D1682628

DIGITAL DESIGN
A Practical Course

Peter Burger
DIGITAL DESIGN
A Practical Course

John Wiley & Sons, Inc.
New York · Chichester · Brisbane · Toronto · Singapore

Copyright © 1988, by John Wiley & Sons, Inc.

All rights reserved. Published simultaneously in Canada.

Reproduction or translation of any part of
this work beyond that permitted by Sections
107 and 108 of the 1976 United States Copyright
Act without the permission of the copyright
owner is unlawful. Requests for permission
or further information should be addressed to
the Permissions Department, John Wiley & Sons.

Library of Congress Cataloging in Publication Data:

Burger, Peter.
   Introduction to digital design/Peter Burger.
     p. cm.
   Bibliography: p.
   Includes index.
   ISBN 0-471-60493-3
   1. Digital electronics.   I. Title.
TK7868.D5B87 1988
621.3815′3—dc19   87-35546

Printed in the United States of America

10 9 8 7 6 5 4 3 2 1

Preface

A NEW KIND OF COURSE

Digital design involves the systematic design of digital circuits and systems from which digital computers are constructed. The term "design" is often used in the sense of artistic design, such as in architecture, landscaping, or fashion, where the design activity indicates an artistically creative endeavor which, starting with an initial concept, creates a pleasing (and useful) product. In architecture, for example, testing and evaluation (public opinion) can occur only after the building has been built.

For engineers, design has a wider meaning because it must include the entire **design cycle,** which consists of specification, implementation, testing, and evaluation. The design cycle, as the term indicates, is not a single design procedure that leads directly from specifications to the final system but consists of numerous modifications (redesign) as constraints in its implementation, or some undesired characteristics in its performance, become evident. For an engineer, a successfully completed design means that a reliable system has been constructed that meets all its specifications and is the *best* according to some given performance criteria.

The purpose of this textbook is to provide an introductory one-semester course that combines theory (lectures) and practice (laboratory) and thus addresses the entire engineering design cycle. Traditional courses treat design as a collection of design tools and principles that are acquired by the student during the term of the course. The student provides his or her designed circuits on paper or passes examinations, proving that he or she has acquired the required tools. The course supported by this textbook asks the student to design, construct, test, and evaluate experiments. The integration of a traditional digital design and a practical digital laboratory course is the result.

Students are presented with two basic requirements:

1. Complete the design, construction, testing, and evaluation of a number of digital circuits or experiments (usually, the number is between 9 and 11) which are described by a set of specifications.
2. The design and construction must be your own work (visible evidence is required of the working circuits and the design methods used).

The obligations of the instructor are:

1. Present and explain design methods that allow the systematic design of the circuits.
2. Provide a physical environment that makes the construction and testing of the circuits possible within the available time limit.
3. Provide help to those students who are unable to determine why a particular circuit does not operate correctly.

This may be the first course in which students are asked to design and construct actual physical circuits; therefore, a large number of important principles will be demonstrated to them. One of the most important principles is that a correctly designed circuit on paper is not equivalent to a working circuit in practice. There are many things that could go wrong (and, according to Murphy, *they will*). The truth emerges that a good designer starts to prove himself only when the circuit does not work and the fault has to be found. Another important principle that should be learned is that systematic, logical, and well-organized testing methods are at least as important as systematic design methods and that the former (systematic testing) can be used only if the latter (systematic design) is practiced. During the latter part of the course it also becomes apparent that whereas some design problems can be successfully tackled by systematic design methods, a search for the *best* solution to other problems may never be found.

There are no required homework exercises in this course and, if rules allow, examinations are not needed either. As we have stated before, the only requirement from the student is that he or she provide a number of complete, well documented, and correctly operating experiments. One may describe this course as a working environment in which design problems are defined, design tools (both theoretical and physical) are provided, and the final products are evaluated. Many of the problems are open-ended, and in some cases trial-and-error methods have to be used because systematic design procedures are not yet available.

No specific prerequisites are needed for this course, but it is assumed that it would be taught at the sophomore level, preferably in the first semester. In this case, a freshman physics course that covers elementary circuits (Ohm's law) and a computer programming course in a higher-level language (Pascal, FORTRAN, etc.) are very helpful to the students in the understanding of IC models, the construction of the circuits, and in the grasping of experiments based on digital computers.

MOTIVATION AND SATISFACTION

One of the most pleasing aspects of teaching this integrated course is the high degree of motivation usually shown by the students. Since the students are pursuing an engineering degree (it is my belief that computer professionals should be considered engineers as well), they will derive satisfaction from working systems that they have designed, built, and tested. At the same time, they will also realize that sometimes a large proportion of the total effort must be spent on the correction of a few faults in a system that works most of the time. This realization will increase the value of designed systems that work all the time.

A further amount of motivation can be added to the course by the calculation of an overall circuit cost for each experiment. The search for the least expensive circuit can introduce a healthy measure of competition among the students. Ample recognition should be given to students who provide the least expensive solution to a design problem, but course grades should not be significantly affected by this contest.

Documentation is the area for which it is the most difficult to generate enthusiasm. All of us remember from our student days the dreadful hours spent on making up (or copying) the required sections of laboratory reports in which we had very little to say. In this course the required reports should be reduced to the essentials: an engineering notebook for record keeping and a concise and well-organized formal report that describes the design procedures and methods of testing. Guidelines on documentation are given in Appendix 2.

It must be emphasized at the outset that, at times, this course will require a great deal of time from most students. This is unavoidable. To construct something that works 100 percent takes time.

ABOUT THIS BOOK

The development of the described course and the subsequent writing of this book were governed by the following requirements:

1. To introduce a course with practical design content into the engineering curriculum.
2. To use more efficient teaching methods (large classes with limited staff resources).
3. To provide a closely coordinated lecture and laboratory course.

These requirements governed the style of this book and made it a self-contained *user's manual* for the students. It was planned that the book should provide all the necessary information for the successful completion of the design experiments. This does not mean that design methods are given in a *"cookbook"* form. The overriding style of the book is teaching by example. First, the principles of several design methods are described, which are followed by a large number of worked-out examples.

The included material, its order of presentation, and the level of detail are all governed by the practical problem that students must complete one experiment per chapter. In some places there are, or may appear to be, unreasonably long and repetitive explanations. These sections were added after the first version of this book, and they were added to topics that appeared difficult to the students. The experiments were chosen in such a way that the necessary "theoretical" material for their completion should cover all the important aspects of digital design. Obviously, because of practical limitations in time and facilities, some desirable material had to be left out. In a few places material was included that does not appear in the experiments, but this was kept to a minimum.

The course material falls naturally into three groups. The first 4 chapters and experiments cover combinational circuits. Since previous experience of digital circuits or practical wiring is not required, these 4 experiments are also used to provide the students with the necessary practical experience in constructing and debugging their circuits. For this reason, 2 weeks should be allowed for the first experiment (Chapter 1), and 1 week for Chapters 2 and 3 each, which appear to be progressing at a slower pace than the rest of the book. Chapter 4, which introduces the technique of K-maps, provides simpler circuits than Chapters 2 and 3, and 1 week should be sufficient to cover it.

The second group of 4 chapters and experiments (Chapters 5–8) introduces and presents several fundamental aspects of sequential circuits. To understand digital memory and sequential circuits from basic combinational principles (by introducing time dependency) is usually a difficult step for the student; therefore, 2 weeks should be allowed for Chapter 5. The detailed understanding of how flip-flops operate is not an absolutely essential requirement for designing synchronous sequential circuits (the topics of Chapters 6–8), but these fundamental aspects of sequential circuits should be included in the education of engineering students. Each of the 3 following chapters (6–8) can be completed in 1 week.

The first 3 of the last 4 chapters (9–11) use already familiar design methods for the design of simple digital computers. Software at the machine-language level is introduced in order to emphasize the fact that when digital computers are designed, both the hardware and software aspects of the computer should be analyzed. The length of time allowed for these experiments will depend on the student's ability, laboratory equipment, and so on. Experiments 9 and 10 could be completed in 1 week under ideal conditions, but experiment 11 will require at least 2 weeks. Finally, Chapter 12 addresses the problems of designing complex digital systems (designing in the large) and introduces programmable logic devices. The material in this chapter can serve as an introduction to more advanced digital courses, such as digital systems, interfacing, VLSI, and so on.

Appendix 1 includes a list of ICs, which makes this textbook self-contained. The list was carefully selected and tested, and it provides a reasonable supply for the first 10 experiments. It is suggested that experiment 11 be constructed by a group of students, in which case several lists of ICs become available. It is equally possible to run the course with an unlimited supply of ICs supported by an engineering design book, and there are good arguments for using both types of IC supplies. I have chosen the limited list (no OR gates, for example) because the experiments are fundamentally simple, and in order to simulate practical design problems, the small IC list will present physical limitations to the students, which will have to be dealt with in some form of trial-and-error methods.

PRACTICAL ISSUES

The success and effectiveness of this course are obviously influenced by the equipment used. I believe that the experimental facilities should allow the students to construct and test the circuits at their convenience. This requires portable self-contained digital breadboard units (we call them digital kits) which, in addition to the circuit board, the ICs, and the power supply, must include switches, debounced digital clock pulse inputs, and test lights. The time periods suggested for the experiments were based on this type of equipment.

Another advantage of providing a digital kit for each student is the efficient way the laboratory period can be used. Each student is required to show the completed circuits to the instructor who then spends 10 minutes with each student to ascertain that the design is the student's own work. If digital equipment (a digital analyzer, for example) is available, then the testing of the circuits can be done automatically.

We have been describing the *ideal* course, but it is also possible to use this book in other ways. One possibility is to use a computer simulation package in place of the digital kits. In this case, the course could be very similar to the one described previously and may even be advantageous if the wiring experience is not required (for computer science students, for example). Finally, the book can be used for a conventional digital design course, since most of the traditional material on digital design principles has been included, and there are a selection of drill problems in each chapter. In this case, the self-contained aspect of the book may be utilized by students for self-teaching, whereas the experiments may be designed on paper.

London, England **Peter Burger**
1987

ACKNOWLEDGMENTS

I first wish to thank the administrators of the University of Lowell, Lowell, Massachusetts, who allowed me to experiment with such an unusual course and provided the funds for the digital kits. Among the many, I would like to thank Professor James E. Powers, who as Chairman of the Electrical Engineering Department during the first year of development played a major role in defining the course objectives, helped me teach it, and provided unfailing encouragement. Many of the original ideas that are still present in this book were suggested by him.

Help from many of my other colleagues is also greatly appreciated, in particular Professors Holmstrom, Laste, Leonard, Tran, and Wade (all at the University of Lowell), who helped me greatly in the latter development and teaching of the course.

I owe many thanks to the more than 400 students who, with their dilligence and enthusiasm, proved to me that in addition to being fun, this course was also a worthwhile undertaking for all concerned.

I would like to thank my daughter, Juliet, for her patient proofreading of the first version of the manuscript and my wife, Vera, for the proofreading of this text. I also would like to thank Vera for her patient and constant encouragement over these last six years, a large portion of which I spent in front of a word processor. Finally, I would like to thank the members of the Department of Computing at Imperial College, London, who provided me with computer facilities and allowed me time for completing this textbook.

Contents

CHAPTER 1
Introduction to Digital Systems, Logic Gates, and Boolean Algebra   1

    1.1 Introduction   1
    1.2 Physical Systems and Their Representation   1
    1.3 Binary Systems   4
    1.4 Basic Binar (Logic) Operations   8
    1.5 Binary (Logic) Gates   11
    1.6 Binary (Logic) Circuits   14
    1.7 Fundamentals of Boolean Algebra   16
    1.8 Physical Realization of Binary Circuits   23
    1.9 Practical Digital Circuits   28

EXPERIMENT 1
Analysis, Construction, and Testing of Binary Circuits   34

    E1.1 Summary and Objectives   34
    E1.2 The Experiment   34
    E1.3 Alternate Experiments   36
    E1.4 Testing and Debugging the Circuit   37
    References   38
    Drill Problems   38

CHAPTER 2
Boolean Simplification, Binary Functions, and Digital (Logic) Circuits   41

    2.1 Introduction   41
    2.2 Set Theory and the Venn Diagram   41
    2.3 Boolean Simplification   42
    2.4 Functions of One Binary Variable   48
    2.5 Functions of Two Binary Variables   49
    2.6 Logic Gates with One and Two Inputs   50
    2.7 Practical Gate Transformations   53

EXPERIMENT 2
Two-Input, Three-Output Boolean Function Generator   58

    E2.1 Summary and Objectives   58
    E2.2 Functional Specifications   58
    E2.3 Results and Testing   59
    E2.4 Design Example   60
    E2.5 Further Experiments   63
    Drill Problems   63

CHAPTER 3
Variable Boolean Function Generators   65

    3.1 Introduction   65
    3.2 Simple Variable Boolean Function Generators   65
    3.3 Digital Multiplexers   68
    3.4 Using a Decoder To Build a Multiplexer   71
    3.5 Using Multiplexers and Decoders for Generating Other Boolean Functions   73

EXPERIMENT 3
Variable Boolean Function Generator   77

    E3.1 Summary and Objectives   77
    E3.2 Functional Specifications   77
    E3.3 Expected Results, Restrictions, and Designing Process   78
    E3.4 Testing   79
    E3.5 Alternate Experiments   80
    Drill Problems   80

CHAPTER 4
Combinatorial Circuit Design and Minimization   82

    4.1 Introduction   82
    4.2 Binary Circuit Design with Minterms and Maxterms   83
    4.3 Two-, Three-, and Four-Variable Karnaugh Maps   88
    4.4 Minimization of Sum-of-Product Expressions Using K-Maps   91
    4.5 Minimization of Product-of-Sum Expressions Using K-Maps   96
    4.6 Simplification of Boolean Expressions   98
    4.7 K-Map Minimization with "Don't Care" Terms   104
    4.8 Minimization with Multiple Outputs   106
    4.9 K-Maps for Five and Six Variables   113

EXPERIMENT 4
Minimized Variable Boolean Function Generator   116

    E4.1 Summary and Objectives   116
    E4.2 Description of Overall Circuit   117
    E4.3 Expected Results, Requirements, and Restrictions   117
    E4.4 Alternate Experiments   119

References 119
Drill Problems 119

CHAPTER 5
Introduction to Sequential Systems 121

5.1 Introduction 121
5.2 Feedback Connections in Binary Circuits 122
5.3 Synchronous Sequential Systems 133
5.4 Practical Flip-Flop Circuits 140

EXPERIMENT 5
Time-Dependent and Sequential Circuits 153

E5.1 Summary and Objectives 153
E5.2 Switch Bounce Eliminator Circuit 153
E5.3 Asynchronous Time-Dependent Digital Signals 153
E5.4 Detection of Unwanted Signal Transitions (Glitches) 154
E5.5 A Synchronous Sequential System 157
E5.6 A Faulty Master–Slave Flip-Flop 158
E5.7 Display of Time-Varying Digital Signals 159
Drill Problems 160

CHAPTER 6
Design of Counters 164

6.1 Introduction 164
6.2 Design of Synchronous Counters with D-Type Flip-Flops 165
6.3 Counter Design Using General Flip-Flop Types 176
6.4 Asynchronous (Ripple) Counters 191
6.5 Synchronous and Asynchronous Control of Practical Counters 194

EXPERIMENT 6
Design of Five-State Up–Down Counter 197

E6.1 Summary and Objectives 197
E6.2 Description of Design Problem 197
E6.3 Selection of Counting Sequence 198
E6.4 Expected Results and Testing 198
E6.5 Design Requirements 199
E6.6 Alternate Experiments 199
Drill Problems 200

CHAPTER 7
Synchronous Controllers and State Assignments 201

7.1 Introduction 201
7.2 State Controllers 202

xvi   CONTENTS

    7.3 Design of Traffic Signal Controller   204
    7.4 One-Flip-Flop-per-State Method   213
    7.5 Bit-Sequence Detector Circuits   217
    7.6 Input and Output Signal Synchronization in Sequential Systems   219
    7.7 Moore Detector Circuits for Bit Sequence 101   222
    7.8 Mealy Detector Circuits for Bit Sequence 101   225
    7.9 State Assignment Selection   228

EXPERIMENT 7
Serially Controlled Combination Lock   233

    E7.1 Summary and Objectives   233
    E7.2 Operation of Combination Lock   233
    E7.3 Bit Sequence and Standard State Transition Diagram   234
    E7.4 Design Requirements   235
    E7.5 Checking and Testing Your Circuits   236
    E7.6 Alternate Experiments   236
    References   237
    Drill Problems   237

CHAPTER 8
Number Representation and Computer Arithmetic   239

    8.1 Introduction   239
    8.2 Registers and Register Transfer Operations   240
    8.3 Unsigned Number Representation and Arithmetic   242
    8.4 Parallel Addition and Subtraction   249
    8.5 Serial Addition   250
    8.6 Sign-Plus-Magnitude Representation   252
    8.7 Two's Complement Representation   255
    8.8 Octal and Hexadecimal Representation of Binary Numbers   260
    8.9 General Data Processor (CPU)   263

EXPERIMENT 8
Serial Data Processor   269

    E8.1 Summary and Objectives   269
    E8.2 Serial Representation of Data   269
    E8.3 Serial Processor Block Diagrams and Operation   271
    E8.4 Design Requirements and Testing   273
    E8.5 Additional Circuits and Alternate Experiment   274
    Drill Problems   275

CHAPTER 9
Serial–Parallel Processing and Conversion   277

    9.1 Introduction   277
    9.2 Shift Registers and Parallel–Serial Data Conversion   279

9.3 Multiple Precision, Shift Operations, and Multiplication   285
9.4 Carry Propagation   290

EXPERIMENT 9
Four-Bit Serial–Parallel Arithmetic Processor   294

E9.1 Summary and Objectives   294
E9.2 Serial Control, Input Data, and Expected Output Signals   295
E9.3 Recommended Internal Organization   296
E9.4 State Controller   300
E9.5 Requirements and Suggestions   303
E9.6 Simplified Processors and/or Input Conditions   303
Drill Problems   304

CHAPTER 10
Introduction to Digital Computers   306

10.1 Introduction   306
10.2 Register Transfers, Data Bus Structures, and Memory Systems   308
10.3 Computer Instructions and Computer Cycles   320
10.4 Data-Handling Instructions   322
10.5 Externally Programmed Computers   327

EXPERIMENT 10
Externally Programmed Digital Computer   328

E10.1 Summary and Objectives   328
E10.2 Simple Data-Sorting Computer   329
E10.3 Sorting Program   330
E10.4 Hardware Requirements   331
E10.5 Instruction Encoding   332
E10.6 Hardware Design   334
E10.7 Alternate Experiments   335
Drill Problems   335

CHAPTER 11
Introduction to Programmable Digital Computers   337

11.1 Introduction   337
11.2 Memory   338
11.3 Data Transfer Operations   341
11.4 Control Instructions   345
11.5 Data-Processing Instructions   348
11.6 Control Circuitry   350
11.7 Selection of Computer's Instruction Set   356
11.8 Programming in Assembly Language   365

EXPERIMENT 11
Design of Four-Bit Computer   367

    E11.1 Summary and Objectives   367
    E11.2 Some Practical Aspects   369
    E11.3 Hardware Design   372
    E11.4 Requirements   374
    E11.5 Software Tasks   375
    Drill Problems   376

CHAPTER 12
Digital System Design   379

    12.1 Introduction   379
    12.2 Design in the Large and in the Small   380
    12.3 ROMs and Programmable Logic Modules   387
    12.4 Design Examples with ROMs and PLAs   395
    12.5 Microprogramming   407

EXPERIMENT 12
Experimental Designs with ROMs and PLAs   415

    E12.1 Summary and Objectives   415
    Drill Problems   415

Appendix 1:
Integrated Circuit Devices   417

    A1.1 Introduction   417
    A1.2 List of Available IC Devices   417
    A1.3 IC Devices   419
    A1.3.1 NAND Gates   420
    A1.3.2 AND Gates and Inverters   421
    A1.3.3 NOR Gates and XOR Gates   422
    A1.3.4 Flip-Flops   422
    A1.3.5 Multiplexers   424
    A1.3.6 Decoders/Demultiplexers   425
    A1.3.7 Binary Counter   426
    A1.3.8 Shift Register   427
    A1.3.9 Comparator   427
    A1.3.10 Random-Access Memory   428
    A1.3.11 Arithmetic-Logic Unit   429

Appendix 2
Documentation   430

    A2.1 Introduction   430
    A2.2 Engineering Notebook   431
    A2.3 Summary Report   432
    A2.4 Example of a Summary Report   433

INDEX   439

CHAPTER 1

Introduction to Digital Systems, Logic Gates, and Boolean Algebra

1.1 INTRODUCTION

Our aim is to study methods that engineers use to design digital circuits. We shall approach this vast subject through the design, construction, and testing of practical digital circuits, which we will call **experiments**. In general, the design of a system means synthesis, the "putting together" of system elements into a working whole, which provides the desired system. We assume that the reader has no prior knowledge of digital systems or their components; hence, it is not possible to start with design at once. A short introductory period of analysis, or "finding out" phase, is necessary. Hence, the first chapter of this book and the first experiment do not deal with actual design but are based on analysis. We start by analyzing the behavior of digital systems and their components known as logic gates. During this finding-out phase we will introduce basic digital methods, such as Boolean algebra, which will be used for all our later work when the designing process may be started in earnest.

Digital systems are very specialized examples of general physical systems; therefore, a good place to start is to examine the characteristics and the behavior of physical systems in general. The treatment of material in this chapter starts with general systems and then proceeds to the specialized digital/binary systems. It covers both the idealized and the practical aspects of digital circuits. After studying this chapter, the reader should be able to complete the first experiment, whose requirements are to build and test a given combinational logic circuit using practical logic gates.

1.2 PHYSICAL SYSTEMS AND THEIR REPRESENTATION

The general concept of a **system** is widely used today. We use electrical power and computer systems, study biological and political systems, and build models of economic, social, and environmental systems. We always formulate a system as a collection of interconnected parts, or **primitive elements** of the system whose combined behavior may be predicted from the known individual behavior of the elements and the given interconnections. What makes a system concept useful is its formal, often mathematical, description, which we shall refer to as its **representation**.

The mathematical representation of a system must be given in terms of **variables** that must be associated with physically measurable quantities in the real, that is, physical, system. A general system may be represented schematically by a **block diagram,** as shown in Figure 1.1. The variables are divided into three classes. Two of these classes contain the **input** and **output** variables. The values of the input variables are determined independently from and are not under the control of the system. The system controls the values of the output variables. The input variables drive the system, whereas the output variables express the system's response to a particular set of input values and operating conditions. The behavior of the system may be specified as an output–input, or transfer, function, which is in fact the functional description or external specification of the system.

Although, for many simple systems, the functional description in terms of how the output variables behave for a given set of input variables suffices, for general dynamic, that is, time-varying systems, a third class of variables must be included. This third class of system variables is called **state variables.** State variables are necessary in systems for which one cannot simply relate the output variable values to the input variables. The following example demonstrates the role that each of the three classes of variables play.

We model a supermarket as a dynamic system. Input variables are the quantities and buying prices of arriving goods: pounds of flour, pints of milk, number of soap bars, and so on. Output variables are the selling prices and quantities of goods bought by the customers. The state variables are the quantities of goods stored on the shelves or in the warehouse.

Some of the input variables (e.g., buying price) affect output variables (e.g., the selling price of goods) in a direct manner. Other input variables, such as the quantities of arriving goods, may have a less direct effect. Some of the state variables may also influence the output variables, as, for example, the length of time perishable items have been on the shelves. The effect of the selling prices and quantities of displayed goods on the quantities sold may be difficult to establish. In general, we may expect that the selling price would increase with increasing buying price and decrease with increasing stock; whereas the quantities of goods sold should increase with decreasing selling prices.

The supermarket as a dynamic system has easily measurable variables,

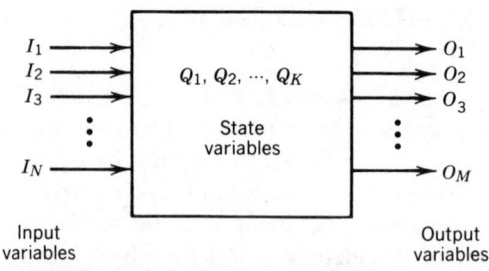

FIGURE 1.1 Block diagram of general system.

1.2 PHYSICAL SYSTEMS AND THEIR REPRESENTATION

and the representation of the system is very close to the physical system. However, some of the system laws are unknown and could be very difficult to establish. We may also notice that the supermarket is a mixed digital and analog system, since some of the goods are measured by the piece, whereas others are measured by continuous quantities.

Let us consider now a purely digital system for which the system laws are clearly defined. In Figure 1.2 the system representation of a chess game is shown. There are only two input variables. These are the times available (in seconds) for the two players. To force the system to be completely digital, digital clocks are assumed. The output variables are all **logic,** or **binary,** variables, that is, they may have one of only two possible values. They represent statements that may be either true of false. The physical system may have a message board on which the true statements are lit up. The state variables are the chess pieces. The measured values for the state variables are the positions of the pieces. Since there are 64 squares on the board, each chess piece may have 1 of 64 possible values. The sixty-fifth value indicates that the piece is off the board.

From the values of the input variables (amount of time left for each player) and the state variables (the current chess position), the values for the output variables can be uniquely determined. This seems to be a well-defined, trouble-free digital system, but we find that the system cannot be measured reliably at all times. What happens if at the particular time we want to measure the system a chess piece is being held by a player in his hand? In this case the system variable describing the piece being held has no valid value, and our representation of the physical system breaks down. In order to bring the representation of the system closer to its physical realization, some corrective action must be taken.

There are two possible remedies. First, the number of system variable values may be increased to 66, in which case the sixty-sixth value could

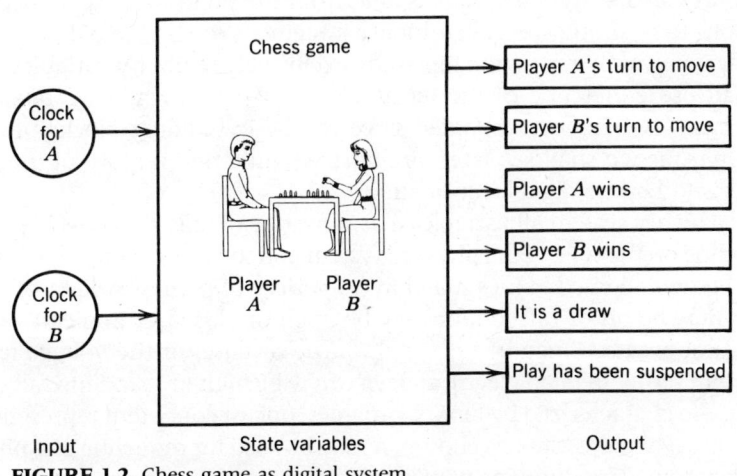

FIGURE 1.2 Chess game as digital system.

indicate that the piece is being moved. A new message could be added to the system that would account for this situation. Another way to handle the moving of a piece is by restricting the times of measurements to those time periods when all pieces are on the board. Either method of introducing changes to our original system improves the correspondence between the physical system and its representation.

The preceding two systems demonstrate how we may build models of physical systems. The most important parts of the model for good correspondence between the physical system and its representation are easily and reliably measured system variables. It is interesting to note that the development of the most successful man-made physical systems, digital computers, followed the inverse of model building. The representation of digital (binary) systems was well established before digital computers could be built. In fact, digital computer circuits are the physical realization of **Boolean** and **switching** algebra representations and not the other way around. The amazing success story of digital computers can be directly attributed to the fact that the behavior of a real computer system follows its idealized binary model extremely well. This is so in spite of the continuous nature of the electrical signals that represent digital (binary) variables in the physical system.

1.3 BINARY SYSTEMS

Our main interest lies in the design of digital systems. The term *digital* means that the variables of the system can take on one of a given number of values. Values of digital system variables jump discontinuously from one allowed value to another. The variables expressing the positions of chess pieces, or the number of eggs on the supermarket shelf, are good examples of digital variables we use every day. When a variable may have one of only two possible values, it is called a **binary variable.**

In practice, the largest and most successful digital systems, digital computers, use binary system variables, and from the point of view of their construction, they should be called binary systems. We shall avoid this inconsistency in terms by restricting our study exclusively to binary variables, while we shall use either one of the terms *binary* or *digital* when we talk about entire systems. This is reasonable since the basic building blocks of digital computers, as we shall see later, are binary, while the purpose of the entire system is to handle digital information.

In a binary system all variables may have one of only two possible values. Depending on the type of the physical system and the roles the binary variables play, the two allowed values may have a variety of names. A light or a light switch may be on or off, a door may be open or closed, a Boolean variable in a Pascal program may be set equal to true or false, or the voltage level at the output terminal of an electrical device may be high or low. Since all binary systems are characterized by binary variables, once a consistent representation for a general binary system is chosen, it may be used for modeling any physical binary system. The different names are convenient to express the functions of the physical system, but it is easier to use only one set when the representation of the system is studied.

We shall adopt the names **Logic-1** and **Logic-0** throughout this book, but in most cases will use the digits **1** and **0** as a shortened and more convenient notation. The names *Logic-1* and *Logic-0* emphasize the fact that in the binary system the digits 1 and 0 do not represent numerical values. This may cause some initial confusion, especially, when in the next section, binary operations are introduced. However, the use of 1's and 0's for binary values is so simple and widely used that this is still the best choice for their representation.

The choice of the word *Logic* also expresses the fact that binary systems are equivalent to logic systems. In fact, our current design tools for binary systems developed originally from the field of classical logic. We shall discuss the equivalence of binary and logic systems in the next section.

Let us examine the simple binary system shown in Figure 1.3. The system is completely specified. The block diagram shows that the system has two input and two output variables. To the right of the block diagram the **truth table** specifies the output–input transfer function, indicating that the output values may be directly calculated if the input values are known (e.g., state variables are not required). In the truth table the output values are tabulated for all combinations of the two possible input values.

In the block diagram the system variables are identified by symbols or variable names. The two input variables are named A and B, and the two output variables are identified as X and Y. In practical binary circuits the system variables are called **signals.** Since there are two output variables, the truth table has two output columns. The number of entries in the truth table (each row in the table is one entry) is equal to the number of possible unique input value combinations, which in this case is equal to 4. Since each input variable may have one of two possible values, in the general case, the number of entries is given by the following expression:

$$\text{Number of entries in truth table} = 2^{\text{number of input variables}}$$

This simplest type of binary system, for which a truth table can be constructed, is called a **combinational** binary system. Since the output values are specified in the truth table for all possible combinations of input values, the truth table

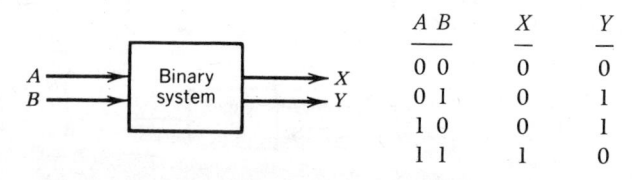

A B	X	Y
0 0	0	0
0 1	0	1
1 0	0	1
1 1	1	0

Block diagram   Truth table

FIGURE 1.3 Block diagram and truth table for two-input, two-output binary system.

6 DIGITAL SYSTEMS, LOGIC GATES, AND BOOLEAN ALGEBRA

represents a **complete** description of the system's behavior. We may use other methods to describe the system's behavior that are equivalent, but possibly more convenient, descriptions; however, nothing can be added to the information contained in the truth table. The table contains "the whole truth, and nothing but the truth" as far as the combinational system is concerned.

We have mentioned before that complex systems may be constructed from suitable primitive components of a system. Any combinational system may be used as a primitive building block, and a more complex system may be built by connecting the outputs of the components to the inputs of other components. The entire system will have independent inputs, which are not connected to any outputs, and some outputs may be designated as the overall output variables of the combined system (see Figures 1.4 and 1.5). The question is: If all components are combinational binary systems, will the overall system be also combinational? Another way to ask the same question is: If we know the truth table of each component, can we determine an overall truth table for the entire system?

The answer to the preceding question is that there are two ways one may connect combinational system elements to form more complex systems. When one uses only **feed-forward** connections (Figure 1.4), the overall system remains combinational. If one or more **feedback** connections are used (Figure 1.5), the system is no longer combinational, and a truth table for the overall system does not exist. In this more complicated case state variables must be added to the system whose behavior can no longer be specified by a simple functional relationship between output and input variables.

We may test a system for the lack of feedback connections by assigning "levels" to its variables (see Figure 1.4). The independent inputs of the system are assigned level 0. All the outputs of a system component are assigned a

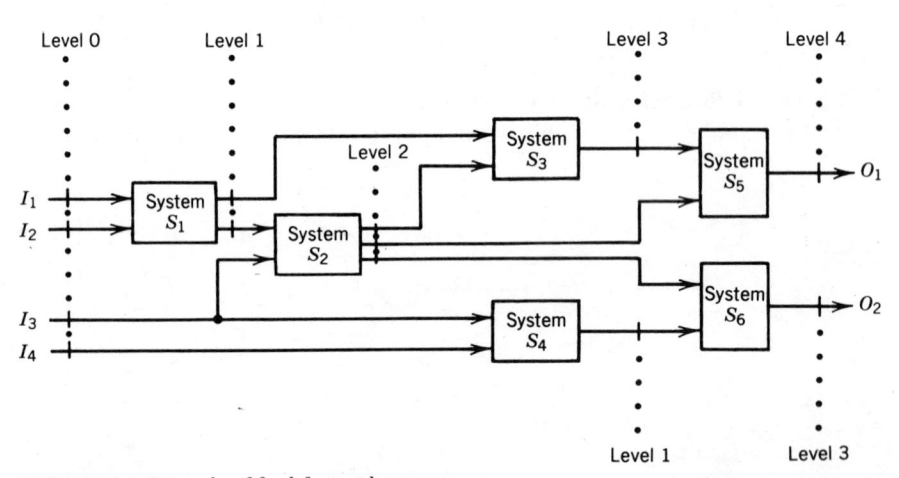

FIGURE 1.4 Example of feed-forward system.

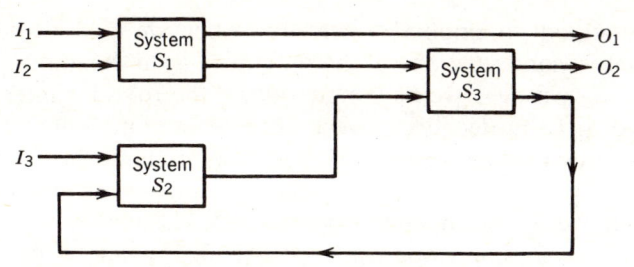

FIGURE 1.5 Example of system with feedback.

level that is one higher than the maximum level found among its input variables. Hence, a combinational block element raises the level by 1, or:

Output level = 1 + max(input levels)

If such an assignment is possible for all system variables, then the system is free of feedback connections, and it is combinational. This is true because if such a level assignment is possible, then the truth table of the overall system may be determined in a systematic way. The calculations are done in as many steps as there are levels.

At the first step a combination of Logic-1's and Logic-0's are specified for the independent inputs, which are at level 0. Now all the outputs at level 1 can be calculated (outputs at level 1 are outputs of combinational components whose inputs are at level 0). This is the first step in the calculations. The second step is to calculate the values for all variables at level 2. Knowing all the values for variables at level 0 and level 1, the output variables of system components whose output variables are at level 2 can be determined because the inputs could only be at level 0 or level 1. Once the second step in the calculation is completed, we know the values of all variables whose level are at 0, 1, or 2.

At the next step the variables at level 3 can be evaluated, and at every succeeding calculation step the variables belonging to one higher level can be determined. In a similar manner, the calculations proceed from levels to higher levels until all variable values are evaluated. Once all variable values are known, the values of the designated system output variables specify one entry in the truth table, namely, the entry that corresponds to the selected combination of the independent input values. Similar calculations for all combinations of input values determine the complete truth table for the overall system. Since the truth table exists, the overall system must be combinational.

A system that uses a feedback connection is shown in Figure 1.5. It is impossible to assign levels to the outputs of system component S_2 and S_3. Assume that the output of system component S_2 is at level N. This sets the level for the outputs of S_3 to be one higher, that is, $N + 1$. The same signal is connected to the input of S_2, which forces the output of component S_2 to be at level $N + 2$. This is two higher than N, which was our original assumption. This shows that it is impossible to assign levels to such a system.

It is important to emphasize here that when we used a feedback connection, that is, connected one of the outputs of S_3 to the input S_2, we have done "nothing wrong"; we have not violated the rules for interconnecting binary circuits. Using feedback, we built a more complicated system that cannot be represented by a truth table. We shall postpone the study of binary feedback, or **sequential,** systems to Chapter 5 and later chapters. In the first four chapters we will study only combinational systems.

The truth table is only one way of specifying the output–input relationship of a combinational system. Examine again the simple two-input, two-output system shown in Figure 1.3. It is also possible to describe the system by words. Examining the truth table, we can state that the output X of the system is equal to Logic-1 if and only if both inputs have values Logic-1. The output variable Y is equal to Logic-0 if and only if the two input variables are equal (both are 0 or both are 1). These two statements constitute different, but equivalent, functional descriptions to the truth table of the system shown in Figure 1.3. Description by words will be called the **functional specification** of the system.

Note that the truth table can always be constructed from a complete functional specification. On the other hand, the truth table may be described by many different (and equivalent) functional specifications. Another functional specification for the system in Figure 1.3 is: The output variable Y is equal to 1 if and only if the number of 1's among the input values is equal to 1.

Functional specifications become more and more important as the complexity of a binary system increases. For a system that has a large number of independent input variables, truth tables are impractical. A large, complex combinational system is divided into smaller systems, as shown in Figure 1.4, where the divisions are governed by functional specifications rather than truth tables. This division into modules is similar to the division of computer programs into subroutines, procedures, or functions. The successive divisions of a large complex system into logically well-defined system modules is the most fundamental technique of modern system design and is always practiced by good system designers. There is a practical reason why this division by modules is fundamental to digital design. Practical digital circuits are designed and constructed by the interconnection of ready-made digital modules, or **integrated circuits (ICs),** when truth tables are rarely used in the overall design process of a more than trivial digital system.

In the first four chapters, when the basics of digital systems are studied, truth tables, algebraic representations, and functional specifications will be equally important; in later chapters, when the overall systems become more complex, functional specifications will be used more and more.

1.4 BASIC BINARY (LOGIC) OPERATIONS

Many people working in the field of digital system design use the terms *logic* or *logical* in place of the term *binary*. Historically, the first formal binary system was used by ancient philosophers who developed the field of propositional logic. A **proposition** or verbal statement may be either **true** or **false;** hence, a proposition may have one of two values. The proposition is the binary variable of logic systems. We have already used simple logical propositions as the output variables of our chess game system (Figure 1.2). For example, the statement "It is player A's turn to move" is a proposition whose value alternates between true and false as the game progresses.

The laws of logic are based on three basic logical operators: **AND, OR,** and **NOT**. The first two are logical connectives that combine two simple propositions into one, more complex proposition. The combined proposition

It is player A's turn to move AND it is player B's turn to move

has always the false value since the combined proposition violates the rules of the game. On the other hand, the proposition

It is player A's turn to move OR it is player B's turn to move

does not always have the true value because the game might have been finished already or it could have been suspended.

The NOT operator inverts the meaning of a proposition. This operator produces the **denial,** or **complement,** of a proposition. If we know that the statement "It is player A's turn to move" is true, then we also know that the more complex statement "It is NOT player A's turn to move" is false. The laws for these three basic logic operators are expressed below where the symbols P1 and P2 are used for two arbitrary propositions:

$$
\begin{array}{l}
\text{(P1 \textbf{AND} P2) is true} \\
\text{(P1 \textbf{OR} P2) is true} \\
\text{(\textbf{NOT} P1) is true}
\end{array}
\left\{ \begin{array}{c} \text{If and} \\ \text{only if} \end{array} \right\}
\begin{array}{l}
\text{both P1 and P2 are true} \\
\text{either P1 or P2 or both are true} \\
\text{P1 is false}
\end{array}
$$

These three laws are fundamental to propositional logic. Any proposition, no matter how complex, may be expressed by the combination of simpler propositions and the application of these three basic logic operators. We postpone the proof of this statement until Chapter 4, when we will show that any complex logical statement (Boolean expression) defined by an arbitrary truth table may be expressed in canonical forms which use only these three basic logical laws (Boolean operators).

In fact, one can easily show that two basic operators are sufficient. If we select the **NOT** operator and, say, the **AND** operator then (as we will prove shortly) the **OR** operator may be expressed in terms of the first two because the expression

(P1 **OR** P2)

is equivalent to the expression

NOT[NOT(P1) **AND NOT**(P2)]

A similar expression can be found for the two basic operators **NOT** and **OR,** since the expression

(P1 **AND** P2)

is equivalent to the logical expression

NOT[NOT(P1) **OR NOT**(P2)]

Nevertheless, propositional logic in particular and logic designers in general maintain the three operators because they provide simpler expressions than if only two operators were used and because of a useful symmetrical relationship between **AND** and the **OR** operators called **duality,** which we will frequently encounter.

The three basic logical laws may be expressed in truth tables. The preceding three statements are functional specifications for the three laws and may be converted to tabular forms by using the letter T for the value of true and the letter F for the value of false. Table 1.1 is a truth table for the three logic operators. (We give up the convenient binary notation of 1's and 0's for a short while in order to avoid the problem of deciding which one should be used for the value of true and which for the value of false. We shall tackle this assignment problem in the next section). The reader should be convinced that the logical postulates and the truth tables express identical definitions for the three basic logic operators.

We have mentioned before that one truth table may be expressed by many different functional specifications. The laws for the three logic operators were expressed by combined statements that were true. An equivalent, but different, set of laws may be stated by choosing combined statements that are false:

(P1 **AND** P2) is false $\quad$ either P1 or P2 or both are false
(P1 **OR** P2) is false $\quad$ If and only if $\quad$ both P1 and P2 are false
(**NOT** P1) is false $\quad$ P1 is true

Note the symmetry between the two sets of logical expressions. In the first set the AND connective is described by the words *both* and *and*, whereas the OR connective uses the words *either, or* and *both.* In the second set the descriptive words are interchanged. The words *either, or* and *both* are used for the AND operator, and the words *both* and *and* are used for the OR operator. The existence of two symmetrical sets of representation for logical variables may be extended to all binary systems. When the values of true and false (or Logic-1 and Logic-0) and the roles of the operators AND and OR are interchanged, we can generate an alternate description of the system. This

TABLE 1.1 Truth Table for Three Basic Logical Operators **AND, OR,** and **NOT** in Terms of TRUE (T) and FALSE (F) Values

P1	P2	P1 **AND** P2	P1 **OR** P2	**NOT** P1
F	F	F	F	T
F	T	F	T	T
T	F	F	T	F
T	T	T	T	F

fundamental property of binary systems, called **duality,** will appear often in our later discussions.

It is now worthwhile to stop for a minute, glance back, and review how much territory we have covered. We started out by stating that all binary systems are equivalent since they all use binary variables. The difference between binary systems is only the names we give to the two different values of the variables. We then showed that a combinational system of any complexity may be constructed from simpler combinational elements as long as feedback connections are not used. Finally, we reviewed formal logic, which postulates that three basic operators are sufficient to express any desired logic function. The three basic operators AND, OR, and NOT can be used now as our combinational primitives, and feed-forward connections can be applied to build combinational systems of any complexity. We are now ready to transfer these operators to general binary systems expressed in terms of logic gates and binary (logic) circuits.

1.5 BINARY (LOGIC) GATES

Since binary (or logic) systems are usually constructed from the three basic operators AND, OR, and NOT, three elementary building blocks can be defined that perform these operations. In practice, these and other simple binary system elements are called **logic gates.** Logic gates are used to construct **binary circuits,** the physical realization of general binary systems. In practical circuits logic gates are given special system block symbols so that they can be easily recognized in a binary circuit diagram. For example, the AND operator may be represented by a traditional block diagram, or the more commonly used AND gate symbol shown in Figure 1.6.

As shown in Figure 1.6, the general rectangular block diagram representation has no orientation to the right or left; therefore, arrows must distinguish between the input and output connections. The AND gate symbol is asymmetrical; its curved right side indicates output, and the flat left side indicates the input side of the gate. Because the input and output sides of the gate are clearly indicated, arrows are not necessary. This is an advantage in a complicated circuit diagram where gates may have many different orientations. When the asymmetrical shapes of gates are used, inputs and outputs become instantly recognizable. There are similar, generally accepted symbols for the two other gates that perform the OR and NOT operations. They are called the OR gate and the INVERTER gate (or INVERTER), respectively.

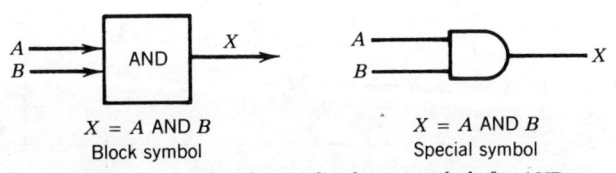

$X = A$ AND B
Block symbol

$X = A$ AND B
Special symbol

FIGURE 1.6 General and specialized gate symbols for AND operation.

12   DIGITAL SYSTEMS, LOGIC GATES, AND BOOLEAN ALGEBRA

Gate symbols, together with their truth tables for these three basic logic operations, are shown in Figure 1.7. Truth tables are expressed in terms of 1's and 0's (i.e., Logic-1's and Logic-0's). They are equivalent to the truth table in Table 1.1, with 1's substituted for the value of true and 0's for the value of false. Using these substitutions, we have followed the so-called positive-logic convention. The term *asserted* is often used in place of the term *true* (*not asserted* in place of *false*); therefore, identifying the value of Logic-1 with the assertion of a desired result or action seems to be the "logical" choice. But, as it often happens in practice, different people have different preferences, and sooner or later somebody will use an alternate to the logical choice. The same happened in the field of digital design. Some digital designers chose the alternate assignment, the negative-logic convention, and substituted the value of Logic-0 for true and Logic-1 for false. Using these alternate substitutions, the truth table for this **dual** representation of the same three basic logic operators can be constructed (see Table 1.2).

In order to compare the truth table for negative logic (Table 1.2) to the more conventional positive-logic representation (Figure 1.7), the entries in Table 1.2 must be rearranged so that they are listed in the same order in the two truth tables. After having rearranged the entries in Table 1.2, we discover that the truth tables for the AND and OR gates become exchanged, whereas the truth table for the INVERTER gate remains the same. This shows that using both positive and negative logic conventions at the same time, which is called the mixed logic convention, could be extremely confusing. It would require the designation of each binary variable whether it is expressed in terms of positive or negative logic. Instead of adding these complications to our circuits, we adopt the positive-logic convention throughout this book.

Type of gate	Symbol	Truth table
AND	 A, B → ⟶ X	A B | X 0 0 | 0 0 1 | 0 1 0 | 0 1 1 | 1
OR	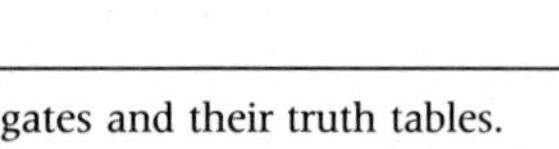 A, B → ⟶ X	A B | X 0 0 | 0 0 1 | 1 1 0 | 1 1 1 | 1
INVERTER	A ──▷○── X	A | X 0 | 1 1 | 0

FIGURE 1.7 Three basic binary gates and their truth tables.

1.5 BINARY (LOGIC) GATES 13

TABLE 1.2 Truth Table for **AND**, **OR**, and **NOT**
Operators Using Negative-logic Convention

P1	P2	P1 **AND** P2	P1 **OR** P2	**NOT** P1
1	1	1	1	0
1	0	1	0	0
0	1	1	0	1
0	0	0	0	1

This does not mean that the negative-logic convention will disappear completely. It will reappear many times when the dual alternative of a method will be investigated. However, we shall adopt the conventional logic gate symbols and truth tables shown in Figure 1.7, where a Logic-1 means true and a Logic-0 means false.

Let us consider an example that demonstrates how logic gates may be used to express propositional logic functions. If gates represent logic operators, then inputs and outputs represent logic propositions. The inputs are simple propositions, whereas the outputs are the combined propositions obtained by the application of logic operators. Considering the example of the chess game (Figure 1.2), we are interested in the new proposition "The game is in progress." How could we represent this new proposition using logic gates? The solution is shown in Figure 1.8. The game is in progress if it is either player A's turn to move, or player B's turn to move. Consequently, we use an OR gate. The binary signals are designated by propositions. A Logic-1 value for a variable indicates assertion, the condition that the statement is true. For example, when the proposition "The game is in progress" is false, then it can be neither player's move, and both input propositions must be false.

This simple example already shows the way logic gates and circuits may be used for practical applications. The output signals from the chess game system may be used to light the appropriate messages on a message board. When the logic signal is asserted, that is, the binary variable has a value 1, the message lights up. We could now "wire" the new message, "The game is in progress," by using a new light and an OR gate and the already wired output lights of the system. If either one of the light for the messages "It is player A's turn to move" or "It is player B's turn to move" lights up, then our new sign will also light up; otherwise, it will remain dark. By adding this new message, we, in fact, designed a binary circuit according to the functional specification of our new system output.

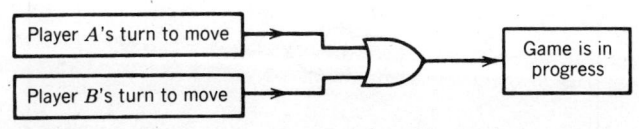

FIGURE 1.8 Two logic assertions combined by OR logic operator.

14 DIGITAL SYSTEMS, LOGIC GATES, AND BOOLEAN ALGEBRA

Logic gates are the lowest level tools of digital circuit designers. The three gates shown so far were borrowed from the classical theory of logic and are sufficient for building any combinational circuit. However, from the digital circuit designer's point of view, two other gates, the **NAND** and the **NOR** gates are just as basic and may even be more important building blocks than the AND, OR, or NOT gate. The NAND (NOT-AND) gate may be constructed by inverting the output of an AND gate, and similarly, the NOR (NOT-OR) gate is an OR gate with its output values inverted. The gate symbols and truth tables for these two new gates are shown in Figure 1.9.

The gate symbols shown in Figures 1.7 and 1.9 are consistent in the sense that a small circle represents the logical operation of inversion. The INVERTER gate consists of a triangle, a symbol that indicates no logical operation at all (in a practical circuit this shape is used for an amplifier, e.g., a "power booster," meaning that the logic value of the output follows exactly the value of the input, no matter how many connections are "loading" the output of the circuit), and a small circle, which indicates inversion. The NAND gate symbol uses an AND gate symbol with a small circle added to it for inversion of the output, as does the NOR gate symbol with respect to the OR gate.

1.6 BINARY (LOGIC) CIRCUITS

Basic laws of logic and combined logic operations provided us with five basic logic gates (Figures 1.7 and 1.9). These gates are representations of elementary binary systems described in terms of binary variables. Just as complex logic propositions may be constructed from simple propositions and logic operations, complex digital circuits may be constructed from basic logic gates and connections between their output and input terminals. As long as the feedforward nature of the connections are observed, a combinational binary system is constructed, and the overall output–input transfer function, that is,

Type of gate	Symbol	Truth table
NAND	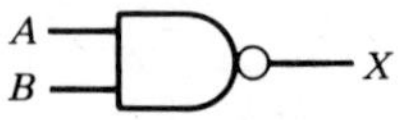 A, B → X	A B / X 0 0 / 1 0 1 / 1 1 0 / 1 1 1 / 0
NOR	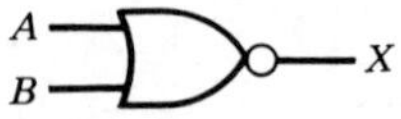 A, B → X	A B / X 0 0 / 1 0 1 / 0 1 0 / 0 1 1 / 0

FIGURE 1.9 NAND and NOR gates and their truth tables.

the truth table of the system, may be determined. A circuit example that uses three types of logic gates is shown in Figure 1.10.

The circuit has three independent input variables labeled A, B, and C, respectively. It has one output variable for which the symbol X is used. Three symbols are added, O_1, O_2, and O_3, which identify "internal" variables, the outputs of the three gates that previously had no labels. The overall truth table of this circuit is given in terms of inputs A, B, and C and output X. The truth table is calculated from level to level, as was shown in Section 1.3.

The eight rows include all possible combinations of the three input values. Values for signal O_1 are determined by the application of the truth table for the AND gate (Figure 1.7) using the given values for A and B as input variables. The values for O_2 are calculated by inverting the values of input C. Now, level 1 is complete. Values for O_3 are determined by inverting the values of O_1, which completes the calculations for level 2. Finally, the values for output X are determined by using the truth table for a NOR gate (Figure 1.9) and variables O_2 and O_3 as inputs.

Once the output–input transfer function of the circuit of Figure 1.10 has been determined, the entire circuit may be replaced by a new system component that has three inputs, one output, and the calculated truth table. Knowing the truth table of this circuit, we may find its functional specification as a complex proposition, namely, that the output is equal to Logic-1 if and only if all three inputs are equal to 1. The same logical statement may be expressed by the combination of two simple propositions which involve only two variables:

Both A and B are equal to Logic-1 and C is equal to Logic-1.

The block diagram and functional specifications of this system are shown in Figure 1.11. From the overall specifications (expressed as a combined prop-

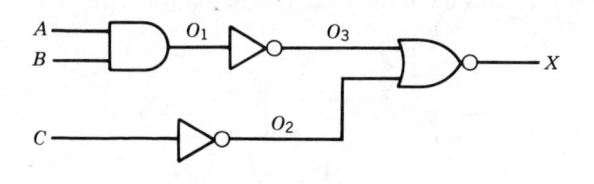

A B C	O_1 = A AND B	O_2 = NOT C	O_3 = NOT O_1	X = O_2 NOR O_3
0 0 0	0	1	1	0
0 0 1	0	0	1	0
0 1 0	0	1	1	0
0 1 1	0	0	1	0
1 0 0	0	1	1	0
1 0 1	0	0	1	0
1 1 0	1	1	0	0
1 1 1	1	0	0	1

FIGURE 1.10 Circuit example using three gate types.

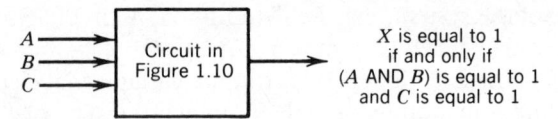

16   DIGITAL SYSTEMS, LOGIC GATES, AND BOOLEAN ALGEBRA

A →
B → Circuit in Figure 1.10 → X is equal to 1 if and only if (A AND B) is equal to 1 and C is equal to 1
C →

FIGURE 1.11 Circuit in Figure 1.10 redefined by logic assertions.

osition) a new binary circuit can be constructed that uses only two-input AND gates. As shown in Figure 1.12, this new circuit produces exactly the same overall truth table as that in Figure 1.10.

The preceding examples demonstrate two important properties of binary circuits. First, many different binary circuits could satisfy the same functional specifications. These circuits are indistinguishable from each other as far as their output–input characteristics are concerned, but they could have vastly different internal structures. One of the major practical problems digital designers face is the search for the least expensive circuit that satisfies a given set of functional specifications.

The second observation we can make is that the use of binary gates and the systematic evaluation of truth tables may be equated to complex logical reasoning. The calculation of the entries in the truth table is much simpler than logical reasoning, and since the two systems are equivalent, it could be used in place of logic. In constructing binary circuits, we have found an important analogous method, a tool, that can be extended to other applications, logic in particular, and to any binary system in general.

1.7 FUNDAMENTALS OF BOOLEAN ALGEBRA

So far we have used only one method for evaluating the output–input function of binary combinational systems. This method involved the use of basic binary gates and their interconnections, which form binary circuits. From the graph-

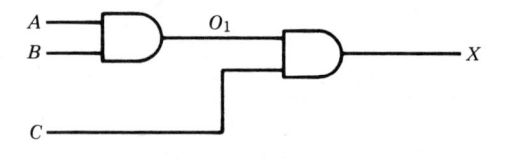

A B C	$O_1 = A$ AND B	$X = O_1$ AND C
0 0 0	0	0
0 0 1	0	0
0 1 0	0	0
0 1 1	0	0
1 0 0	0	0
1 0 1	0	0
1 1 0	1	0
1 1 1	1	1

FIGURE 1.12 Simplified circuit replacement for circuit in Figure 1.10.

ical representation of the circuit diagram and the known truth tables of the gates, the output values of the circuit could be calculated for any given combination of input values. The same calculations may be done by a special algebraic method called **Boolean algebra.** It is often the case that there are parallel, complementary methods of design. In many fields of design graphical and written (algebraic) methods exist side by side. Both of these methods are self-sufficient, that is, they can be used to calculate the complete answer. Nevertheless, it is useful to have a number of methods. In some cases graphical methods are more useful, whereas in others more formal, algebraic methods are easier to use. People handle graphical information extremely well; computers prefer algebra. Another advantage of having several design methods is that each method may provide a different insight into the fundamental properties of the design process. For example, in the design of computer programs the program itself plays the role of "algebra," whereas flowcharts or data flow diagrams use graphical methods. The two representations complement each other and help in the understanding of the structure of the program. In digital systems we started with the graphical method. Let us now look at the algebraic representation.

The definitions for Boolean algebra were first formalized by mathematician George Boole [1] in the middle of the nineteenth century. For a long time Boolean algebra was applied only to logic. With growing interest in two-valued digital, so-called switching systems, Shannon [2] extended this algebra in the early 1900s to binary gates, circuits, and binary time-varying systems. In the twentieth century the rules of Boolean algebra were applied more and more to the design of binary circuits. Today Boolean algebra forms the formal foundation of digital circuit design.

It is possible to present a formal algebraic system that applies to both binary and multivalued digital variables; however, we shall restrict our discussions to two-valued binary variables only. We will not use strictly formal definitions because we use Boolean algebra as a design tool, not as a formal theorem-proving system. The axiomatic, formal way of defining Boolean algebra uses postulates formulated by E. U. Huntington [3] in 1904. We shall use these postulates but without the usual formal axiomatic language of mathematics. Instead, we shall build the rules of the algebra in a systematic way, presenting the simplest ones here, while postponing the presentation of a few more complex rules until the next chapter.

The use of an algebraic method requires symbols, operators, and rules that transform one algebraic expression into another. We have already used symbols for binary variables and defined three basic operators. We have to assign algebraic symbols to these operators and then find the useful rules of the algebra. There is controversy about which symbols to use for the AND and OR operators. Logic uses the special symbol $\wedge$ (wedge), set theory uses $\cap$ (intersection), and most digital designers use the centered period, · (multiplication), for the AND operator. For the OR operator the symbols $\vee$ (vee), $\cup$ (union), and the plus sign, + (addition), are used.

The multiplication and addition symbols are more commonly used for digital design and will be adopted throughout this book. The problem with these common symbols is that they are the same as the symbols for the multiplication and addition operators of common algebra. This is not an accident. We shall often refer to the AND operator as *logical multiplication* and to the OR operator as *logical addition* because many of the basic Boolean rules are similar to their common algebraic equivalents. Similarities between Boolean operators and ordinary algebraic addition and multiplication may be confusing at first, especially when the rules of the two algebra are different. The confusion may be compounded by the use of the symbols 1 and 0 in place of Logic-1 and Logic-0. It is best to approach these two basic logic operators as if ordinary multiplication and addition did not exist. Rules of algebra, such as the use of parentheses, substitution, simplification, factoring, and general properties such as commutation, association, and distribution will be very useful. However, the danger in relying on experience with ordinary algebra is that the familiar rules will be used more often, and the rules of Boolean algebra, which have no similar rules in regular algebra, will be ignored. This introduces an unnecessary bias in our algebraic methods, which may hinder us in finding the simplest solution to a problem.

We have no similar problems with NOT, the complement operator, because it has no equivalent in regular algebra. There are two commonly used symbols. One is the prime (') and the other is the overbar ($\overline{}$). When the prime is attached to a symbol or a Boolean expression, the variable or the entire expression is complemented. The symbols A' and A may be considered as two different variables with the constraint that if A is equal to 0 then A' is equal to 1 (and vice versa). The symbols $\overline{A}$ and A' are equivalent. Since both of these complement notations are used in practice, we shall alternate between using one or the other. Table 1.3 summarizes the operator symbols used throughout this text.

We already know that any binary variable may take one of two possible values: 0 or 1. One is the **null,** the other the **identity** element, but their roles are exchanged with respect to the two basic operators AND and OR. These two basic elements of algebra are defined by a given operator as follows: When a variable is operated on using the null element, the result is equal to the null element; when the indentity element is used, the result is equal to the variable. For the AND operator we have the following two basic rules:

$$A \cdot 0 = 0 \quad \text{and} \quad A \cdot 1 = A$$

TABLE 1.3 Operators and Their Symbols for Boolean Algebra

Logic operator	Logic statement	Boolean operator	Boolean expression
AND	P1 **AND** P2	$\cdot$	$A \cdot B$ or AB
OR	P1 **OR** P2	$+$	$A + B$
NOT	**NOT** P1	$'$, $\overline{}$	A' or $\overline{A}$

and we can see that the 0 value is the null, whereas the 1 is the identity element. (It does look like regular algebra, doesn't it?)

The two fundamental rules for the OR operator are:

$$A + 0 = A \quad \text{and} \quad A + 1 = 1$$

where the roles of the identity and null elements are interchanged. The last rule does not hold for regular algebra, of course, but it makes Boolean algebra symmetrical as far as the two operators AND and OR are concerned. We can also see that for the OR operator the 0 takes on the role of the identity element, whereas 1 behaves as the null element.

How can we prove that these equalities are correct? Constructing an appropriate truth table is a simple and fundamental method with which one may prove conclusively that a Boolean equality is correct. Since the truth table contains all possibilities, the proof is a complete proof. The preceding four equalities contain only one variable, A; therefore, the truth table will have only two entries. The truth table for the four equalities follows:

A	$A \cdot 0$	$A \cdot 1$	$A + 0$	$A + 1$
0	0	0	0	1
1	0	1	1	1

The values in the truth table are determined by substituting the correct value for A (0 for the first row and 1 for the second) and by the appropriate entries in the truth table given for the AND and the OR operations (Fig. 1.7). For example, in the second column and first row we have $0 \cdot 1$, which is equal to 0, and in the second column, second row we have $1 \cdot 1$, which is equal to 1. Once the truth table is known, the results may be reinterpreted as rules. The first column is always 0 (a constant), all entries of the second and third column are equal to the values of the variable A, and the fourth column is equal to 1.

There are two other basic equalities that contain only one variable and the AND or the OR operator. These are

$$A \cdot A = A \quad \text{and} \quad A + A = A$$

So far we have concentrated on the AND and OR operators. We now look for simple rules that contain the NOT operator as well. Remembering that the null element for the AND operator is 0, whereas for the OR operator it is 1, the two equalities

$$A \cdot \overline{A} = 0 \quad \text{and} \quad A + \overline{A} = 1$$

show that when either the AND or the OR operator operates on a variable and its complement, the result is equal to the null element for the operator (0 for AND and 1 for OR). There are no "inverse" operations in Boolean algebra (in the sense that subtraction is the inverse operation of addition in regular algebra), but these two equalities do express the fact that A' is the inverse of A with respect to both basic operations. Finally, the nature of the NOT operator (complement) indicates that the equality

$$(A')' = A$$

must be true. The truth table proof of this equality requires two columns:

A	A'	$(A')'$
0	1	0
1	0	1

These nine rules express the simplest and most fundamental Boolean equalities. The next group of rules express various properties of the operators.

An operator is **commutative** if the roles of the two variables used for the operation may be exchanged. Both the AND and OR operators are commutative; therefore, the following two equalities are true:

$$A \cdot B = B \cdot A \quad \text{and} \quad A + B = B + A$$

Of course, these equalities may also be proven by a truth table, which, in this case, has four entries, since two variables are involved.

The **distributive** property requires three variables and applies to two different operators (in this case to the operators AND and OR). It helps in untangling parenthesized expressions by *factoring* or *multiplying out*. In Boolean algebra both the AND operator is distributive over the OR operator and the OR operator over the AND operator:

$$A \cdot (B + C) = (A \cdot B) + (A \cdot C) \quad \text{and} \quad A + (B \cdot C) = (A + B) \cdot (A + C)$$

The parentheses on the right side of the first equality are superfluous because the AND operator is of higher precedence than the OR operator. We included them to demonstrate the symmetry of the two equalities. Normally, in the more concise form, the same two expressions are stated as

$$A(B + C) = AB + AC \quad \text{and} \quad A + BC = (A + B)(A + C)$$

The first expression is similar to regular algebra and allows factorization and multiplying out parenthesized expressions. The second expression is unique to Boolean algebra and will take some time to get used to.

Let us demonstrate once again the use of truth tables for proving equalities. The first of the preceding equalities may be proven by a truth table that contains five columns and eight entries. This is shown in Table 1.4. The truth table entries are determined in the same way as before. Using the required values for the variables as they appear in the first column of the truth table and the truth tables of the basic operators AND and OR, the results can be calculated for each horizontal entry, progressing from left to right. Finally, when the entire table is filled, the values in the two columns labeled $A \cdot (B + C)$ and $A \cdot B + A \cdot C$ are compared. The values in these two columns are equal, and the proof is complete.

TABLE 1.4 Using Truth Tables to Prove Equality $A \cdot (B + C) = A \cdot B + A \cdot C$

A B C	B + C	$A \cdot (B + C)$	$A \cdot B$	$A \cdot C$	$A \cdot B + A \cdot C$
0 0 0	0	0	0	0	0
0 0 1	1	0	0	0	0
0 1 0	1	0	0	0	0
0 1 1	1	0	0	0	0
1 0 0	0	0	0	0	0
1 0 1	1	1	0	1	1
1 1 0	1	1	1	0	1
1 1 1	1	1	1	1	1

The **associative** property applies to the repeated application of one operator and expresses the property that the order of performing the operations is immaterial. Three variables are involved. Both the AND and the OR operators are associative; therefore,

$$A \cdot (B \cdot C) = (A \cdot B) \cdot C \quad \text{and} \quad A + (B + C) = (A + B) + C$$

Since the order in which the operations are performed is immaterial, the parentheses are superfluous. Both sides of the first equality may be written as $A \cdot B \cdot C$ (or ABC) and of the second as $A + B + C$. The associative property may be applied repeatedly to terms containing more than three variables when only one type of operator is involved. In these cases the parentheses are always dropped.

We turn our attention now to basic rules that contain all three operators. We find only two more simple rules, which are called De Morgan's laws. They are

$$\overline{A \cdot B} = \overline{A} + \overline{B} \quad \text{and} \quad \overline{A + B} = \overline{A} \cdot \overline{B}$$

We can notice the symmetry of these two equalities. De Morgan's laws are often used when entire expressions are complemented, since these are the only rules that can be used to simplify a complemented expression by separating it into individual terms. In fact, De Morgan's laws apply to any number of variables:

$$(A + B + C + D + \cdots)' = A' \cdot B' \cdot C' \cdot D' \cdots \quad \text{and}$$
$$(A \cdot B \cdot C \cdot D \cdots)' = A' + B' + C' + D' + \cdots$$

This concludes the list of the simplest, most basic rules of Boolean algebra. They are collected in Table 1.5 as AND rules, OR rules, and NOT rules. We can see that Boolean algebra is very "regular" indeed. Both basic operators satisfy all three useful properties. There is a noticeable symmetry between AND rules and OR rules. In fact, all OR rules may be derived from the AND rules by the application of the principle of **duality.** The duality principle states that a Boolean equality remains valid if all AND operators are changed to OR operators, all OR operators to AND operators, all 0's to 1's, and all 1's

TABLE 1.5 Summary of Basic Boolean Equalities

AND *rules*	OR *rules*	NOT *rules*	Name of rule
$A \cdot 0 = 0$	$A + 1 = 1$	$0' = 1$	
$A \cdot 1 = A$	$A + 0 = A$	$1' = 0$	
$A \cdot A = A$	$A + A = A$	$(A')' = A$	
$A \cdot A' = 0$	$A + A' = 1$		
$A \cdot B = B \cdot A$	$A + B = B + A$		Commutative
$A \cdot (B \cdot C) = (A \cdot B) \cdot C$	$A + (B + C) = (A + B) + C$		Associative
$A \cdot (B + C) = A \cdot B + A \cdot C$	$A + (B \cdot C) = (A + B) \cdot (A + C)$		Distributive
$(A \cdot B)' = A' + B'$	$(A + B)' = A' \cdot B'$		De Morgan

to 0's. The rules are arranged in Table 1.5 so that the two dual expressions are placed next to each other in the AND rules and OR rules columns.

Even though the rules in Table 1.5 are given in terms of variables A, B, and C, it is important to realize that any Boolean expression may be substituted in place of any one of these variables. Of course, methods of substitution are valid for algebras in general, and in fact, they are the most important algebraic methods that make algebra really useful. To demonstrate how substitution may be used, let us derive De Morgan's law for three variables, X, Y, and Z:

$$(X \cdot Y \cdot Z)' = ?$$

Using the substitutions $X = A$ and $Y \cdot Z = B$, we get

$$(A \cdot B)' = A' + B' = X' + (Y \cdot Z)' = X' + Y' + Z'$$

Boolean algebra is used to transform Boolean expressions from a given form to another form that is more useful to the designer. Often, the purpose is simplification. In order to use Boolean algebra for a circuit, a Boolean expression must be derived from the circuit diagram. If all input and output variables of the gates are labeled, and only basic gates are used, then a pair of parentheses may be used for each gate, and the appropriate operator symbol for the gate type inserted into the expression. This may be demonstrated for the circuit shown in Figure 1.10 on Page 15. We get the following binary gate equations:

$$O_1 = (A \cdot B)$$
$$O_2 = (C)'$$
$$O_3 = (O_1)'$$
$$X = (O_2 + O_3)'$$

where we used the fact that the NOR gate is equivalent to an OR gate with a complemented output.

If we substitute the first equation into the third and then the second and third into the fourth, we get

$$X = [(C)' + (A \cdot B)']'$$

which looks complicated but can be simplified using De Morgan's laws. The inverted OR expression on the right side is transformed by De Morgan's law, $(A + B)' = A' \cdot B'$:

$$[(C)' + (A \cdot B)']' = [(C)']' \cdot [(A \cdot B)']'$$

and since complementing any expression twice results in the original expression, we get

$$X = (C) \cdot (A \cdot B)$$

where the AND operators and the parentheses can be dropped and the order of the variables rearranged (commutative property!), so, finally, we get

$$X = ABC$$

This is the same result that we found by interpreting the overall truth table for the circuit.

We have seen two methods (one graphical, the other algebraic), with which binary circuits may be analyzed. In the next two sections we shall examine how these circuits may be built in the laboratory.

1.8 PHYSICAL REALIZATION OF BINARY CIRCUITS

The useful application of our analysis and design of binary circuits depends on our ability to build physical systems that obey similar rules to our system representation. A physical binary system must have physical variables that can be in one of two physical states. As an example of a physical binary system, the see-saw, a mechanical device, clearly has two well-defined states. A battery, a switch, and a light forms an electrical system that also has two states (Figure 1.13). While in our system representation variables may have only two states, physical variables are often continuous, analog quantities. If

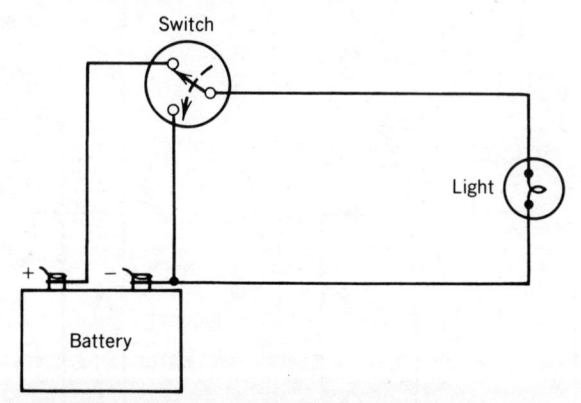

FIGURE 1.13 Electrical switch and light as two examples of simple binary devices.

the two states of the see-saw are defined as one of the two ends of the see-saw touching the ground, then while this mechanical system changes from one state to the other, it remains in an ambiguous state, which is neither 1 nor 0. This "flipping" of states may take a long time.

The electrical system seems to emulate a binary system more closely. The light is either **on** or it is **off**. However, if we want to decide the state of the light by measuring the light intensity, we find that it varies continuously from zero (off) to a given maximum value. In this case it may be very difficult to decide which intensity values should represent the on state, since different light bulbs may produce very different light intensities.

These examples demonstrate two important difficulties that arise in the physical realization of binary systems. First, physical measurements must be able to decide reliably the state of each variable. Second, the changing of states takes time, during which the states of the changing variables are undefined. Hence, the measurements must be made only at those time periods when the system is in a quiescent state, that is, all measured variables are settled down to one state or the other (this is similar to the problem we encountered in the chess game when a chess piece was being moved).

Our main interest is in the design and construction of practical digital circuits that are all based on electrical devices. Two such electrical devices (the switch and the light) are shown in Figure 1.13. The switch, an electromechanical input device, may be placed in its 0 or 1 state by flipping a lever from one position to another. The light, an optoelectrical output device, indicates the state of a physical variable, the voltage across the light bulb, by producing light when the voltage is high. The circuit may be tested by setting switches into their required states and by observing or recording the states of the lights. It is possible to arrange lights and switches in such a way that they perform the two basic Boolean operations AND and OR. In Figure 1.14 both AND and OR operations are demonstrated. Light L_1 is on only if both switches

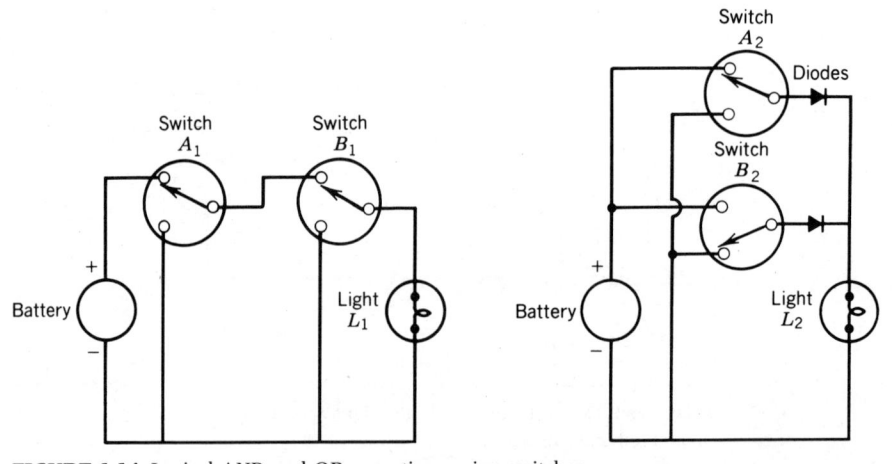

FIGURE 1.14 Logical AND and OR operations using switches.

A_1 and B_1 are on, whereas light L_2 is on if either switch A_2 or B_2 is on. In the case of the OR circuit we could not simply connect the outputs of A_2 and B_2 together, because the battery could be short-circuited. In this case diodes are used that allow the flow of current only in one direction.

The simple OR circuit already shows problems with using switches to realize Boolean operations. In addition to these and similar difficulties, switches and lights are not much use for building binary circuits since they are operated and "measured" (observed) by the operator, not by the system. We need electrically operated "switches" (gates) that perform Boolean operations. These are readily available today in the form of digital integrated circuits, or ICs.

The internal construction of an IC device may be very complex. We are not interested in the details of its electrical construction but would like to present a model for its operation, which exhibits most of its important physical characteristics. It would be possible to consider an IC device equivalent to the symbols of binary gates and to ignore the differences between the analytical representation and its physical realization. Today most IC devices work so reliably that they could be considered equivalent to their Boolean representation. Still, it is worthwhile to examine the main principles of their operation since this will give us some insight into what makes them so reliable.

In Figure 1.15 the schematic diagram of a single IC INVERTER gate is shown. Two separate circuits are identified. Both of these are constructed from electrical devices such as diodes, transistors, resistors, and so on, and both of them require electrical power for their operation. Such an INVERTER gate may be one of a very large number of IC types, called **IC families,** but most of them use the voltage level at their terminals to decide the state of their inputs. The sensing circuit determines whether the input voltage level belongs to the **low-** or **high-voltage** state and sends a control signal to the output driver circuit. The output driver circuit generates the necessary output

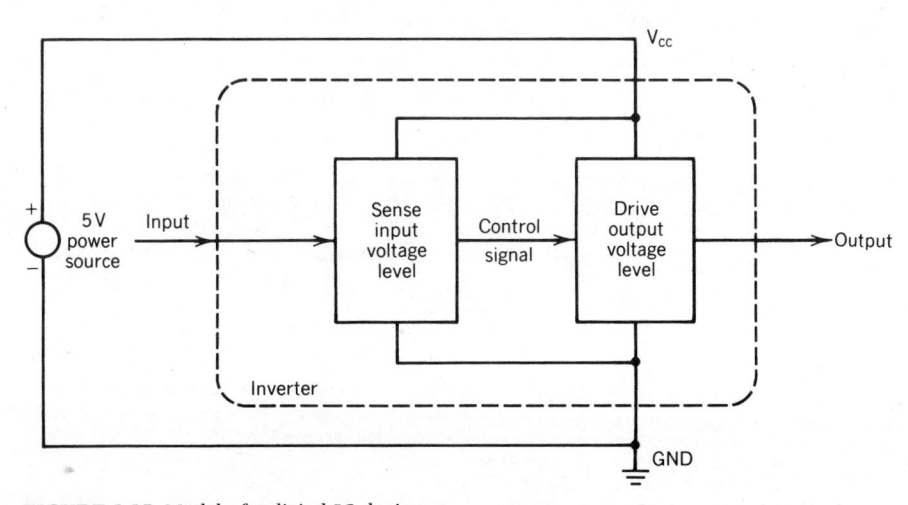

FIGURE 1.15 Model of a digital IC device.

26 DIGITAL SYSTEMS, LOGIC GATES, AND BOOLEAN ALGEBRA

voltage level for the device and provides sufficient power for driving 10 or 20 input terminals of similar IC devices. This model emphasizes the fact that the IC device, in a way, isolates its output from its input and requires an outside electrical power source for its operation. Students of digital design who use IC devices the first time often have the wrong impression that the outputs are somehow directly connected to the inputs.

Digital ICs are inherently analog devices. The output voltage is a continuous (though nonlinear) function of the input voltage. The actual voltage levels associated with an IC device depends on its type. For simplicity, we shall consider only the so-called TTL (transistor–transistor logic) device family for the demonstration of the principles of IC device operation. The TTL family requires only one 5-V power supply, and the voltage levels of all signals are between 0 and 5 V.

The binary nature of the voltage level is defined in the following manner: Whenever a voltage level is above 2 V, the respective signal is in its **high** state; whenever it is below 0.8 V, the signal is in its **low** state (see Figure 1.16). A typical TTL IC device may provide as high as 3.5 V when its output is in the high state and is lightly loaded, and approximately 0.2 V when it is in its low state without much loading. This arrangement of threshold voltages provides a very large degree of reliability. The actual voltage levels will be a function of loading and change from device to device, but there is a margin of 1.5 V for the high state and 0.6 V for the low state in this case within which the signals may vary without any effect on the correct operation of the device. Since real electrical signals exhibit noise (random variations of their voltage levels in time), these threshold voltages provide **noise margin.** This means that noise variations of 0.6 V or less will not affect the operation of TTL circuits. This is one of the basic principles that made computer circuits such unbelievably reliable physical systems. No other man-made systems come even close to the reliability of digital ICs.

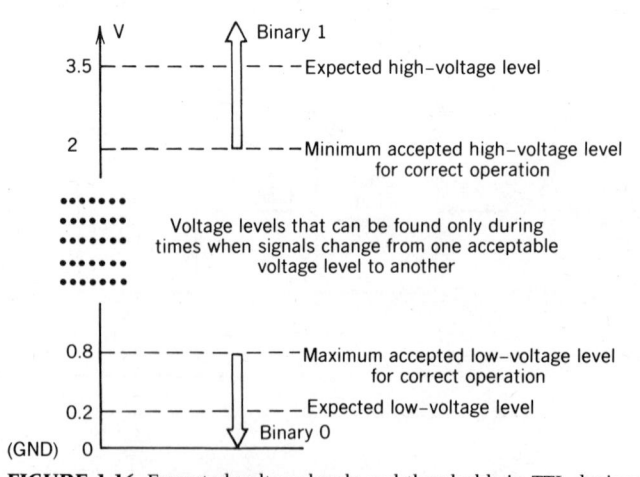

FIGURE 1.16 Expected voltage levels and thresholds in TTL devices.

When a practical IC is heavily loaded (a large number of inputs are connected to its output), its output voltage tends to decrease when it produces a high output and it becomes higher when it is in the low state. IC manufacturers limit the maximum allowable loading of gates by assigning a minimum high (called V_{OH}) and a maximum low output (called V_{OL}). For the TTL family of ICs, V_{OH} = 2.4 V and V_{OL} = 0.4 V. Under these extreme conditions the noise margin is 0.4 V for both the high and the low states.

We have already mentioned that in reality the digital IC is an analog device. This is demonstrated in Figure 1.17, where a typical output–input voltage function of a TTL INVERTER gate is shown. We can see that the "flipping" of the output voltage occurs within a relatively narrow input voltage range of around 1 V. From this curve we can also clearly see that if the input voltage is in the low state (e.g., it is around 0.2 V), then it may increase up to 0.8 V (a difference of 0.6 V) without any significant changes occurring in the output voltage. Similarly, if the input voltage is in the high state (3.5 V), then it may decrease as much as 2 V (down to 1.5 V) before the output voltage changes. The output–input voltage characteristics of the IC device allows input voltage variations (noise) without disturbing the correct binary output level of the device. The interconnection of such components yields a very reliable digital system.

The selection of proper voltage threshold levels allows the assignment of binary states to analog signals and handles the first mentioned difficulties of physical systems very well. The second difficulty, the fact that during transitions signals have undefined states, is not a serious concern for combinational circuits. The independent inputs of a combinational system may be provided by the setting of switches. While the switches are operated, the system variables (the voltage levels in the circuit) are also changing. After the last independent input signal has been changed, the circuit must be allowed to settle. Since practical IC gates have settling times of a fraction of a microsecond, the settling of a practical combinational circuit occurs automatically. We can postpone our worries about the time dependency of signal state

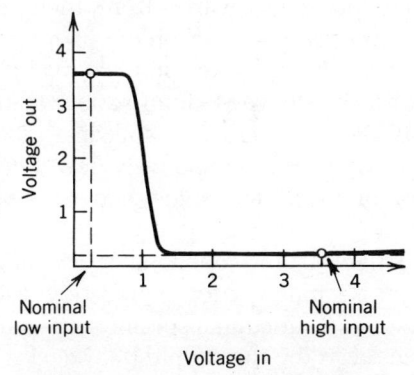

FIGURE 1.17 Voltage output–input characteristics of TTL INVERTER gate.

28 DIGITAL SYSTEMS, LOGIC GATES, AND BOOLEAN ALGEBRA

transitions until Chapter 5, where we will begin the study of sequential systems.

This is as much detail as we want to go into concerning the physical operation of IC devices. For different IC families the voltages, thresholds, and voltage curves are all different, but the principles of operation are the same. As long as devices of only one family are interconnected, they should work reliably together. If the interconnection of devices from different families is required, the detailed electrical characteristics of the ICs must be taken into account, and often interface circuits need to be used. We will simplify our problems in design and construction by selecting one family, the already mentioned TTL family of IC devices. Other technologies, (NMOS, CMOS, ECL, etc.), are also widely used today, but the study of their detailed electrical characteristics is beyond our aim.

We are getting nearer to our principal goal, the design and construction of a practical digital circuit. By choosing devices from one family of ICs, the drawing of a Boolean circuit diagram and the construction of the physical circuit become two very closely related activities. Physical circuits may have a few problems that do not occur in circuit diagrams drawn on paper. These are discussed in the next section.

1.9 PRACTICAL DIGITAL CIRCUITS

Practical IC devices are made in various physical forms. A photograph of the most common type, the dual in-line IC device is shown in Figure 1.18. The active circuit, a tiny silicon microchip, is encased in a rectangular plastic or ceramic case to which metallic conductors, called IC pins, are fixed. The pins are connected internally to the microcircuit. The digital circuit is constructed by providing good electrical connections to the IC pins. The connections are either between pins, or between the power supply terminals, switches, or lights, and the pins of ICs.

For each IC device family a large number of different gate types are made. The number of IC pins per device varies from 14 to 80, and the internal complexity of the devices vary by five orders of magnitude. The simplest devices have less than ten binary gates and are manufactured with small-scale integration (SSI) technology. Medium-scale integration (MSI) devices hold ten to hundreds of gates, whereas large-scale integration (LSI) can produce devices with thousands of gates. The relatively new, but rapidly developing, very large scale integration (VLSI) technology can pack as many as one million gates into one IC device.

The electrical, mechanical, and functional specifications of a family of IC devices are collected in a handbook for engineering design [4]. In order to use a particular IC device, its functional specification and its pin assignments must be known. Often, even the the simplest SSI device contains more than one gate. One of the most frequently used small-scale TTL device, the type 7400 device, has 14 IC pins and contains 4 NAND gates. The functional characteristics of the IC device are often shown in a graphical form, which

FIGURE 1.18 Photograph of dual in-Line IC device.

defines both its functional characteristics and its pin assignments (see Figure 1.19). The pins of the physical device are numbered counterclockwise when the device is viewed from the top. A notch, or a small dot, identifies pin 1. In addition to the eight input and four output connections for the four NAND gates, one pin, in this case pin 7, must be connected to the ground (GND), whereas pin 14 is connected to the +5-V power supply terminal (V_{CC}). Some IC devices contain unconnected pins (UC) as well. A useful list of pin assignments for TTL devices are shown in Appendix 1. Data for other devices

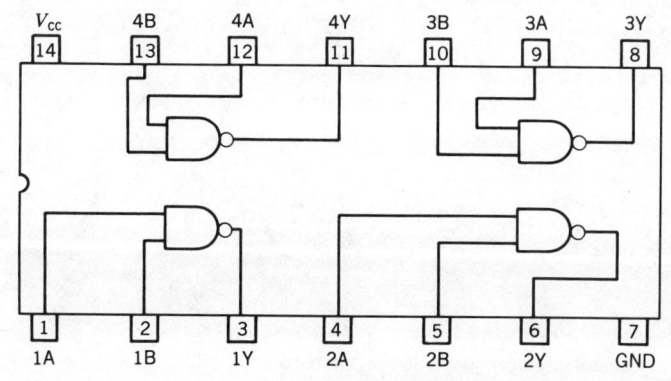

FIGURE 1.19 Pin assignments for TTL type 7400 IC device.

may be found in engineering design handbooks of most major IC manufacturers [4].

As mentioned before, connections between the IC pins, the power supply terminals, switches, and lights must be made by metallic wires or strips. With respect to the physical ruggedness of the circuit, there are three major ways by which digital circuits can be constructed. The most temporary method of construction uses an IC circuit board and wires (see Figure 1.20). The dual in-line IC devices are inserted into the circuit board. Small sockets accept wires, and an electrical contact is established between an IC pin and the inserted wire next to it. This method of construction is used for the so-called digital breadboard, which is usually the first circuit constructed by a designer. It is relatively easy to modify such a circuit, but obviously, physically, the circuit is not very rugged, the connecting wires may get loose, cause electrical shorts, and so on.

The second method of construction, what may be called a semipermanent circuit, is wire wrapping (see Figure 1.21). The IC devices are inserted into special IC sockets that have thin but strong metallic pins protruding from below the sockets. A special wire-wrapping tool is used to wrap wires around the pins, which provide both good electrical contacts and a reasonably strong physical bond. Prototype systems often contain wire-wrapped circuit boards. It is a good compromise between solid physical construction and semiper-

FIGURE 1.20 Digital circuits constructed on circuit board.

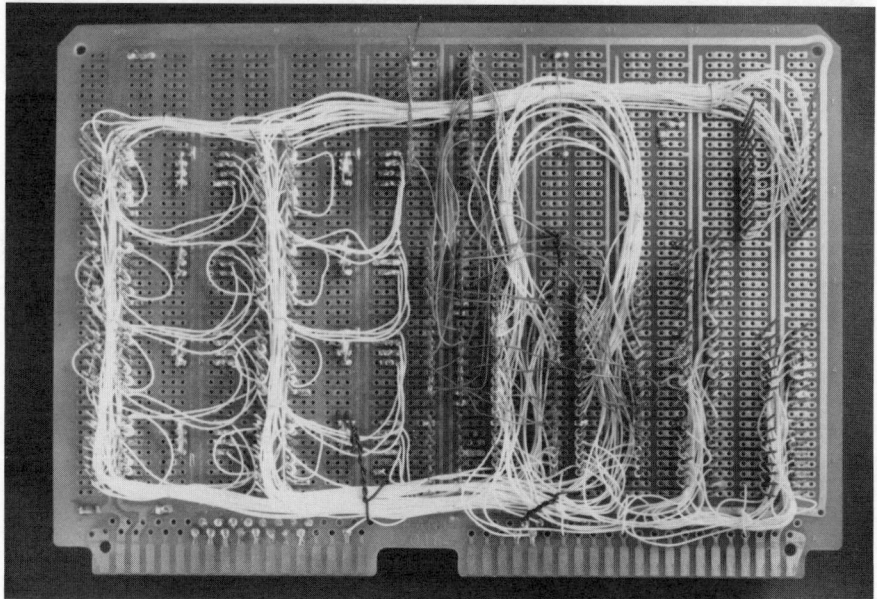

FIGURE 1.21 Digital circuit using wire-wrapped connections.

32 DIGITAL SYSTEMS, LOGIC GATES, AND BOOLEAN ALGEBRA

manency of electrical connections. Errors found in such a circuit board may be corrected with relative ease.

The third, most permanent type of construction uses printed circuit boards (PCBs). In most cases, the IC devices are solidly soldered into the board, and electrical connections are provided by thin metallic strips on either side of the PCB (see Figure 1.22). Sometimes IC sockets rather than the devices themselves are soldered into the board, in which case a number of different digital circuits may be provided that use the same electrical connections but could accommodate different IC device types.

Circuit construction, using any one of these three methods, must start

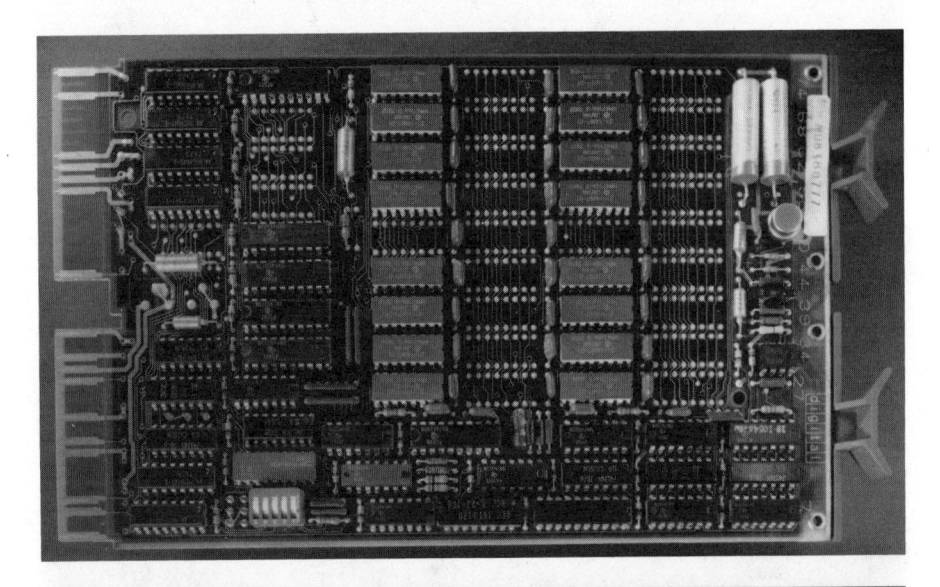

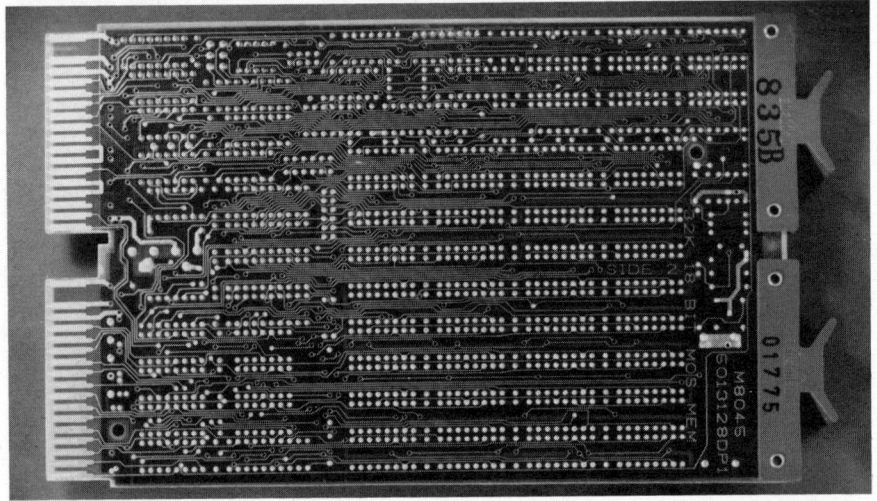

FIGURE 1.22 Printed-circuit board (PCB).

1.9 PRACTICAL DIGITAL CIRCUITS

with a carefully drawn circuit diagram. The circuit diagram is used not only during the construction of the circuit but also, sometimes more importantly, during its testing and maintenance. In either case, if erroneous connections must be found or modifications must be made, an accurate and easily interpreted circuit diagram is imperative for success. A practical circuit diagram is similar to the binary circuit discussed in Section 1.5, but in addition to the gate symbols and wire connections, it must contain information that establishes correspondence between the diagram and the physical circuit. The physical circuit is described in terms of IC devices, their physical placement in the circuit board, and IC pin numbers. A practical circuit diagram must also contain the same information. If there is only one row of IC devices used, then a consistent numbering of the ICs (e.g., left to right) is sufficient. If the devices are placed in rows and columns, both physical locations must be identified. Usually, this IC identification number is placed inside the gate symbol; hence, for the type 7400 device four NAND gates will have the same number.

The IC number indicates where in the board the IC device is placed. The pin numbers for each IC device must also appear in a practical circuit diagram (see Figure 1.23). These numbers are placed near the input–output connections of gates. In addition to these numbers, a list of IC types is necessary, which identifies the IC type used for each IC identification number. A simple practical circuit diagram is shown in Figure 1.23, where two type 7400 IC devices are used, each containing four two-input NAND gates. A more complex circuit is shown in Appendix 2.

Before we proceed to the first experiment, a few practical hints should be in order. It may be a challenge to draw the first correct circuit diagram on paper, but entirely different problems turn up during its construction. Observe the following warnings:

1. Never take pin assignments for granted. Make sure that pin numbers are correctly placed on the circuit diagram.
2. Carefully insert the IC device into the circuit board or socket, making certain that pin 1 is placed where it should be. The easiest way to damage an IC is to plug it in backward.

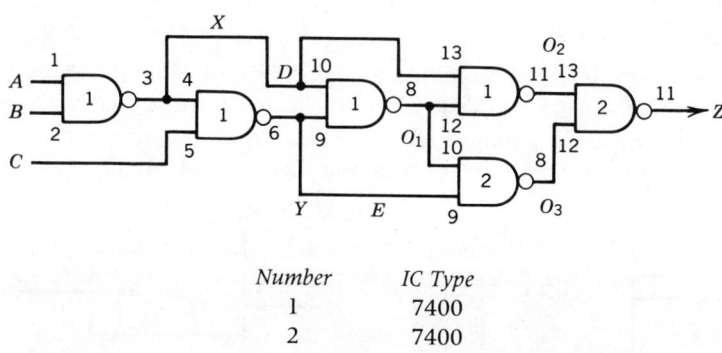

Number	IC Type
1	7400
2	7400

FIGURE 1.23 Practical circuit diagram.

34 DIGITAL SYSTEMS, LOGIC GATES, AND BOOLEAN ALGEBRA

3. Make absolutely certain that the ground (GND) and voltage supply (V_{CC}) connections are correctly connected to every IC device in the circuit.
4. Never leave any input connections of a used gate unconnected. Different IC types assume different voltage levels for unconnected inputs. It is wrong to assume that an unconnected input is equivalent to a Logic-0 (or to a Logic-1). If Logic-0 is required, connect the input to the GND terminal; if the input should be a Logic-1, connect it to V_{CC}.
5. The most difficult error to find is when outputs of gates are shorted to each other. In addition to the danger of damaging the IC devices, it is never correct to connect the output of one gate to the output of another (in a much later experiment there will be exceptions).
6. After the circuit is constructed, check that all pin numbers for which connections are shown in the circuit diagram have physical connections.
7. As long as no more than ten wires are connected to an output terminal, you do not have to worry about **fan-out,** the ability of IC devices to drive a number of input terminals. When this number approaches or exceeds 10, be aware that problems may start to occur.

EXPERIMENT 1
Analysis, Construction, and Testing of Binary Circuits

E1.1 Summary and Objectives

For this experiment two combinational binary circuits are specified. One has three inputs and two outputs; the other has two inputs and one output. After both circuits are constructed and tested, the two circuits are connected, which results in one three-input, one-output combinational circuit. The overall truth table of the combined circuit is evaluated, and the circuit is tested. The main objectives of this experiment are to provide experience in drawing practical circuit diagrams, constructing the circuits, calculating truth tables, evaluating Boolean expressions, and testing combinational circuits according to their truth tables.

E1.2 The Experiment

The block diagrams of two combinational circuits are shown in Figure E1.1. The logic circuit diagram for circuit 1 is selected from the circuit diagrams shown in Figure E1.2 (circuits 1A–1D). The circuit diagram for circuit 2 is shown in Figure E1.3. Before constructing the physical circuits, calculate the

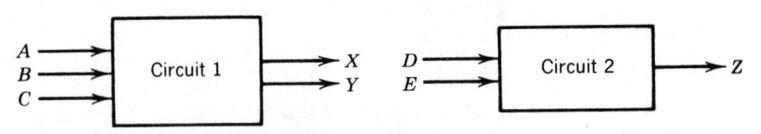

FIGURE E1.1 Block diagrams for Experiment 1.

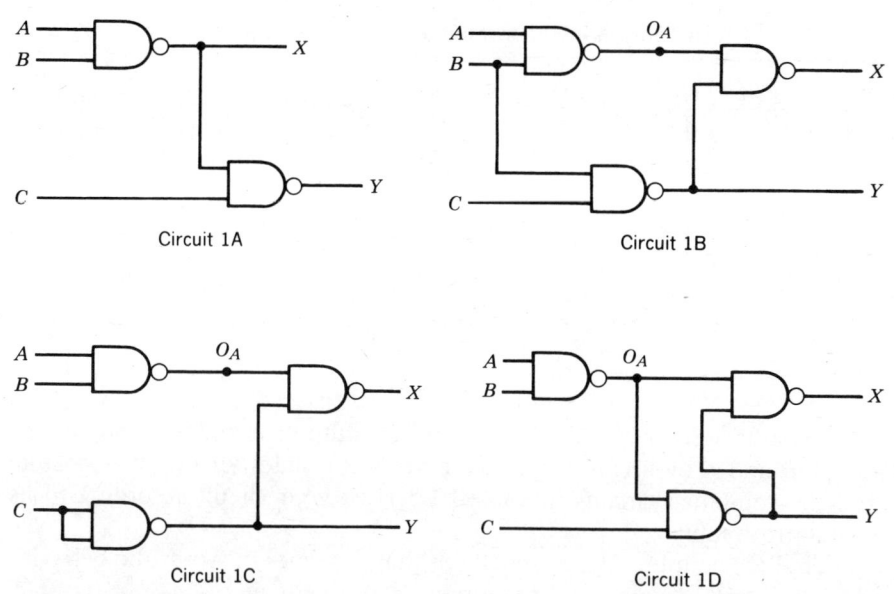

FIGURE E1.2 Four possible variations for circuit 1.

truth tables for both circuits 1 and 2 evaluating columns from left to right in Table E1.1.

Derive Boolean expressions for both circuits. Use De Morgan's laws to expand complemented expressions, and the distributive law to multiply out expressions so that the final expressions contain sums of products, Boolean terms ORed together, where each term is an ANDed expression containing variables and/or complemented variables. Calculate the truth tables for each term, and then calculate the truth tables for the outputs of the two circuits by the appropriate OR expressions. Use the tables provided in Table E1.2. Compare your results with those in Table E1.1.

Since each type 7400 IC device contains four two-input NAND gates, one IC device may be used for each circuit. Construct the circuits and test them separately using switches for independent inputs and lights for outputs.

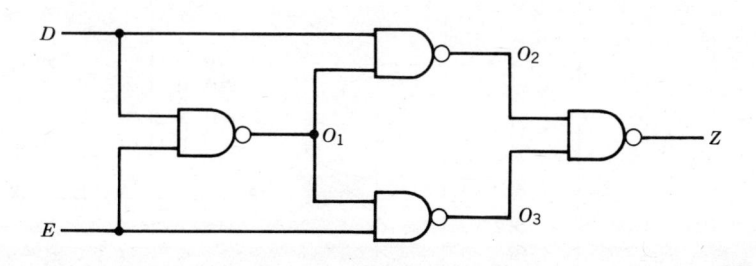

FIGURE E1.3 Four NAND gates used for circuit 2.

36 DIGITAL SYSTEMS, LOGIC GATES, AND BOOLEAN ALGEBRA

TABLE E1.1 Truth Tables for Experiment 1

A B C	O_A	X	Y
0 0 0			
0 0 1			
0 1 0			
0 1 1			
1 0 0			
1 0 1			
1 1 0			
1 1 1			

D E	O_1	O_2	O_3	Z
0 0				
0 1				
1 0				
1 1				

If the results agree with Table E1.1, connect outputs X and Y of circuit 1 to inputs D and E of circuit 2. Complete the truth tables for the two separate and the combined circuits in Table E1.3. Test your circuit according to its overall truth table.

Find the simplest logical statement that describes in words the behavior of your overall three-input, one-output Boolean circuit.

E1.3 Alternate Experiments

In Appendix 1 the pin assignments of a large number of TTL devices are listed. Note that the TTL type 7402 device contains four NOR gates. Use either two 7402 type devices by changing each NAND gate in Figure E1.2 to NOR gates or one type for one of the circuits and another for the other circuit (e.g., circuit 1 uses NOR gates and circuit 2 uses NAND gates). There are 4 combinations and 4 possible combined circuits; therefore, 16 different experiments may be tried.

TABLE E1.2 Truth Tables for Independent Check of Outputs

Boolean expression: _____

A B C	Terms		X	Y
0 0 0				
0 0 1				
0 1 0				
0 1 1				
1 0 0				
1 0 1				
1 1 0				
1 1 1				

D E	Terms		Z
0 0			
0 1			
1 0			
1 1			

TABLE E1.3 Combined Truth Table for Experiment 1

A B C	D	E	X
0 0 0			
0 0 1			
0 1 0			
0 1 1			
1 0 0			
1 0 1			
1 1 0			
1 1 1			

E1.4 Testing and Debugging the Circuit

So far we have assumed that everything has worked perfectly, and anyone who follows these directions will end up with correct truth tables and correct circuits. Since we know that in practice this is hardly ever the case, the efficient testing and debugging of the experiments are at least as important as their design and construction. For this reason, we have suggested two independent methods of calculating the truth tables. If the two results agree, then the probability of an error in the overall truth tables is greatly minimized. The most frustrating situation is when a correct circuit is tested against an erroneous truth table. It may take many alterations in the circuit before the experimenter realizes that the circuit was correct in its original form.

Once one has obtained a reasonable amount of confidence in the truth tables, the circuit must be tested against them. It is extremely important that the truth tables contain the expected output values for all gates (not only for the required outputs) because if the circuit does not work properly, then these internal values must be used to find the location of the error. It is also extremely important that a neat, easily readable practical circuit diagram is prepared that corresponds exactly to the physical circuit. It should be easy to look up any input or output terminal pins in the circuit diagram and to find the corresponding physical IC pins in the circuit. The efficient debugging of digital circuits is based solely on this strict correspondence between the circuit diagram and the circuit.

If an error is found during testing, that is, an output value is the opposite of the expected value for a given combination of inputs, the error may be found by the point-to-point tracing of signal values. When the output of a gate shows an incorrect value, then all the input values of the same gate should be checked. They could be checked physically by using a light as a testing light. Most likely, one or more of the input values will also be incorrect. These input values must be traced back to output values of other gates, which should also be checked at this stage. All this checking is done with the help of the circuit diagram, which shows how the signals are related as well as the physical locations of the pins to be tested. Once this technique is acquired, the testing and debugging proceed very rapidly, even for the more complex circuits of later experiments.

If correct input values provide an incorrect output value, then the error must be related to the gate itself. It is very rarely a "hardware" fault, meaning that the IC device is faulty. In most cases the wiring is at fault. The most frequent problems are unconnected V_{CC} or GND leads, ICs plugged in upside down or in the wrong place (pin 1 does not correspond to its expected physical position), the output of the gate is connected to a switch or to another gate output terminal, wrong pin number assignment in the circuit diagram, or a wrong IC device type.

All the mentioned problems, and numerous unmentioned ones, almost certainly guarantee that experiments will not work correctly the first time. To plan for testing and to be good at testing are often more important than being extremely careful and meticulous during design and construction. The parallels to generating computer software are very clear. It is often the case that the debugging and maintaining of complex software products are many times more expensive than their generation. You will save time in the long run if you acquire a fast, efficient hardware debugging skill while the experiments are relatively simple.

References

1. Boole, G. *An Investigation of the Laws of Thought* (1854). New York: Dover Publications (reprint), 1954.

2. Shannon, C. E. A symbolic analysis of relay and switching circuits. *Trans. AIEE,* **57,** 713–723 (1938).

3. Huntington, E. V. Sets of independent postulates for the algebra of logic. *Trans. Am. Math. Soc.*, **5,** 288–309 (1904).

4. *The TTL Data Book for Design Engineers.* Dallas Texas: Texas Instruments, 1981.

Drill Problems

1.1. Describe a railroad crossing in terms of a combinational digital system.

1.2. Describe a railroad station in terms of a digital system. Include railroad tracks, platforms, switches, and signals in your description.

1.3. Twenty-five balls are divided into five sets of five balls. Each set contains different-colored balls (red, blue, green, yellow, and black) numbered from 1 to 5. There are ten simple propositions, such as the "ball is red" or "the number of the ball is 3." State the simplest combined propositions that define the following set of balls:
  (a) The blue or green number 4 ball.
  (b) All black even-numbered balls.
  (c) Balls that are not yellow and whose numbers are larger than 3.
  (d) All red, green, or black balls except the ones with number 5 on them.

EXPERIMENT 1   39

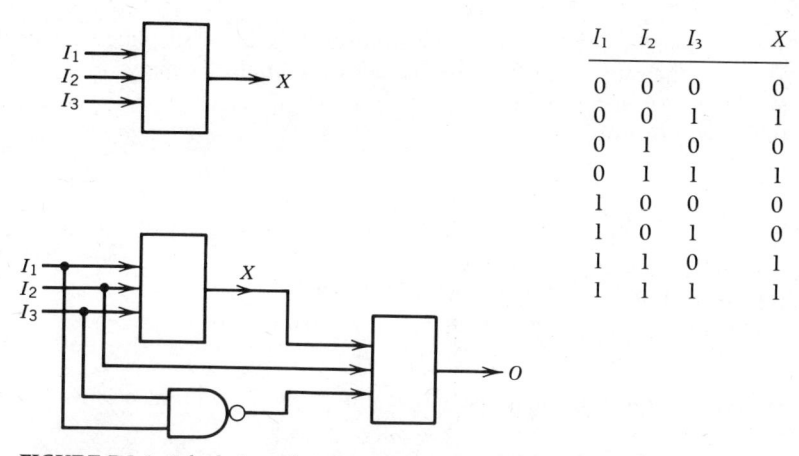

I_1	I_2	I_3	X
0	0	0	0
0	0	1	1
0	1	0	0
0	1	1	1
1	0	0	0
1	0	1	0
1	1	0	1
1	1	1	1

FIGURE D1.1 Calculation of truth table for a combinational circuit.

1.4. The truth table of a three-input, one-output binary device is shown in Figure D1.1. Calculate the truth table of the overall circuit shown also in Figure D1.1 that uses two of the same device.

1.5. Calculate the truth tables for the two circuits shown in Figure D1.2

1.6. Draw an equivalent circuit, calculate the truth table, and if possible, by restating

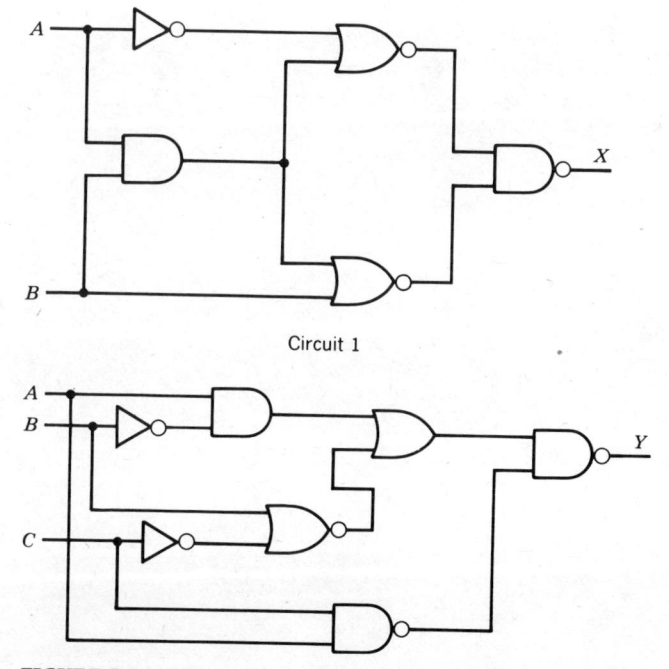

FIGURE D1.2 Two circuits using various logic gates.

the functional specifications for the truth table, show a simplified circuit that generates the same truth table for the following Boolean expressions:
(a) $A \cdot B \cdot (B + C')'$
(b) $A \cdot B \cdot (B \cdot C')'$
(c) $A \cdot B' \cdot C' + A' \cdot B' + A \cdot B' \cdot C$
(d) $A \cdot B + A' \cdot B + A \cdot B'$
(e) $(A + B) \cdot (A \cdot B + B' \cdot C + A)$

1.7. Using truth tables, prove that
(a) $(A + B + C) \cdot (A' + B) \cdot (B + C') = B$
(b) If $A \cdot B = A \cdot C$ and $A + B = A + C$, then $B = C$
(c) $(A' \cdot B' + A \cdot B)' = A \cdot B' + A' \cdot B$

1.8. Show that the NAND and the NOR operators do not have the associative property using
(a) Truth tables.
(b) Boolean algebra.

CHAPTER 2

Boolean Simplification, Binary Functions, and Digital (Logic) Circuits

2.1 INTRODUCTION

In the last chapter we introduced basic binary (logic) gates, Boolean algebra, and practical digital circuits. Our main objectives in this chapter are to expand our familiarity with these topics, acquire experience in simplifying Boolean expressions, and design and build a simple combinational digital circuit from its functional specifications. Since algebraic simplification is based on the application of a few simplifying rules, a few more rules will be added to the basic rules we studied in Chapter 1. The skill of manipulating Boolean expressions is similar to manipulating any type of algebraic expressions, which is acquired only through experience. To gain experience, the simplification of a few algebraic expressions will be demonstrated. In addition, one- and two-variable binary functions will be presented, with emphasis on their functional aspects.

Since Experiment 2 involves a practical design problem when only a few gate types can be used, the possibility of providing a variety of logic operations with a given specific gate type will be examined. This will lead to gate transformations that are necessary to satisfy the requirements of Experiment 2, the design of a two-input, three-output Boolean function generator circuit.

2.2 SET THEORY AND THE VENN DIAGRAM

Before continuing the study of Boolean algebra, we use set theory, which can be presented with similar terminology to binary systems. Set theory will provide us with a graphical technique called the Venn diagram. With this technique Boolean relationships can be illustrated using two-dimensional diagrams. As we have mentioned before, a new representation can add new insights that will ultimately help us in the design process.

A set is a collection of elements, or objects. When sets are considered as binary variables, we seek an answer to the question: does one set of objects is equal to another set? The answer can be yes (true) or no (false), itself a binary variable. A logical equality states that two sets are equal (that is, they contain the same elements). A set variable is a subset of elements where the **universal set** contains all possible objects and is analogous to the binary 1 value. The **null set** contains no objects and represents the binary 0. The binary variables are analogous to possible subsets that contain a given number of objects. The **intersection** of two sets is defined as the set of objects that belong to both sets and is clearly analogous to the binary **AND** operation. It is easy to see from the set theory point of view that the intersection of a set with the universal set is equal to the set itself ($A \cdot 1 = A$), whereas the intersection with the null set is equal to the null set ($A \cdot 0 = 0$). The **union** of two sets is the set of objects that belong to either set. This is equivalent to the binary **OR** operation. The union of the universal set with any set is equal to the universal set ($A + 1 = 1$), whereas the union of any set with the null set is equal to the set itself ($A + 0 = A$). Finally, the binary **NOT** operation is equivalent to the **complement** of a set, which is equal to the set of objects that do not belong to the set.

A graphical method for the visualization of set operations (or the equivalent Boolean operations) is the Venn diagram. As shown in Figure 2.1, a rectangle (or square) is drawn that contains the **universe,** that is, all points inside the rectangle. A circle drawn inside the rectangle represents an arbitrary set, or a Boolean variable. The set of points inside a circle represents objects belonging to the set. Shading of areas demonstrates subsets, such as the binary equivalent operations for the intersection, union, or complement of a set, as shown in Figure 2.1.

In order to represent the intersection and the union of sets (AND and OR operations for binary variables) by shaded regions in the diagram, the circles must overlap. Therefore, three is the maximum number of variables that can be used in a Venn diagram. Venn diagrams may be used to illustrate the validity of Boolean rules. For example, De Morgan's rule for three variables is illustrated in Figure 2.2. The expression $A \cdot B \cdot C$ is inverted, and the complement set is shaded. The sets A', B', and C' are indicated with three different shading patterns. The union of these three sets, $A' + B' + C'$, is the collection of shaded areas, which is equivalent to the shaded area for the $(A \cdot B \cdot C)'$ expression.

2.3 BOOLEAN SIMPLIFICATION

Before we discuss Boolean simplification methods, we should decide how we can determine whether one Boolean expression is simpler than another. From our earlier experience with regular algebra, we could count the number of literals in the expression as the deciding factor. In most cases the purpose of

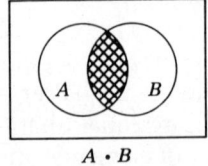

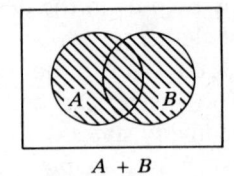

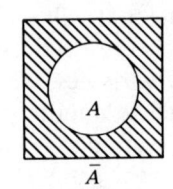

FIGURE 2.1 Venn diagram for three Boolean operators AND, OR, and NOT.

2.3 BOOLEAN SIMPLIFICATION   43

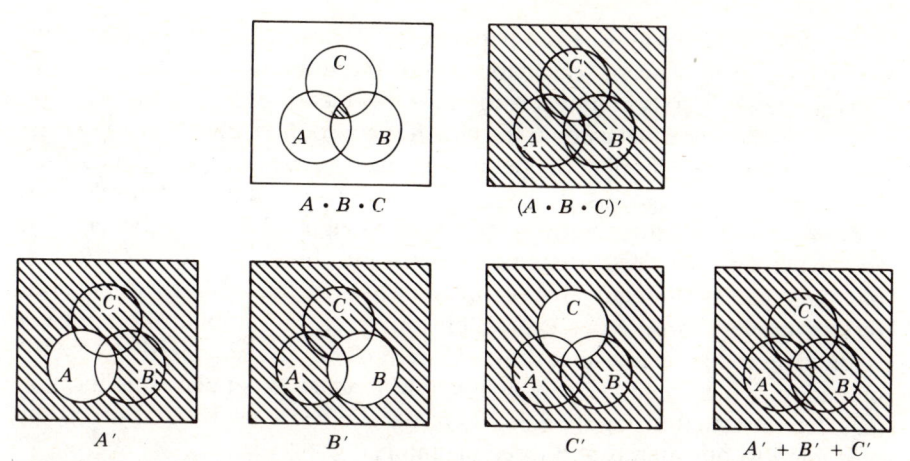

FIGURE 2.2 Proof of equality $(A \cdot B \cdot C)' = A' + B' + C'$ using Venn diagrams.

expressing a logic function as a Boolean expression is to build a digital circuit according to the derived Boolean expression. If this is our aim, then simplification has a more practical aspect. The simplest (the **minimized**) expression should provide the simplest circuit, that is, the fewest number of logic gates with the fewest number of interconnections.

In Figure 2.3 two Boolean expressions and their circuit realizations are shown. In both expressions there are four literals and three operators. In one expression parentheses are used, whereas in the other the associative law is applied, and the parentheses are dropped. According to the number of literals or the number of operators, the two expressions are of similar complexity. However, the two expressions result in different circuits because the four-input AND gate is one of the available IC devices. Which circuit is better? Unfortunately, the answer is not easy. If the number of gates is the criterion, then circuits may be compared only if they use the same gate types. There is another problem. When we say "better," what exactly do we mean? The criterion for deciding may be that the circuit is cheaper, or uses less space, or uses less power, or is more reliable, and so on.

We can see that it is very difficult to select the best circuit when all these practical problems occur. If we would like to have a measure by which two

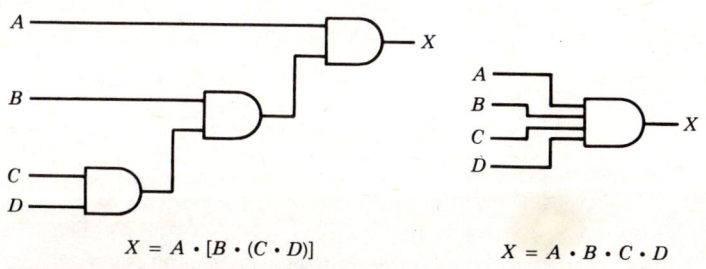

$X = A \cdot [B \cdot (C \cdot D)]$ $\qquad$ $X = A \cdot B \cdot C \cdot D$

FIGURE 2.3 Two possible realizations for Boolean expression $A \cdot B \cdot C \cdot D$.

44 BOOLEAN SIMPLIFICATION, BINARY FUNCTIONS, AND DIGITAL CIRCUITS

Boolean expressions may be compared, some of these practical problems must be ignored. It may seem reasonable to use the number of operators and/or the number of literals in an expression for this measure. The operators AND, OR, and NOT appear as gates in the circuit that produces the Boolean expression; therefore, the number of operators in the expression is a good indication of the number of gates used. Even this simple rule is not valid at all times. From De Morgan's rule the two expressions in Figure 2.4 are equivalent. The first expression, $(A' + B')$ has two literals and three operators. The circuit has three gates. The equivalent expression $(A \cdot B)'$ has only two operators, so it is simpler than the first expression. However, the simplification in gates is much more significant since the expression $(A \cdot B)'$ is recognized as a NAND operation, which requires only one gate in practice. A similar simplification was demonstrated in Figure 2.3, where we showed a circuit that used only one gate for four literals and three operators.

We shall divide the problem of deciding on a measure of circuit complexity into two problems: Boolean simplification and circuit simplification. First, we shall concentrate on Boolean simplification when our aim will be to minimize the number of operators and/or literals in the expression. In the next section we shall examine the circuits generated according to a Boolean expression with possibilities of rearranging the gates so that a minimum number of IC devices and/or gates are used.

Algebraic simplification uses simplifying rules, rules for which one side of the equation contains fewer literals than the other side. Many of our basic rules (Table 1.4) show possibilities of significant reductions. For example, the rule $A + 1 = 1$ eliminates a literal (or an expression, if A stands for a Boolean expression), as does the rule $A \cdot 0 = 0$. The rules $A + A' = 1$ and $A \cdot A' = 0$ eliminate two literals and two operators. The four other basic rules,

$$A + 0 = A \cdot 1 = A + A = A \cdot A = A$$

also reduce the number of literals and the number of operators. The distributive law also shows a different number of literals on the respective sides of the equations and can be used for Boolean simplification. In addition to these basic rules, six useful simplifying rules are shown in Table 2.1, which can be derived from the basic rules in Table 1.4 or can be proven by truth tables. The first OR rule in Table 2.1, for example, may be proven by the following

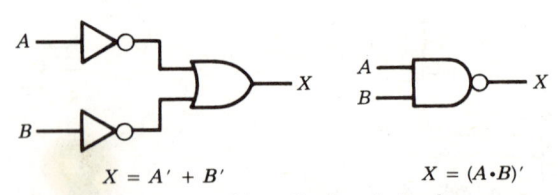

$X = A' + B'$ $\qquad\qquad$ $X = (A \cdot B)'$

FIGURE 2.4 Two possible realizations for Boolean expression $A' + B'$.

2.3 BOOLEAN SIMPLIFICATION

TABLE 2.1 Summary of Simplifying Rules

AND *rules*	OR *rules*
$A \cdot (A + B) = A$	$A + A \cdot B = A$
$(A + B) \cdot (A + B') = A$	$A \cdot B + A \cdot B' = A$
$A \cdot (A' + B) = A \cdot B$	$A + A' \cdot B = A + B$

substitutions of basic rules:

$$A + A \cdot B \underset{\text{Distributive property}}{=} A \cdot (1 + B) \underset{\text{Basic rules}}{= A \cdot 1 = A}$$

The first AND rule can also be easily proven:

$$A \cdot (A + B) \underset{\text{``multiply out''}}{= A + A \cdot B} \underset{\text{Use first OR rule above}}{= A}$$

The two rules in the second row and the third AND simplifying rule in Table 2.1 can be similarly proven. One of these proofs is illustrated by using the Venn diagram in Figure 2.5. The proof of the third OR rule using Boolean algebra shows a feature of Boolean simplification worth demonstrating. The left side of this rule is

$$A + A' \cdot B$$

Since we know that $A + A \cdot B = A$, we can add the term $A \cdot B$ to the term A without affecting the expression and get

$$A + A \cdot B + A' \cdot B$$

which can be "factored" and then evaluated:

$$A + A \cdot B + A' \cdot B = A + (A + A') \cdot B = A + (1) \cdot B = A + B$$

We will now show three examples that should demonstrate the "flavor" of Boolean algebraic manipulations and simplifications.

Simplify the Expression $A \cdot B \cdot C + A \cdot B' + A \cdot C'$

Using the distributive law to factor A out of the last two terms, we get

$$A \cdot B \cdot C + A \cdot (B' + C')$$

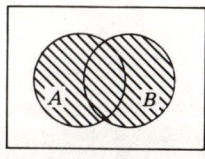

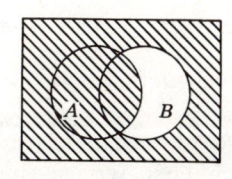

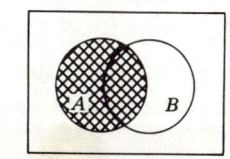

FIGURE 2.5 Proof of equality $(A + B) \cdot (A + B') = A$ using Venn diagrams.

46 BOOLEAN SIMPLIFICATION, BINARY FUNCTIONS, AND DIGITAL CIRCUITS

De Morgan's rule for the expression in parentheses gives us

$$A \cdot B \cdot C + A \cdot (B \cdot C)'$$

which may be written as $A \cdot (B \cdot C) + A \cdot (B \cdot C)'$. If we recognize that the term $B \cdot C$ may be considered as one literal, say, $X = B \cdot C$, then the expression is equivalent to the simplified expression $A \cdot X + A \cdot X'$, which, according to our second simplifying OR rule, is equal to the single literal A. Alternately, we can show that

$$A \cdot (B \cdot C) + A \cdot (B \cdot C)' = A \cdot [(B \cdot C) + (B \cdot C)'] = A \cdot (X + X') = A \cdot (1) = A$$

Simplify the Expression:
$A \cdot B' \cdot D + A \cdot B \cdot C \cdot D' + B' \cdot C \cdot D + A' \cdot B \cdot C' \cdot D'$

For this expression the distributive property may be used to factor the expression in two ways. Factoring out term A, we get

$$A \cdot (B' \cdot D + B \cdot C \cdot D') + B' \cdot C \cdot D + A' \cdot B \cdot C' \cdot D'$$

which has 13 literals and 18 operators; whereas factoring the same expression in a different way, we get

$$B' \cdot D \cdot (A + C) + B \cdot D' \cdot (A \cdot C + A' \cdot C')$$

which contains 10 literals and 13 operators. The second expression is clearly the simpler one. We have been using the number of operators as a measure of complexity for a Boolean expression because they are realized as logic gates. In the first factored expression we counted the NOT operator for B' twice, since B' appeared in both the first and second terms. Similarly, D' appeared in the first and third terms. This gives us an unnecessarily pessimistic measure because in a circuit, once an INVERTER gate is used to produce the output B' from the signal B, any further use of the term B' utilizes the same output and another gate is not required. Therefore, the more realistic calculation for the number of operators is to count the number of AND and OR operators and add to it the number of required inversions. For the first factored expression the number of required inversions is 4 (all four literals appear in their inverted form); therefore, the total operator count is equal to 16. For the second expression the number of operators is equal to 12, which also includes four inversions.

Simplify the Expression:
$(A + B' + C) \cdot (A' + B + C) \cdot (B + C')$

We can use the OR distributive rule:

$$(X + Y) \cdot (X + Z) = X + Y \cdot Z$$

for simplification in two ways. First, we use the rule for the first two parenthesized terms and get

$$[C + (A + B') \cdot (A' + B)] \cdot (B + C')$$

2.3 BOOLEAN SIMPLIFICATION

The term $(A + B') \cdot (A' + B)$ may be "multiplied out" to yield

$$(C + A \cdot B + A' \cdot B') \cdot (B + C')$$

because the terms $A \cdot A'$ and $B \cdot B'$ produce 0, which may be dropped. The last expression may be further multipled out, and we get

$$A \cdot B + B \cdot C + A' \cdot B' \cdot C'$$

where, again, terms like $B \cdot B'$ and $C \cdot C'$ have been dropped, and the expression $AB + ABC'$ is equivalent to the simpler expression AB. The original expression contained eight literals and ten operators. All the simplified expressions contain seven literals and nine operators. The distributive property can be used for the last expression to produce an even simpler one:

$$B \cdot (A + C) + A' \cdot B' \cdot C'$$

which contains only six literals and eight operators.

We now return to the original expression and use the OR distributive rule for the last two parenthesized terms, so we get

$$(A + B' + C) \cdot [B + (A' + C) \cdot C']$$

The term $(A' + C) \cdot C'$ is equivalent to the term $A' \cdot C'$ because the term $C \cdot C'$ is equal to 0 and may be dropped, so we get

$$(A + B' + C) \cdot (B + A' \cdot C')$$

which has six literals and eight operators. If this expression is multipled out, we again get

$$A \cdot B + B \cdot C + A' \cdot B' \cdot C'$$

which is equal to the expression we had before and demonstrates a good way of checking the answers.

The last expression, which is in the form of a **logical sum of logical products,** contains no parentheses. Any Boolean expression may be expressed as the logical sum of terms containing logical products only, since the multiplying out of expressions work similarly to that of ordinary algebra. It would be convenient if this form would yield the simplest expression, but unfortunately, this is not generally the case. In fact, if we use De Morgan's rule for the last term of the preceding expression, we get

$$A \cdot B + B \cdot C + (A + B + C)'$$

and factoring the variable B from the first two terms gives us

$$B \cdot (A + C) + (A + B + C)'$$

which has only six literals and six operators and is clearly the simplest expression we have found so far.

Even these few examples show that there is an element of "black magic" in Boolean simplification procedures. It is difficult to design a systematic method by which algebraic expressions may be simplified. This is true for all types of algebra. One may attempt to restrict the final expression to special

48 BOOLEAN SIMPLIFICATION, BINARY FUNCTIONS, AND DIGITAL CIRCUITS

forms (like logical sum of product terms), and one may set up special rules, like not counting inversion operators at all, that result in more uniform methods of simplification. We shall not pursue these more theoretical aspects of Boolean simplification for two reasons. First, in Chapter 4 we return to the problem of minimization, when more general methods will be used. Second, we are interested in the practical aspects of design, which often involve elements of skill and experience that are difficult to describe by strict rules. Since, ultimately, we are interested in minimized circuits rather than minimized Boolean expressions, it would not be very useful to set up special restrictions or rules that help in the systematic minimization of algebraic expressions but provide little help in finding simpler circuits.

We have been concentrating on the algebraic aspects of Boolean expressions. Now we shall change our viewpoint and consider their **functional** aspects. Boolean expressions viewed as Boolean **functions** will be the topic of the next two sections.

2.4 FUNCTIONS OF ONE BINARY VARIABLE

The simplest digital system is a combinational binary system that has one input and one output (Figure 2.6). The system F transforms variable A (the input) into variable X (the output); hence, the output X may be interpreted as a function of A, or $F(A)$. Any specific function $F(\ )$ may be defined by a truth table that has two entries because for one input variable only two different input values are possible.

One may enumerate all such possible functions since there are only four ways by which the two values 0 and 1 may be arranged in a two-entry truth table. Table 2.2 is the truth table of the four functions F_0, F_1, F_2, and F_3.

For F_0 and F_3 the output $F(A)$ is not a function of the input variable A. This means that the functional form $F(A)$ produces a constant, the same way as the continuous function $y = 5$ expresses a constant for the general functional representation $y(x)$. Obviously, there are only two such constant functions for binary variables, one with value 0, the other with value 1. The other two functions express identity, $F(A) = A$, and inversion, $F(A) = A'$, respectively. A practical circuit that generates all four possible functions of one binary variable is shown in Figure 2.7. This circuit requires only one INVERTER gate.

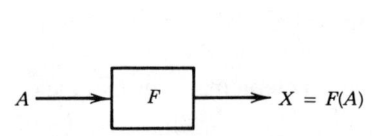

FIGURE 2.6 General block diagram of Boolean function of one variable.

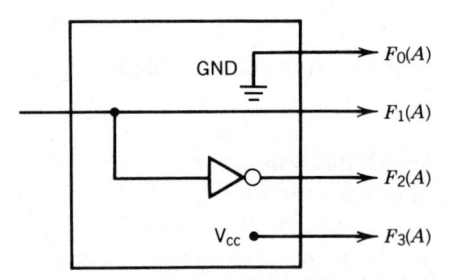

FIGURE 2.7 Circuits that generate all possible functions of one Boolean variable.

TABLE 2.2 Truth Tables for the Four Functions of One Binary Variable

A	F_0	F_1	F_2	F_3
0	0	0	1	1
1	0	1	0	1
Name	ZERO	Identity	Inversion	ONE
Expr.	0	A	A'	1

2.5 FUNCTIONS OF TWO BINARY VARIABLES

The block diagram of a two-input, one-output combinational digital circuit is shown in Figure 2.8. The truth table for a given function $G(A, B)$ has four entries, and there are 16 ways by which 1's and 0's may be arranged in four-entry truth tables. These 16 functions are G_0–G_{15} and are shown in Tables 2.3 and 2.4.

Six of these functions, G_0, G_3, G_5, G_{10}, G_{12}, and G_{15}, are constants or functions of one variable only, which were discussed in the previous section. From the remaining ten functions, four, G_1 (AND), G_7 (OR), G_8 (NOR), and G_{14} (NAND), are already familiar. The function G_6 is called the **Exclusive OR**, or **XOR** function. It has its own defined operation symbol, $\oplus$, and its unique gate symbol is shown in Figure 2.9. This function is called *Exclusive OR* because it is similar to the OR function (which is often called *Inclusive OR*), but the generation of a Logic-1 for the output excludes the case when the two inputs are both Logic-1.

The XOR gate with an inverted output is often called the **Exclusive NOR** gate, but we will refer to it as the **Equivalence** gate, or **EQU** gate, because its output is equal to Logic-1 (true for positive logic) if and only if its two input values are equal (both are 0 or both are 1). It also has its own defined operation symbol, $\odot$, and gate symbol. The symbol for the EQU gate is consistent with our graphical representation of inversion since a small circle is added to the output of the XOR gate (see Figure 2.9).

The remaining four functions, G_2, G_4, G_{11}, and G_{13}, are not used as simple gates. The reason for this is that these functions are not symmetrical with respect to their two input variables. From a practical point of view, the output of any one of the six basic two-input binary gates remains the same if the two inputs are interchanged. In other words, all these six Boolean operations satisfy the commutative property. The remaining four functions do not satisfy

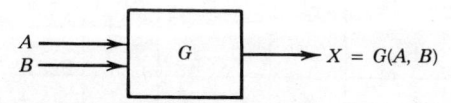

FIGURE 2.8 General block diagram of Boolean function of two variables.

TABLE 2.3 Truth Tables for the First Eight Functions of Two Binary Variables

A B	G_0	G_1	G_2	G_3	G_4	G_5	G_6	G_7
0 0	0	0	0	0	0	0	0	0
0 1	0	0	0	0	1	1	1	1
1 0	0	0	1	1	0	0	1	1
1 1	0	1	0	1	0	1	0	1
Name	ZERO	AND		ID		ID	XOR	OR
Expr.	0	$A \cdot B$	$A \cdot B'$	A	$A' \cdot B$	B	$A \oplus B$	$A + B$

commutation; therefore, they have not been found useful as practical binary building blocks. The expression $A' + B$ (G_{13}) may be interpreted by the logic statement "A implies B." If we know that A is equal to 1, then the expression is true (equal to 1) if and only if variable B is equal to 1. The assertion of A implies the assertion of B as long as the expression itself is asserted. The other function, $A + B'$, is interpreted as "B implies A." The term **implication** is used for these two functions. The two remaining functions, $G_2 = A \cdot B'$ and $G_4 = A' \cdot B$, are the complements of the implication functions and are called **inhibition** functions.

2.6 LOGIC GATES WITH ONE AND TWO INPUTS

We have already seen that the symbols for the INVERTER, NAND, NOR, and EQU gates use small circles to indicate inversion. So far, inversion was applied only to the output of a gate. The inversion operation may also be applied to the input of a gate, in which case the small circle appears at the input side of the gate symbol. In case of the INVERTER gate, logically, there is no difference. In fact, as shown in Figure 2.10, the circle itself, without the gate symbol, can indicate the inversion operation. The convention is that the triangle indicates a physical gate, whereas the small circle is used for the logical function of inversion.

TABLE 2.4 Truth Tables for the Last Eight Functions of Two Binary Variables

A B	G_8	G_9	G_{10}	G_{11}	G_{12}	G_{13}	G_{14}	G_{15}
0 0	1	1	1	1	1	1	1	1
0 1	0	0	0	0	1	1	1	1
1 0	0	0	1	1	0	0	1	1
1 1	0	1	0	1	0	1	0	1
Name	NOR	EQU	INV		INV		NAND	ONE
Expr.	$(A + B)'$	$A \odot B$	B'	$A + B'$	A'	$A' + B$	$(A \cdot B)'$	1

2.6 LOGIC GATES WITH ONE AND TWO INPUTS

$A \oplus B = A' \cdot B + A \cdot B'$ $A \odot B = A \cdot B + A' \cdot B'$

A B	$A \oplus B$
0 0	0
0 1	1
1 0	1
1 1	0

XOR gate

A B	$A \odot B$
0 0	1
0 1	0
1 0	0
1 1	1

EQU gate

FIGURE 2.9 Exclusive-OR (XOR) and Equivalence (EQU) gates and their truth tables.

Using De Morgan's rule, we know that

$$A + B = (A' \cdot B')'$$

which means that a NAND gate with inverted inputs is equivalent to an OR operation (Figure 2.11). Using De Morgan's rule in another way,

$$(A \cdot B)' = A' + B'$$

shows that a NAND operation may be executed with an OR gate whose inputs are inverted. Similar (dual) relationships exist between the NOR operation and an AND gate. As shown in Figure 2.11, when circles (inversion operators) are attached to the inputs of gate symbols, they change the logical function of the gate. This may cause quite a lot of confusion.

Difficulties arise in practice because digital designers who work with negative or mixed logic conventions use these alternate gate symbols. Instead of having only four gate symbols for the AND, NAND, OR, and NOR operations, they may use all eight possibilities. Following the positive-logic convention, we shall use basic gate symbols that have no inversion circles at their inputs for any practical circuit diagram. During the designing process of circuit diagrams, small circles that indicate inversion operations may appear. For the final circuit, however, these inverted inputs will be eliminated either by changing the gate type or by the introduction of additional INVERTER gates. These practical gate transformations will be discussed in the next section.

We have seen that classical logic is based on the three fundamental binary

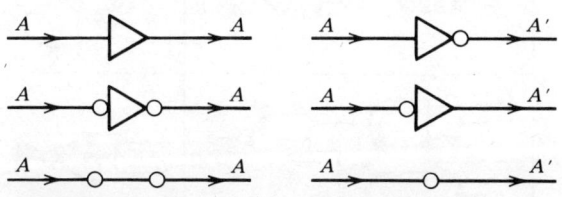

FIGURE 2.10 Using small circles to indicate logical inversion (NOT) operations.

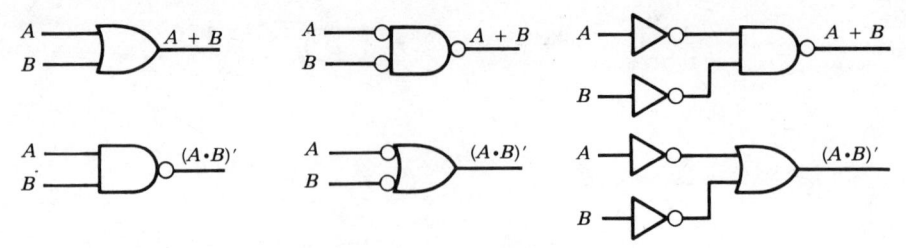

52 BOOLEAN SIMPLIFICATION, BINARY FUNCTIONS, AND DIGITAL CIRCUITS

FIGURE 2.11 OR–NAND and NAND–OR gate transformations.

operations AND, OR, and NOT. In contrast, in the earliest days of practical digital circuits, designers found it useful to use only one single gate type as much as possible. At that time hardware was very expensive, and the repeated use of a single gate type with highly optimized circuitry and minimum cost was very economical. The two-input NAND gate became the most popular gate type at that time. Entire computer systems were designed overwhelmingly from two-input NAND gates.

It can be shown that the two-input NAND gate (or the two-input NOR gate) can be used as a "universal" gate for digital circuits because all three basic operations may be realized by this one gate type. Since

$$(A \cdot A)' = A'$$

a two-input NAND gate is equivalent to an INVERTER gate when its two inputs are connected to each other. The same holds for a NOR gate. To provide other logic functions, we already know that NAND gates may be converted to OR gates (and vice versa) by inverting their inputs and outputs (Figure 2.11).

Boolean function	NAND gates	NOR gates
INVERSION		
AND		
OR		

FIGURE 2.12 Three basic Boolean operations constructed from NAND and NOR gates only.

In Figure 2.12 the three fundamental Boolean operations are realized using only two-input NAND or NOR gates, respectively. These circuits may seem to complicate the design process, but they do find practical applications when it is desirable to use one gate type to execute a different logic operation. Although, "theoretically," there may be no need for such gate transformations, the complexity of practical circuits may be reduced when such techniques are used.

2.7 PRACTICAL GATE TRANSFORMATIONS

Practical digital circuits are constructed from available digital IC devices. As we have seen, it is rare that an IC device contains only one logic gate. Having one, or only a very few, gates per IC is highly uneconomical, since the cost of the packaging, the IC socket, or the area of the printed circuit board used for the device are relatively high. When there are many gates per device, the cost of the package is divided among the large number of logic functions provided. At the same time, the variety of available IC devices is limited. We have already used types 7400 and 7402 devices, which contain four NAND or NOR gates, respectively, but there is no IC device that contains, say, two NAND gates and two OR gates. If our logic requirement calls for one NAND and one OR operation, we would have to use two different IC types (resulting in two different IC devices) or could use the three extra NAND gates of the type 7400 IC to perform the OR operation (Figure 2.13). In practical circuits the minimization of IC devices is more important than the minimization of logic gates.

A different type of practical gate transformation is necessary when the required number of inputs for a logic function does not match the number of inputs of an available logic gate. The simplest case is when the available number of inputs is larger than the one required. When only simple gates are involved, the interconnection of inputs can be used to decrease the number of available inputs. As shown in Figure 2.14, it is also possible to connect selected inputs to GND (Logic-0) or to V_{cc} (Logic-1) that provide similar reductions of available input terminals. These input terminal reduction methods use the four Boolean equalities

$$A \cdot A = A + A = A \cdot 1 = A + 0 = A$$

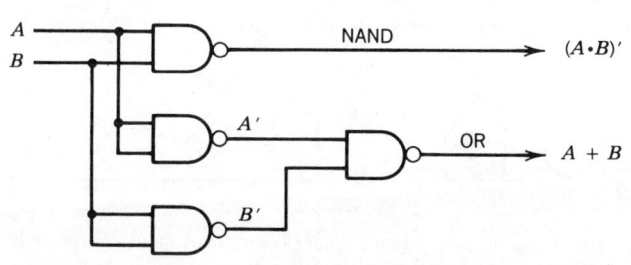

FIGURE 2.13 Generating both NAND and OR functions with NAND gates only.

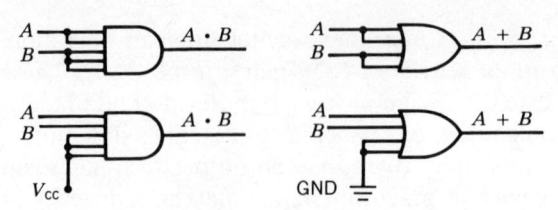

54 BOOLEAN SIMPLIFICATION, BINARY FUNCTIONS, AND DIGITAL CIRCUITS

FIGURE 2.14 Using fewer than available number of gate inputs.

When inputs are connected together, as shown in Figure 2.14, the loading of the output driver of the preceding gate is also affected.

If the number of available inputs is smaller than that required by the circuit, then cascading of similar gates is necessary. The simple cascading of gates works only for gate types that have noninverted outputs (AND, OR, and XOR). For gates with inverted outputs (NAND, NOR, and EQU), IN-VERTER gates must be used. Examples for four-input AND and NOR functions that use two-input gates are shown on Figure 2.15.

In practical digital circuits that use all five gate types (INVERTER, AND, NAND, OR, and NOR), gate transformations may be used to reduce the number of required IC devices because otherwise unused gates can be utilized to provide various logic functions. Similarly, a practical requirement may be to realize various logic functions using a small number of gate types. To help in the gate transformations, a summary of logic function and gate type equivalences are given in Figure 2.16.

Small circles (not INVERTER gates) are shown for the inversion function in Figure 2.16. We must distinguish between circles attached to the output of logic gate symbols that indicate that a NAND or a NOR gate is used and the circles separated from gate symbols, or attached to gate inputs, which indicate logical inversion. The detached circles and circles attached to inputs are used only for the gate transformation process when intermediate logic circuit diagrams are constructed. When the final, practical logic circuit diagram is drawn, circles that are not attached to gate symbol outputs must be converted to INVERTER gates.

The gate transformation procedure begins with the original logic circuit diagram derived from a given Boolean expression. In this diagram, for each AND operation in the Boolean expression there is an equivalent AND gate, for each OR operation an OR gate, and for each inversion an INVERTER gate. Gate type transformations begin with the addition of pairs of circles next to

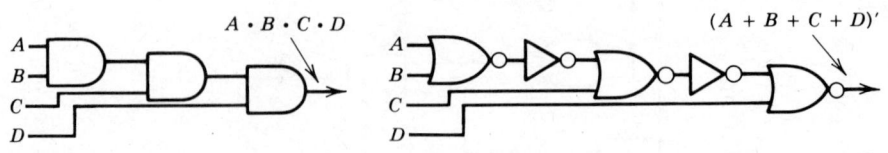

FIGURE 2.15 Cascading of gates when number of inputs is larger than number of inputs of gates.

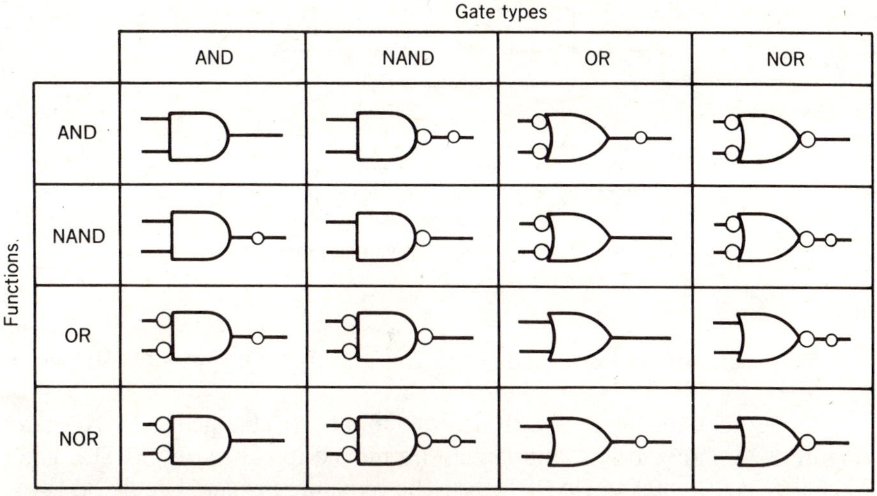

2.7 PRACTICAL GATE TRANSFORMATIONS 55

FIGURE 2.16 Summary of gate-type transformations.

each other. These additional "double" inversions may be added to inputs or outputs of gates, as required. They do not affect the logic function of the circuit in any way, since if inversion is applied twice to any Boolean variable, the original Boolean variable is reproduced (see Figure 2.17).

The circles may be freely moved along lines until they meet a gate symbol, another circle, or a connecting point where another line branches out. By moving selected circles around, they may be detached from gate outputs, which transforms NAND and NOR gates to AND and OR gates, or attached to gate outputs, which produces the opposite results. When a circle is moved past a connecting point (where two lines meet), circles must be added to both lines. This is demonstrated in Figure 2.18, where the symbols show that the illustrated methods do not alter the functional aspects of the circuit. When circles are attached to inputs, they produce gate type changes (see Figure 2.16). When all required gate transformations have been completed, pairs of circles may be eliminated (since they do not affect the logic functions at all), and isolated circles are changed to INVERTER gates. The three examples that follow demonstrate this procedure.

Use NAND and INVERTER Gates to Realize Boolean Expression
$A' \cdot B \cdot C' + (A + B') \cdot C' \cdot D'$

The logic diagram derived from the required Boolean expression is shown in Figure 2.19.

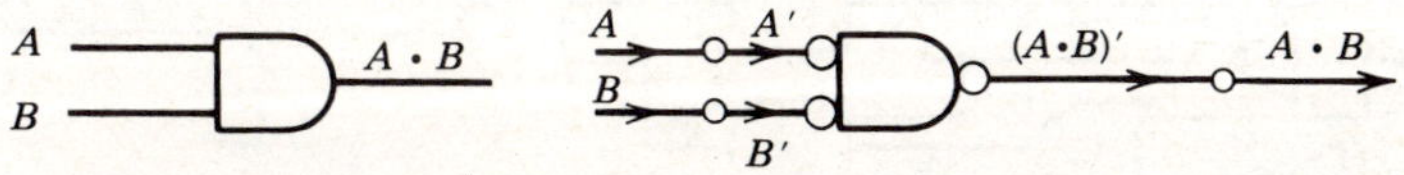

FIGURE 2.17 Addition of pairs of inverting circles for gate-type transformations.

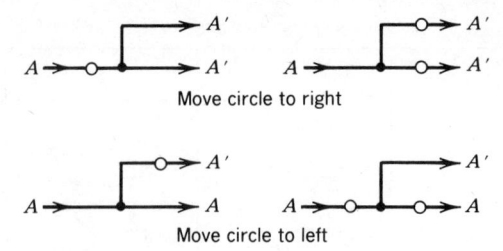

56 BOOLEAN SIMPLIFICATION, BINARY FUNCTIONS, AND DIGITAL CIRCUITS

FIGURE 2.18 Moving of inverting circle across junction of wires.

First, pairs of circles are added to the inputs of OR gates, as shown in Figure 2.20.

In Figure 2.20 OR gates with inverted inputs are changed to NAND gates (Figure 2.16 shows how), and circles are moved to cancel INVERTER gates or change AND gates to NAND gates. The remaining isolated circle on the A input line must be changed to an INVERTER gate, but since an INVERTER gate is already present, the input A' is already available. The final circuit diagram is shown in Figure 2.21.

Use NOR and INVERTER Gates to Realize Same Boolean Function

The original circuit diagram is shown in Figure 2.19. Pairs of circles are added now to inputs of AND gates and outputs of OR gates. The intermediate logic diagram is shown in Figure 2.22.

AND gates with inverted inputs are changed to NOR gates, and circles are moved around to cancel INVERTER gates or to change OR gates to NOR gates. The isolated circle in the B input line is changed to an INVERTER gate. The resulting circuit diagram, which contains only NOR and INVERTER gates, is shown in Figure 2.23.

AND–OR and OR–AND Circuits

When a Boolean expression is in the form of a logical sum of logical product terms, the resulting digital circuit contains AND gates connected to one multi-input OR gate. Such circuits may be easily transformed to circuits that contain NAND and INVERTER gates only. This is shown in Figure 2.24. In Figure 2.25 the dual of the AND–OR circuits is used where the Boolean expression

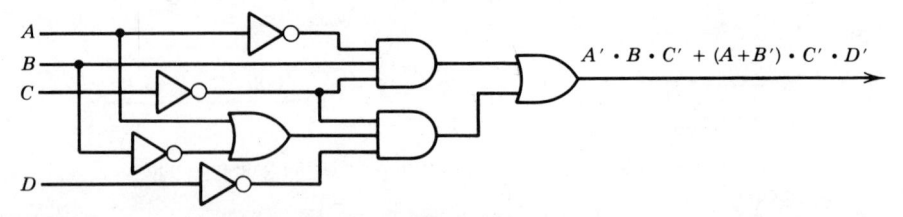

FIGURE 2.19 Original circuit before gate-type transformations.

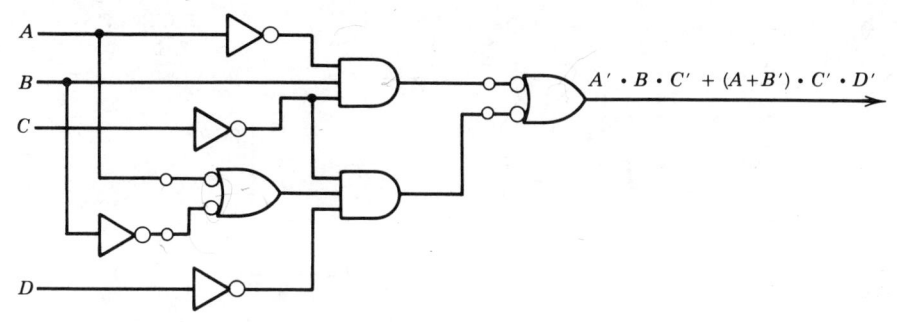

FIGURE 2.20 Addition of pairs of inverting circles to inputs of OR–NOR gates.

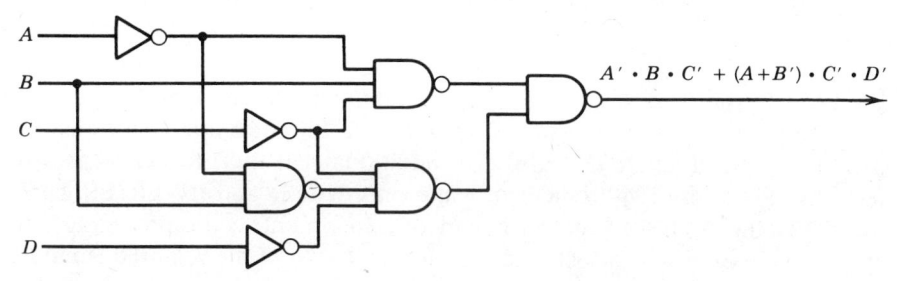

FIGURE 2.21 Final circuit that contains NAND and INVERTER gates only.

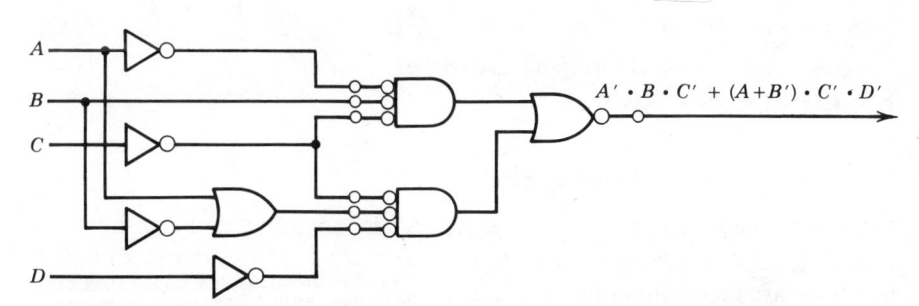

FIGURE 2.22 Addition of pairs of inverting circles to inputs of AND–NAND gates.

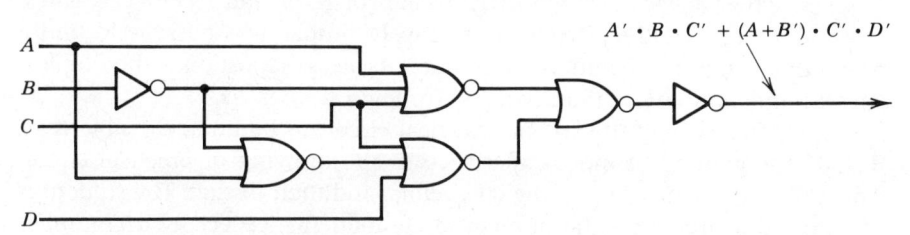

FIGURE 2.23 Final circuit with NOR and INVERTER gates only.

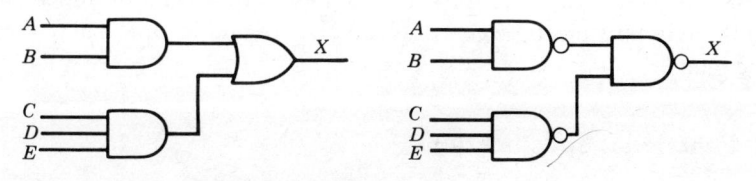

$$X = A \cdot B + C \cdot D \cdot E$$

FIGURE 2.24 Transformation of AND–OR circuit to pure NAND circuit.

57

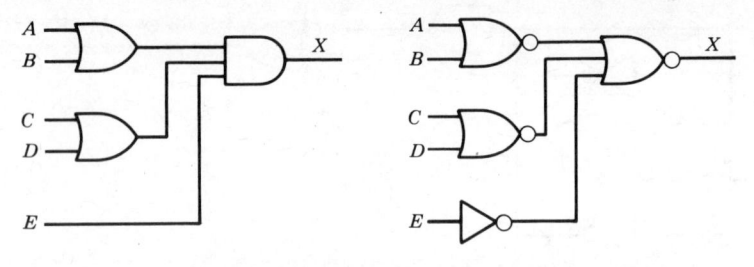

58 BOOLEAN SIMPLIFICATION, BINARY FUNCTIONS, AND DIGITAL CIRCUITS

$$X = (A+B) \cdot (C+D) \cdot E$$

FIGURE 2.25 Transformation of OR–AND circuit to NOR–INVERTER circuit.

is a logical product of logical sum terms and yield OR–AND circuits. In this case the final circuit contains NOR and INVERTER gates only. Note that in the case of the final NOR circuit, an additional INVERTER gate was also required. Since the requirement of using only NAND and INVERTER gates (or NOR and INVERTER gates) is part of Experiment 2, the preceding examples were especially selected to help in the design of the required circuits.

EXPERIMENT 2
Two-Input, Three-Output Boolean Function Generator

E2.1 Summary and Objectives

The functional specifications for this experiment require that a circuit is built that generates three Boolean functions of two variables. Three groups of functions are listed, and one function from each group must be selected. The internal construction of the circuit is constrained by allowing only two types of gates: either NAND and INVERTER gates or NOR and INVERTER gates. Since in either case the circuits are relatively simple, it is possible to find a circuit that contains the minimum number of gates. In most cases the simplest circuit requires no more than two SSI IC devices.

Since this is only the second practical circuit to be built, the objectives include the gaining of more experience in the drawing of practical circuit diagrams, the construction of digital circuits, and their testing. The student is encouraged to minimize the number of components, especially the number of required IC devices. Gate transformations are necessary for the final circuits. The main objective of using these techniques is to gain more understanding about the relationships between Boolean algebra and its practical realization, logic circuits.

E2.2 Functional Specifications

In Figure E2.1 the block diagram for this experiment is shown. The three outputs O_1, O_2, and O_3 generate three different functions of the two Boolean

```
         A  ────▶  ┌─────────┐  ────▶  O₁
                   │ Circuit │  ────▶  O₂
         B  ────▶  │   E2    │  ────▶  O₃
                   └─────────┘
```

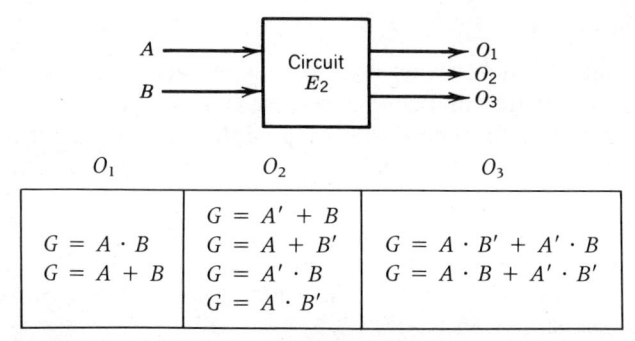

O_1	O_2	O_3
	$G = A' + B$	
$G = A \cdot B$	$G = A + B'$	$G = A \cdot B' + A' \cdot B$
$G = A + B$	$G = A' \cdot B$	$G = A \cdot B + A' \cdot B'$
	$G = A \cdot B'$	

FIGURE E2.1 Block diagram and available functions for Experiment 2.

variables A and B. Groups of possible functions are given for the three outputs, and one function per group must be selected. The internal construction of the circuits is restricted to the use of only two types of IC devices. If the designer chooses NAND and INVERTER gates, IC types 7400 and 7404 can be used (see Appendix 1 for the pin assignment and logic functions of IC devices). For NOR and INVERTER gates the two allowed IC types are 7402 and 7404. A reasonable amount of time should be spent to find the simplest circuit that satisfies the functional specifications. Both the requirement of finding a minimized circuit and the restrictions on the types of gates used may be satisfied by using gate transformations. A good designer always tries to find the most economical design. In addition to the benefit of reducing cost, a simplified circuit is easier to construct and to debug.

E2.3 Results and Testing

Before the circuits are designed, the truth table for the three selected functions should be determined and the results listed in Table E2.1. After the circuits have been found and appropriate gate transformations have been made, a practical circuit diagram should be drawn that shows IC device types, physical device positions, and pin assignments. Before testing the circuit with test lights and switches, the correctness of the circuit diagram should be tested by the generation of Boolean expressions from the circuit diagram and the evaluation of the output values according to these expressions. The output values should be compared to the completed truth table in Table E2.1.

TABLE E2.1 Table for Expected Results

A B	O_1	O_2	O_3
0 0			
0 1			
1 0			
1 1			

60 BOOLEAN SIMPLIFICATION, BINARY FUNCTIONS, AND DIGITAL CIRCUITS

E2.4 Design Example

To demonstrate the methods used for this experiment, we select three functions, two of which are not included in the given list of functions. In this way we do not provide information that may be directly copied for any version of the required experiment. Assume that the three required functions are the NAND, NOR, and EQU functions. Figure E2.2 shows three separate circuits, the first one uses a NAND gate, the second a NOR gate, and the third, which generates the Boolean function $A \cdot B + A' \cdot B'$, uses two AND gates, an OR gate, and two INVERTER gates. The truth table corresponding to these three functions is also shown in Figure E2.2.

If we connect the three input terminals designated as A and the other three terminals designated as B, the functional specifications of the required circuit are satisfied. This is shown schematically in Figure E2.3. In order to satisfy the internal constraints, the gates in these three circuits must be changed to NAND and INVERTER gates (or NOR and INVERTER gates). This can be done by using gate transformations, as shown in Section 2.7.

Before we apply gate transformations, which change the given gates to required gate types, let us examine the circuit to determine if simplifications are possible. The required internal functions are indicated in Figure E2.2. If we examine the Boolean expression for the EQU function, it contains two terms. The first term is equal to $A \cdot B$. We know that for the first output

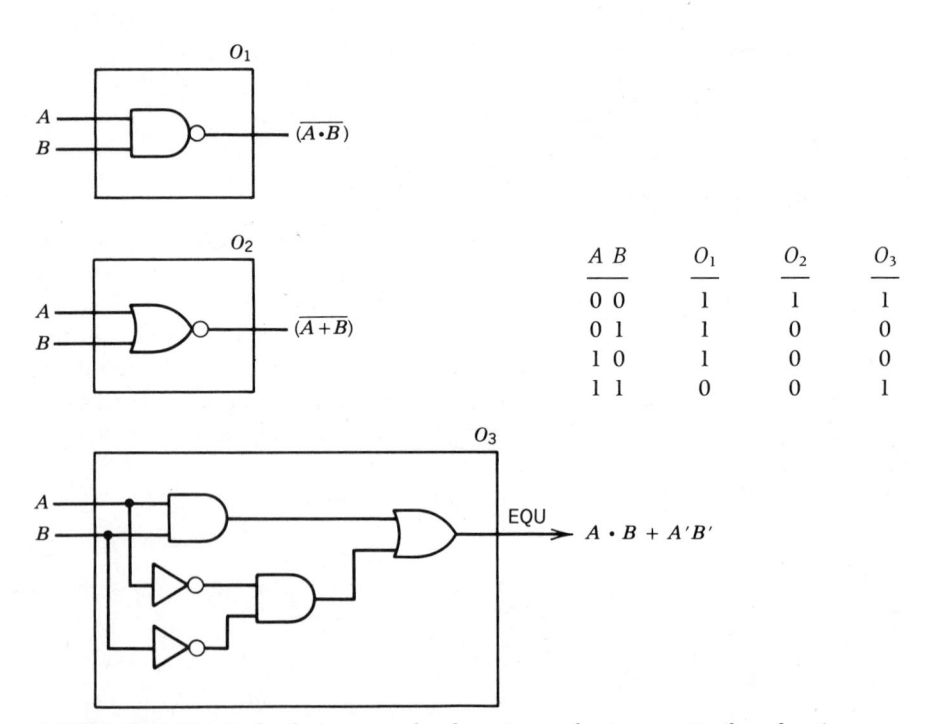

A B	O_1	O_2	O_3
0 0	1	1	1
0 1	1	0	0
1 0	1	0	0
1 1	0	0	1

FIGURE E2.2 Circuits for design example of two-input, three-output Boolean function generator circuit.

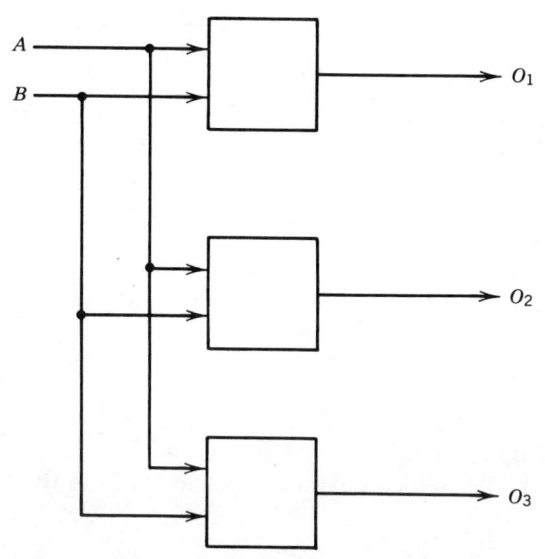

FIGURE E2.3 Combined circuits.

the NAND function is generated, which is equal to the Boolean expression $(A \cdot B)'$. Hence, if we generate the function $A \cdot B$ for the EQU function, a simple inversion of its output can generate the first output. For the second output the required function is equal to a NOR function, which is expressed as $(A + B)'$. Using De Morgan's rule, the NOR function is also equal to the expression $A' \cdot B'$. This is the same as the second term in the evaluation of the EQU function. By sharing combined Boolean functions between the required outputs, a substantial reduction of the number of gates is possible. The reduced logic diagram is shown in Figure E2.4.

Now, we use gate transformations and change all gates to the desired gate types. We demonstrate the gate transformations for both NAND and NOR gates. In Figure E2.5 the circuit is changed to all NAND and INVERTER gates, whereas in Figure E2.6 it is changed to all NOR and INVERTER gates. An alternate design method is to use gate transformations first (for the circuits shown in Figure E2.2 and E2.3) and do the simplification of the circuit afterward.

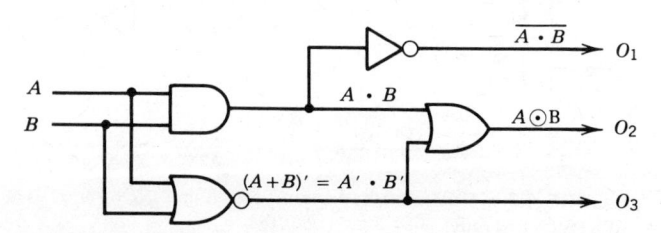

FIGURE E2.4 Circuit diagram for design example.

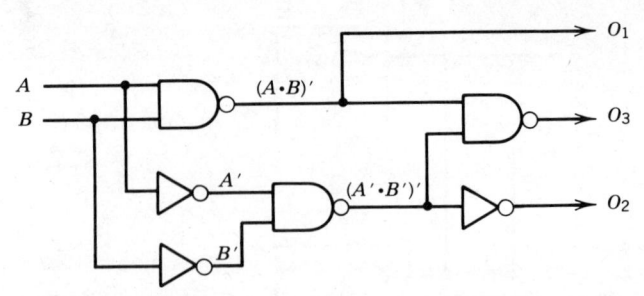

62 BOOLEAN SIMPLIFICATION, BINARY FUNCTIONS, AND DIGITAL CIRCUITS

FIGURE E2.5 Final circuit diagram containing NAND and IN-VERTER gates only.

We check our results by deriving Boolean expressions from the final circuit diagrams. The Boolean expressions derived from the circuit in Figure E2.5 are

$$O_1 = (A \cdot B)'$$
$$O_2 = [(A' \cdot B')']' = (A' \cdot B') = (A + B)'$$
$$O_3 = [(A \cdot B)' \cdot (A' \cdot B')']' = A \cdot B + A' \cdot B' = A \odot B$$

Clearly, O_1 and O_2 are equal to the required NAND and NOR functions. The output values for O_3 are evaluated, as shown in Table E2.2, and are found to be equal to the EQU function.

The Boolean expressions derived from the circuit in Figure E2.6,

$$O_1 = [(A' + B')']' = A' + B' = (A \cdot B)'$$
$$O_2 = (A + B)'$$
$$O_3 = [((A' + B')' + (A + B)')']' = (A' + B')' + (A + B)' = A \cdot B + A' \cdot B'$$

are the correct expressions for the circuit.

FIGURE E2.6 Final circuit diagram containing NOR and IN-VERTER gates only.

TABLE E2.2 Boolean Algebraic Check for Design Example

A B	$(A \cdot B)'$	$A' \cdot B'$	$(A' \cdot B')'$	$(A \cdot B)' \cdot (A' \cdot B')'$	$[(A \cdot B)' \cdot (A' \cdot B')']'$
0 0	1	1	0	0	1
0 1	1	0	1	1	0
1 0	1	0	1	1	0
1 1	0	0	1	0	1

E2.5 Further Experiments

Expand your circuit by adding one more function from any group in Figure E2.1. The expanded circuit generates four different functions, and in most likelihood, additional circuitry will be required. Find the functions that can be added with minimum circuitry, especially with circuitry that does not require more IC devices. Keep the restriction of only two gate types for this extended circuit.

Design a circuit that generates all 8 functions given in Figure E2.1. This circuit may generate all possible 16 functions of two binary variables. If not, how many more gates do you need to generate all 16 functions?

Drill Problems

2.1. Use Venn diagrams to prove the simplifying rules shown in Table 2.1.

2.2. Prove that the following Boolean equalities are true:
   (a) $A' \cdot B' + A \cdot B' + A'B = A' + B'$
   (b) $A \cdot B \cdot C + B \cdot C' = B \cdot (A + C')$
   (c) $(A + B)' \cdot B \cdot C = 0$
   (d) $A' \cdot B' \cdot C + A \cdot B \cdot C' + B \cdot C = A' \cdot C + A \cdot B$
   (e) $(B + C) \cdot (A' + B + C') \cdot (A + B' + C) = (A' + B) \cdot (A + C) = A' \cdot C + A \cdot B$
   (f) $(A \cdot B' + A' \cdot B)' = A' \cdot B' + A \cdot B$
   (g) $[(A + B \cdot (C + A')]' = A' \cdot B'$
   Use Venn diagrams and/or algebraic simplification.

2.3. Use algebraic simplification and factoring to reduce the following Boolean expressions to a form that contains the smallest number of operators.
   (a) $[(A' + B \cdot C') \cdot D']'$
   (b) $\{(A \cdot B + C) \cdot [(AB)' + C']\}'$
   (c) $A' \cdot B' \cdot C + B' \cdot C \cdot D + A \cdot B' \cdot C \cdot D$
   (d) $A \cdot B \cdot C' + A' \cdot B \cdot C' + A' \cdot B' \cdot C'$
   (e) $(B + C \cdot D) \cdot (A \cdot D + B + B \cdot C' + C \cdot D')$

2.4. A new computer firm, CCC Inc. (Crazy Computer Company), invented a process with which two-input INHIBITION gates can be manufactured at an incredibly low price. The gate symbol is shown in Figure D2.1. Determine and draw the simplest circuits that generate the seven logical functions AND, NAND, OR, NOR, XOR, EQU, and IMPLICATION using only INHIBITION gates.

64 BOOLEAN SIMPLIFICATION, BINARY FUNCTIONS, AND DIGITAL CIRCUITS

$$X = A' \cdot B$$

FIGURE D2.1 Implication gate.

2.5. Draw the circuit diagrams that generate the following Boolean expressions. Use gate transformations to produce circuits that contain only two-input NAND and INVERTER gates (or NOR and INVERTER gates). Derive Boolean expressions for your transformed circuits and check your results by using Boolean algebra to transform them back to their original form.
  (a) $A \cdot B' \cdot C + (B \cdot C' \cdot D)'$
  (b) $(A \odot B) \odot C$
  (c) $(A \cdot B \cdot C + A' \cdot B' \cdot C') \cdot D'$
  (d) $A \cdot B + C \cdot (A + B)$
  (e) $\{[(A \cdot B) \cdot C'] + (A + B)' \cdot C\}'$

2.6. Use theorems of Boolean algebra to prove that if $A \cdot B = A \cdot C$ and $A + B = A + C$, then $B = C$ (difficult). Show the circuit that tests the first two postulates:

$$[(A \cdot B) \odot (A \cdot C)] \cdot [(A + B) \odot (A + C)]$$

and show that it is equivalent to the circuit that tests the simple statement $B = C$, which is: $B \odot C$

2.7. Show the simplest circuits that generate the five logic functions INVERTION, AND, NAND, OR, and NOR:
  (a) Using XOR gates only.
  (b) Using EQU gates only.

2.8. It is possible to connect four two-input NAND gates in such a way that the output of the circuit is equal to the XOR function of its two input variables (if four two-input NOR gates are used, then the EQU function is generated). Find these circuits and prove by algebraic manipulations that the XOR (or EQU) functions are generated by the output.

CHAPTER 3

Variable Boolean Function Generators

3.1 INTRODUCTION

In Experiment 2 we designed and built a binary circuit that contained a number of outputs. These outputs produced various functions of two binary variables. For our design techniques we used Boolean algebra, basic binary gates, and gate transformations.

In Experiment 3 we will turn our attention to a variable function generator circuit. This type of circuit has only one output, but the Boolean function generated at this output varies depending on the values of some additional, so-called **control** variables.

Even though we have produced binary functions and used functional specifications, until now our design techniques were based on truth tables, Boolean algebra, and gates, not on functional specifications. In this chapter, and in Experiment 3, we will concentrate more on the "functional" descriptions of binary circuits. The variable function generator is chosen as an example, first, because it demonstrates how one can think in terms of functions rather than gates and, second, because it is an important element of most digital systems.

Two useful binary circuits, multiplexers and decoders, will be introduced, both because they are helpful in building variable function generators and because they themselves are excellent examples of Boolean functional circuit elements. The study of these two circuits will help us to achieve our main objective, which is to obtain a new view of our digital circuits, which, on the basic level, are still constructed from binary gates. This new view recognizes groups of gates that work together to give them special, well-defined functional identities.

3.2 SIMPLE VARIABLE BOOLEAN FUNCTION GENERATORS

The block diagram of the simplest variable function generator is shown in Figure 3.1. As before, the output of a binary circuit, X, is interpreted as some function of a single input variable I. A control variable C is added to the circuit and, depending on the value of this variable, the function $F(A)$ is changed. The application of this circuit may be an "intelligent" function generator.

VARIABLE BOOLEAN FUNCTION GENERATORS

$$X = \begin{cases} F_0(I) & \text{if } C = 0 \\ F_1(I) & \text{if } C = 1 \end{cases}$$

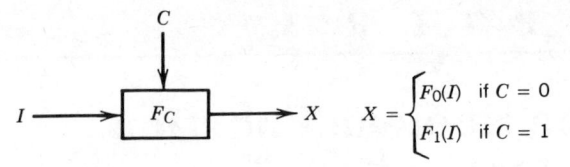

FIGURE 3.1 Block diagram of simple variable function generator.

Under some circumstances, when the variable C is equal to 0, we get one function, at other times, when C is equal to 1, we get another function.

As shown in Figure 3.1, there are two input variables and one output variable. The three variables of the circuit are of two types. The input variable I and the output variable X are data variables (they carry information). The input variable C is a control variable (it carries control information). The normal operation of the circuit assumes that control input C is set to a given value, and the circuit is used as a simple one-input function generator. Even though input C is interpreted differently, it is an ordinary binary variable, just like any other variables in the circuit. It is **our interpretation** that adds a new dimension to the circuit. Let us demonstrate this important fact by revisiting and reinterpreting two familiar basic gates.

The AND Gate as a Variable Function Generator

In Figure 3.2 the input variables of a two-input AND gate are separated, so that one is interpreted as a data input, and the other is interpreted as a control input. The appearance of the truth table of the AND gate is also changed. There are still four entries, but they are separated into two groups of two entries. One entry is for the case when $C = 0$, and the other for the case when $C = 1$. Examination of these two groups of outputs reveals that when the control variable is equal to 0, then nothing "gets through" the circuit (the input variable is "blocked"), and the output remains Logic-0 regardless of the value of the input variable. When the control variable is equal to 1, then the "gate" is opened, and variable I is passed through this circuit unchanged, that is, $X = I$. This interpretation of the AND gate explains the name *gate* splendidly and provides the simple AND gate with a functional importance it well deserves.

$$X = \begin{cases} 0 & \text{if } C = 0 \\ I & \text{if } C = 1 \end{cases}$$

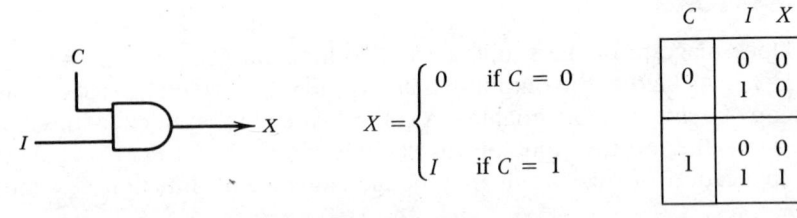

C	I	X
0	0	0
0	1	0
1	0	0
1	1	1

FIGURE 3.2 AND gate as variable function generator circuit.

3.2 SIMPLE VARIABLE BOOLEAN FUNCTION GENERATORS 67

$$X = \begin{cases} I & \text{if } C = 0 \\ I' & \text{if } C = 1 \end{cases}$$

C	I	X
0	0 0	0
	1 1	1
1	0 1	1
	1 0	0

FIGURE 3.3 XOR gate as variable function generator circuit.

The XOR Gate as a Variable Function Generator

The same analysis for an XOR gate reveals another useful and interesting variable function generator circuit. As shown in Figure 3.3, the rearrangement of the truth table reveals that the output of the XOR gate is equal to the unchanged input X when the control input value is equal to Logic-0, and it generates the complement of X, X', when the value of C is equal to Logic-1. Thus, the simple XOR gate can be interpreted as a controllable complementer. Depending on the value of a control variable, the circuit may or may not complement a Boolean variable.

A General One-Input Variable Function Generator Circuit

With the success of using the AND and XOR gates as function generators, it seems reasonable to search for a circuit that uses these gates for the generation of all possible functions of one Boolean variable. We know already that there are four such functions. In the two preceding examples one function was selected from two possible functions; therefore, one control variable was sufficient. If there are four possible functions, then two control variables are required. The number of possible combinations of 1's and 0's for the two control input values is equal to 4, and each particular setting of the control variables selects one function. After a little experimentation, we find the simple circuit shown in Figure 3.4, which contains one AND gate and one

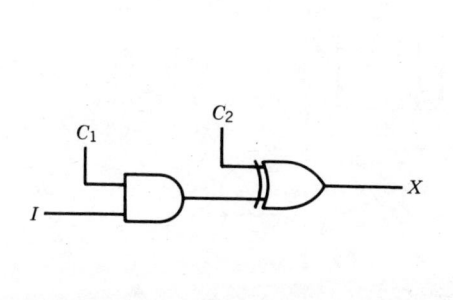

C_1 C_2	I	X	Function
0 0	0 0	0 0	0
	1	0	
0 1	0 1	0 1	1
	1	1 1	
1 0	0 0	0 0	I
	1	1 1	
1 1	0 1	0 1	I'
	1	1 0	

FIGURE 3.4 Generation of all four possible Boolean functions of one variable.

68 VARIABLE BOOLEAN FUNCTION GENERATORS

XOR gate. This circuit satisfies the specifications for the most general one-input function generator. As shown in Figure 3.4, it can generate any one of the possible four functions of one binary variable.

3.3 DIGITAL MULTIPLEXERS

It is possible to construct a general controllable function generator with the aid of a special digital circuit, called a multiplexer. A digital multiplexer has n control inputs, m data, or signal inputs ($m \leq 2^n$), and one output. For each unique combination of 1's and 0's of the control inputs the output becomes equal to one of the signal inputs. Hence, the multiplexer behaves as a switch, as shown schematically in Figure 3.5. The position of the switch is set according to the values of the control inputs. Since the control inputs **select** the particular signal input connected to the output, the control inputs are often referred to as **selection** inputs.

One of the many applications of multiplexers is to use them for the construction of variable function generator circuits. To begin, one needs a Boolean function generator circuit, similar to the circuit we built for Experiment 2. This circuit has many outputs, and a different Boolean function is generated by each of them. By the addition of a suitable multiplexer, the combined circuit becomes a variable function generator.

In Figure 3.6 we demonstrate a specific example where a circuit generates four different functions, and a 4-to-1 multiplexer is used. The signal inputs to the multiplexer are renamed. The indices are changed, and the signal inputs become I_{00}, I_{01}, I_{10}, and I_{11}. The same indices are shown inside the multiplexer block symbol. With these new indices the selection process is more clearly indicated. The indices of an input signal show the combination of 1's and 0's of the control variables C_1 and C_2 for which a particular input signal is selected. When the values of the two control inputs are 00, the output is equal to I_{00}; when they are equal to 01, then the I_{01} input is selected, and so forth. As shown in Figure 3.6, when the inputs C_1 and C_2 are set to values 00, X becomes equal to O_1, the first Boolean function. The values 01 select the second function, O_2, 10 selects O_3, and finally, values 11 select O_4. The truth

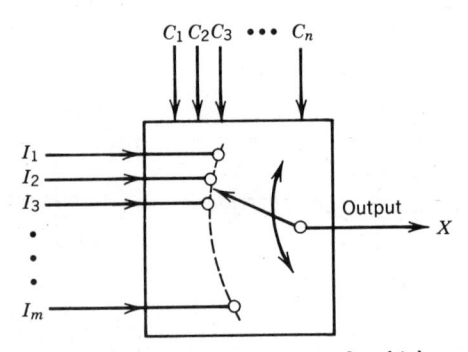

FIGURE 3.5 Schematic diagram of multiplexer.

3.3 DIGITAL MULTIPLEXERS

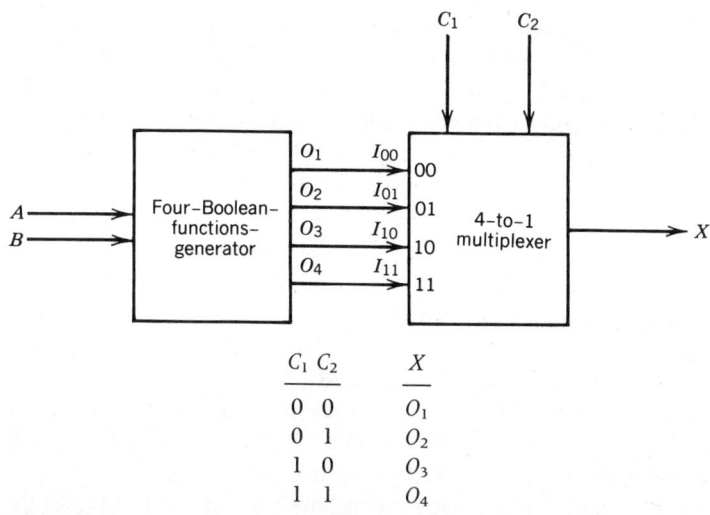

FIGURE 3.6 Construction of variable function generator circuit using multiplexer.

table shown in Figure 3.6 expresses the same functional operation of this circuit. Note that the truth table is given not in terms of 1's and 0's, as in the past, but in terms of other Boolean variables, O_1–O_4. In fact, the truth table defines the functional operation of the multiplexer without specifying the actual values for some of its inputs. Hence, when we contrasted "functional" representation and description by truth tables, we did not mention the fact that truth tables are very useful in functional specifications as well. The use of truth tables for functional specifications is a very efficient technique for describing complex digital circuits. The technique is based on separating control inputs, whose values are specified by 1's and 0's, and signal inputs, which are used in functional expressions.

The circuits in Figure 3.6 can be easily generalized for any number of functions. If there are more than four functions, then at least three control inputs are required. The overall block diagram of the circuit remains the same. On the left a combinational circuit generates m outputs, or m functions, whereas on the right a multiplexer with an appropriate number of control inputs selects one of the m functions for its output. It should be pointed out that such simple descriptions for generalizing specific circuit examples are another important feature of functional descriptions.

Let us proceed to the internal structures of multiplexers and examine how they might be built from basic logic gates. The familiar 4-to-1 multiplexer is considered first. Once a suitable logic circuit has been found for it, the basic ideas will be extended to general m-to-1 multiplexer circuits. A Boolean expression for the 4-to-1 multiplexer circuit in Figure 3.6 is

$$X = C_1' \cdot C_2' \cdot I_{00} + C_1' \cdot C_2 \cdot I_{01} + C_1 \cdot C_2' \cdot I_{10} + C_1 \cdot C_2 \cdot I_{11}$$

This expression is a logical sum of logical product terms. At this time it is not worth worrying about how this expression was derived because in the next chapter we will learn a general method for generating such expressions, and anyone who will study Chapter 4 will be able to derive this expression.

Let us examine this expression closely. There are four logical product (AND) terms. The output X is equal to Logic-1 if any one of the four expressions is equal to Logic-1, since the OR operator is used between all four terms. Of course, if more than one term has Logic-1 values, then the result is still equal to 1. Each product term contains both control variables, C_1 and C_2. The complementing of either or both control variables is arranged in such a way that for any combination of 1's and 0's three out of the four product terms become equal to Logic-0. For example, if $C_1 = 0$ and $C_2 = 0$, the first product term in the preceding expression is equal to $1 \cdot 1 \cdot I_{00} = I_{00}$, the second is equal to $1 \cdot 0 \cdot I_{01} = 0$, the third is equal to $0 \cdot 1 \cdot I_{10} = 0$, and the fourth is equal to $0 \cdot 0 \cdot I_{11} = 0$. The four terms ORed together are equal to I_{00}, that is, the first function, O_1. Other combinations of 1's and 0's in the values of C_1 and C_2 also eliminate three terms. The three terms that are made equal to Logic-0 are selected in such a way that the result, X, becomes equal to I_{01}, I_{10}, and I_{11}. We can see that the Boolean expression indeed satisfies the functional requirements of a 4-to-1 multiplexer.

A digital circuit designed on the basis of the preceding Boolean expression is shown in Figure 3.7. It contains two INVERTER, four three-input AND, and one four-input OR gates. If we generalize this circuit to an m-to-1 multiplexer, which has n control inputs ($m \leq 2^n$), it will have n inverters (since all control variables appear in their inverted form), m AND gates, each having $n + 1$ inputs, and one OR gate with m inputs. Each ($n + 1$)-input AND gate takes up the role of one product term in the Boolean expression of the multiplexer.

The description of a general m-to-1 multiplexer demonstrates that a circuit described in terms of its functional behavior often exhibits features in its circuit diagram that can be easily related to its functional description. If gate

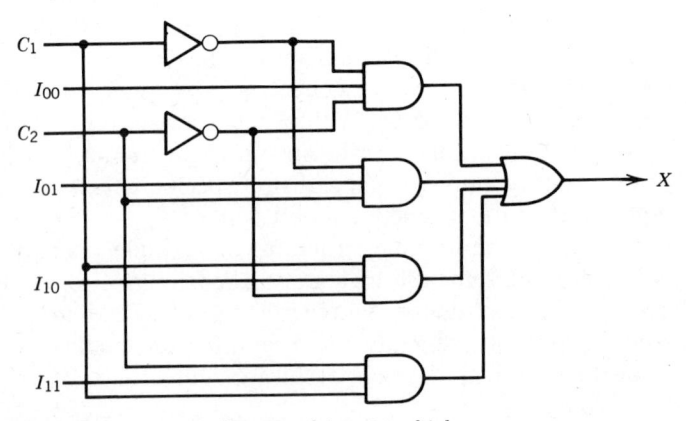

FIGURE 3.7 Circuit diagram of 4-to-1 multiplexer.

transformations were used for the circuit in Figure 3.7 and, say, all gates were changed to NOR gates, the correspondence between the circuit diagram and the functional description of the circuit could be lost. In the same manner, the circuit may be changed by factoring the Boolean expression in Figure 3.7, in which case a simpler, less expensive circuit may result; however, the correspondence between functional description and circuit elements would be lost. This is a frequent dilemma for a designer. Should the functional aspects of the circuits be kept, or should the least expensive circuit be used, which may not show any correspondence to the functional behavior of the circuit? In practice, both design methods are used. When the number of circuits to be built is moderate, the functional approach is used. This approach has many advantages. The main advantages are that it is easier to design the circuit as well as to test, maintain, and modify it. The question of economics for the least expensive version of the circuit becomes important when the circuit is manufactured in large quantities.

Circuit minimization, gate transformations, and Boolean algebraic manipulations were very important in the past when hardware was expensive. Recent advances in VLSI technology and the rapidly decreasing cost of digital hardware have made the functional approach the primary concern of modern digital designers. In the next chapter important relationships between the functional and the minimizing design methods will become much clearer, when the circuit of Experiment 3 will be redesigned using minimization techniques.

3.4 USING A DECODER TO BUILD A MULTIPLEXER

Let us reexamine the Boolean expression for the 4-to-1 multiplexer shown in Section 3.3. We substitute four new variables, D_{00}, D_{01}, D_{10}, and D_{11}, in place of the control variables and get

$$X = D_{00} \cdot I_{00} + D_{01} \cdot I_{01} + D_{10} \cdot I_{10} + D_{11} \cdot I_{11}$$

Each new variable is a function of the control variables C_1 and C_2. Following the analysis of this expression in the last section, we know that for a particular combination of 1's and 0's of the control variables, only one of the four new variables is equal to 1, whereas the other three are equal to 0. For example, when both C_1 and C_2 are equal to 0, D_{00} is equal to Logic-1, whereas the other variables (D_{01}, D_{10}, and D_{11}) are equal to Logic-0. In this case the output is equal to I_{00}. Since the four new variables are functions of the two control variables C_1 and C_2, one may construct a combinational circuit with two input and four output variables. This circuit, called a 2-to-4 decoder, is shown in Figure 3.8. Even though we use the decoder here to construct a multiplexer, the decoder is a frequently used digital circuit element in its own right.

The truth table for the 2-to-4 decoder is also shown in Figure 3.8. The truth table can be easily derived from the functional specifications of the decoder. For each combination of input values only one output is equal to 1; the others are equal to 0. The selected output, the output that becomes

72 VARIABLE BOOLEAN FUNCTION GENERATORS

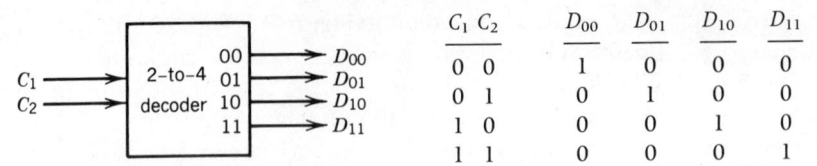

$C_1\ C_2$	D_{00}	D_{01}	D_{10}	D_{11}
0 0	1	0	0	0
0 1	0	1	0	0
1 0	0	0	1	0
1 1	0	0	0	1

FIGURE 3.8 A 2-to-4 decoder and its truth table.

Logic-1, is determined by the values of the control inputs, similarly to the multiplexer. Input values 00 select output D_{00} to be equal to 1, input values 01 select output D_{01}, and so on.

The Boolean expressions for the four outputs can be easily derived from the Boolean expression of the multiplexer in Section 3.3. They are

$$D_{00} = C_1' \cdot C_2' \qquad D_{01} = C_1' \cdot C_2$$
$$D_{10} = C_1 \cdot C_2' \qquad D_{11} = C_1 \cdot C_2$$

Using these expressions, a digital circuit for the 4-to-1 decoder can be constructed from four two-input AND gates and two inverters. This circuit is shown in Figure 3.9. Finally, the 4-to-1 multiplexer circuit constructed with a 2-to-1 decoder is shown in Figure 3.10. The combination of AND and OR gates on the right is called an AND–OR circuit, which is another frequently used functional unit. When a logical sum of logical product terms is needed, an AND–OR circuit may be used in practice.

The digital circuit diagram in Figure 3.10 demonstrates a very close correspondence between the functional description of the circuit and its construction. Remembering that a two-input AND gate is a variable function generator that either blocks or passes a digital variable, the circuit's operation becomes clear. The output is the combination (OR operation) of four possible input functions, but the decoder allows only one of the four functions to pass through at any one time. This function passes through the circuit unchanged and appears at the output, making X equal to I_{00} at one time, I_{01} at another, and so forth.

The generalization of the decoder circuit for an n-input, m-output decoder ($m \leq 2^n$) is reasonably straightforward. The circuit consists of n inverters (since all selector inputs appear in their inverted form), and m n-input AND gates. Each AND gate receives n inputs because each product term in the

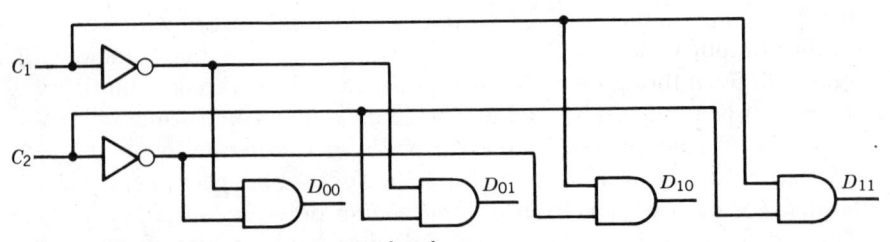

FIGURE 3.9 Circuit diagram for 2-to-4 decoder.

3.5 MULTIPLEXERS AND DECODERS FOR OTHER BOOLEAN FUNCTIONS

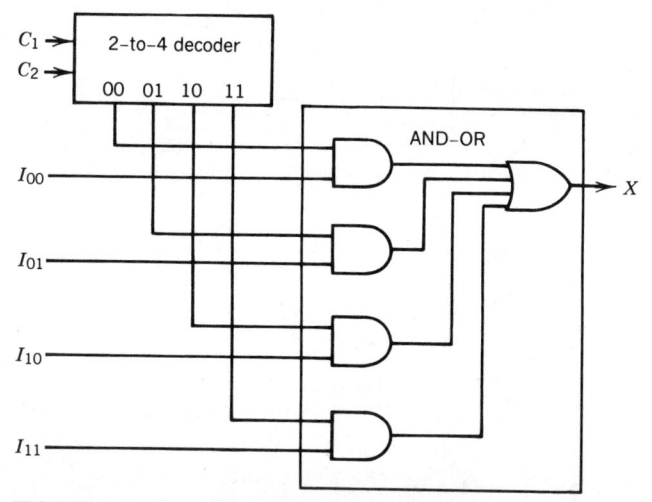

FIGURE 3.10 Circuit diagram of 4-to-1 multiplexer that uses 2-to-4 decoder.

Boolean expression of a decoder contains all the decoder input variables, some of them in normal form, others in inverted form. Logical product terms that contain all the input variables will be very important in our general method of minimizing circuits, which will be the topic of Chapter 4.

Multiplexers, decoders, and AND–OR circuits are all available commercially as digital SSI and MSI devices. In Appendix 1 the pin assignments of a few devices of these types are shown. A larger variety may be found in the engineering design handbook of any major digital IC manufacturer. Since the actual devices often contain additional inputs, or are not exactly in the form presented here, we shall postpone their description and their use until later chapters. It would be very unusual, indeed, to build a multiplexer from basic gates for a commercial digital circuit. In Experiment 3 this exercise is included, first, to provide further experience with constructing digital circuits; second, to gain experience in designing and testing a circuit that is described by a functional description; and third, to be able to compare the functional circuit to the minimized circuit that will be built in Experiment 4.

3.5 USING MULTIPLEXERS AND DECODERS FOR GENERATING OTHER BOOLEAN FUNCTIONS

We have seen in Chapter 2 that basic Boolean gates can generate several different logic functions when the inversion operation is applied to their inputs and/or outputs. Extending the same types of logic function transformations to more complex digital circuits, it may not be surprising that multiplexers and decoders can also be used for various logic functions when INVERTER or other basic gate types are added to them. A 2-to-1 multiplexer circuit is shown in Figure 3.11, where the inputs of the multiplexer are renamed as

74 VARIABLE BOOLEAN FUNCTION GENERATORS

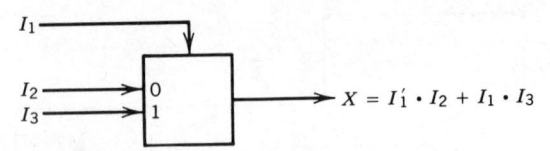

FIGURE 3.11 Boolean equation for 2-to-1 multiplexer.

I_1, I_2, and I_3. The Boolean expression for the multiplexer in terms of these input variables is equal to

$$X = I_1' \cdot I_2 + I_1 \cdot I_3$$

Examining the Boolean expression for the output of the multiplexer, X, we can see that it is similar to an XOR operation, which for the two variables A and B is equal to $A \cdot B' + A' \cdot B$. If we substitute A in place of I_1, B in place of I_3, and B' in place of I_2, then the two expressions become equivalent. We need one INVERTER gate to make a 2-to-1 multiplexer generate the XOR function, as shown in Figure 3.12.

It is somewhat more difficult to produce an OR function of two variables ($A + B$) using the same 2-to-1 multiplexer. From one of the simplifying Boolean expressions we know that the OR expression $A + B$ is equivalent to the more complex expression $A' \cdot B + A$, which can be expanded further to yield

$$A + B = A' \cdot B + A \cdot A$$

We can make the substitutions $I_1 = A$, $I_2 = B$, and $I_3 = A$ and generate the preceding expression. The equivalent circuit with its truth table is shown in Figure 3.13. The truth table may be derived from the functional description of the multiplexer. When the selection input is equal to Logic-0 ($A = 0$, the first two entries in the truth table), the output is equal to B; therefore the truth table entries are 0 and 1. When the selection input is equal to logic-1 (the last two entries in the truth table), the output is equal to A, which provides the last two 1 entries in the truth table. The result is a simple OR function.

Similar analysis yields a systematic method by which it becomes possible to generate an arbitrary Boolean function of two variables with a 2-to-1 multiplexer and, at most, one additional INVERTER gate. One of the two input variables, say, A, is connected to the control input of the multiplexer. Setting input A equal to 0, depending on the value of the other input variable

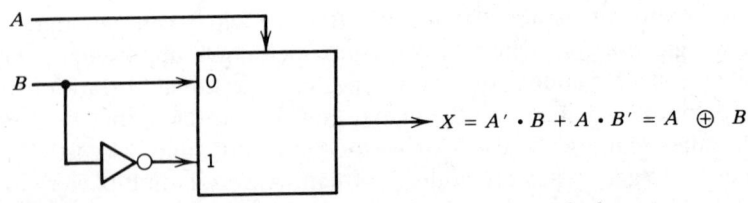

FIGURE 3.12 Generation of XOR function using 2-to-1 multiplexer.

3.5 MULTIPLEXERS AND DECODERS FOR OTHER BOOLEAN FUNCTIONS

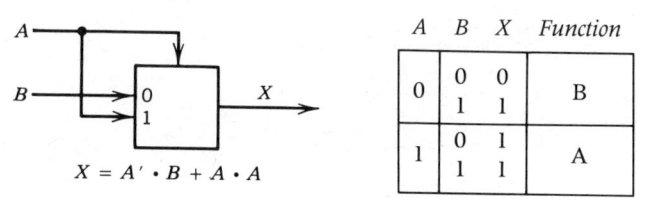

$X = A' \cdot B + A \cdot A$

A	B	X	Function
0	0	0	B
0	1	1	
1	0	1	A
1	1	1	

FIGURE 3.13 Generation of OR function using 2-to-1 multiplexer.

B, the output can be equal to one of four possibilities. It can be equal to the constant 0 (the output is equal to 0 regardless of the value of B); it can be equal to the constant 1 (the output is equal to 1 for either value of B); it can be equal to B (for $B = 0$ the output is equal to 0, for $B = 1$ the output is equal to 1); or it can be equal to B' (for $B = 0$ the output is equal to 1, for $B = 1$ the output is equal to 0). Hence, depending on the output values in the truth table for $A = 0$, the input I_0 is connected to constants 0 or 1, to the input B, or using an INVERTER gate, to B'. The same analysis for $A = 1$ determines the necessary input for I_1, and the required Boolean function is generated at the output of the multiplexer. Of course, it may not be an advantage in practical circuits to use complex logical functions, such as the multiplexer, to generate simple logic operations. We have used these examples to show how easily one logic function may be transformed to another. This is a general principle true for digital logic and hence for all binary systems.

In general, an arbitrary Boolean function with n variables may be generated with a multiplexer with $n - 1$ selection inputs (which has 2^{n-1} signal inputs) and one additional INVERTER gate. First, $n - 1$ signals of the n variables are connected to the selection inputs. For each combination of 1's and 0's for the connected $n - 1$ variables the Boolean function output may be expressed as a function of only one, the unconnected variable. The function of one variable may be only one of four: the constants 0 or 1, or the functions V or V', where the unconnected variable is named V. The input of the multiplexer selected by the particular combination of 1's and 0's is then connected to ground if the calculated function $F(V)$ is equal to 0, to V_{CC} if it is 1, and to either to V or to V' in the other two cases. Thus, an arbitrary function of two variables may be generated by a 2-to-1, of three variables by a 4-to-1, of four variables by an 8-to-1 multiplexer, and so on.

Boolean functions may be also generated using a 2-to-4 decoder and a two-input or three-input OR gate in a systematic way. The generation of the XOR function using a decoder is shown in Figure 3.14. The outputs of the decoder are labeled with logical product terms. Each product term is a function of both input variables, and they are either in their normal or inverted forms. The decoder output that becomes Logic-1 when the two input values are 00 is labeled with the product term $A' \cdot B'$. Since in this case $A' = 1$, and $B' = 1$, the product term is equal to 1, whereas all other product terms are equal to 0. These product terms may be used to describe the decoder in terms of logical propositions. Each output of the decoder indicates a given product

76 VARIABLE BOOLEAN FUNCTION GENERATORS

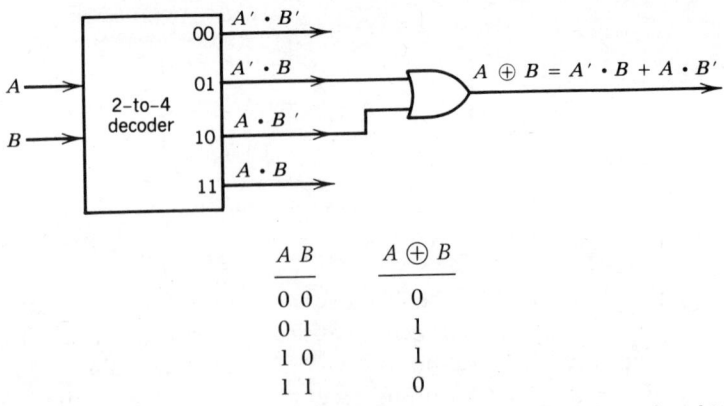

FIGURE 3.14 Generation of XOR function using 2-to-4 decoder and multi-input OR gate.

term that is asserted (equal to Logic-1 for positive logic). We notice that no matter what values A and B take on, exactly one output of the decoder is equal to Logic-1. The Boolean expression of the XOR function is equal to $A \cdot B' + A' \cdot B$, which can be interpreted as a combined logical proposition; that is, the expression is true if the product term $A \cdot B'$ is true or the term $A' \cdot B$ is true. With the addition of the two-input OR gate we can construct an equivalent digital circuit to the combined logical proposition. One of the inputs to the OR gate becomes Logic-1 when the input values are equal to 10 (product term $A \cdot B'$) or equal to 01 (product term $A' \cdot B$); they remain Logic-0 in the two other cases. This satisfies the truth table for the XOR function.

Another way to interpret the same circuit is by examining the truth table for the generated function. We notice that exactly two 1's appear in the truth table. These two 1's point to the two logical product terms asserted in the Boolean expression. This interpretation gives us a simple method for the generation of any Boolean function of two variables. The inputs of an OR gate are connected to those outputs of the decoder for which there are corresponding Logic-1 entries in the truth table. For the XOR function there are two such inputs.

In Figure 3.15 an OR function is provided using a 2-to-4 decoder and a three-input OR gate. The three inputs of the OR gate are connected to the decoder outputs labeled 01, 10, and 11, since there are three 1's in the truth table of the OR function. The three 1's appear in rows labeled with the same three input value combinations, 01, 10, and 11. Boolean algebra can also be used to show that the result is equal to an OR function. Forming the logical sum of the appropriate product terms, we get

$$A' \cdot B + A \cdot B' + A \cdot B = A' \cdot B + A \cdot (B + B') = A' \cdot B + A = A + B$$

Obviously, the circuit shown in Figure 3.15 is not a useful practical circuit. A complex circuit, the decoder, and a three-input OR gate are wasted to

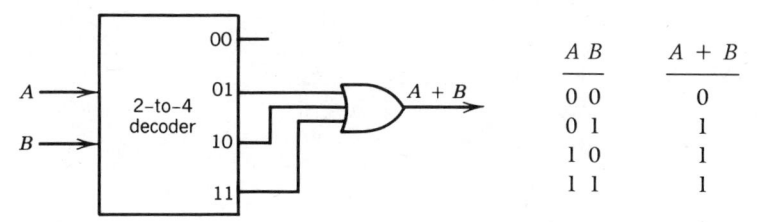

FIGURE 3.15 Generation of OR function using 2-to-4 decoder.

provide the equivalent logic function of a two-input OR gate. However, we used this circuit to demonstrate a general method. Given the truth table of an arbitrary Boolean function, a decoder and an OR gate can be used to generate the specified function. The ideas behind this method will be generalized and extensively used in Chapter 4.

EXPERIMENT 3
VARIABLE BOOLEAN FUNCTION GENERATOR

E3.1 Summary and Objectives

The objective of this experiment is to design a variable Boolean function generator circuit that provides one of three Boolean functions of two variables. Two control inputs are used; therefore, for one combination of input values (00), no output specifications are given. The function generator is constructed by the addition of a multiplexer to the circuit of Experiment 2 that provides three Boolean functions. Gate types are restricted to INVERTER and two-input NAND, NOR, and AND gates.

This experiment provides further practice in the construction and testing of reasonably complex circuits. It should demonstrate the advantages in designing and building digital circuits by the division of the whole task into smaller and simpler subcircuits.

E3.2 Functional Specifications

The block diagram and functional truth table for this experiment are shown in Figure E3.1. There are two data (signal) inputs and two control inputs. The output of the circuit is equal to one of three given functions of the two data inputs (A and B). The selected function is determined by values of the two control inputs C_1 and C_2. The three functions are the same as those specified in Experiment 2; therefore, the circuits built for Experiment 2 are reused. These circuits are expanded with an additional 3-to-1 multiplexer circuit that transforms the circuit of Experiment 2 into a variable function generator circuit (see Figure E3.2).

As shown in Figure E3.1, for each given value of the control inputs C_1

78 VARIABLE BOOLEAN FUNCTION GENERATORS

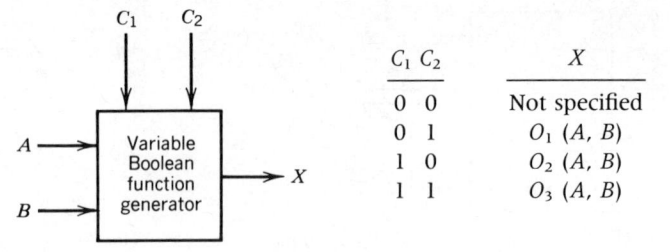

C_1 C_2	X
0 0	Not specified
0 1	$O_1 (A, B)$
1 0	$O_2 (A, B)$
1 1	$O_3 (A, B)$

FIGURE E3.1 Block diagram and functional specifications for Experiment 3.

and C_2, except values 00, the circuit behaves as a simple two-input (A and B), one-output (X) Boolean function generator. No output specifications are given for the case when the values of C_1 and C_2 are 00. During the designing process it should be assumed that variables C_1 and C_2 are never equal to 00.

E3.3 Expected Results, Restrictions, and Designing Process

The expected results should be tabulated in four truth tables, as shown in Table E3.1. For the three specified cases the values will be equal to those of Experiment 2. Even though for the control input values 00 there are no output specifications, once the circuit is designed, the expected values for this case can be determined. The values for this fourth case ($C_1C_2 = 00$) should also be entered in Table E3.1.

Gate types are restricted to INVERTERs and two-input NAND, NOR, and AND gates. The use of different gate types within the stated restrictions is encouraged if that simplifies the circuit for the multiplexer. If desired, the original circuitry built for Experiment 2 may be also redesigned, using these new, less severe restrictions. In this way more gates may be shared between the circuit for Experiment 2 and the multiplexer. The main objective is to keep the number of IC devices as small as possible. For this experiment the use of OR gates is not allowed.

Designing of the multiplexer can start with the Boolean expression given in Section 3.3. Factorization may be used to reduce the number of required

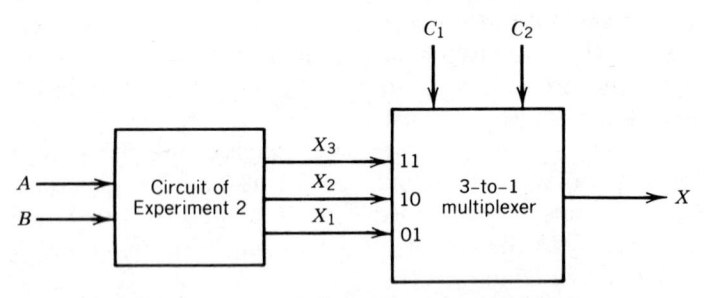

FIGURE E3.2 Construction of Experiment 3 using multiplexer.

TABLE E3.1 Expected Results for Experiment 3

C_1 C_2	A B	X
0 0	0 0 0 1 1 0 1 1	
0 1	0 0 0 1 1 0 1 1	
1 0	0 0 0 1 1 0 1 1	
1 1	0 0 0 1 1 0 1 1	

gates. Remember, there are two ways to factorize Boolean expressions, since there are two distributive rules (see Table 1.5). The two methods may result in different types and numbers of gates. At the end, gate transformations may be required to force the final circuit to contain only allowed gate types.

Another approach is to use the 2-to-4 decoder and a 3 × 2-input AND–OR circuit (three two-input AND gates and a three-input OR gate). You may use either method, or, if you like, you may use both, and then choose the simpler circuit for construction.

E3.4 Testing

Unless you are already an experienced circuit builder, it will be useful to use the functional aspects of the circuit for testing it **during** construction. Before the multiplexer circuit is connected to the circuit of Experiment 2, test the multiplexer circuit on its own. If you used a decoder circuit, test the decoder circuit first. This is analogous, of course, to the testing of individual procedures or subroutines in a computer program before the entire program is tested. And it has the same benefit: simpler testing procedures. Unfortunately, it may still happen that having successfully tested each subfunction, the combined circuit does not work according to specifications. Such is real life (as opposed to designing a circuit on paper). In this case the benefit of pretesting subcircuits is that in most cases the error is found among the few connections added after the initial tests were made.

E3.5 Alternate Experiments

The circuits may be redesigned with the gate-type restrictions relaxed so that no restrictions are placed on the number of inputs of gates, and all basic gate types (INVERTER, AND, NAND, OR, and NOR) are allowed. Gate types are restricted to available practical IC devices, listed in Appendix 1. The circuits, including those of Experiment 2, should be redesigned with these relaxed restrictions. The most important design criterium is to satisfy the functional specifications with the smallest number of IC devices. In addition, the number of gates used for the circuit should also be minimized. Report the savings in the number of IC devices and/or gates (if any) that are due to the relaxation of the gate-type constraints.

You may relax the gate-type restrictions further by allowing the use of the IC device type 74153 (see Appendix 1), which contains two 4-to-1 multiplexers. Compare the total component costs for your circuits built with various component restrictions.

Drill Problems

3.1. Describe the EQU gate as a variable function generator of one input variable.

3.2. Describe the INHIBITION gate shown in Figure D2.1 as a variable function generator of one input variable in two ways. First, use input A as a control input and input B as data input; second, reverse the roles of the two input variables.

3.3. Consider a three-input NOR gate with inputs A, B, and C and output X as a variable function generator. Determine the two two-variable functions $G_i(A, B)$ generated when input C is used as a control input. If both B and C are used as control inputs, which functions $F_i(A)$ are generated for the four possible control input settings?

3.4. Use two 2-to-1 multiplexers and the fewest possible number of basic gates (AND, NAND, OR, or NOR) to build one 4-to-1 multiplexer. Generalize your results to show how one may construct a $2m$-to-1 multiplexer from two m-to-1 multiplexers.

3.5. Show how each of the following functions can be generated using a 2-to-1 multiplexer and, if required, an INVERTER gate:
  (a) NAND
  (b) $F(A, B) = A' \cdot B$
  (c) $F(A, B) = A + B'$
  (d) EQU

3.6. Extend the method of generating two-variable functions with multiplexers to three-variable functions by using 4-to-1 multiplexers. Consider the entries in the truth table as functions of one variable only with the other two input variables used as control variables. Generate the three-input majority circuit (output is equal to 1 when at least two of the input variables are equal to 1) using a 4-to-1 multiplexer.

3.7. Use two 2-to-4 decoders and the fewest possible number of basic gates (AND, NAND, OR, or NOR) to build one 3-to-8 decoder. Generalize your results to show how one may construct a $(m + 1)$-to-$2n$ decoder from two m-to-n decoders.

3.8. Use two INVERTER gates and four two-input NOR gates to build a 2-to-4 decoder. Draw the circuit and label the outputs with the appropriate output symbols from D_{00} to D_{11}.

3.9. Generate the two-variable Boolean functions in Drill Problem 3.5 using a 2-to-4 decoder and a multi-input OR gate.

CHAPTER 4

Combinational Circuit Design and Minimization

4.1 INTRODUCTION

In the three previous chapters we introduced simple combinational circuits and their truth tables and studied logic gates, Boolean algebra, functional specifications of digital circuits, and gate transformations. So far, our approach has been mainly analytical. We observed how digital circuits behave, how Boolean algebra can help in their simplification, and how functional descriptions can break down a complex circuit into simpler, more easily managed subcircuits.

Having studied and constructed several binary circuits, we acquired an initial understanding of binary systems. However, we have not yet seen a general design method with which any combinational system may be designed from its functional specifications. The main purpose of this chapter is to present such a design method.

In this chapter we will solve the problem of finding a logic circuit that operates according to a given truth table. Gate transformations and algebraic manipulations have already shown that many different logic circuits can generate the same truth table. Therefore, the general design problem of finding a circuit from a given truth table does not have a unique solution. If possible, we should try to find a design method that can produce the "best" logic circuit, where *best* is determined by some given design criteria. If the criterium for the best design is that the number of gates should be at a minimum, then our aim is to **minimize** the circuit.

Digital (logic) circuit minimization has been the greatest concern of logic designers for the past 30 years. Consequently, the amount of research and the number of practical methods in this field are enormous. We present here the simplest and most appealing method, Karnaugh maps. The main appeal of Karnaugh maps is that they use graphics. Graphical methods have the advantage that once learned, they are rarely forgotten. Humans seem to have a far superior memory for remembering graphical than numerical information.

We have already mentioned that the decreasing cost of digital hardware and rapid advances in VLSI technology have significantly decreased the importance of digital design techniques, for example, Karnaugh maps, that employ truth tables and gates rather than functional specifications. The importance of Karnaugh maps may have decreased in practice, but they demonstrate a highly successful design method. In addition to giving some further insight into the understanding of binary systems, it shows how the discovery of a useful (often graphical) technique can greatly advance the designer's ability to produce efficient systems.

Logic design techniques based on Karnaugh maps start to become awkward when the number of input variables is larger than 4 and become impossible to use when the number of variables is larger than 6. Various minimization techniques have been developed that can handle large number of variables, both input and output. These techniques use numerical methods, and many computer programs have been developed that solve the minimization problem of general combinational circuits in a more or less satisfactory manner. Many digital design textbooks cover the easiest and most popular variations of these techniques [1, 2]. Since our principal goal is to complete specific experiments for which the number of variables will never be larger than 6, the only minimization technique we will discuss is the one based on Karnaugh maps.

First, we will solve the general problem of finding at least one Boolean expression for an arbitrary truth table (in fact, due to the duality principle, two circuits will be found). We will use the so-called canonical forms, minterms, and maxterms for solving this problem. Then we will examine how a general Boolean expression of not more than six variables may be minimized using Karnaugh maps. Finally, the functional specifications for Experiment 3 will be revisited. For Experiment 4 a circuit will be designed and constructed that has essentially the same functional specifications as those of Experiment 3. The design method used at this time is Karnaugh maps, and hopefully, a much simpler circuit will be found. A comparison of the two design techniques and the two resulting circuits provides a useful model for these two types of design methods: functional design and gate design.

4.2 BINARY CIRCUIT DESIGN WITH MINTERMS AND MAXTERMS

We already know that the behavior of any binary combinational circuit may be defined by, or any complete functional specification converted to, a truth table. Therefore, the truth table is a convenient starting point for the design of a binary circuit. The problem of designing a circuit for a given truth table is equivalent to finding the Boolean expression that generates the same truth table because a logic circuit can always be constructed from a given Boolean expression.

The three-input majority circuit will be used as an example. A majority circuit may be compared to a voting machine. The inputs are the voters, a Logic-1 is a yes vote, a Logic-0 means nay. We assume that the number of inputs is odd. The output is equal to 1 (yes) if the majority of inputs is also 1. The block diagram and the truth table of the three-input majority circuit is shown in Figure 4.1. The output is equal to 1 when at least two of its inputs are equal to 1.

84 COMBINATIONAL CIRCUIT DESIGN AND MINIMIZATION

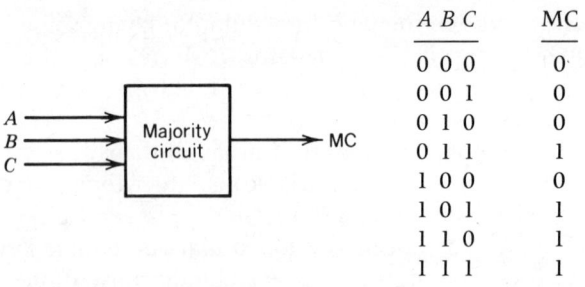

A B C	MC
0 0 0	0
0 0 1	0
0 1 0	0
0 1 1	1
1 0 0	0
1 0 1	1
1 1 0	1
1 1 1	1

FIGURE 4.1 Block diagram and truth table of three-input majority circuit.

We would like to produce a logic circuit consisting of basic gates that produces the truth table shown in Figure 4.1. In Section 3.6 we have already seen a "functional" method with which the required Boolean function can be generated. Using a 3-to-8 decoder and a multi-input OR gate, those outputs of the decoder must be used as inputs to the OR gate, which correspond to 1's in the truth table. As shown in Figure 4.2, there are four such inputs. The first 1 occurs for input values of 011; therefore, the decoder output labeled 011 is used. This output becomes 1 when $A = 0$, $B = 1$, and $C = 1$. In Figure 4.2 (similar to our decoder circuit in Section 3.6), we labeled the four decoder outputs with logical product terms that become 1 when the combination of input values is equal to the labels of the four outputs. Using these product terms, we can assemble the following logical sum for the majority circuit (MC):

$$MC(A, B, C) = A' \cdot B \cdot C + A \cdot B' \cdot C + A \cdot B \cdot C' + A \cdot B \cdot C$$

This Boolean expression generates the correct truth table because for any one of the four cases when there is a 1 in the truth table, one of the product terms becomes a 1. For the other four cases the four terms are all 0 and, therefore, the entire expression is equal to 0. Once the Boolean expression is known, of course, an equivalent logic circuit may be constructed. In this case the

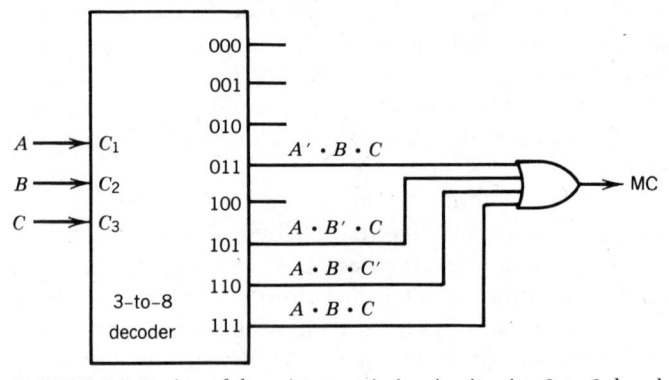

FIGURE 4.2 Design of three-input majority circuit using 3-to-8 decoder.

circuit contains three INVERTER, four three-input AND, and one four-input OR gates.

The preceding Boolean expression for the majority gate is in one of the so-called **canonical** forms and is in fact a logical **sum of minterms**. What makes this expression a canonical expression is that all the logical product terms contain all three input variables. Product terms that contain all the input variables of a Boolean function are called **minterms**. Minterms are constructed in such a way that only one minterm becomes a logic 1 for any one combination of 1's and 0's of the variables of the function.

There is, of course, a **dual canonical** representation for an arbitrary Boolean expression. Using the rules of duality, all OR operations must be changed to AND operations and vice versa. The dual terms become logical sums, and the overall expression becomes a logical product of logical sum terms, or **maxterms**. A maxterm is a logical sum that contains all the input variables, and the canonical expression is a logical **product of maxterms**. For each combination of 1's and 0's of the input variables only one maxterm is equal to 0; all other maxterms are equal to Logic-1. We can see this, for example, if we examine the maxterm $A + B + C$. For this expression to be equal to 0, all three input variables must be equal to 0. All other maxterms contain at least one input variable in its inverted form. If $A = B = C = 0$, then the inverted variable is equal to Logic-1, which is logically ORed to the other terms; hence, all the other maxterms become Logic-1.

How can we find this dual expression? There are two ways in which we may proceed. First, we can determine those maxterms for which 0's are entered in the truth table. On the left in Table 4.1 selected minterms (for which there are 1's in the truth table) and selected maxterms (for which there are 0's) are shown. There are four 0's in the truth table; therefore, there are four maxterms. The first 0 is for the input values of 000 that select maxterm $A + B + C$. The second 0 is for input values 001, for which the maxterm is equal to $A + B + C'$, and so on. The complete canonical expression is given by

$$MC(A, B, C) = (A + B + C) \cdot (A + B + C') \cdot (A + B' + C) \cdot (A' + B + C)$$

TABLE 4.1 Minterms and Maxterms for Three-Input Majority Circuit (MC) and Its Complement Function (MC')

\multicolumn{4}{c}{Truth table for MC}	\multicolumn{4}{c}{Truth table for MC'}
ABC	MC	Minterms	Maxterms	ABC	MC'	Minterms	Maxterms
000	0		$A + B + C$	000	1	$A' \cdot B' \cdot C'$	
001	0		$A + B + C'$	001	1	$A' \cdot B' \cdot C$	
010	0		$A + B' + C$	010	1	$A' \cdot B \cdot C'$	
011	1	$A' \cdot B \cdot C$		011	0		$A + B' + C'$
100	0		$A' + B + C$	100	1	$A \cdot B' \cdot C'$	
101	1	$A \cdot B' \cdot C$		101	0		$A' + B + C'$
110	1	$A \cdot B \cdot C'$		110	0		$A' + B' + C$
111	1	$A \cdot B \cdot C$		111	0		$A' + B' + C'$

86 COMBINATIONAL CIRCUIT DESIGN AND MINIMIZATION

The preceding expression was determined from maxterms that corresponded to 0's in the truth table. The other method uses the 1's (minterms) of the inverted function $MC'(A, B, C)$. For the inverted function all 1's are changed to 0's and 0's to 1's (see Table 4.1). The sum of minterms, which is equivalent to the Boolean expression for the inverted function $MC'(A, B, C)$ is equal to

$$MC'(A, B, C) = A' \cdot B' \cdot C' + A' \cdot B' \cdot C + A' \cdot B \cdot C' + A \cdot B' \cdot C'$$

When the preceding expression is inverted, the application of De Morgan's rule gives us

$$MC(A, B, C) = (A + B + C) \cdot (A + B + C') \cdot (A + B' + C) \cdot (A' + B + C)$$

which is equivalent to the expression we calculated using the 0's of the original truth table. If this expression is not factored, then the corresponding circuit yields three INVERTER, four three-input OR, and one four-input AND gates.

Similar to the majority circuit, the truth table of any Boolean function of three variables may be used to produce two canonical expressions and hence two circuits. There are eight minterms and eight maxterms for the three variables shown in Table 4.2. Selecting any row in the table, the substitution of the indicated input values into the corresponding minterm yields Logic-1, whereas the substitution of these values into the maxterm expression yields Logic-0. The same input values substituted in the minterms of any other row produces 0 values, whereas these values substituted for the corresponding maxterms yields 1 values. Hence, the logical sum of those minterms for which the function is equal to Logic-1 or the logical product of those maxterms for which the function has Logic-0 values yield the two required canonical expressions.

It is easy to generalize the concepts of minterms, maxterms, and canonical expressions for n variables. There are 2^n minterms and 2^n maxterms for n input variables. Each minterm is a logical product of all n input variables, and they appear in normal or inverted forms so that the minterm is equal to Logic-1 when the corresponding input values are substituted into the min-

TABLE 4.2 Minterms and Maxterms for Three-Variable Boolean Function

Input variables ABC	Minterm symbol	Minterm expression	Maxterm symbol	Maxterm expression
000	m_0	$A' \cdot B' \cdot C'$	M_0	$A + B + C$
001	m_1	$A' \cdot B' \cdot C$	M_1	$A + B + C'$
010	m_2	$A' \cdot B \cdot C'$	M_2	$A + B' + C$
011	m_3	$A' \cdot B \cdot C$	M_3	$A + B' + C'$
100	m_4	$A \cdot B' \cdot C'$	M_4	$A' + B + C$
101	m_5	$A \cdot B' \cdot C$	M_5	$A' + B + C'$
110	m_6	$A \cdot B \cdot C'$	M_6	$A' + B' + C$
111	m_7	$A \cdot B \cdot C$	M_7	$A' + B' + C'$

term's Boolean expression. Similarly, the maxterms are logical sums of all n input variables, and similar substitutions yield Logic-0 values for the maxterms. The canonical expressions for a Boolean function of n variables are constructed similarly to those for functions of three input variables.

In decoder and multiplexer circuits indices were chosen that clearly pointed to the values of the selection variables. For example, the decoder output D_{000} became Logic-1 when all three inputs were equal to Logic-0. We could adopt the same convention for minterms and maxterms and, for example, use the symbols m_{000} or M_{000} for the same entry in the truth table. However, the indices would become rather unwieldy for a large number of input variables; therefore, the input variable values are interpreted as a positive binary number of n bits rather than a string of 1's and 0's, and the index is expressed as a decimal integer in the range from 0 to $2^n - 1$. In Table 4.2 the standard minterm and maxterm symbols for three variables are shown. The range of indices are from 0 to 7.

The easiest way to calculate the numerical value of a binary number is to label its bits b_0, b_1, and so on. The rightmost bit, that is, the least significant bit (LSB), is labeled b_0 (in Table 4.2, input variable C is equal to b_0). The bits to the left of b_0 are labeled b_1, b_2, and so on, until the leftmost bit, the most significant bit (MSB), is labeled b_{n-1} (variable A is the MSB, which is equal to b_2). The numerical value of the binary number is given by

$$\text{Numerical value} = \sum_{i=0}^{n-1} 2^i b_i$$

where b_i is the numerical value associated with a bit of the binary number, 0 for Logic-0 and 1 for Logic-1. This numerical value is expressed as a decimal integer and is used for the indices of the minterms and maxterms.

In the truth table the input variable values are arranged in increasing numerical order of the minterm and maxterm indices. Table 4.2 shows that for three variables the first entry is for input values 000, or numerical value 0, whereas the last entry is for 111, or numerical value 7. In practice, especially when there are a large number of variables, a shorthand notation is used for canonical expressions. The letters m for minterms or M for maxterms are dropped, only the sigma for logical sums (or the pi for logical products) and the indices are shown. The two canonical expressions for the majority circuit in this shorthand notation are

$$MC(A, B, C) = \Sigma\,(3, 5, 6, 7) \quad \text{or} \quad MC(A, B, C) = \Pi\,(0, 1, 2, 4)$$

On examining the minterm and maxterm expressions in this simple shorthand notation, it becomes clear that the indices appearing in one of the canonical expressions are missing in the other expression (and vice versa). This is true in general and may be stated as follows: The set of indices in the minterm expression is equal to the complement set of indices in the maxterm expression. In what follows we derive this relationship between minterm and maxterm indices using the majority gate as an example.

For any minterm and maxterm belonging to the same input variables, the following equality holds:

$$(m_i)' = M_i$$

This equality must be true since the minterm is equal to 1, whereas the maxterm is equal to 0 for the given input variable values, and they are equal to 0 and 1, respectively, for all other input values; hence, if the truth table for minterm m_i is inverted, the truth table for the maxterm M_i is produced.

Knowing the minterm expression for a function, the minterm expression for the inverted function may be written by inspection. The inverted function contains all minterms that were not included in the logical sum of the minterm expression for the original function (since the 1's of the inverted function are the 0's of the original function). It follows that the minterm expression for the inverted majority circuit is

$$MC'(A, B, C) = \Sigma\ (0, 1, 2, 4)$$

When the logical sum expression for $MC'(A, B, C)$ is inverted, we get $MC(A, B, C)$. According to De Morgan's rule, the logical sum becomes the logical product of the inverted terms, and the inverted minterms become maxterms with the same indices. Therefore, in case of the majority circuit

$$MC(A, B, C) = \Pi\ (0, 1, 2, 4)$$

which is the identical maxterm expression to that calculated before.

4.3 TWO-, THREE-, AND FOUR-VARIABLE KARNAUGH MAPS

We have succeeded in providing two canonical expressions for a given truth table. If we are interested in a minimized expression, Boolean algebra may be used to simplify the expressions. But we have seen that simplifications with Boolean algebra is a rather haphazard method. Even after expending a large amount of time and energy on algebraic manipulations, one cannot be certain that the final expressions are the simplest ones.

As long as the number of variables is 6 or less, with the use of a graphical method based on the so-called **Karnaugh map** (also called the Veitch diagram) [3, 4], the simplest algebraic expression for a given truth table can be found. The "simplest" is interpreted as the fewest literals and/or operators in a logical sum of product terms (or logical product of sum terms for the dual method). The method is based on the repeated application of the simplifying rule

$$A \cdot B' + A \cdot B = A$$

It can be shown that if this simplifying rule is applied for all possible combinations of the terms in an expression, the simplest expression according to the preceding criteria may be found. In many cases there will be more than one simplest expression, that is, different expressions that have the same number of literals and operators and produce the same truth table.

4.3 TWO-, THREE-, AND FOUR-VARIABLE KARNAUGH MAPS

The Karnaugh map is a rectangle (or square) divided into a number of small rectangles. Each small rectangular region corresponds to one entry in the truth table, that is, one specific combination of input variable values. In Figure 4.3 two-, three-, and four-variable Karnaugh maps (or K-maps) are introduced. For the purpose of indicating which input values belong to which rectangular region, the specific combinations of 1's and 0's of the input variables are shown inside each small rectangle. This is used only for these introductory discussions because, as we shall see shortly, for ordinary K-maps the output values are the ones that appear in the small rectangular regions.

The input values inside the small rectangular regions are redundant in Figure 4.3 because the same input values are indicated outside the large rectangles along the horizontal and vertical borders. For the two-variable map the names of the two input variables A and B appear at the upper left corner of the K-map. From their positions it can be deduced that variable A varies along the vertical border, whereas variable B takes on different values along the horizontal border. This means that in the first row of the two-variable K-map the value of A is equal to 0, whereas in the second row it is equal to 1. Similarly, in the first column B is equal to 0, whereas in the second column B is equal to 1.

In the three-variable K-map variable A varies along the vertical border, as before, but along the horizontal border two variables, B and C, change their values. The horizontal border is labeled with both values (e.g., 00, 01, 11, and 10), which refer to the two input variables B and C. For example,

Two-Variable Map

A \ B	0	1
0	00	01
1	10	11

Three-Variable Map

A \ BC	00	01	11	10
0	000	001	011	010
1	100	101	111	110

Four-Variable Map

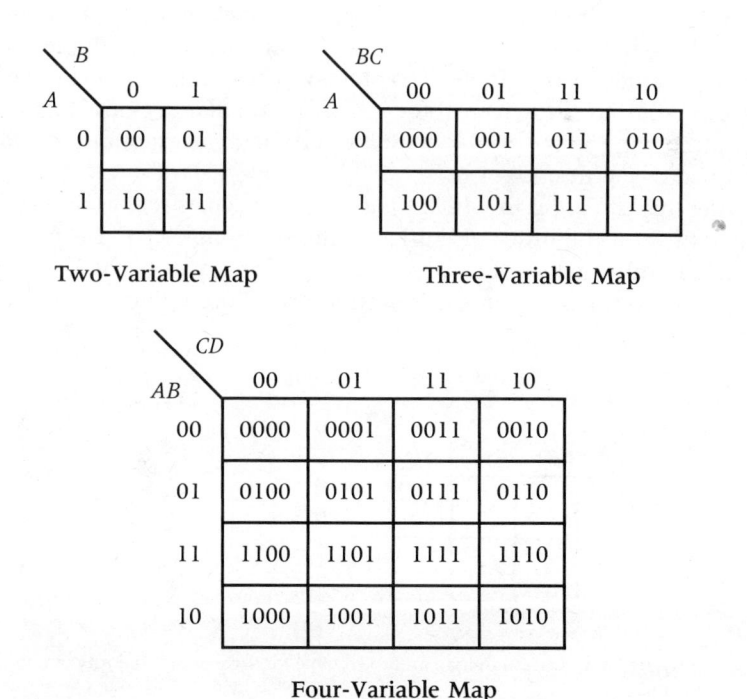

AB \ CD	00	01	11	10
00	0000	0001	0011	0010
01	0100	0101	0111	0110
11	1100	1101	1111	1110
10	1000	1001	1011	1010

FIGURE 4.3 Two-, three-, and four-variable K-maps.

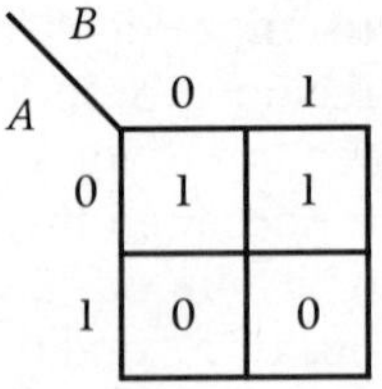

90 COMBINATIONAL CIRCUIT DESIGN AND MINIMIZATION

	B=0	B=1
A=0	1	1
A=1	0	0

A B	F(A, B)
0 0	1
0 1	1
1 0	0
1 1	0

FIGURE 4.4 Karnaugh map and truth table for two-variable function.

the small rectangular region in the lower right corner of the three-variable K-map refers to input variable values 110, since it is in the second row, which indicates that $A = 1$, and in the fourth column, which is labeled with input values $B = 1$ and $C = 0$ (values 10 are shown at the top of the horizontal border).

The four-variable K-map is labeled similarly. Two of the four variables, A and B, change along the vertical border, whereas the remaining two variables, C and D, vary along the horizontal direction. In one row of the four-variable K-map the values of the variables A and B remain constant. The same is true for variables C and D with reference to columns. For example, in the third row of the second column the input values are equal to 1101, since in the third row both A and B are equal to 1 (values 11 are shown), whereas in the second column $C = 0$ and $D = 1$ (values 01 are shown).

The input variable values outside the borders of the K-map uniquely identify each small rectangular region and associate it with one entry in the truth table. When the K-map is used for the simplification of Boolean expressions, the values of the Boolean expression corresponding to the entries in the truth table are shown in these small rectangular regions. Examples for two-, three-, and four-variable K-maps and their corresponding truth tables are shown in Figures 4.4–4.6.

Note that the order in which the small rectangular regions are laid down is different from the order of entries in the truth table. The K-map minimization procedure relies on this specific order. This will become apparent in the next section, where the minimization procedure will be discussed.

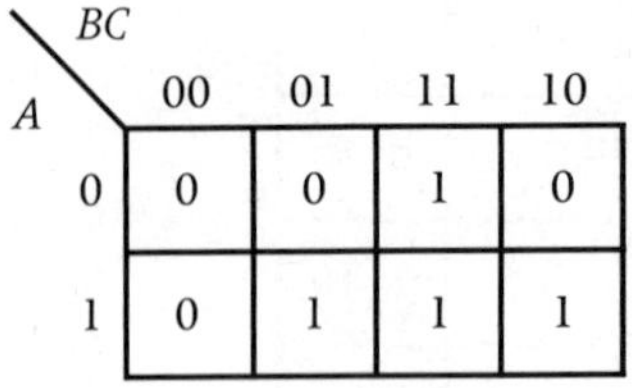

A\BC	00	01	11	10
0	0	0	1	0
1	0	1	1	1

A B C	F(A, B, C) = MC
0 0 0	0
0 0 1	0
0 1 0	0
0 1 1	1
1 0 0	0
1 0 1	1
1 1 0	1
1 1 1	1

FIGURE 4.5 Karnaugh map and truth table for three-input majority circuit (MC).

4.4 MINIMIZATION OF SUM-OF-PRODUCT EXPRESSIONS USING K-MAPS

AB \ CD	00	01	11	10
00	0	0	1	0
01	0	1	1	1
11	1	1	1	1
10	0	1	1	1

A B C D	F(A, B, C, D)
0 0 0 0	0
0 0 0 1	0
0 0 1 0	0
0 0 1 1	1
0 1 0 0	0
0 1 0 1	1
0 1 1 0	1
0 1 1 1	1
1 0 0 0	0
1 0 0 1	1
1 0 1 0	1
1 0 1 1	1
1 1 0 0	1
1 1 0 1	1
1 1 1 0	1
1 1 1 1	1

FIGURE 4.6 Karnaugh map and truth table for four-variable function.

4.4 MINIMIZATION OF SUM-OF-PRODUCT EXPRESSIONS USING K-MAPS

The small rectangular regions in the K-map are laid down in a special order. For any two neighboring rectangular regions the values of the input variables differ only in one variable. For this reason, the order of input variable values are listed in the order 00, 01, 11, and 10 whenever two variables are involved. Note that the first pair of values, 00, may be considered "next" to the last pair, 10, since they differ only in the value of the first variable. (This fact is important, and we will make use of it when we study regions located next to the border.)

The input variable values determine a specific entry and its associated minterm expression in the truth table. For example, as shown in Figure 4.4, values $A = 0$ and $B = 0$ indicate logical products $A' \cdot B'$, whereas values $A = 1$ and $B = 0$ are associated with the term $A \cdot B'$. Since neighboring small rectangular regions differ only by the value of one input variable, the associated minterms differ only by the inversion operator for one input variable. If both of these minterms belong to the Boolean expression (two 1's appear in the truth table for neighboring rectangular regions), then the combined expression, which is the logical sum of the two minterms, is

$$A' \cdot B' + A \cdot B' = (A' + A) \cdot B' = (1) \cdot B' = B'$$

and significant simplification occurs.

Let us examine how this simplification appears in a four-variable K-map. In Figure 4.7 two neighboring 1's are shown. Each 1 refers to a corresponding 1 in the truth table and to the respective minterm in the canonical "sum-of-minterms" expression. The 1 in the second row and second column is in the

92 COMBINATIONAL CIRCUIT DESIGN AND MINIMIZATION

```
         CD
  AB\    00  01  11  10
  00  [  ][  ][  ][  ]
  01  [  ][ 1][ 1][  ]
  11  [  ][  ][  ][  ]
  10  [  ][  ][  ][  ]
```

A B C D	Minterm
0 1 0 1	$A' \cdot B \cdot C' \cdot D$
0 1 1 1	$A' \cdot B \cdot C \cdot D$

$$A' \cdot B \cdot C' \cdot D + A' \cdot B \cdot C \cdot D = A' \cdot B \cdot D$$

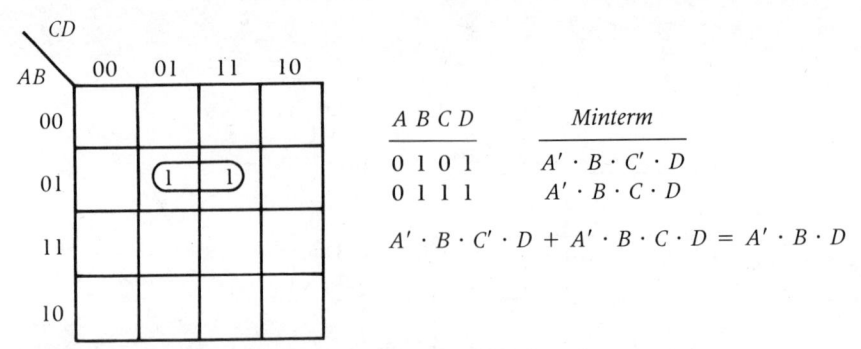

FIGURE 4.7 Two neighboring 1's and their minimized Boolean expression.

region with input variable values $AB = 01$ (meaning, of course that $A = 0$ and $B = 1$) and $CD = 01$. The minterm for this entry in the truth table is equal to $A' \cdot B \cdot C' \cdot D$. Similarly, for the 1 in the second row and third column the input variable values $ABCD$ are equal to 0111, and the minterm is equal to $A' \cdot B \cdot C \cdot D$. The logical sum of these two minterms is

$$A' \cdot B \cdot C' \cdot D + A' \cdot B \cdot C \cdot D = (A' \cdot B \cdot D) \cdot C' + (A' \cdot B \cdot D) \cdot C = A' \cdot B \cdot D$$

where the variable C, whose value changed in the two truth table entries, disappeared from the final simplified expression.

This is the heart of the simplification procedure. The simplification of the original Boolean expression that contains two minterms occurs because the two minterms differ only in the inversion of one variable. Since all neighboring small rectangular regions in the K-map have this same property, all neighboring 1's in the K-map provide similar simplifications. The simplification produces a term for which the number of variables is reduced by 1. In the four-variable case single isolated 1's in the K-map indicate minterms, which are logical products of all four input variables, whereas the combination of two neighboring 1's produce only one product term that contains three variables.

The same simplification procedure for neighboring 1's may be extended to neighboring "pairs" of 1's, which provide recognizable groups of four 1's in the K-map. Two different patterns of four 1's are shown in Figure 4.8. As expected, four 1's reduce the number of variables by 2, resulting in logical product terms of only two variables. In the first row we indicate two pairs of neighboring 1's. These two groups are associated with product terms $A' \cdot B' \cdot C'$ (on the left) and $A' \cdot B' \cdot C$ (on the right). The combined OR expression for all four 1's yield

$$A' \cdot B' \cdot C' + A' \cdot B' \cdot C = (A' \cdot B') \cdot C' + (A' \cdot B') \cdot C = A' \cdot B'$$

which is the minimized expression for the top group of four 1's shown, and it contains only two variables.

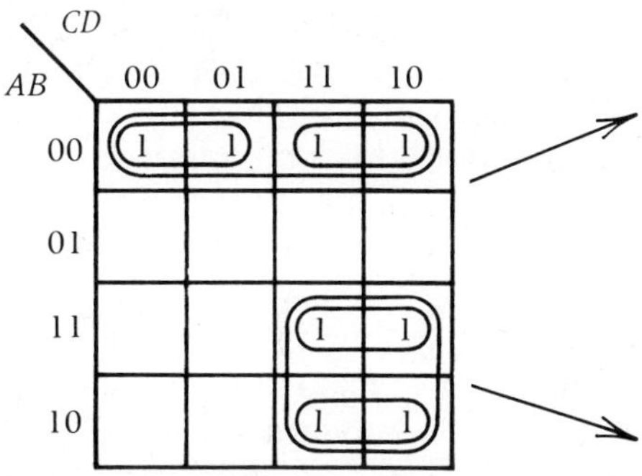

4.4 MINIMIZATION OF SUM-OF-PRODUCT EXPRESSIONS USING K-MAPS

A B C D	Minterms	Reduced terms
0 0 0 0	$A'B'C'D'$	
0 0 0 1	$A'B'C'D$	$A'B'C'$
0 0 1 1	$A'B'CD$	
0 0 1 0	$A'B'CD'$	$A'B'C$
1 1 1 1	$ABCD$	
1 1 1 0	$ABCD'$	ABC
1 0 1 1	$AB'CD$	
1 0 1 0	$AB'CD'$	$AB'C$

FIGURE 4.8 Two groups of four 1's and their associated Boolean functions.

The group of four 1's, shown in the lower right corner, also contains two groups of two neighboring 1's. The upper pair of 1's yields the product term $A \cdot B \cdot C$, whereas the lower pair is associated with the simplified expression $A \cdot B' \cdot C$. The logical sum of these two terms is

$$A \cdot B \cdot C + A \cdot B' \cdot C = (A \cdot C) \cdot B + (A \cdot C) \cdot B' = A \cdot C$$

which, as before, contains only two variables.

The Boolean product terms of three or two variables may be found directly from the groups of two or four 1's in the K-map. For any group of neighboring 1's the input variables are divided into two groups. In the first group are the variables whose values change. These variables are dropped; they do not appear in the simplified expression. In the second group are the variables whose values remain constant for all the regions that belong to the group of 1's under consideration. In the final expression these variables appear in their normal form if the input value is equal to 1, whereas they appear in their inverted form if the input variable value is equal to 0.

We will demonstrate this simple method for the lower group of four 1's in Figure 4.8: The input variable values for these four 1's are 1111, 1110, 1011, and 1010. The second and fourth variables (B and D) change their values and are therefore dropped. Both the first variable (A) and the third variable (C) are equal to 1 for all four 1's in the K-map; hence, the simplified product term is equal to $A \cdot C$.

So far we have seen how 1's and groups of neighboring 1's produce logical product terms in simplified logical-sum-of-product expressions. It is easy to establish the same correspondence between logical product terms and 1's appearing in the K-map. This is called *mapping* a given Boolean function. The mapping is done separately for each product term. Each term produces a pattern of 1's in the K-map. It is possible that more than one term produces a 1 in the same region. Of course, once a 1 appears in a rectangular region, other product terms have no effect on the same region.

The method of mapping is quite simple. First, each product term is converted to a set of input variable values. For example, in Figure 4.8 the term

$A' \cdot B'$ is converted to input variable values $AB = 00$ (the same rules apply as before; 0 values are associated with inverted variables, 1 values are associated with normal variables). Once the input variable values are known, 1's are entered for **all** regions in the K-map for which the input variable values are true. In the four-variable K-map of Figure 4.8, all four regions in the first row have input variable values $A = 0$ and $B = 0$; hence, four 1's are entered into these four regions.

We may pursue groups that contain more than four 1's. Since groups of four neighboring 1's produce product terms with two literals for four-variable functions, we should expect that groups of eight 1's produce terms with only one literal. There are only eight such terms $(A, A', B, B', \cdots, D')$ for a four-variable function. Instead of examining groups of eight 1's in the K-map, let us map three functions, $F(A, B, C, D) = A', D,$ and B' and study the patterns of 1's produced by the mapping process. The results for the selected three functions are shown in Figure 4.9. Indeed, eight 1's appear in the K-map for each function. However, for the function $F(A, B, C, D) = B'$ the eight 1's do not seem to form a group of eight 1's but appear as two groups of four 1's.

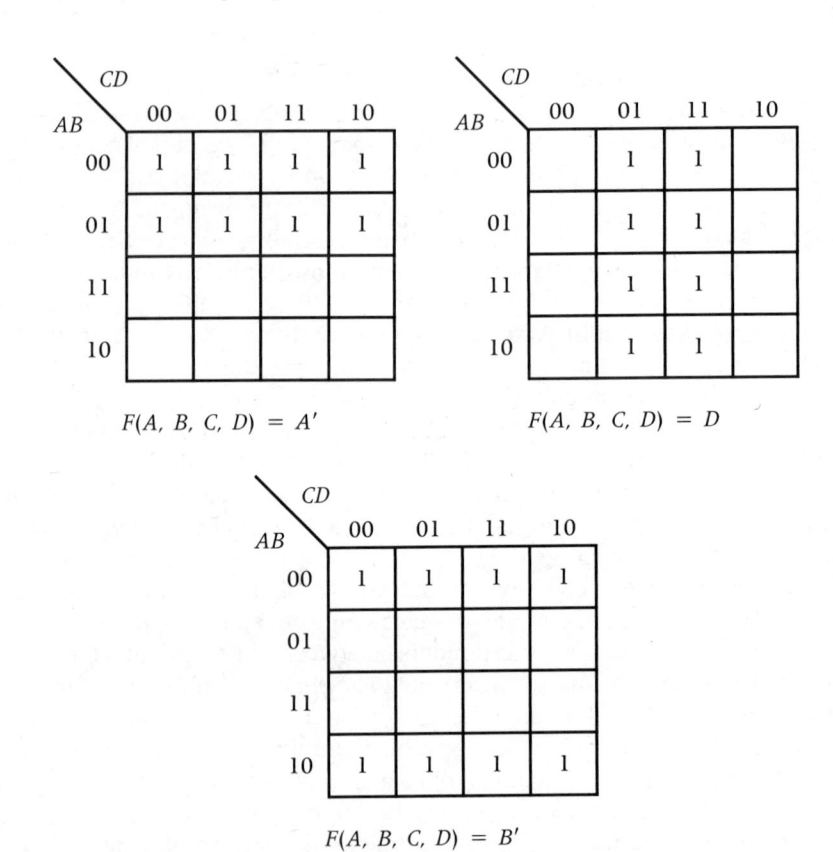

$F(A, B, C, D) = A'$

$F(A, B, C, D) = D$

$F(A, B, C, D) = B'$

FIGURE 4.9 Groups of eight 1's.

4.4 MINIMIZATION OF SUM-OF-PRODUCT EXPRESSIONS USING K-MAPS

In order to see that this group is equivalent to a group of eight 1's, we must deal with 1's located next to the borders of the K-map as special cases.

Simplification at Borders and Corners

We have shown how groups of neighboring 1's yield simplified Boolean expressions. The concept of "neighboring regions" may be extended to the border and corner squares of the K-map. The first column in the four-variable K-map has input variable values $CD = 00$. The fourth column has input variable values $CD = 10$. Since these two sets of input values differ only in the change of the C variable, the first and fourth columns in the K-map contain neighboring squares. The same can be shown for the first and fourth rows and for the four corners. The four corner regions have input variable values 0000, 0010, 1000, and 1010, which produce a group of four 1's with the first and third variables changing and with $B = 0$ and $D = 0$.

A relatively simple way of recognizing how 1's at the border form groups of neighboring 1's is to draw the K-map as a periodic diagram. The fourth column (with input values 01) is drawn to the left of the first column, and the fourth row (also with input values 01) is drawn at the top of the first row. The periodic nature of the K-map is demonstrated in Figure 4.10 by showing the input variable values in the small rectangular regions, similar to Figure 4.3.

In Figure 4.11 we demonstrate the simplification procedure for a pattern of 1's next to the borders. Since the map is periodic, the 1's in the fourth row and fourth column of the original K-map (within the heavier drawn borders) are copied into the top row and left column of the extended diagram. The lower right corner square value is copied into the upper left corner of the extended diagram, which in this case is equal to 0, so it has no effect on the minimization procedure.

AB \ CD	10	00	01	11	10
10	1010	1000	1001	1011	1010
00	0010	0000	0001	0011	0010
01	0110	0100	0101	0111	0110
11	1110	1100	1101	1111	1110
10	1010	1000	1001	1011	1010

FIGURE 4.10 Extended K-map that shows its periodic nature.

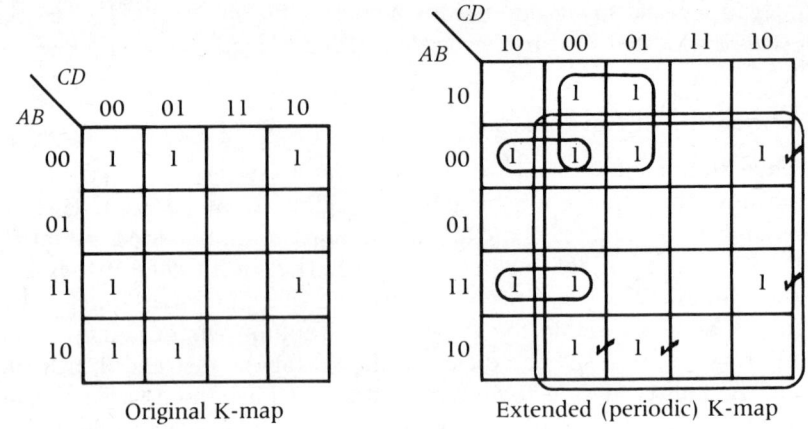

96   COMBINATIONAL CIRCUIT DESIGN AND MINIMIZATION

FIGURE 4.11 Simplification at borders.

The extended K-map shows a group of four 1's and two groups of two 1's. The 1's at the border form these groups, and these are indicated in the extended K-map. When a 1 located in the top row or left column of the extended diagram is included in a group of 1's, it is checked off in the original K-map. This is to indicate that the 1's in the top row are equivalent to the 1's in the bottom row, since they have the same input variable values ($AB = 10$). The same relationship is true between the leftmost and the rightmost columns ($CD = 01$). The checked-off 1's do not produce additional product terms, since they are already covered by the terms associated with the groups at the borders.

After some experience in recognizing groups of neighboring 1's, both within K-maps and at the borders or corners, the groups can be easily identified without drawing the extended periodic map. In Section 4.6 various examples will demonstrate simplification with two-, three-, and four-variable K-maps for which simplifications at the borders and corners will be included.

4.5 MINIMIZATION OF PRODUCT-OF-SUM EXPRESSIONS USING K-MAPS

We have already seen that it is worth exploring duality since it often produces another, quite independent method of solution with a very small amount of additional work. This is also true for minimization procedures using K-maps. We always find the dual method by replacing OR operations with AND operations, AND operations with OR operations, 1's with 0's, and 0's with 1's. In the case of K-maps we also replace minterms by maxterms.

For the dual method we concentrate on the 0 entries in the truth table. Each 0 in the truth table corresponds to a maxterm, which is a logical sum that contains all the input variables. In the K-map the 0's are entered, and the regions that would contain 1's are left blank. Since neighboring regions

4.5 MINIMIZATION OF PRODUCT-OF-SUM EXPRESSIONS USING K-MAPS

have input variable values for which only one variable changes, the logical product of the respective maxterms yields a very much simplified logical sum expression that contains one less variable than the maxterms contained.

In Figure 4.12 a Boolean function of three variables is shown. The 0 in the upper left corner is in the region with input variable values $ABC = 000$ and maxterm $A + B + C$. The 0 to the right is in the region with $ABC = 001$ and maxterm $A + B + C'$. Forming the logical product of these two maxterms, we get

$$(A + B + C) \cdot (A + B + C') = [(A + B) + C] \cdot [(A + B) + C']$$
$$= A + B + C \cdot C' = A + B$$

where we used the OR distributive law for the simplification of the Boolean expression (Table 1.5). The results are similar to the simplification one gets for minterms. The third variable (C) changes; therefore, it is missing in the final expression. The other two variables have constant values, $A = 0$ and $B = 0$, which produce the maxterm $A + B$. For the other group of 0's in Figure 4.12, the input variable values are equal to 010 and 110, which correspond to maxterms $A + B' + C$ and $A' + B' + C$. From these two maxterms we get

$$(A + B' + C) \cdot (A' + B' + C) = B' + C + A \cdot A' = B' + C$$

and the combined minimized expression as a logical-sum-of-product terms is given by

$$F(A, B, C) = \Pi\,(0, 1, 2, 6) = (A + B) \cdot (B' + C)$$

The techniques of simplification discussed in the previous section for 1's in the K-maps are applicable to 0's as well. The only difference is in the final expression, which becomes a logical-product-of-logical-sum expression.

Before we continue our study of minimization techniques by showing various examples of three- and four-variable functions, let us examine the simplified three-variable function for the K-map shown in Figure 4.12.

The final product-of-sum expression has four literals and four operators. A binary circuit built according to this expression contains two two-input OR gates, one two-input AND gate, and one INVERTER gate. We can see that if

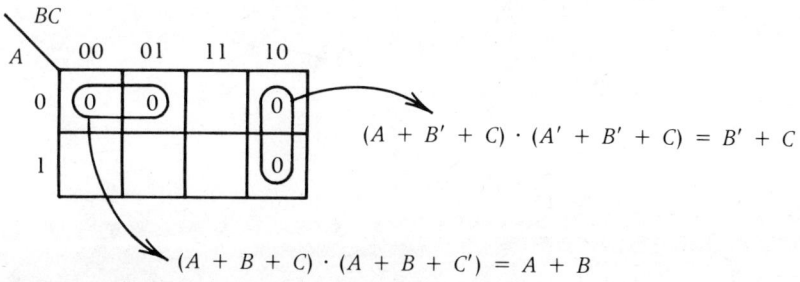

FIGURE 4.12 Minimization for groups of 0's.

98 COMBINATIONAL CIRCUIT DESIGN AND MINIMIZATION

this expression is multiplied out, the following sum-of-product expression is produced:

$$F(A, B, C) = A \cdot B' + A \cdot C + B \cdot C$$

which contains six operators. It is not immediately obvious how one may simplify the preceding expression using Boolean algebra and the simplifying expressions we have studied so far. On the other hand, if we use the original K-map and find minimum product-of-sum expressions (see Figure 4.13), we find that the simplified product-of-sum expression is given by

$$F(A, B, C) = A \cdot B' + B \cdot C$$

which contains four literals and four operators and can be realized with a circuit containing two two-input AND gates, one two-input OR gate, and one INVERTER gate. The simplification of this function from three terms to two terms is important enough to have acquired a name and is called the **consensus** theorem. It may be memorized and added to the collection of simplifying rules shown in Chapters 1 and 2. For those of us who do not like to memorize a large number of formulas, the K-map technique, which clearly shows the application of this theorem, is a more attractive alternative.

This simple example demonstrates that in order to find the circuit with the smallest number of gates, both methods of minimization should be tried. It is difficult to predict in advance which method will yield the simpler expression. The circuits found for these two methods contained an equal number of logic gates, but the distribution of gate types differed. When practical IC devices are used and other parts of a circuit are considered, one minimized version may produce a less expensive overall circuit than the other.

4.6 SIMPLIFICATION OF BOOLEAN EXPRESSIONS

We have studied two methods with which simplified sum-of-product and product-of-sum expressions may be found for Boolean functions of up to four variables. Before considering further extensions to these two techniques, a number of examples using these two methods will be demonstrated. It does not matter very much how the Boolean function is defined initially. As a first step, the initial function definition is converted to a K-map. The decision

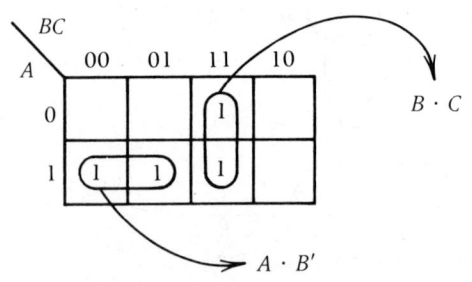

FIGURE 4.13 Minimization for groups of 1's.

whether a truth table should be generated before the K-map is constructed or the K-map generated from the functional specifications does not at all affect the minimization procedures.

Once the K-map has been determined, minimization can proceed. Obviously, a certain amount of experience is needed before one becomes proficient in recognizing the correct groups of 1's or 0's in a K-map. We have been presenting the K-map minimization procedures as if they were simple and straightforward. They are, indeed, relatively simple. However, they are not straightforward because many different groupings of 1's or 0's can be used that produce different circuits.

When one is faced with a pattern of 1's or 0's in a K-map, one's immediate reaction is to try to find the largest possible groups first. This turns out to be the wrong strategy. Since the final expression must account for all 1's (or 0's) in the K-map, the best strategy is to take care of the most isolated 1's (0's) first. This is demonstrated with an example in Figure 4.14. If we assign a product term to the group of four 1's in the middle of the K-map (product term $B \cdot D$), there are still four isolated 1's that are not accounted for. These 1's are parts of groups of two 1's, and therefore, the final expression will contain the term $B \cdot D$ plus four terms with three variables each. On the other hand, if we find the terms for the 1's that do not belong to the middle group of four 1's, as shown in the K-map on the right in Figure 4.14, the four terms include **all** the 1's in the K-map and no other terms are required. The requirement is to include all 1's (or 0's) in the K-map by at least one group (i.e., one term). In general, many 1's or 0's may belong to more than one group.

Even when one follows the strategy of first selecting groups for the more isolated 1's or 0's in the K-map, one may find that the selection of larger groups is not unique. More than one arrangement may produce equally simplified expressions. In this case one cannot decide which selection is better; therefore, all equally minimized expressions should be considered. Boolean factorization operations and gate transformations may affect the final circuit diagram, and preference of one expression over the other at these later stages

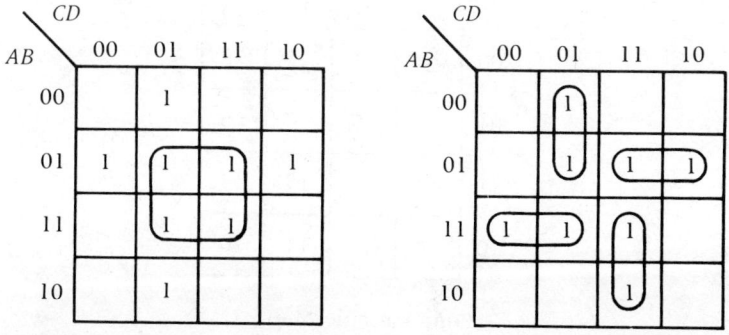

FIGURE 4.14 Possible groups of four 1's and two 1's in four-variable K-map.

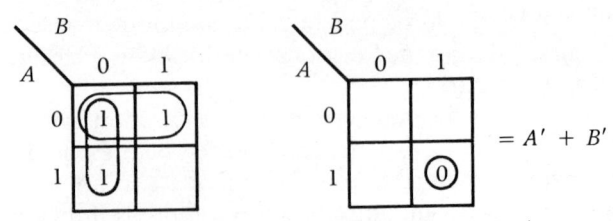

100 COMBINATIONAL CIRCUIT DESIGN AND MINIMIZATION

FIGURE 4.15 Simplification for $F(A, B) = A' \cdot B + B'$.

of digital circuit design may decide which expression should be selected for the final circuit.

Karnaugh map minimization techniques produce sum-of-product and product-of-sum expressions. If the circuit is to be constructed from basic logic gates, the minimized expressions may not produce minimized circuits. Factorization often reduces the number of literals and operators in an expression. Other algebraic manipulations may also improve the results. Some of these techniques were demonstrated in Chapter 2. These less systematic methods indicate that K-map minimization is not the final word in producing the best possible circuit. However, minimization techniques based on the K-map are very effective as a first step in producing minimized expressions in standard forms.

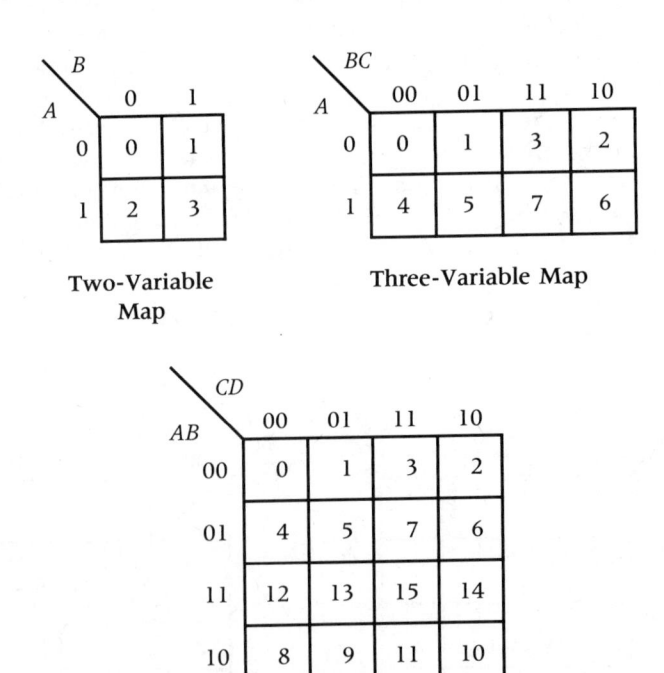

Two-Variable Map

Three-Variable Map

Four-Variable Map

FIGURE 4.16 Minterm and maxterm indices in various K-maps.

4.6 SIMPLIFICATION OF BOOLEAN EXPRESSIONS 101

$$= A \cdot B + A \cdot C + B \cdot C$$

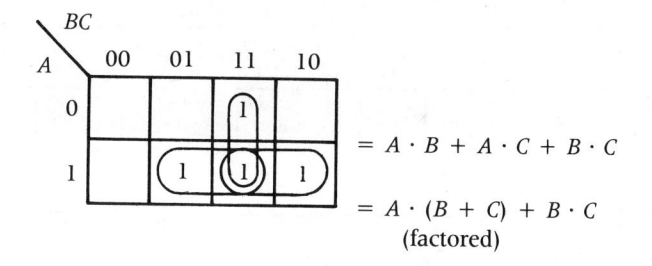

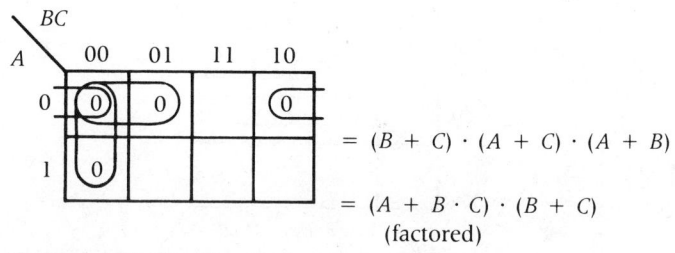

$$= A \cdot (B + C) + B \cdot C$$
(factored)

$$= (B + C) \cdot (A + C) \cdot (A + B)$$

$$= (A + B \cdot C) \cdot (B + C)$$
(factored)

FIGURE 4.17 Simplification for $F(A, B, C) = \Sigma\ (3, 5, 6, 7)$.

Examples of K-map minimization procedures follow. For each case the function is specified, the original K-map is shown, and then both sum-of-product and product-of-sum expressions are generated. Factorizations, when possible, are also executed. The encircled groups in the original K-map show the 1's or 0's that were mapped directly from the given expression. The final groups show the selected minimized expressions.

Simplify $F(A, B) = A' \cdot B + B'$

As shown in Figure 4.15, both minimization methods produce the same result.

Find an Expression for Majority Circuit $F(A, B, C) = \Sigma\ (3, 5, 6, 7)$

In Figure 4.16 the minterm and maxterm indices are shown in the small rectangular regions of the two-, three-, and four-variable K-maps. These diagrams are very helpful in the translation of canonical expressions to a K-map.

The minimization is carried out and shown in Figure 4.17.

Simplify the $F(A, B, C) = B' \cdot C' + A \cdot B' \cdot C + A \cdot B$

Minimization for both 1's and 0's are shown in Figure 4.18.

Simplify $F(A, B, C) = C + A \cdot C' + B' \cdot C' + A' \cdot B$

As shown in Figure 4.19, the original K-map indicates that this function is not a function at all but is equal to the constant 1 (eight 1's in the K-map).

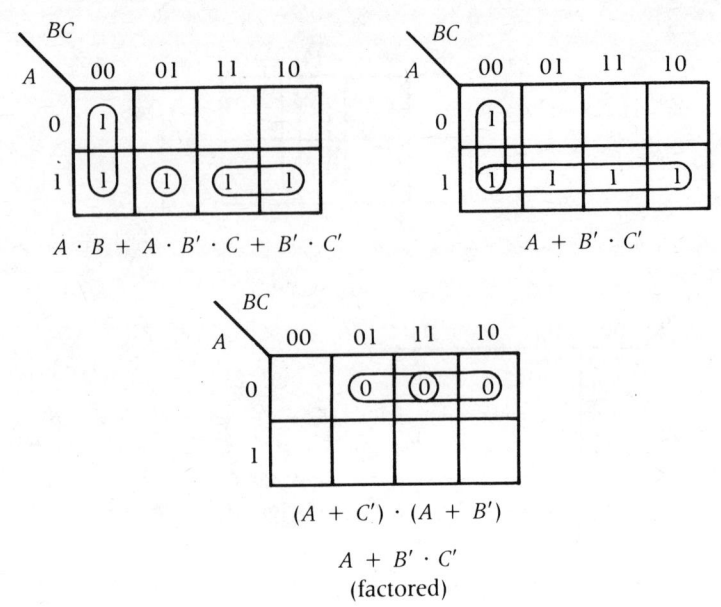

102 COMBINATIONAL CIRCUIT DESIGN AND MINIMIZATION

$A \cdot B + A \cdot B' \cdot C + B' \cdot C'$

$A + B' \cdot C'$

$(A + C') \cdot (A + B')$

$A + B' \cdot C'$
(factored)

FIGURE 4.18 Simplification for $F(A, B, C) = B' \cdot C' + A \cdot B' \cdot C + A \cdot B$.

Determine Boolean Function for Truth Table in Figure 4.6

In Figure 4.20 the results for both minimization methods are shown. Six groups of four 1's can be found. Many different factored expressions may be produced, but only two are shown.

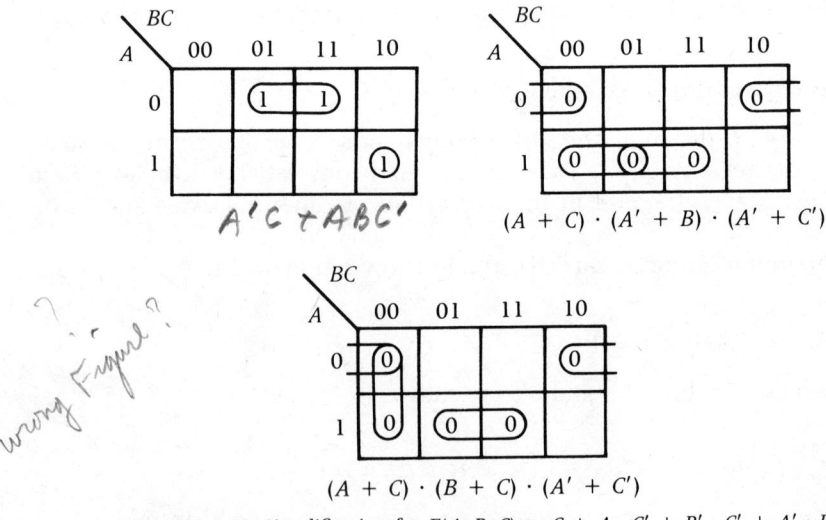

$A'C + ABC'$

$(A + C) \cdot (A' + B) \cdot (A' + C')$

$(A + C) \cdot (B + C) \cdot (A' + C')$

FIGURE 4.19 Simplification for $F(A, B, C) = C + A \cdot C' + B' \cdot C' + A' \cdot B$.

$= A \cdot B + C \cdot D + B \cdot D + B \cdot C + A \cdot D + A \cdot C$

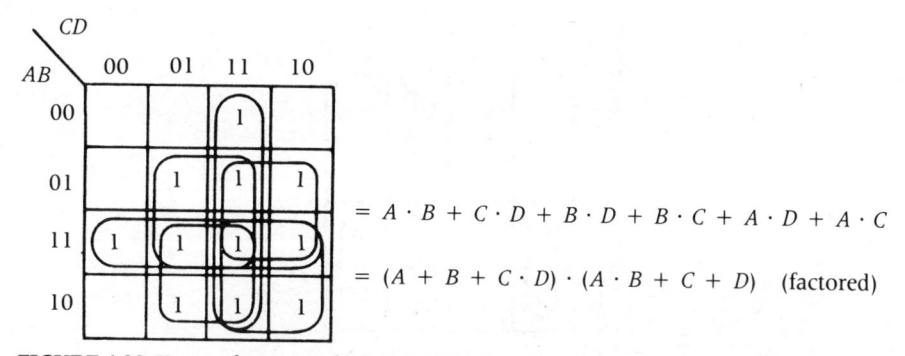

$= (A + B + C \cdot D) \cdot (A \cdot B + C + D)$ (factored)

FIGURE 4.20 Karnaugh maps and minimized Boolean expressions for truth table in Figure 4.6.

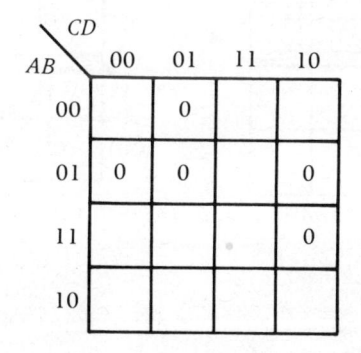

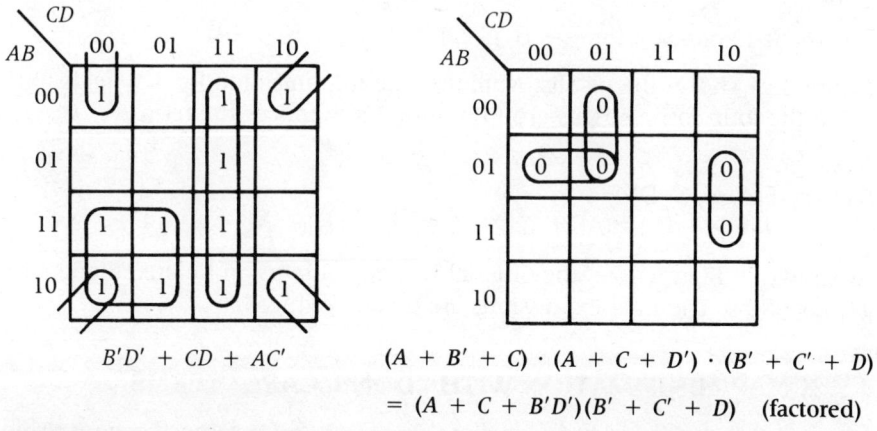

$B'D' + CD + AC'$

$(A + B' + C) \cdot (A + C + D') \cdot (B' + C' + D)$

$= (A + C + B'D')(B' + C' + D)$ (factored)

FIGURE 4.21 Karnaugh map and minimized expressions for $F(A, B, C, D) = \Pi(1, 4, 5, 6, 14)$.

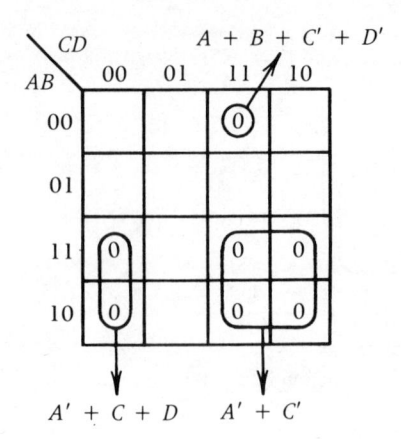

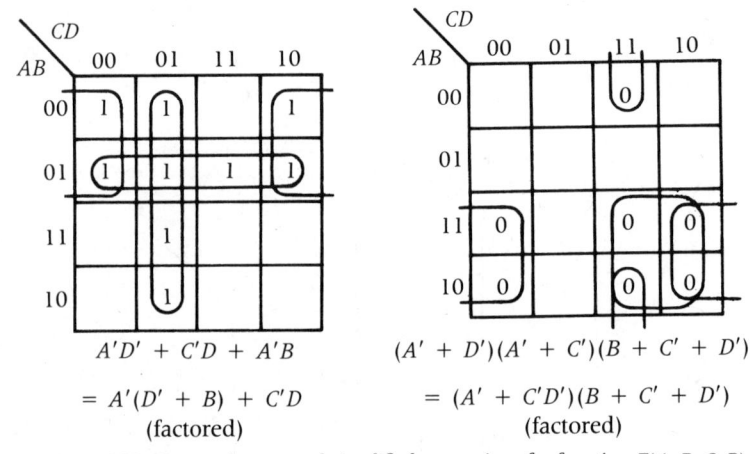

FIGURE 4.22 Karnaugh map and simplified expressions for function $F(A, B, C, D) = (A + B + C' + D') \cdot (A' + C') \cdot (A' + C + D)$.

Determine Expression for $F(A, B, C, D) = \Pi (1, 4, 5, 6, 14)$

Figure 4.21 shows the results. Note that the minimization for 1's yields four 1's in the four corners. This group of four 1's produces the term $B' \cdot D'$.

Simplify $F(A, B, C, D) = (A + B + C' + D') \cdot (A' + C') \cdot (A' + C + D)$

As shown in Figure 4.22, the original K-map contains an isolated 0 and two groups of 0's. The final expressions are greatly reduced.

4.7 K-MAP MINIMIZATION WITH "DON'T CARE" TERMS

Many digital design problems include don't care output specifications. This means that for some combinations of input variable values the circuit's re-

4.7 K-MAP MINIMIZATION WITH "DON'T CARE" TERMS

quired output value is not specified. Don't care outputs arise in situations where not all input value combinations are used. For example, the so-called binary-coded decimal (BCD) digits use 4 bits to represent the decimal digits from 0 to 9. There are 16 possible combinations of values for 4 bits, but the BCD digits use only 10 of these 16 combinations. Six combinations of input value are not used and become don't care cases.

Don't care outputs are marked with the letter X in the truth table and they appear in the K-map for both minimization methods. Since required values for the don't care outputs are not specified, the designer can select either a 1 or a 0 value for them. The selection of a value may be made in such a way that the simplest Boolean expression is produced for the output function. For both minimization methods (when simplification is carried out for 1's and for 0's), the don't care terms appear in the K-map. It is possible that different combinations of 1's and 0's are selected for the two minimization procedures. This means that the resulting two circuits generate different truth tables. For the required output values they both generate the same output, but they differ for the don't care cases.

The best way to demonstrate these procedures is to give an example. In Figure 4.23 the truth table and the K-map of a four-variable Boolean function are shown. In addition to the sum of minterms that produces the required 1's in the truth table, a sum of indices for the don't care cases are also shown. These indices are used for both minterms and maxterms. Figure 4.24 shows the K-maps and the final Boolean expression for the minimization procedure when the minterms are used. The first map shows the 1's and the don't care terms. The selected groups of 1's are encircled. Inside these groups all don't care terms are changed to 1 values. The don't care terms outside the selected groups remain 0. The second map on the right shows the K-map for the

A B C D	F(A, B, C, D)
0 0 0 0	1
0 0 0 1	1
0 0 1 0	1
0 0 1 1	1
0 1 0 0	1
0 1 0 1	Don't care
0 1 1 0	1
0 1 1 1	0
1 0 0 0	0
1 0 0 1	0
1 0 1 0	Don't care
1 0 1 1	Don't care
1 1 0 0	1
1 1 0 1	Don't care
1 1 1 0	0
1 1 1 1	0

AB \ CD	00	01	11	10
00	1	1	1	1
01	1	X	0	1
11	1	X	0	0
10	0	0	X	X

$F(A, B, C, D) = \Sigma\ (0, 1, 2, 3, 4, 6, 12)$
Don't care $= \Sigma\ (5, 10, 11, 13)$

FIGURE 4.23 Truth table and K-map for four-variable Boolean function with don't care terms.

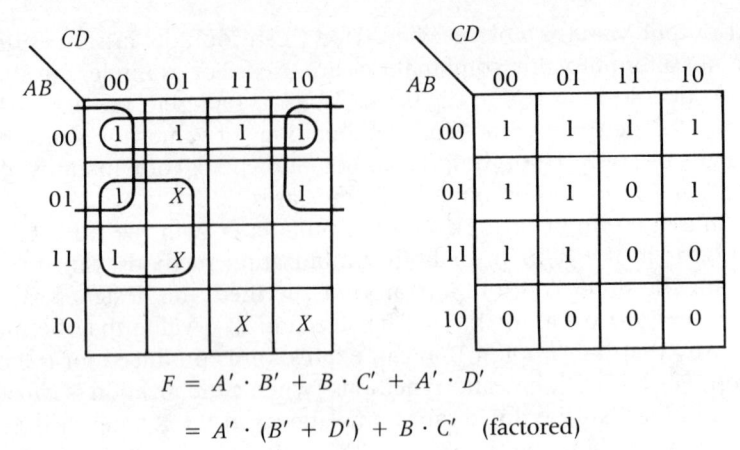

$$F = A' \cdot B' + B \cdot C' + A' \cdot D'$$
$$= A' \cdot (B' + D') + B \cdot C' \quad \text{(factored)}$$

FIGURE 4.24 Minimization for 1's for K-map in Figure 4.23.

output of the circuit in which the don't care outputs are exchanged with the actual 1's and 0's generated by the simplified Boolean expression.

The minimization procedure for maxterms is shown in Figure 4.25. The K-map with 0's and don't care terms is shown on the left, whereas the K-map on the right shows the actual output values generated by the simplified expression for maxterms. Note that the two final K-maps are not the same. However, they both satisfy the functional requirements of the posed problem.

4.8 MINIMIZATION WITH MULTIPLE OUTPUTS

We have seen how K-maps are generated from Boolean expressions and, with the help of these maps, how the expressions are minimized. So far we have considered only one Boolean function at a time, but many practical circuits have more than one output. For circuits with multiple outputs the

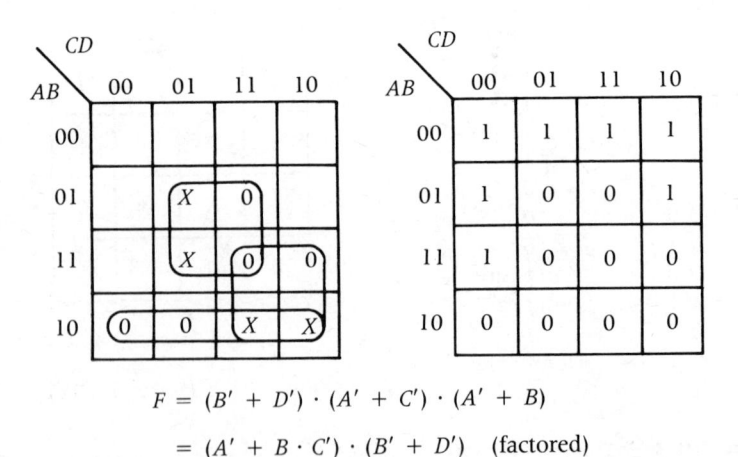

$$F = (B' + D') \cdot (A' + C') \cdot (A' + B)$$
$$= (A' + B \cdot C') \cdot (B' + D') \quad \text{(factored)}$$

FIGURE 4.25 Minimization for 0's for K-map in Figure 4.23.

design problem is more complex because the circuit provides all the required outputs simultaneously, and the sharing of different parts of the circuit among the outputs become possible.

When multiple outputs are required, as many K-maps are produced as there are outputs. Each K-map may be minimized on its own. However, it is not certain that the resulting circuits provide the best overall solution to the multiple-output design problem. By using other than minimized circuits for some of the outputs, it may be possible to share a considerably larger portion of the circuit between the outputs than that provided by the individually minimized circuits. The reduced cost arising from sharing parts of the circuit for different outputs may be more significant than the independent minimization of each circuit for each output.

Many researchers have tackled this very complex problem, but none can claim a solution that works for all cases. This is one of the areas of digital design where it is difficult to find general systematic design methods and where designers with experience (i.e., the experts), outperform automated design programs.

As we have mentioned before, the importance of minimization techniques on the logic gate level is decreasing; therefore, we do not want to attribute too much importance to this subject. Two design examples with multiple outputs are presented in what follows that should demonstrate some of the techniques one may use to achieve simplified circuits for multiple-output problems. We will stop there and will not explore this very complicated field any further.

Design Example 1 Octal Digit Incrementer

For the first problem an octal digit incrementer circuit will be designed. An octal digit has 3 bits, and the digits have unsigned binary values from 0 (000) to 7 (111). The incrementer circuit has three inputs and three outputs. The output bits provide an octal digit whose value is equal to the input digit plus 1. Figure 4.26 shows the block diagram of the octal incrementer and its truth table.

When the digit is equal to 7 (111), then incrementing the digit should be equal to 8, but the value 8 cannot be expressed with 3 bits (the binary number for 8 is equal to 1000 which has 4 bits). If we assume that the octal incrementer produces the lowest 3 bits of the result, then incrementing the digit for 7 (111) becomes equal to digit 0 (000).

The output functions for the three output variables are shown in Figure 4.27 in three three-variable K-maps. For each K-map the simplification procedures for 1's and 0's provide the following minimized Boolean expressions: for minterms

$$R_2 = D_2' \cdot D_1 \cdot D_0 + D_2 \cdot D_1' + D_2 \cdot D_0' = D_2' \cdot D_1 \cdot D_0 + D_2 \cdot (D_1' + D_0')$$

$$R_1 = D_1' \cdot D_0 + D_1 \cdot D_0'$$

$$R_0 = D_0'$$

108 COMBINATIONAL CIRCUIT DESIGN AND MINIMIZATION

FIGURE 4.26 Block diagram and truth tables for octal incrementer.

and for maxterms

$$R_2 = (D_2 + D_0) \cdot (D_2 + D_1) \cdot (D_2' + D_1' + D_0') = (D_2 + D_1 \cdot D_0) \cdot (D_2' + D_1' + D_0')$$

$$R_1 = (D_1 + D_0) \cdot (D_1' + D_0')$$

$$R_0 = D_0'$$

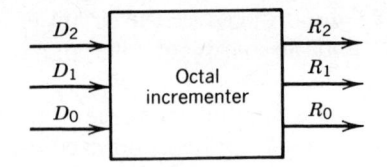

If we examine the two groups of minimized expressions separately, there are no terms that include similar expressions. However, if we use the expres-

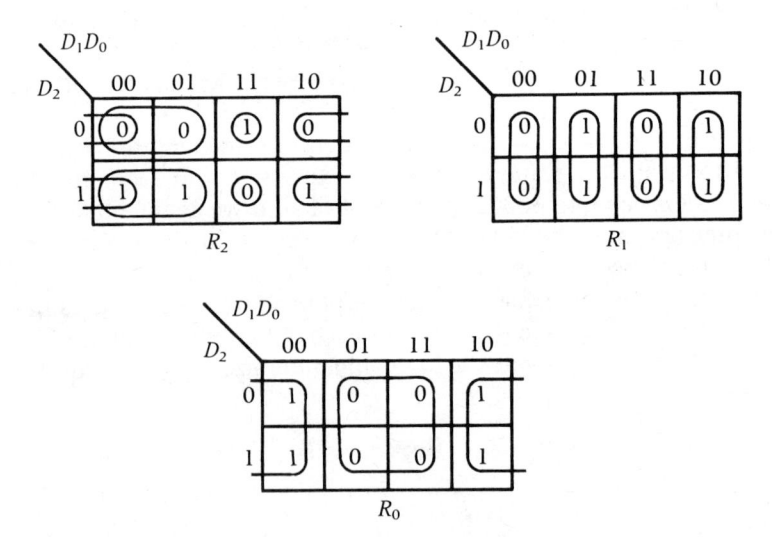

FIGURE 4.27 Karnaugh maps for octal incrementer.

sion for R_2 for minterms and the expression for R_1 for maxterms, the term $D_1' + D_0'$ becomes common to both expressions. Furthermore, the term $D_1' + D_0'$ can also be expressed as $(D_1 \cdot D_0)'$, which becomes a single NAND gate.

The resulting Boolean expressions for the octal incrementer become

$$R_2 = D_2' \cdot (D_1 \cdot D_0) + D_2 \cdot (D_1 \cdot D_0)'$$

$$R_1 = (D_1 + D_0) \cdot (D_1 \cdot D_0)'$$

$$R_0 = D_0'$$

The circuits constructed according to these last three expressions are shown in Figure 4.28.

Design Example 2 Decimal Decrementer Circuit

For the second example we design a circuit that decrements a decimal digit. The decimal digits 0–9 require at least 4 bits for their representation. The bits for the input are called D_3, D_2, D_1, and D_0. The outputs of the decrementer also consist of 4 bits. These are R_3, R_2, R_1, and R_0.

Since for 4 bits there are 16 combinations of input values, 6 combinations will not be used and become don't care cases. There are many different ways in which one may select 10 input combinations for the digits out of the possible 16. We will use a popular selection, called the **Excess-3** binary representation. For this representation binary value 3 (0011) represents decimal digit 0, binary value 4 (0100) digit 1, and so on. Binary value 12 (1100) represents decimal digit 9.

The decrementer produces an output digit (represented by outputs R_0–R_3) equal to 1 less than the input digit (inputs D_0–D_3). If the digit 0 is decremented (digit 0 is equivalent to inputs 0011), the result is equal to digit 9 (1100). This result is consistent with the decrement operation because the outputs

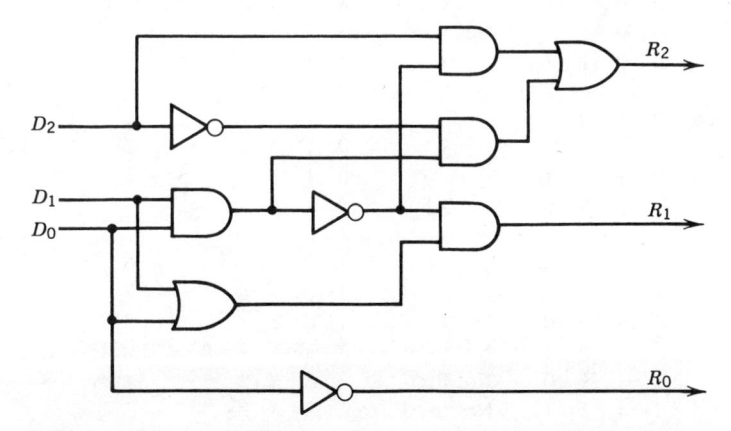

FIGURE 4.28 Completed circuit diagram for octal incrementer.

110 COMBINATIONAL CIRCUIT DESIGN AND MINIMIZATION

represent the four lowest order bits of the result, similar to the octal incrementer. The block diagram and truth table for the digital decrementer are shown in Figure 4.29.

We find that the LSB of the result, R_0, is equal to D_0', similar to the octal incrementer circuit. The K-maps for minterms and don't care outputs for the other three outputs are shown in Figure 4.30. They yield the following results for the output bits:

$$R_3 = D_3 \cdot D_2 + D_2' \cdot D_1 + D_3 \cdot D_0 = D_3 \cdot (D_2 + D_0) + D_2' \cdot D_1$$

$$R_2 = D_3' \cdot D_1 + D_2 \cdot D_0 + D_2' \cdot D_1' \cdot D_0'$$

$$R_1 = D_1' \cdot D_0' + D_2 \cdot D_1 \cdot D_0 + D_3 \cdot D_1 \cdot D_0 = D_1' \cdot D_0' + D_1 \cdot D_0 \cdot (D_2 + D_3)$$

The minimization for maxterms for the same three outputs is shown in Figure 4.31, and the resulting expressions are

$$R_3 = (D_3 + D_2') \cdot (D_2 + D_1 + D_0)$$

$$R_2 = (D_2' + D_1 + D_0) \cdot (D_3' + D_0') \cdot (D_3' + D_1') + (D_2' + D_1 + D_0) \cdot (D_3' + D_1' \cdot D_0')$$

$$R_1 = (D_3 + D_2) \cdot (D_1 + D_0') \cdot (D_1' + D_0)$$

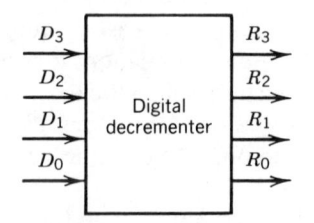

Input bits D_3 D_2 D_1 D_0	Digit value	Output bits R_3 R_2 R_1 R_0	Output value
0 0 0 0	Not used	X X X X	Don't
0 0 0 1	Not used	X X X X	cares
0 0 1 0	Not used	X X X X	
0 0 1 1	0	1 1 0 0	9
0 1 0 0	1	0 0 1 1	0
0 1 0 1	2	0 1 0 0	1
0 1 1 0	3	0 1 0 1	2
0 1 1 1	4	0 1 1 0	3
1 0 0 0	5	0 1 1 1	4
1 0 0 1	6	1 0 0 0	5
1 0 1 0	7	1 0 0 1	6
1 0 1 1	8	1 0 1 0	7
1 1 0 0	9	1 0 1 1	8
1 1 0 1	Not used	X X X X	Don't
1 1 1 0	Not used	X X X X	cares
1 1 1 1	Not used	X X X X	

FIGURE 4.29 Block diagram and truth tables for decimal decrementer circuit using Excess-3 representation.

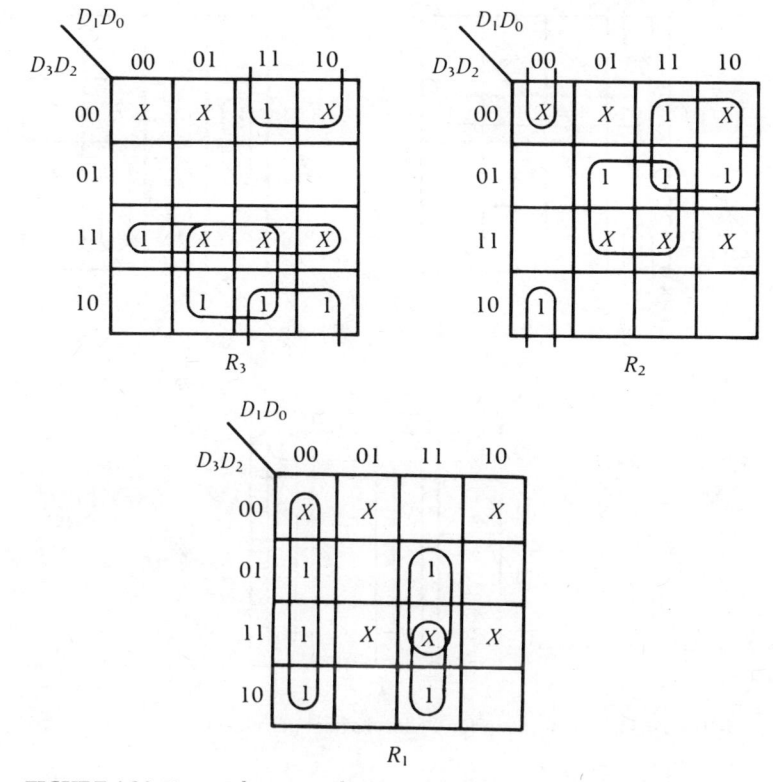

FIGURE 4.30 Karnaugh maps and minimization for 1's for decimal decrementer circuit.

These may be the simplest expressions for the decimal decrementer circuit, but it is also possible that by the selection of different values for the don't care terms or the selection of different groups of 1's and 0's for the minimized expressions, more common terms could have been found. This example demonstrates the fact that even for four input variables, when multiple outputs are required, the design process becomes very complicated. The "best" design may be found only after an exhaustive search for all the different possibilities. This exhaustive search may not be feasible because of time constraints, in which case the best solution may never be found.

In the six preceding expressions the common term $D_1' \cdot D_0' = (D_1 + D_0)'$ may be used to yield the following four expressions for the output bits of the decrementer:

$$R_3 = (D_3 + D_2') \cdot \{D_2 + [(D_1 + D_0)']\}$$

$$R_2 = D_3' \cdot D_1 + D_2 \cdot D_0 + D_2' \cdot (D_1 + D_0)'$$

$$R_1 = (D_1 + D_0)' + D_1 \cdot D_0 \cdot (D_2 + D_3)$$

$$R_0 = D_0'$$

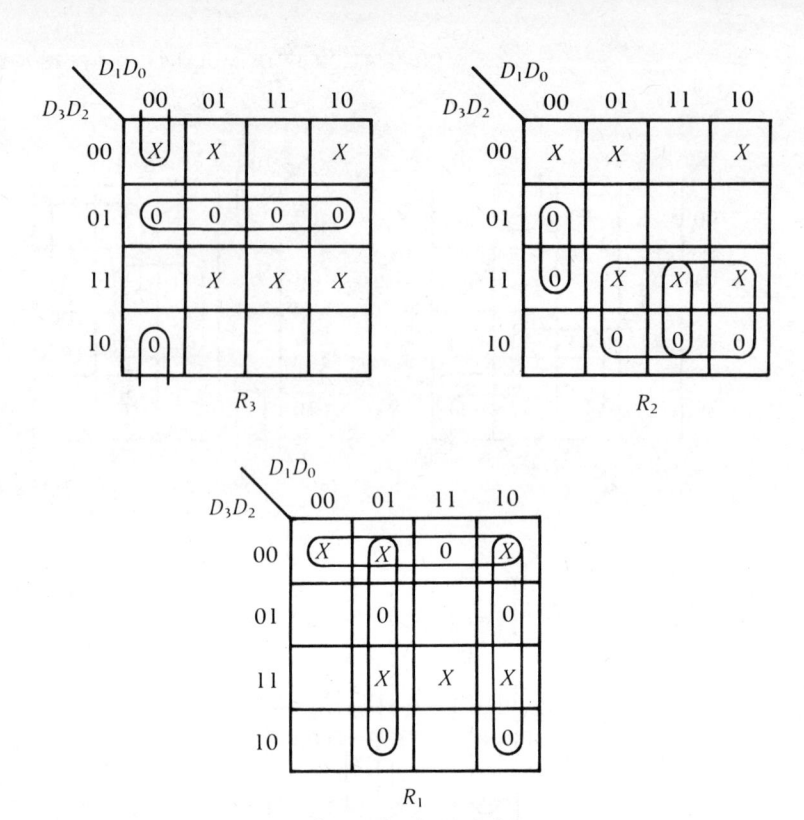

FIGURE 4.31 Minimization for 0's for decimal decrementer.

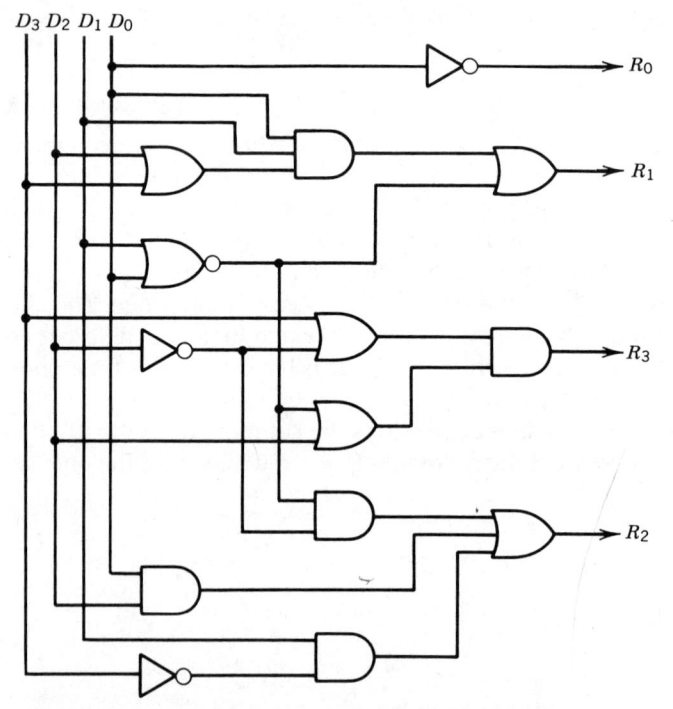

FIGURE 4.32 Completed circuit for decimal decrementer.

112

The circuits built according to these four expressions are shown in Figure 4.32.

4.9 K-MAPS FOR FIVE AND SIX VARIABLES

We stated at the beginning of this chapter that K-maps are useful for Boolean expressions of up to six variables. So far we have shown minimization techniques and K-maps only for two, three, and four variables. It has become quite apparent that circuits with only four variables, especially those with multiple outputs, already produce complex minimization problems. We should expect that the complexities grow considerably larger when five or six inputs are involved.

Since it is possible to use K-maps for five and six variables and the extension of K-map minimization techniques for a larger number of variables is not difficult, we include their description here. This is more for the sake of completeness than for practical use because in the following chapters problems are designed in such a way that in most cases the use of four-variable K-maps will be sufficient.

As shown in Figure 4.33, two four-variable K-maps are used for five variables. From the five input variables ($ABCDE$), four ($BCDE$) are used for the K-maps. The K-map on the left is for input values when $A = 0$, whereas the map on the right is for $A = 1$. There are 32 entries in a five-variable truth table, and the minterm (or maxterm) indices (0–31) are also indicated in Figure 4.33.

Minimization can proceed now with groups of 1's or 0's in the two maps. However, neighboring rectangular regions occur not only within the two separate maps, as before, but also between the two maps. The easiest way to "see" the neighboring regions between the two maps is to imagine that one map is placed on top of the other and that the touching rectangles form neighbors as well. For example, minterm m_0 (00000) neighbors minterm m_{16} (10000), and minterm m_{15} (01111) is neighbor to m_{31} (11111). In a similar

$F(A, B, C, D, E)$

BC\DE	00	01	11	10
00	0	1	3	2
01	4	5	7	6
11	12	13	15	14
10	8	9	11	10

$A = 0$

BC\DE	00	01	11	10
00	16	17	19	18
01	20	21	23	22
11	28	29	31	30
10	24	25	27	26

$A = 1$

FIGURE 4.33 Karnaugh maps with minterm and maxterm indices for functions of five variables.

way, the four minterms m_5, m_{13}, m_{21}, and m_{29} form a group of four 1's with changing values for variables A and B and constant values for variables $C = 1$, $D = 0$, and $E = 1$. Hence, the product term for this group of four minterms is given by

$$F(A, B, C, D, E) = \Sigma (5, 13, 21, 29) = C \cdot D' \cdot E$$

The four four-variable K-maps in Figure 4.34 are similarly arranged for functions that contain six input variables. The K-maps represent variables C, D, E, and F, whereas for the four different maps the values of A and B are equal to 00, 01, 11, and 10, respectively. Each neighboring K-map contains neighboring minterms (or maxterms) that can be visualized again as placing one map on top of the other map. The map in the upper left corner (minterms m_0–m_{15}) is neighbor to both the map in the right upper corner (minterms m_{16}–m_{31}) and to that in the lower left corner (minterms m_{32}–m_{47}). The same is true for the other K-maps. The maps are neighbors because the values of the two variables A and B are different only in one variable, whereas the values for the other four variables remain the same.

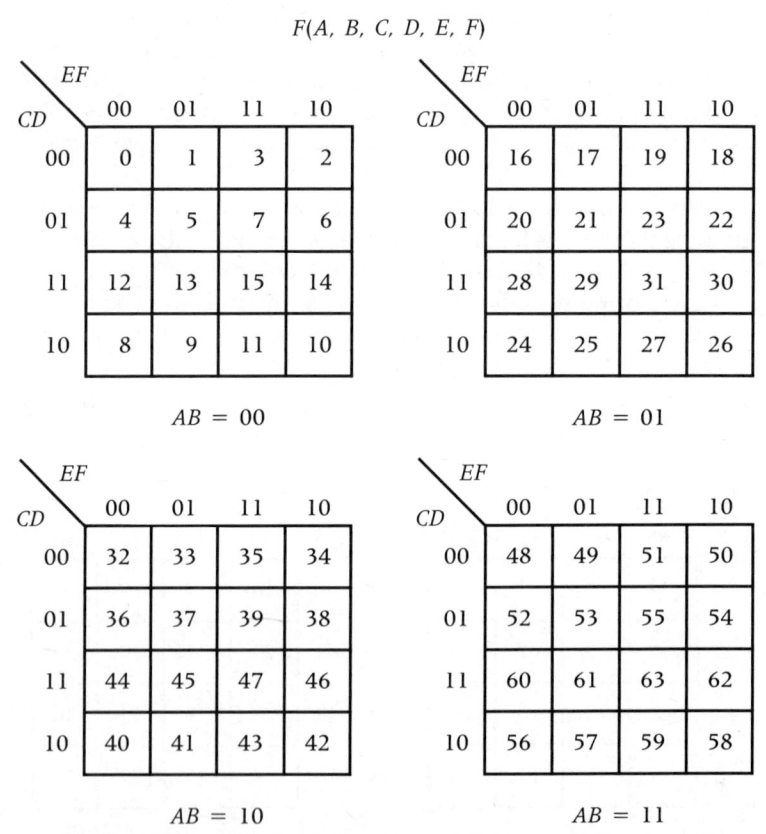

$F(A, B, C, D, E, F)$

CD\EF	00	01	11	10
00	0	1	3	2
01	4	5	7	6
11	12	13	15	14
10	8	9	11	10

$AB = 00$

CD\EF	00	01	11	10
00	16	17	19	18
01	20	21	23	22
11	28	29	31	30
10	24	25	27	26

$AB = 01$

CD\EF	00	01	11	10
00	32	33	35	34
01	36	37	39	38
11	44	45	47	46
10	40	41	43	42

$AB = 10$

CD\EF	00	01	11	10
00	48	49	51	50
01	52	53	55	54
11	60	61	63	62
10	56	57	59	58

$AB = 11$

FIGURE 4.34 Karnaugh maps with minterm and maxterm indices for functions of six variables.

4.9 K-MAPS FOR FIVE AND SIX VARIABLES

When all four maps are placed on top of one another and a minterm (or maxterm) appears in each map in exactly the same location, a group of four 1's (or 0's) is formed. For example, the minterms m_4, m_{20}, m_{36}, and m_{52} form a group of four neighboring regions; therefore,

$$F(A, B, C, D, E, F) = \Sigma (4, 20, 36, 52) = C' \cdot D \cdot E' \cdot F'$$

where both variables A and B change, whereas $C = 0$, $D = 1$, $E = 0$, and $F = 0$. The same simplification occurs between neighboring maps when groups of 1's or 0's are located in identical positions. The following example demonstrates how simplification occurs in the six-variable K-maps (refer to Figure 4.35).

Simplify $F(A, B, C, D, E, F) =$
$\Sigma (5, 7, 13, 15, 21, 23, 29, 31, 37, 39, 45, 47, 55, 63)$

Examining the group of two 1's in the lower right map, we find that a similar group of 1's in the other maps forms a group of eight 1's, for which

$$F(A, B, C, D, E, F) = \Sigma (7, 15, 23, 31, 39, 47, 55, 63) = D \cdot E \cdot F$$

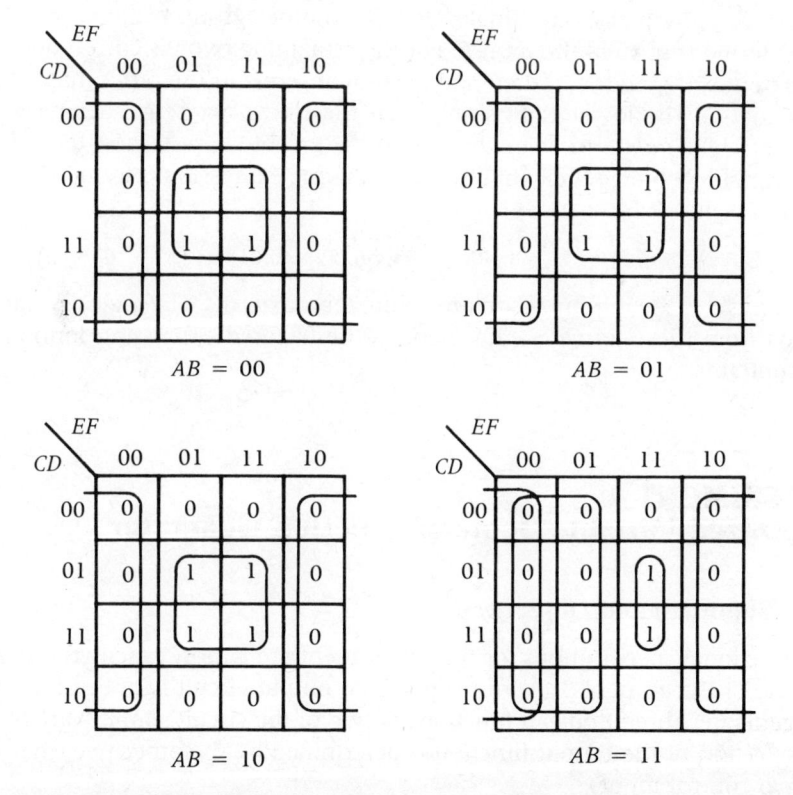

$F(A, B, C, D, E, F) = \Sigma (5, 7, 13, 15, 21, 23, 29, 31, 37, 39, 45, 47, 55, 63)$

FIGURE 4.35 Minimization for function of six variables.

The group of four 1's in the upper left map can be matched with similar patterns either in the upper right or in the lower left map. For these group of eight 1's we get

$$F(A, B, C, D, E, F) = \Sigma\ (5, 7, 13, 15, 21, 23, 29, 31) = A' \cdot D \cdot F$$

and

$$F(A, B, C, D, E, F) = \Sigma\ (5, 7, 13, 15, 37, 39, 45, 47) = B' \cdot D \cdot F$$

The three terms cover all 1's in the four maps, and the complete (factored) expression is

$$F(A, B, C, D, E, F) = D \cdot F \cdot (A' + B' + E)$$

When minimization is carried out for the maxterms, the same expression should be the result since it is already in the form of a logical-product-of-logical-sum terms. For the maxterms we find a group of thirty-two 0's (group of eight 0's in all four maps), which are given by the following product:

$$\Pi\ (0, 1, 2, 3, 8, 9, 10, 11, 16, 17, 18, 19, 24, 25, 26, 27, 32, 33, 34, 35, 40, 41,$$
$$42, 43, 48, 49, 50, 51, 56, 57, 58, 59)$$

and which are equal to the single term D. Another group of thirty-two 0's can be found that yield the term F. For the remaining two 0's corresponding to maxterms M_{53} and M_{61}, there are no similar terms in the other maps, but the K-map in the lower right shows that these two maxterms belong to a group of eight maxterms. For this group, $AB = 11$, $E = 0$, and the other three variables change their values. The following expression is found for this group of eight 0's:

$$F(A, B, C, D, E, F) = \Pi\ (48, 49, 52, 53, 56, 57, 60, 61) = (A' + B' + E)$$

This completes the minimization procedure for maxterms. The logical product formed from these terms is equivalent to the factored expression found for the minterms.

EXPERIMENT 4
Minimized Variable Boolean Function Generator

E4.1 Summary and Objectives

The functional specifications for this experiment are almost exactly the same as those of Experiment 3. A four-input, one-output circuit is to be designed that generates three Boolean functions of two of the circuit's input variables. The selection of the output function is determined by the other two, that is, the two control inputs.

There are two major differences between Experiment 3 and Experiment 4. First, it is not specified which control input settings generate which required output function. Second, the required design method uses K-maps (i.e., overall circuit minimization), not the functional approach of Experiment 3.

The main objective of this experiment is to gain experience in applying circuit minimization with K-maps to a design problem. Comparisons between this experiment and Experiment 3 should be very instructive. For the functional approach, the values of the control variables for each function selection had to be specified, but these values affected the final circuit very little. For the minimization procedures the freedom of choice in the control variable values may result in further simplification of the circuits.

Restrictions on IC types are further relaxed. All available basic gate types (INVERTER, AND, OR, NAND, and NOR gates) are allowed as components. This larger choice of gate types should also help in finding a correctly working circuit that consists of much fewer gates than that of Experiment 3.

E4.2 Description of Overall Circuit

The overall block diagram of the required circuit is shown in Figure E4.1. (This block diagram is the same as that of Experiment 3.) The three output functions are defined in terms of the two signal inputs A and B, and they are equivalent to the functions selected for Experiment 3. Values for the control inputs are not shown because it is the designer's option to choose the three control input settings for the three required functions. Similar to Experiment 3, for each selected control input setting, the circuit behaves as a two-input, one-output Boolean function generator circuit. The output has don't care values for the fourth, or unused, settings of the control inputs. It is the designer's task to select the control input values, minimize the appropriate Boolean expressions, construct the circuit, and demonstrate that it satisfies the stated functional specifications.

E4.3 Expected Results, Requirements, and Restrictions

Since general circuit minimization is used for this experiment, the results should be tabulated in a truth table that has four input variables. Values for the output can be shown only after the selection of the control input settings

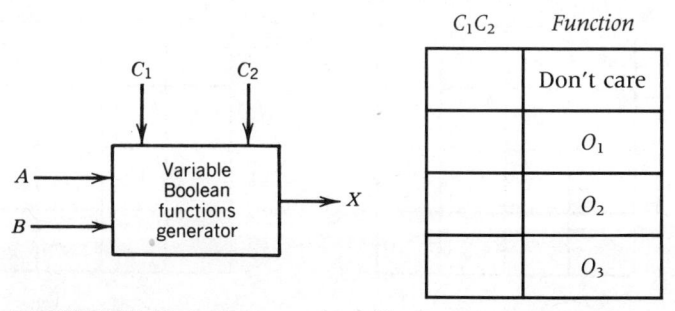

C_1C_2	Function
	Don't care
	O_1
	O_2
	O_3

FIGURE E4.1 Block diagram and specifications for Experiment 4.

118   COMBINATIONAL CIRCUIT DESIGN AND MINIMIZATION

for each required output function. The requirement of this experiment is that at least two different control input settings must be tried, neither of which is equal to the one used for Experiment 3. The two truth tables and K-maps shown in Figure E4.2 should be filled in with the correct output values. Both minimization methods (one for minterms, the other for maxterms) should be tried once, one for each K-map. If time permits, it may be worthwhile to try both minimization methods for both K-maps, since further reductions in the final circuit may occur.

Once the minimized Boolean expressions are determined, they should be factored (if possible) in order to reduce the required number of gates. From the expression that is judged to be the simplest, a circuit diagram should be

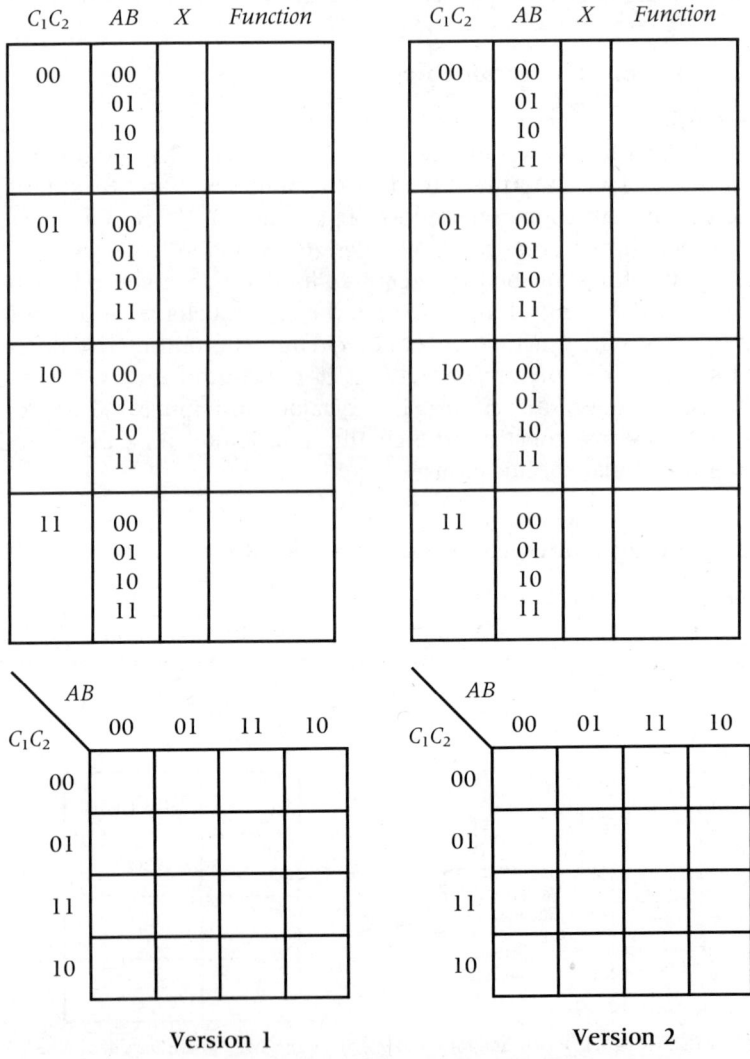

C_1C_2	AB	X	Function
00	00		
	01		
	10		
	11		
01	00		
	01		
	10		
	11		
10	00		
	01		
	10		
	11		
11	00		
	01		
	10		
	11		

C_1C_2	AB	X	Function
00	00		
	01		
	10		
	11		
01	00		
	01		
	10		
	11		
10	00		
	01		
	10		
	11		
11	00		
	01		
	10		
	11		

C_1C_2 \ AB	00	01	11	10
00				
01				
11				
10				

C_1C_2 \ AB	00	01	11	10
00				
01				
11				
10				

Version 1   Version 2

FIGURE E4.2 Karnaugh maps and truth tables for two design methods.

drawn. Depending on available IC types, gate transformations may be necessary. The IC types are restricted to basic gates (INVERTER, AND, NAND, OR, and NOR gates), but the number of inputs for these gates is not restricted. The number of IC devices should be kept to a minimum.

Since several operations and transformations occur between the required truth table and the drawing of the final circuit, it is also a requirement of this experiment that before the circuit is constructed, an independent check of the circuit should be carried out. From the proposed circuit a Boolean expression should be derived and multiplied out so that it is in the form of a logical sum of logical product terms. The expression should be mapped on a four-variable K-map and compared to the completed K-map in Figure E3.2. Then, from the K-map, a truth table should be derived and checked whether it satisfies the functional specifications of this design problem. Of course, this truth table contains only 1's and 0's (there are no X entries) which show the actual output values generated for the don't care as well as for the required input value combinations.

E4.4 Alternate Experiments

After the minimized circuit has been designed, explore the possibility of generating an additional Boolean function with a slightly expanded circuit. This new function should be generated as output when the control inputs are set to the so-far unused selection values. Estimate how much additional circuitry would be required for each new function selected from the eight functions listed in Figure E2.1. Compare the work involved in expanding this circuit to that of Experiment 3, which used a multiplexer.

Use five-variable K-maps to design a circuit that has three control inputs and two data inputs. For each control input setting the circuit generates one of the eight Boolean functions listed in Figure E2.1.

References

1. Mano, M. M. *Digital Logic and Computer Design.* Englewood Cliffs, NJ: Prentice-Hall, 1979, pp. 102–110.
2. Lewin, D. *Design of Logic Systems.* London: Van Nostrand Reinhold, 1985, pp. 70–97.
3. Veitch, E. W., A chart method for simplifying truth functions. *Proc. ACM,* May, 127–133 (1952).
4. Karnaugh, M., A map method for synthesis of combinational logic circuits. *Trans. AIEE, Comm. Electron.,* **72** (Part I) 593–599 (1953).

Drill Problems

4.1. Minimize the following three-variable functions using K-maps both for minterms and maxterms. Factor the minimized expressions and draw the circuit diagram for the least expensive circuit for each case.
  (a) $F(A, B, C) = \Sigma\ (0, 3, 4, 5, 6, 7)$
  (b) $F(A, B, C) = \Pi\ (3, 5, 7)$
  (c) $F(A, B, C) = \Sigma\ (0, 2, 5, 7)$

4.2. Minimize the following three-variable functions that include don't care terms. Determine and draw the circuit diagram for the the least expensive circuit in each case.
  (a) $F(A, B, C) = \Sigma (0, 4, 6)$, don't care $= \Sigma (1, 3)$
  (b) $F(A, B, C) = \Pi (3, 5)$, don't care $= \Sigma (0, 1, 6, 7)$
  (c) $F(A, B, C) = \Pi (3, 5)$, don't care $= \Sigma (1, 2)$

4.3. Map the following three-variable functions and then reduce them using K-map minimization techniques for both minterms and maxterms. Show that the same minimizations may be achieved using Boolean algebra. For each function find the least expensive circuit and draw its circuit diagram.
  (a) $F(A, B, C) = B \cdot C \cdot (A + B' + C') \cdot (A' + C')$
  (b) $F(A, B, C) = B' \cdot C' + A \cdot B' + A' \cdot B \cdot C' + A' \cdot B' \cdot C$
  (c) $F(A, B, C) = A' \cdot B' + A \cdot C' + B' \cdot C' + A' \cdot B \cdot C$

4.4. This problem is the same as Problem 4.1, but four-variable functions are used.
  (a) $F(A, B, C, D) = \Sigma (1, 5, 6, 7, 11, 12, 13, 15)$
  (b) $F(A, B, C, D) = \Pi (3, 6, 7, 12, 14, 15)$
  (c) $F(A, B, C, D) = \Pi (0, 2, 3, 8, 10)$

4.5. Follow the requirements for Problem 4.2 for these four-variable functions with don't care terms.
  (a) $F(A, B, C, D) = \Pi (3, 4, 12, 15)$, don't care $= \Sigma (2, 5, 6, 9, 10, 14)$
  (b) $F(A, B, C, D) = \Sigma (1, 3, 5, 6, 7, 15)$, don't care $= \Sigma (0, 4, 8, 10, 11)$
  (c) $F(A, B, C, D) = \Sigma (4, 7, 12, 13)$, don't care $= \Sigma (0, 1, 2, 6, 7, 9)$

4.6. This problem is the same as Problem 4.3, but four-variable functions are used.
  (a) $A \cdot C + A' \cdot B' \cdot C \cdot D' + A' \cdot B \cdot C \cdot D + A \cdot B' \cdot C' \cdot D + A \cdot B \cdot C' \cdot D'$
  (b) $(A + C + D') \cdot (A' + D') \cdot (A + B' + C' + D') \cdot (B + C' + D')$
  (c) $A' \cdot C' + B' \cdot C' \cdot D + A' \cdot C \cdot D' + A \cdot B \cdot C \cdot D' + A \cdot B \cdot D$

4.7. Design a circuit that multiplies two 2-bit binary numbers (in the range of $00 = 0$ to $11 = 3$). The four output bits (R_0, R_1, R_2, and R_3) are the bits of a decimal digit, which is expressed in the Excess-3 representation.

4.8. Design a five-input majority circuit. The output is equal to logic-1 when at least three of the input variables are equal to logic-1. Try to find the least expensive circuit and draw the circuit diagram.

CHAPTER 5

Introduction to Sequential Systems

5.1 INTRODUCTION

In the last four chapters we limited our exploration of binary systems to combinational circuits. Because a combinational circuit does not contain feedback connections, the outputs of such circuits can always be expressed as Boolean functions of their input variables. The output–input specifications can be tabulated in truth tables, which define the behavior of the circuit for all possible combinations of input values. Even though one may build very complex and sophisticated combinational systems, still another type of building block called a **sequential system** is necessary for the operation of digital computers.

The difference between the operation of sequential and combinational systems is the introduction of **time** as a new variable. Of course, time plays an important role in all physical systems; however, it is possible to eliminate the effect of time in the representation of combinational systems. The behavior of sequential systems depends not only on the current values of the input variables but also on the sequence of input values that occurred in the past. A digital combination lock is a good example of a practical sequential system. The output of a digital lock (whether the lock is closed or open) depends entirely on the sequence of input numbers.

Since the output of a sequential system depends on past events, it must "remember" what happened in the past, or, in other words, it must have **memory.** The two examples of general physical systems discussed in Chapter 1 both contained memory. In the supermarket the items were stored on the shelves, whereas in the chess game the positions of the pieces were stored on the chess board. We called the stored elements of the system **state variables.** The collection of all the stored elements of a sequential system defines its **state.** Since state variables in our digital systems are binary variables, the possible combinations of 1's and 0's for the state variables determine the number of states a digital system may have. For this reason, sequential systems are often called **finite-state** machines.

Hence, by analyzing the operation of practical digital systems, we find that "sequential systems" are required that, by necessity, contain "memory" elements. We have also indicated that noncombinational digital systems differ from combinational systems in that they contain feedback connections. On the basis of our present knowledge of binary systems, it is not at all clear

122 INTRODUCTION TO SEQUENTIAL SYSTEMS

how the two new concepts, memory and feedback, are related. Therefore, instead of introducing digital memory simply as a new type of circuit element, we shall show how the addition of feedback connections leads naturally to the concepts of memory and sequential systems. At the same time we will show that the addition of arbitrary feedback connections in a digital circuit may cause severe difficulties in predicting and/or controlling the behavior of the circuit. These difficulties arise from the fact that sequential systems are time-varying systems. To alleviate these difficulties, we will examine in detail how digital signals may be measured reliably under time-varying conditions. It will be shown that the simplest method involves the introduction of a new binary input variable called **SYSTEM CLOCK,** which controls the operation of the sequential system at all times. The use of SYSTEM CLOCK will allow us to define **synchronous** sequential systems that will be used throughout the latter part of this book.

Having defined the characteristics of synchronous sequential systems, we will show how the memory elements of such systems may be constructed from simple logic gates and feedback connections. Based on the model of a memory element, the general model of a synchronous sequential system will be developed and, finally, practical memory circuits will be studied.

5.2 FEEDBACK CONNECTIONS IN BINARY CIRCUITS

We had no difficulty analyzing binary circuits so long as feedback connections were absent. Our next task is to try to determine the behavior of circuits that contain feedback. We start with the simple circuit shown in Figure 5.1, which is constructed from one AND gate. The circuit contains one independent input (I) and one output (Q), as indicated by its block diagram. Let us try to determine the output Q as a Boolean function of the input variable. The equivalent Boolean equation for this circuit is

$$Q = I \cdot Q$$

which is an **implicit** equation in Q, and since we do not have an "inverse" for the AND operation, it cannot be solved for Q using Boolean algebra. On the other hand, all variables may be equal only to one of two possible values; therefore, we can easily examine all possible cases. First, let us assume that the independent input variable I is equal to Logic-0. In this case

$$Q = 0 \cdot Q = 0$$

which means that if the input is equal to Logic-0, then the circuit behaves as a simple combinational circuit; that is, the output is uniquely determined

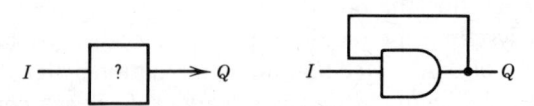

FIGURE 5.1 AND gate with one feedback connection.

and is found to be equal to Logic-0 as well. By examining the simple circuit without considering Boolean algebra, we arrive at the same conclusion. A Logic-0 value at input I will force the output of the AND gate to be 0 regardless of the value of its other input. We notice that the effect of the feedback connection is **blocked.**

Now let us consider the case when the input variable is set to be equal to Logic-1. Substituting this value into the circuit equation gives us

$$Q = 1 \cdot Q = Q$$

which is an equation that contains no information about the value of the output Q, but it is satisfied in either case when Q is equal to Logic-0 or to Logic-1. By examining the circuit for the same input condition, we may reason in the following way: Let us assume that the output is equal to Logic-0, in which case the two input values give us $1 \cdot 0 = 0$ for the output. Since we **assumed** that the output was equal to 0 and then found that according to the circuit equations it **is** equal to 0, we have found a **stable** (self-consistent) solution for the circuit. Assuming that the output is equal to 1 yields another self-consistent solution, since the output of the AND gate, according to its input values, is equal to $1 \cdot 1 = 1$. In fact, we have found **two** stable solutions that allow the circuit to exhibit a **bistable** state.

In the preceding analysis of the circuit of Figure 5.1, we assumed a value for the output Q, and then calculated a "new" output with the assumed output value "fed back" to the input of the circuit. It may be difficult to accept this line of reasoning for the model of the AND gate shown in Figure 5.1 (where one of the input variables is solidly connected to the output), but it can be justified by the examination of the physical operation of the AND gate. Our physical model of the AND gate is drawn in Figure 5.2 with an element of time delay inserted between the sense input values and the drive output blocks. Time delays are inherent in all physical devices, however small they may be in practice. We can view the AND gate now as if its inputs and output were separated (decoupled) by the time delay block; therefore, we may assume a value for the output without consideration for the output–input transfer function of the AND gate. The assumed output value is fed back immediately to the input-sensing circuitry because, in this model, we assume that there is no time delay in the wire connection. With the introduction of

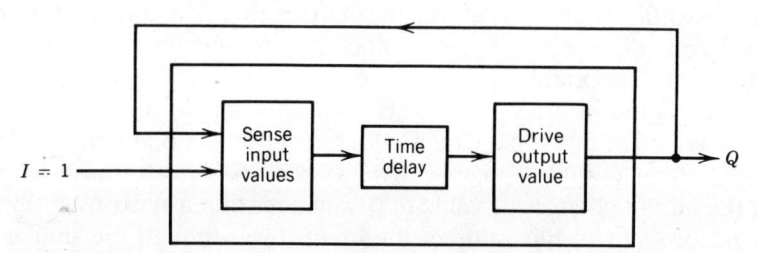

FIGURE 5.2 Physical model of AND gate with one feedback connection.

124 INTRODUCTION TO SEQUENTIAL SYSTEMS

the time delay element in the model of the AND gate, the output of the gate does not change immediately after the input values have changed. The new output, if it is different from the assumed one, will appear only after a small amount of time delay. At this new time, when the output has already changed, this output value is again sensed by the input circuitry, but the output value that reflects these new input conditions is held up again by the internal delay in the model. In this way new output values may be calculated step-by-step, and the time-dependent operation of the logic gate with feedback connections may be traced.

It is true that in a practical circuit the voltage levels change continuously from their low to their high values (and vice versa). Therefore, our model does not represent the true physical system exactly. The determination of the exact details of the time-varying and analog (continuous) signals could be very difficult. Our model is useful for following the time-dependent behavior of the circuit considering only 1's and 0's, and as long as the signal values settle to some final values, it provides the correct results.

Let us summarize what we know about this circuit so far: If the input is equal to 0, then the output is equal to 0. If the input is equal to 1, then the output is either equal to 0 or is equal to 1 and **remains so** until input I changes its value to 0. Could this circuit be of any practical use? Let us assume that at some time we test the output of this circuit and find that it is equal to 1. From this we conclude that its input must also be equal to 1. At some time later we test the circuit again and find that its output is equal to 0. We cannot deduce the present value of the input from this measurement because a 0 output can occur for either a 1 or a 0 input, but we can conclude that in the time period between the two measurements the input must have exhibited a **Logic-1-to-Logic-0 transition** at least once. The input may have returned to its original Logic-1 value, or it could have exhibited several 0-to-1 and 1-to-0 transitions, the output would have remained at the Logic-0 level (look at the circuit in Figure 5.1 and convince yourself that if the output is equal to 0, then the input to the circuit has no effect whatsoever). Hence, the output of this circuit indicates the presence (or absence) of an event in the past. It is a type of memory that "remembers" whether a 1-to-0 transition has occurred at the input of the circuit.

We have found all the important elements of a sequential system in this simple example. The circuit exhibited the operation of **memory** for which **sequences** (or transitions) of logic levels rather than the final logic levels of the input variables are significant. In a circuit that is a combination of sequential and combinational system blocks, both logic levels and logic level transitions are important.

Let us demonstrate how we may use this simple feedback circuit in practice. We want to build a circuit with one input and one output. Initially, both input and output are equal to 1. At the first 1-to-0 transition of the input the output changes its value to 0, and after that it remains at this logic level independent of the value of the input. Even though the simple AND gate circuit we analyzed in the preceding detects a 1-to-0 transition, it is not

5.2 FEEDBACK CONNECTIONS IN BINARY CIRCUITS

a practical circuit because we are unable to initialize it. When the system is first turned on (electrical power is applied to the IC devices) and the input is made equal to 1, the output could be either at the Logic-1 or the Logic-0 level (it is bistable). We must provide another input that initializes the circuit output to the Logic-1 level. With the addition of two OR gates the initialization of the circuit is solved. The new circuit is shown in Figure 5.3. The circuit equation is equal to

$$Q = (INITIALIZE + I) \cdot (INITIALIZE + Q)$$

Normally, the new input labeled **INITIALIZE** is at the Logic-0 level, and the circuit behaves exactly the same way as before (the circuit equation becomes $Q = I \cdot Q$). When the INITIALIZE input is set to Logic-1, then the output is forced to be equal to 1. An analysis of the circuit shows that the output remains at level Logic-1 when the INITIALIZE input is changed to 0 if input I is held at the Logic-1 level. These are the conditions we assumed for the initial state of the circuit. In this form the circuit may be used as an indicator of at least one 1-to-0 transition by the I input variable.

We have seen that with only a few gates and one feedback connection we can construct a useful sequential circuit. Before pursuing this line of analysis further, let us examine a similarly simple circuit example that produces an entirely different result. This circuit example uses a NAND gate and one feedback connection, as shown in Figure 5.4. The implicit circuit equation for this circuit is

$$Q = (I \cdot Q)'$$

When the independent input variable I is equal to 0, the output is equal to $(0 \cdot Q)' = 1$, and this circuit behaves like a combinational circuit. But when the input is equal to 1, the circuit equation becomes

$$Q = Q'$$

which is a contradiction; that is, this equation is not satisfied for either value of 1 or 0 for Q. The physical operation of the circuit can be more easily understood from the model shown in Figure 5.2. If we assume that the output is equal to 0, then the input values for the sensing circuit are 0 and 1, which, because of the NAND function, command the output driver to change the output of the NAND gate to the equivalent voltage level of Logic-1. A small

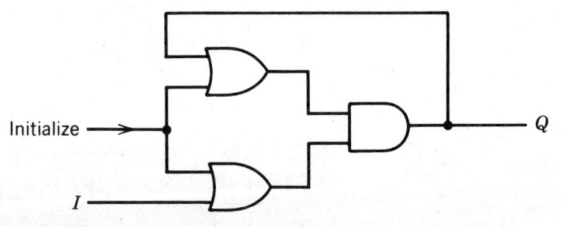

FIGURE 5.3 Practical circuit that indicates that a 1-to-0 transition occured at input I.

126 INTRODUCTION TO SEQUENTIAL SYSTEMS

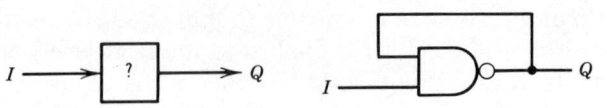

FIGURE 5.4 NAND gate with one feedback connection.

time delay later the output does change to its Logic-1 value. Now the two new input values (two 1's) are sensed by the input circuitry, which for the NAND function commands the output to become 0 again. The cycle continues indefinitely, and the output of the NAND gate exhibits **oscillations.** The frequency of oscillations is controlled by the internal delays in the physical IC devices and could be as high as hundreds of millions of cycles per second.

The analysis of these three simple circuits with feedback connections has shown that output variables of sequential circuits may exhibit three types of behavior when the independent inputs are set to specific combinations of 1's and 0's. First, they can be deterministic (1 or 0); for example, similar to combinational circuits, the output values are uniquely determined by the input conditions. Second, they can be bistable (B), which means that the output is either Logic-1 or Logic-0 and stays unchanged as long as input conditions remain unchanged; and finally, they can be unstable (U), that is, output values alternate between Logic-1 and Logic-0 values with time periods that are determined by the internal delay times of the physical devices used in the circuit. Whereas bistable type of outputs produce memory that is the basic building block of sequential systems, oscillatory type of outputs must be avoided because they produce unpredictable results. In Section 5.3 we will discuss the problems associated with changing signal values and their measurements in digital systems, but before concluding this section, we will present the analysis of an important elementary sequential circuit constructed from two cross-coupled NAND gates, shown in Figure 5.5.

Using the circuit in Figure 5.5 as an example, we present a systematic method by which any logic circuit with feedback connections may be analyzed. The analysis must be done for all combinations of independent input

Assumed	Gate inputs	Calculated
Q_1 Q_2	I_1 X_1 I_2 X_2	Q_1 Q_2
0 0	0 0 0 0	1 1

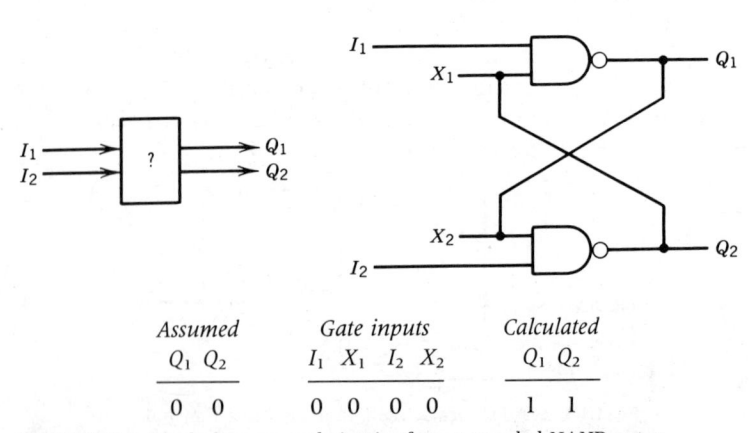

FIGURE 5.5 Block diagram and circuit of cross-coupled NAND gates.

5.2 FEEDBACK CONNECTIONS IN BINARY CIRCUITS   127

values. Since in this case there are two independent inputs (I_1 and I_2), there are four possible input value combinations. The method of analysis is based on assigning "assumed" logic levels to the output of gates, determining the inputs to all gates from the known independent input values and the assumed gate output values and then calculating the "new" output values, which are consistent with the gate types used in the circuit. In order to make this analysis more systematic, variable names are assigned to all unnamed gate inputs. As shown in Figure 5.5, variable names X_1 and X_2 are added to the circuit. With the help of these input variables, a truth table is constructed for a given set of independent input values. The table contains the assumed output values in the leftmost column. In the middle of the table the gate input variables are listed, which, for this example, are I_1 and X_1 for the first NAND gate and I_2 and X_2 for the second NAND gate. The values of I_1 and I_2 are constant for the entire table, whereas X_1 is equal to Q_2, whose value is taken from the assumed output columns, and, similarly, X_2 is equal to Q_1. In the right columns of the table the calculated "new" output values are listed. These output values are calculated according to the gate types used for the circuit and the values of the input variables listed in the middle of the table. We may interpret such tables as lists of "transitions" from assumed output values to "new," or calculated, ones. Figures 5.6–5.9 show the tabulated results for the four cases.

The tabulated results can be interpreted with the help of so-called **state transition** diagrams. Since each entry of the tabulated results start with a set of assumed output values and yields a set of new output values, the tables express how assumed output values change into new values. (We justified this analysis by assuming internal time delays within the physical logic devices.) The outputs of the gates for this analysis are the **state variables** of the circuit. Since there are two state variables (Q_1 and Q_2), there are four possible **states** (the four combinations of 1's and 0's for the two state variables Q_1 and Q_2). These four states are shown in the state transition diagrams as four circles in which the output values are entered. Arrows indicate the transition of states. In Figure 5.6, for example, the assumed state 01 yields the new state 11. An arrow is drawn from the circle indicating state 01 to

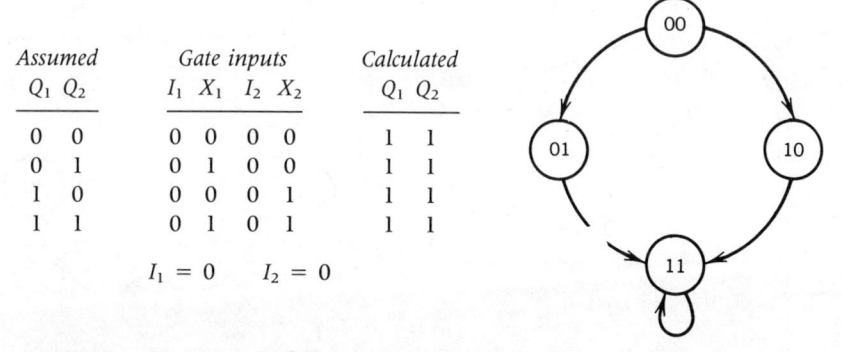

| Assumed | Gate inputs | Calculated |
Q_1 Q_2	I_1 X_1 I_2 X_2	Q_1 Q_2
0 0	0 0 0 0	1 1
0 1	0 1 0 0	1 1
1 0	0 0 0 1	1 1
1 1	0 1 0 1	1 1

$I_1 = 0$ $I_2 = 0$

FIGURE 5.6 The STT and STD for cross-coupled NAND gates with input values 00.

128 INTRODUCTION TO SEQUENTIAL SYSTEMS

Assumed $Q_1\ Q_2$	Gate inputs $I_1\ X_1\ I_2\ X_2$	Calculated $Q_1\ Q_2$
0 0	0 0 1 0	1 1
0 1	0 1 1 0	1 1
1 0	0 0 1 1	1 0
1 1	0 1 1 1	1 0

$I_1 = 0 \quad I_2 = 1$

FIGURE 5.7 The STT and STD for cross-coupled NAND gates with input values 01.

that of state 11. The assumed state 11 also yields the calculated state 11, and this transition appears as an arrow both originating and terminating at the circle of state 11. Thus, state 11 in this diagram is a **stable** state (the circuit will remain in this state as long as the independent input values do not change).

In the case of the assumed state 00 in Figure 5.6, the tabulated results and the transition diagram do not agree. The calculated new state for the assumed state 00 is equal to state 11. This transition requires the simultaneous change of both outputs from their Logic-0 value to their Logic-1 value. In practice, it is essentially impossible for output values of 00 to change to values 11 because this would require that the two outputs change at exactly the same time. The exact time of change depends on internal time delays that are functions of temperature, the precise values of the electrical components, circuit connections, and a large number of other intractable causes. It is unrealistic to expect that two signals could change at exactly the same time; therefore, state 00 must first change either into state 01 or state 10. Since it is impossible to determine which transition will occur in practice, both transitions must be considered, and therefore, two arrows appear in the state transition diagram. Both arrows originate from state 00. The transition diagram in Figure 5.6 clearly shows that whichever is the starting state of the

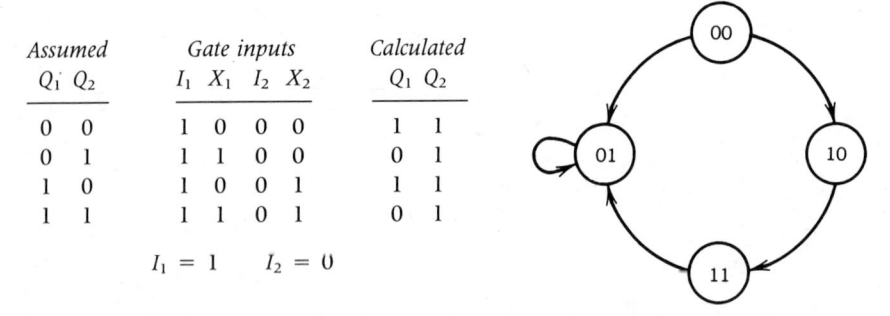

Assumed $Q_1\ Q_2$	Gate inputs $I_1\ X_1\ I_2\ X_2$	Calculated $Q_1\ Q_2$
0 0	1 0 0 0	1 1
0 1	1 1 0 0	0 1
1 0	1 0 0 1	1 1
1 1	1 1 0 1	0 1

$I_1 = 1 \quad I_2 = 0$

FIGURE 5.8 The STT and STD for cross-coupled NAND gates with input values 10.

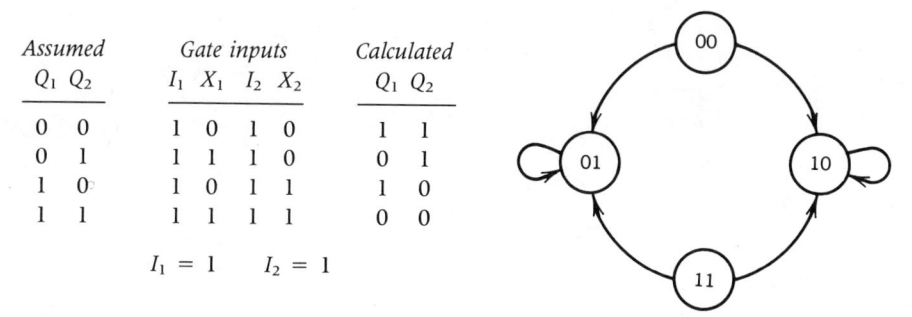

5.2 FEEDBACK CONNECTIONS IN BINARY CIRCUITS 129

Assumed $Q_1\ Q_2$	Gate inputs $I_1\ X_1\ I_2\ X_2$	Calculated $Q_1\ Q_2$
0 0	1 0 1 0	1 1
0 1	1 1 1 0	0 1
1 0	1 0 1 1	1 0
1 1	1 1 1 1	0 0

$I_1 = 1 \qquad I_2 = 1$

FIGURE 5.9 The STT and STD for cross-coupled NAND gates with input values 11.

circuit, its final state is state 11. This trivial result can also be easily deduced from the tabulated values in Figure 5.6. The conclusion from this analysis, for the case when the input values are both equal to Logic-0, is that the output values are deterministic and are equal to 11.

The case when the input values are equal to 01 is shown in Figure 5.7. Similar analysis to the first case shows that no matter from which state the circuit starts, it will settle to the unique state 10. The input conditions $I_1 = 1$ and $I_2 = 0$ produce similar results. The transition diagram in Figure 5.8 clearly indicates that in this case the deterministic outputs are equal to 01. Finally, the case for which the input variable values are equal to 11 yields both the tabulated results and transition diagram shown in Figure 5.9. This is the most interesting case. The tabulated results indicate that the assumed state 00 changes to the new state 11, whereas the assumed state 11 changes to the new state 00. Without our understanding of the physical processes underlying our analysis of these sequential circuits, an oscillatory behavior could be deduced from these results. We already know that state 00 can only change to state 01 or to state 10 and, similarly, state 11 cannot change directly to state 00. Therefore, as the transition diagram clearly shows, the circuit always settles into either the 01 or the 10 state, no matter what state it started from. Hence, both outputs Q_1 and Q_2 are bistable and, in fact, become the complement of each other. Measurements for a practical cross-coupled NAND gate bear out these results.

Our analysis showed that for all four cases of independent input values the cross-coupled NAND circuit produces stable states. This is necessary for a usable sequential circuit. Instability, that is, oscillations, could occur in circuits in which, following the arrows in the transition diagram, there is a loop that contains more than one state. Such circuits cannot be used in sequential systems because the signal values change uncontrollably and the results become unpredictable. As an example, a circuit is constructed from the cross-coupled NAND gates with output Q_1 connected to one of the independent inputs (see Figure 5.10). The analysis of this circuit shows that oscillations are possible in both cases when input I is equal to 0 or to 1. In the $I = 1$ case there are both a stable state (state 10) and an oscillatory state

130 INTRODUCTION TO SEQUENTIAL SYSTEMS

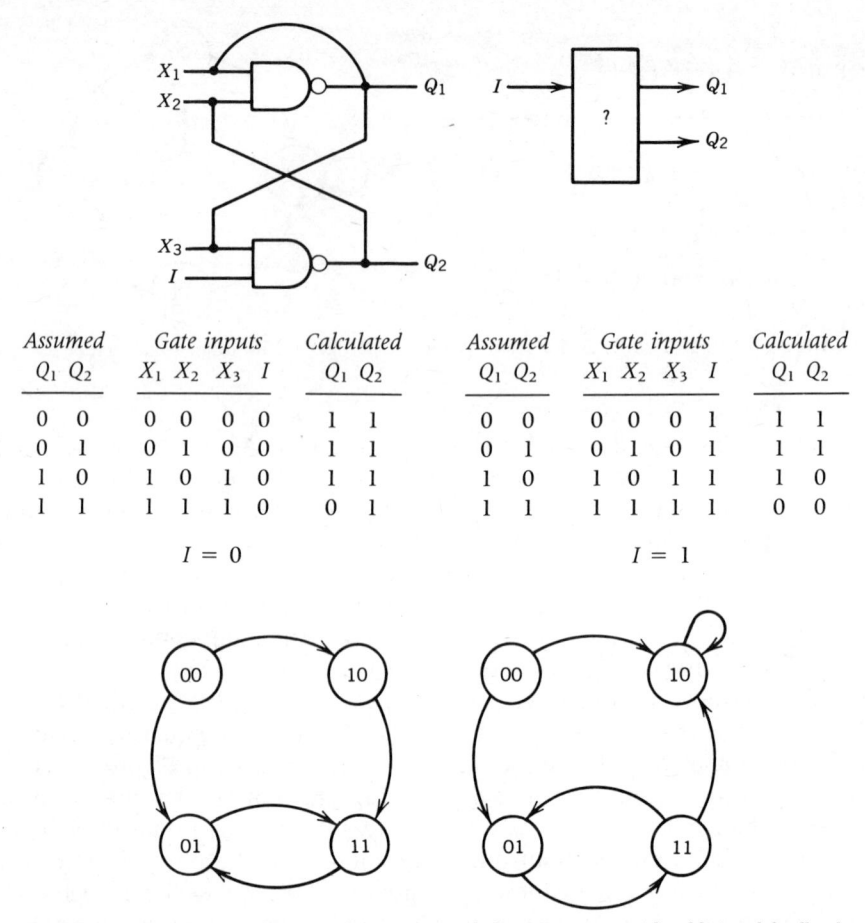

| Assumed | Gate inputs | Calculated | | Assumed | Gate inputs | Calculated |
$Q_1\ Q_2$	$X_1\ X_2\ X_3\ I$	$Q_1\ Q_2$		$Q_1\ Q_2$	$X_1\ X_2\ X_3\ I$	$Q_1\ Q_2$
0 0	0 0 0 0	1 1		0 0	0 0 0 1	1 1
0 1	0 1 0 0	1 1		0 1	0 1 0 1	1 1
1 0	1 0 1 0	1 1		1 0	1 0 1 1	1 0
1 1	1 1 1 0	0 1		1 1	1 1 1 1	0 0

$I = 0 \qquad\qquad\qquad I = 1$

FIGURE 5.10 The STTs and STDs for cross-coupled NAND gates with additional feedback connection.

(oscillating between states 01 and 11). Depending on internal delays, the circuit may have a stable output or can be oscillating.

With the additional bistable (*B*) and unstable (*U*) output types it is possible to characterize the behavior of the output variables of a sequential circuit for each combination of independent input variables, similar to truth tables. Since a truth table does not exist for a sequential circuit, we shall call this tabulated result an **output table.** Table 5.1 is an output table of the five circuits analyzed in this section. Whenever the circuit behaves as a combinational circuit, the deterministic outputs are listed in the table. Bistable outputs are designated with the letter *B* and oscillatory or unstable outputs are shown with the letter *U*. The circuits in Figures 5.5 and 5.10 have two outputs; the others have only one output.

Circuits similar to the cross-coupled NAND gates produce stable output values for any combinations of independent input values; therefore, they are

5.2 FEEDBACK CONNECTIONS IN BINARY CIRCUITS

TABLE 5.1 Output Tables for Circuits in Figures 5.1, 5.3, 5.4, 5.5, and 5.10

I	Figure 5.1	Figure 5.4	Figure 5.10	$I_1 I_2$	Figure 5.3	Figure 5.5 Q_1 Q_2
0	0	1	U	00	0	1 1
1	B	U	01/U	01	1	1 0
				10	B	0 1
				11	1	B B

proper sequential circuits. Since the actual time instances when the transition of the states occur are determined by the signal values of the independent inputs and are not controlled by any external time-dependent digital signals, these types of circuits are "free running" and are called **asynchronous** sequential circuits. Other, much more restricted, sequential circuits belong to the class of **synchronous** sequential systems, which will be discussed in the next section.

From the analysis of a few simple asynchronous circuits we have seen how bistable and unstable outputs arise from feedback connections. Historically, because of the complex problems encountered in the control and testing of asynchronous circuits, synchronous sequential circuits have gained prominence. For this reason, we will concentrate our subsequent studies mainly on synchronous sequential circuits, but as a final example, we will show a practical application for the cross-coupled NAND gate.

The circuit shown in Figure 5.11 is called a "switch debouncer" circuit. Physical switches exhibit contact bounce. This means that when the switch position is changed, the contacts meet and separate many times before their mechanical motion stops and they provide a stable connection. The time-dependent behavior of input signals I_1 and I_2 demonstrates contact bounce in Figure 5.11. The sweeper of the switch is connected to the ground terminal. When the position of the switch is changed, the sweeper "hits" the contact and then bounces and separates from the contact several times, which produces the waveform shown in the diagram for both I_1 and I_2. The output of the circuit, Q_1, eliminates this unpredictable initial behavior of the switch. It shows only one transition for each switch position change. The operation of the switch debouncer circuit is described in what follows.

Whenever the switch sweeper touches one of the contacts, the input to one of the two NAND gates is set to ground potential (Logic-0), and the output of the same NAND gate becomes Logic-1. Assume that at the start the switch is in the indicated position; therefore, $I_2 = 0$ and $I_1 = 1$. Since I_2 is equal to 0, Q_2 is equal to 1, and with both Q_2 and I_1 equal to 1 the output of the NAND gate, Q_1, is set to Logic-0. When the switch position is changed, the input I_1 is made equal to Logic-0 (ground), and this will set Q_1 equal to

132 INTRODUCTION TO SEQUENTIAL SYSTEMS

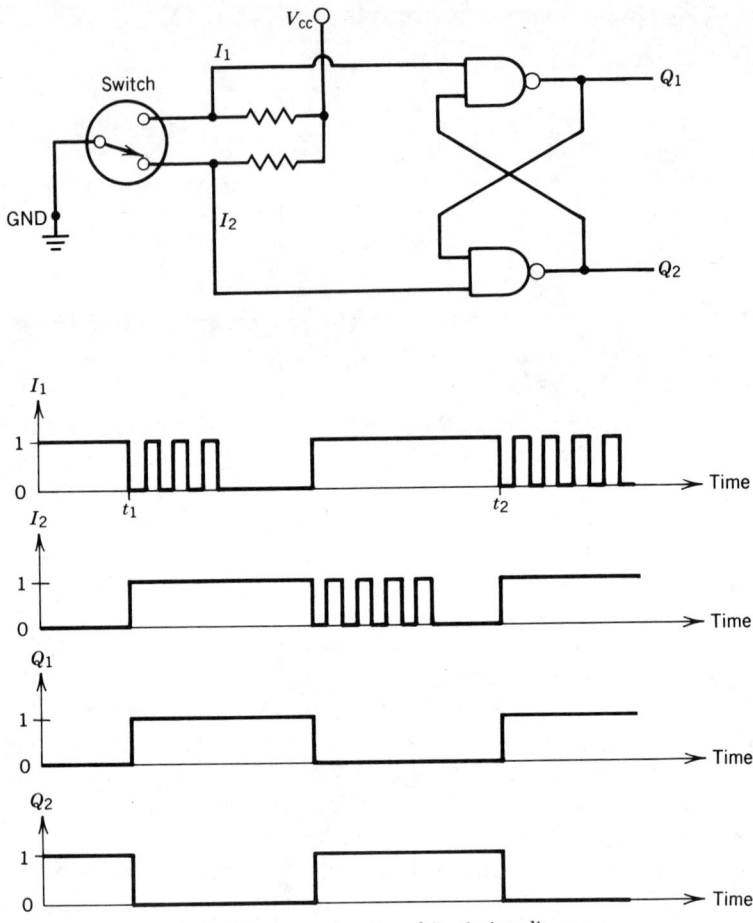

FIGURE 5.11 Switch-debouncer circuit and its timing diagrams.

Logic-1. Since I_2 is also equal to 1, Q_2 becomes equal to Logic-0. This transition is shown at time t_1 on the timing diagram in Figure 5.11. As long as Q_2 is equal to Logic-0, the output of the NAND gate, Q_1, is equal to Logic-1 and is not affected by the value of input I_1, which oscillates between 0 and 1. The output of the debouncing circuit becomes a single 0-to-1 transition. A similar analysis shows that when the switch position is changed back to its original state, only one 1-to-0 transition occurs at the Q_1 output of the debouncing circuit.

For combinational circuits a mechanical switch is adequate because, eventually, the switch output does settle down to its correct value. For sequential circuits, on the other hand, the mechanical switch provides an uncontrolled, unpredictable "sequential" behavior since the operation of sequential circuits depends on the history of transitions of signals, not on their final binary value.

5.3 SYNCHRONOUS SEQUENTIAL SYSTEMS

Sequential systems are useful because they are **dynamic** systems, that is, systems in which signals vary as functions of time. We are interested only in binary systems. The introduction of this new variable, time, causes difficulties because time is a continuously changing variable, and at first it is difficult to see how it could be incorporated into a binary system. There is also an already mentioned practical problem with time-varying "binary" signals, since signals in practice are also continuous variables.

We have solved the latter problem by arranging our physical system in such a way that the voltage levels of all the system variables could be clearly sorted into two classes: signals whose voltage levels are lower than 0.8 V and signals whose voltage levels are higher than 2 V. For combinational circuits it was only a matter of waiting long enough (a few microseconds) before all signal output voltages satisfied the stated binary conditions. In sequential systems the changing of output values plays an essential role; therefore, changes occur all the time, and one cannot wait until all system variables settle to constant values. In fact, it is the changes of values that must be controlled in a systematic way. We must ensure that when the values of variables are measured, all voltage levels correspond to either a Logic-0 or Logic-1 binary value. In other words, time periods must be identified when signal voltages are constant (when signal values may be measured) and others when signal values are changing (when variables may not be measured).

Both problems, continuously changing voltage levels and time as a continuous variable, can be solved by the introduction of a special binary variable called SYSTEM CLOCK. The system clock is an externally generated binary signal connected to all sequential circuits in the system. A 0-to-1 transition of SYSTEM CLOCK defines a time instance indicated in Figure 5.12 as t_n. At some time later the signal returns to its logic-0 level, that is, it exhibits a 1-to-0 transition, after which it can once again register a time instance, which is indicated in Figure 5.12 as t_{n+1}. (The time periods from the 0-to-1 to the 1-to-0 transitions are called **clock pulses**.) The behavior of the sequential system is described in terms of the following time instances: t_0, the initial

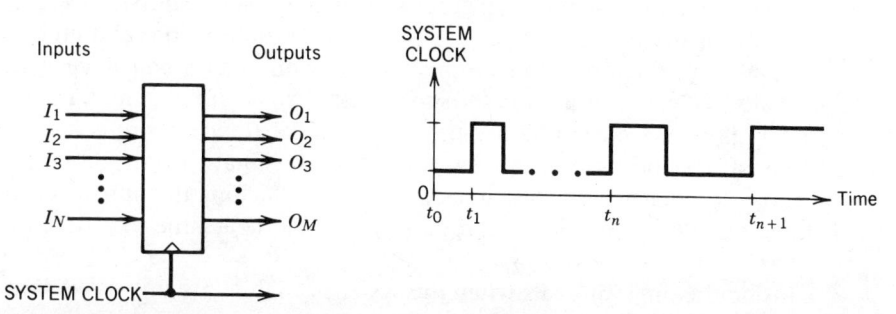

FIGURE 5.12 Block diagram and SYSTEM CLOCK signal for general synchronous sequential system.

time; t_1, the first 0-to-1 transition; t_n, the nth transition; and so forth. All elements of a synchronous sequential system are constructed in such a way that within small time periods around the 0-to-1 transitions of SYSTEM CLOCK all signals are stable and have settled to one of their valid binary values.

There is no special reason why we selected the 0-to-1 transition of SYSTEM CLOCK (which is also known as the positive transition, the positive edge, or the leading edge of the clock pulse) as the significant time instance. In more general systems the complement of SYSTEM CLOCK, or $\overline{\text{SYSTEM CLOCK}}$, is used as well, in which case the the 1-to-0 transitions of the original SYSTEM CLOCK (the negative transition, negative edge, or trailing edge of the clock pulse) also become important time instances. In this case signals must achieve their correct binary values in small time periods around the negative transitions of SYSTEM CLOCK as well. The important fact is that the system is described only at selected time instances (or small time periods), when its variables are all within the allowed physical ranges of voltage levels. With these physical restrictions the reliability of binary systems has been extended to sequential systems, and the analog nature of time has been eliminated. Time enters simply as a counting variable, an integer, the nth index of the SYSTEM CLOCK variable for time instance t_n.

We have restricted the time-dependent behavior of the variables of synchronous sequential systems. In the following section we will demonstrate that such systems can be constructed from simple logic gates. But, of course, the restrictions can be valid only for signals that are controlled by the system. The independent input variables of the system are not under the system's control, and in many practical situations their behavior cannot be similarly restricted. The affect of "uncontrolled," or asynchronous, input signals on the operation of synchronous systems will be discussed later, when general synchronous sequential systems will be studied. For the present we shall assume that the input variables also satisfy the restriction that they must have stable valid binary values within small time periods around the significant transitions of SYSTEM CLOCK.

The operation of these strictly synchronous digital systems may be better understood by the application of the physical model of a logic gate with a small delay element (shown in Figure 5.2) to the system shown in Figure 5.12. Within a small time period that begins some time before and ends at time instance t_n, all input and output variables are stable and have valid binary values. These values are **sensed** (measured) at time t_n, and a short time delay later the output values begin to change. Hence, there is a small time interval around time instance t_n for which all signals have valid binary values. We designate these valid binary values by attaching the time instance t_n to the variables. The value of an output variable O at time instance t_n is shown as $O(t_n)$.

Outside the time intervals when the system variables are measured, both the output and the input variables of the system are allowed to change. Ultimately, if no oscillatory outputs are present in the system, the output and input values settle down to stable, valid binary voltage levels and the system is ready for the "next" time instance, which occurs at t_{n+1}. The behavior of the system is described by the valid binary voltage levels at the significant time instances. Observe that real time does not enter into the description of a synchronous sequential system. It is not possible to determine how much time elapsed between two logic level transitions of SYSTEM CLOCK. The only physical requirement is that the transitions are placed far enough in time so that all signals have settled to stable, valid binary values at each significant SYSTEM CLOCK transition. This restriction places a physical upper limit on the frequency of clock pulses and, hence, on the ultimate speed of the system.

The description of a synchronous sequential system is given in terms of its state (stored) variables at time t_n. The state of the system at the "next" time instance (i.e., at t_{n+1}) depends only on its state variable and input variable values at time t_n. It can be shown that for any practical logic circuit that can be defined as a synchronous sequential system, the complete behavior of the system may be expressed as a Boolean relationship between its state at one time instance and its state and input values at the previous significant time instance.

Before discussing more complex synchronous sequential systems, a general 1-bit memory, the fundamental building block of sequential systems will be described. This system has only one output and is a good example since the 1-bit memory is itself a synchronous sequential system.

A General Synchronous 1-Bit Memory

The block diagram of a synchronous 1-bit memory is shown in Figure 5.13. The system may have many inputs but has only one output, Q, the stored bit of the memory. At any time the **state** of the system is defined by the stored bit value Q. As we have shown before, at time instance t_n the state of the system is defined by the value of the physical variable Q in a small time interval centered at this time instance. This value is indicated as $Q(t_n)$. The physical constraints ensure that the value $Q(t_n)$ is a proper binary value. The value of Q at the next time instance, t_{n+1}, is a function of the input variable values $I_i(t_n)$ and the state of the system, $Q(t_n)$ (all these variables are measured

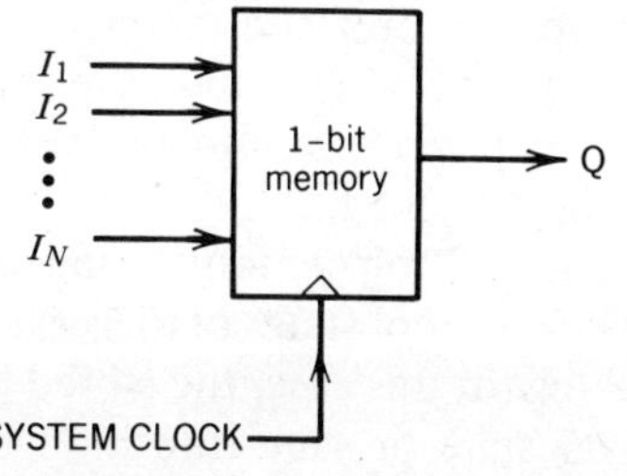

Output at time t_n	Output at time t_{n+1}	Function
$Q(t_n)$	0	RESET
	1	SET
	$Q(t_n)$	No change
	$Q'(t_n)$	TOGGLE

FIGURE 5.13 General 1-bit memory as synchronous sequential system.

at time instance t_n). This functional relationship may be expressed as a Boolean function of $N + 1$ variables (there are N input variables):

$$Q(t_{n+1}) = F[I_1, I_2, \ldots, I_N, Q(t_n)]$$

This Boolean equation is called the **state transition equation** (STE) of the system. If we consider the two variables $Q(t_n)$ and $Q(t_{n+1})$ as two different variables, then we can generate a "truth table" for the STE, which is called the **state transition table** (STT), and a Karnaugh map, which we will call the **state transition Karnaugh map** (STKM). For the 1-bit memory the independent input variables are augmented with the value of the state variable at time t_n, and both the STT and the STKM contain $N + 1$ independent variables.

There is an alternative form of description for the same circuit. Instead of considering the effect of both input and output variables on the new value of the stored bit $[Q(t_{n+1})]$, it is also possible to express the effect of the input variables on the stored bit in a functional manner. There are four possibilities. The output at time t_{n+1} may be set independently of its value at time t_n $[Q(t_n)]$, in which case it may be set to Logic-0 (which we call the **RESET** function) or to Logic-1 (which we call the **SET** function). The output may also be left unchanged $[Q(t_{n+1}) = Q(t_n)]$ or complemented $[Q(t_{n+1}) = Q'(t_n)]$, the latter of which is called the **TOGGLE** function. Hence, the behavior of this one-bit memory may be presented in a table, similarly to the output table for asynchronous sequential systems, where for each combination of input values (at time t_n) one of the four functions is specified. In this case there are only N independent input variables and 2^N entries in the table.

The D-Type Flip-Flop

The D-type flip-flop is the simplest form of a 1-bit memory. It has one independent input (designated with the letter D) and one output (Q, the stored bit). Depending on the value of the input, the memory is either RESET (input value of 0) or SET (input value of 1). The transition equation for the D-type flip-flop is given by:

$$Q(t_{n+1}) = D(t_n)$$

In general, the state of a sequential system at time t_{n+1} is a function of both input and state variables, but for the D-type flip-flop only the independent input variable enters into the equation. The accepted block symbol, the state transition table, the state transition Karnaugh map, and the functional description for the D-type flip-flop are shown in Figure 5.14.

State transition diagrams (STDs) are very useful for demonstrating the behavior of sequential synchronous systems. The different states of the system are designated by the values of the state variables (in this case, the stored bit Q). Depending on the input values, the "next" state or state transition for each "current" state may be determined. On STDs arrows indicate state tran-

5.3 SYNCHRONOUS SEQUENTIAL SYSTEMS   137

Block Symbol

State Transition Table

(t_n) D	(t_n) $Q(t_n)$	(t_{n+1}) $Q(t_{n+1})$
0	0	0
0	1	0
1	0	1
1	1	1

State Transition K-Map

Functional Description

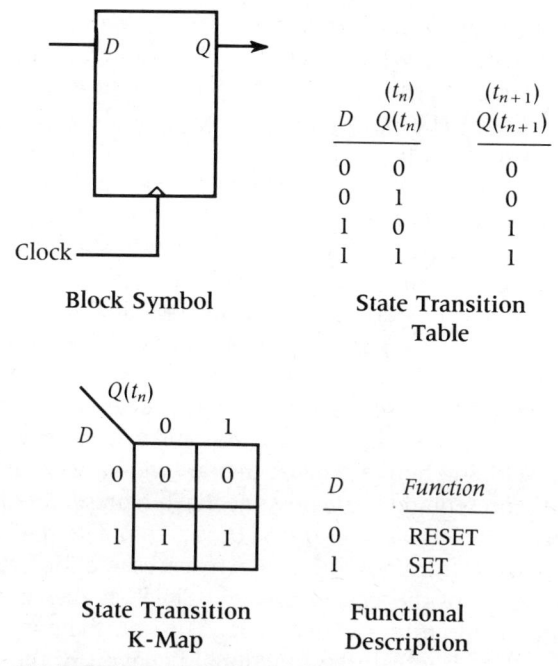

D	Function
0	RESET
1	SET

FIGURE 5.14 The D-type flip-flop.

sitions, and the arrows are labeled with the input values that produce the transitions. The STD for the D-type flip-flop is shown in Figure 5.15. When the input variable is equal to 1, both arrows point to the state for which the stored bit, Q, is equal to 1 (the SET operation). When the input is equal to 0, the arrows point to the other state, when Q is made equal to 0 (the RESET operation).

Even though STDs used for demonstrating the stable and unstable operation of digital circuits with feedback connections and the diagrams used here for synchronous sequential systems look similar, their use is very different from their earlier application. Here the diagram shows the next states of the system (the states at time instance t_{n+1}), which are determined from known current states (the circles from where the arrows originate) and the values of the independent input variables (the input values appear as labels near the arrows). Actual state transitions occur only when SYSTEM CLOCK exhibits significant transitions.

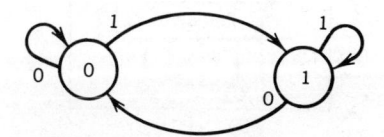

FIGURE 5.15 The STD for D-type flip-flop.

138 INTRODUCTION TO SEQUENTIAL SYSTEMS

In our earlier use of STDs the exact times when logic gate outputs changed were not controlled; they were governed only by internal time delays, and a complete STD was constructed for each combination of independent input values. When we use STDs for synchronous sequential systems, the transitions occur only at given clock times, and the input variable values also appear in the diagram. The number of arrows leaving each state is equal to the number of possible combinations of input values (two in this case). There are no inconsistent solutions here, and internal delays can be ignored because SYSTEM CLOCK transitions are placed far enough for the system to settle to its consistent quiescent state, and no uncontrolled oscillations are allowed in the system. It is important to remember that, so far, synchronous sequential systems have been presented as model systems. No physical realization of these systems have been discussed yet. We will deal with the physical realization of strict synchronized sequential systems in a later section.

We are developing here a reliable digital time-varying model for which many different representations are possible. Each representation, for example, the STE, the STT, the STKM, and the STD, is a complete representation, and each may be derived from any other. Each representation can provide help in a different aspect of the design process. The K-maps, as we have seen, provide help in the minimization of circuits. The state diagram is a convenient graphical method with which the functional behavior of the circuit is demonstrated. Transition tables will find application in the design process when general flip-flop circuits will be used.

General Synchronous Sequential System Representations

With the help of 1-bit memories the schematic diagram of a general synchronous sequential circuit may be constructed. This is shown in Figure 5.16. The state of the system is stored in a collection of K 1-bit memories. The stored bits are designated as $Q_1, Q_2, \ldots, Q_K$. There are N inputs and M outputs. The state of the system at the next time transition, t_{n+1}, is a function of the input and stored bit values at time t_n. The Boolean equations expressing these functional relationships are the STEs. There is one equation for each 1-bit memory, and they all have the same form:

$$Q_i(t_{n+1}) = F_i[I_1(t_n), I_2(t_n), \ldots, I_N(t_n), Q_1(t_n), Q_2(t_n), \ldots, Q_K(t_n)]$$

for $i = 1, 2, \ldots, K$

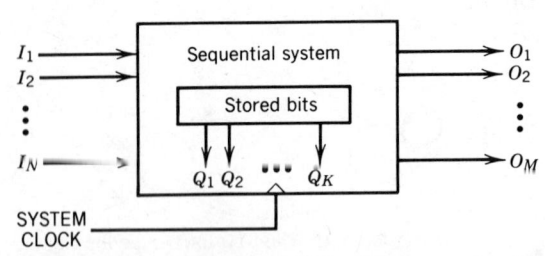

FIGURE 5.16 Model of general synchronous sequential system.

The equations for the output variables are also Boolean functions. There is one **output equation** for each output variable, and there are two types of synchronous sequential systems depending on the form of the output equations. The output equations of Moore [1] systems depend only on the state variables and have the form

$$O_i(t_n) = G_i[Q_1(t_n), Q_2(t_n), \ldots, Q_K(t_n)] \quad \text{for } i = 1, 2, \ldots, M$$

These equations may be handled as simple combinational systems. Once the state of the sequential system is known and as long as inputs have stable binary values around times t_n, the system states can be uniquely determined from the transition equations.

The output equations for Mealy [2] systems depend on the input variables as well:

$$O_i(t_n) = G_i[Q_1(t_n), Q_2(t_n), \ldots, Q_K(t_n), I_1(t_n), I_2(t_n), \ldots, I_N(t_n)]$$
$$\text{for } i = 1, 2, \ldots, M$$

Mealy-type systems are not strictly synchronous systems because the outputs are influenced by externally controlled signals that may not be synchronized with SYSTEM CLOCK. In a later chapter we will use both Moore- and Mealy-type circuits.

D-type flip-flops can be used to construct a general Moore-type synchronous sequential system, as shown in Figure 5.17. The STEs define a combinational function $F(\ )$ that has $N + K$ inputs (there are N independent input and K state variables) and K outputs. Each output of this combinational circuit is connected to the input of a D-type flip-flop. The output equations

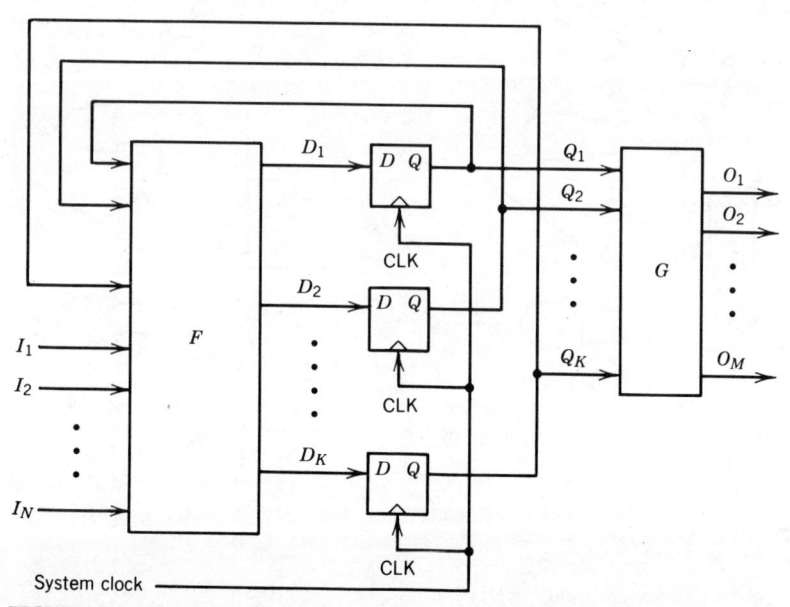

FIGURE 5.17 Construction of general synchronous sequential system using D-type flip-flops.

define another combinational function, $G(\ )$, which has K inputs (the state variables) and M outputs (there are M outputs for the system). Figure 5.17 shows the most general Moore-type circuit that can be designed by methods we have already studied, such as K-maps, since only the design of combinational circuits is involved. The next state of the system is determined by its current state using the STEs:

$$Q_i(t_{n+1}) = D_i(t_n) = F_i[Q_j(t_n), I_k(t_n)]$$

where $i = 1, 2, \ldots, K$;
$j = 1, 2, \ldots, K$;
$k = 1, 2, \ldots, N$

We have reduced the design of a general synchronous sequential system to the design of a synchronous D-type flip-flop and the design of combinational circuits. This assumes that the system is defined by state transition and output equations. In the next section we will show how logic gates may be used to design practical flip-flop circuits, and in later chapters we will study how practical problems can be transformed into minimized sets of state transition and output equations.

5.4 PRACTICAL FLIP-FLOP CIRCUITS

The term *flip-flop* designates 1-bit memories in practice. The cross-coupled NAND gates and their dual circuit, the cross-coupled NOR gates, form the basis of practical flip-flop circuits. These are shown in Figure 5.18. Since there is no SYSTEM CLOCK that controls the state transition of these circuits, both of these are asynchronous and require careful analysis. We have already

$\bar{S}$	$\bar{R}$	Q	$\bar{Q}$	Action
0	0	1	1	—
0	1	1	0	SET
1	0	0	1	RESET
1	1	B	B	None

Cross-coupled NAND

S	R	Q	$\bar{Q}$	Action
0	0	B	B	None
0	1	0	1	RESET
1	0	1	0	SET
1	1	0	0	—

Cross-coupled NOR

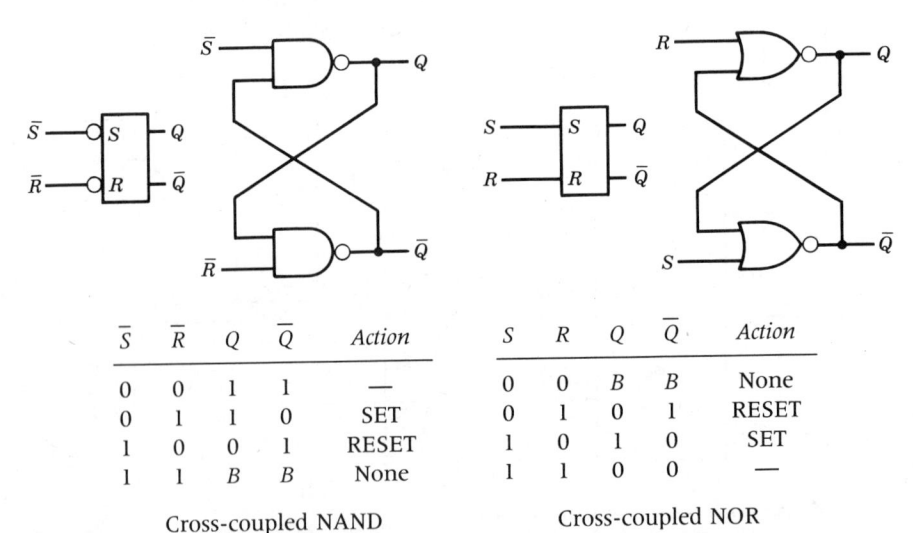

FIGURE 5.18 Asynchronous cross-coupled NAND and cross-coupled NOR circuits.

shown that oscillations do not occur for the cross-coupled NAND gates, and a similar asynchronous analysis shows that the cross-coupled NOR gates do not have oscillatory states either. How can these circuits be used as memories? Examining their output tables, it becomes apparent that for the cross-coupled NAND gates the output is bistable when the two inputs are in the 11 state. The same is true for the cross-coupled NOR gates with input values 00. Hence, we define these input conditions as the **quiescent,** or **passive,** states. These are the conditions under which the stored bit is not influenced directly by the input values. Since we use positive logic convention, the passive state of a signal is equated by its Logic-0 level. This is the reason why the complement of symbols S and R are shown as inputs to the cross-coupled NAND gates. In this case it is not possible to avoid the use of a mixed logic type of signal convention, and small circles indicate that the *active* periods for both signals S and R are the periods when these signals are at their Logic-0 level. The block diagram representation for the cross-coupled NAND gates uses two small circles for the two input signals. These logic devices are called **SET–RESET,** or **SR, latches.**

When we "observe," or measure, the system, it is in its passive state; therefore, both the $\overline{S}$ and $\overline{R}$ inputs are equal to Logic-1 (of course, both S and R are equal to 0). If we do not know the history of the operation of these circuits, the stored bits (Q) may be equal either to 1 or 0. Under these initial conditions the output designated as $\overline{Q}$ is equal to the complement value of Q. The letter B indicates that the output is bistable (meaning, unknown). An "action" occurs by a 0-to-1 transition of either input signals S or R, but in order to measure the stored bit value, the circuit should return to its passive state. Therefore, a complete pulse of both 0-to-1 and 1-to-0 transitions should be considered as a completed action. If a pulse appears at the input designated as S the stored bit value is set to 1 (SET operation). A pulse for input R forces Q to become 0 (RESET operation). The time-dependent signal levels of the two input and two output variables for two cross-coupled NOR gates are shown in Figure 5.19. Before the first pulse appears for the S input variable, the diagram shows "unknown" values for outputs Q and $\overline{Q}$ since they could be equal either to 1 or to 0. As the time-dependent values of these signals show, the cross-coupled gates serve as a memory that remembers whether the latest action in time was a SET or a RESET operation.

Now, let us proceed with the design of a D-type flip-flop circuit, which is based on these cross-coupled gates. Since a D-type flip-flop is controlled by a clock signal, we must add another input to the circuit connected to the SYSTEM CLOCK variable. When the clock signal is passive (equal to Logic-0), the S and R inputs to the cross-coupled gates should be also passive. This can be achieved by the addition of two NAND gates for the cross-coupled NAND circuit and two AND gates for the cross-coupled NOR circuit, as shown in Figure 5.20. Since the D-type flip-flop has only one input, both the SET and RESET operations are provided by the D input. When D is equal to 1, the SET operation should result; when it is equal to 0, the RESET operation should be the result. This can be achieved by a single INVERTER gate that

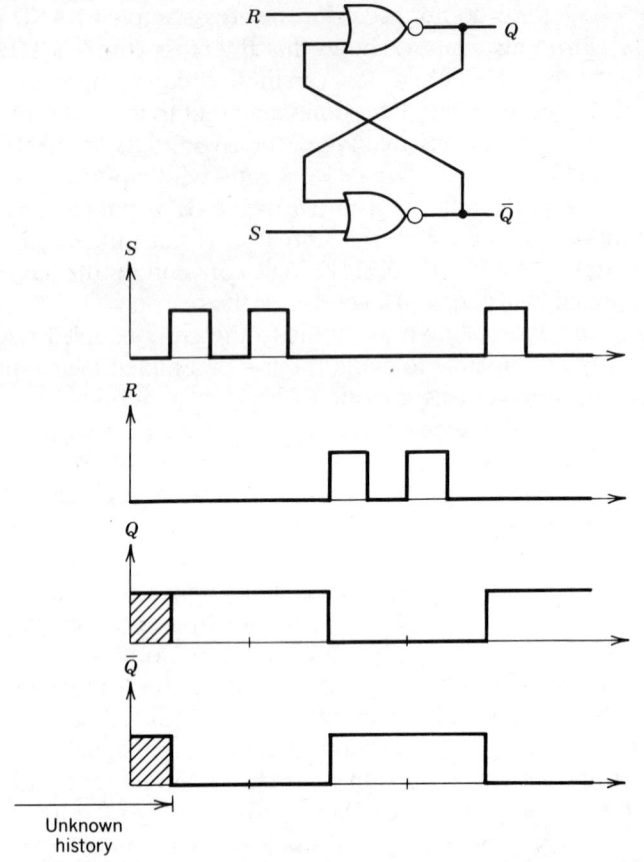

FIGURE 5.19 Operation of asynchronous cross-coupled NOR circuit.

directs the D signal to the S input and the $\overline{D}$ signal to the R input of the cross-coupled gates.

As shown in Figure 5.21, the modified circuit using the coupled OR gates behaves as a D-type flip-flop. When the clock signal has a Logic-0 value, the cross-coupled OR gates have bistable states ($S = R = 0$). When the clock

D-type flip-flop with NAND gates   D-type flip-flop with NOR gates

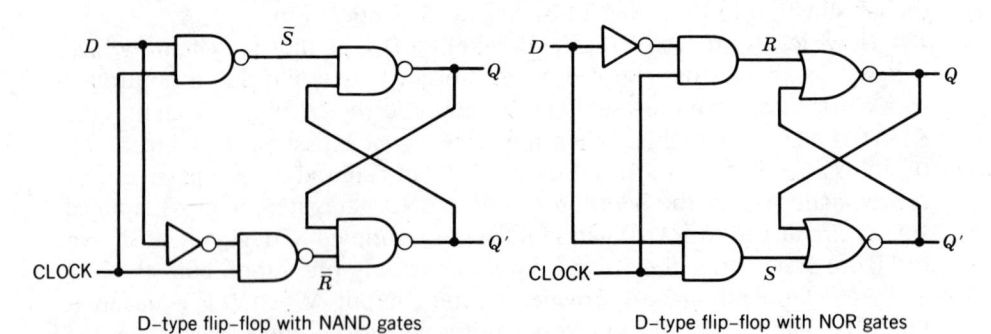

FIGURE 5.20 Cross-coupled NAND and NOR circuits with CLOCK control.

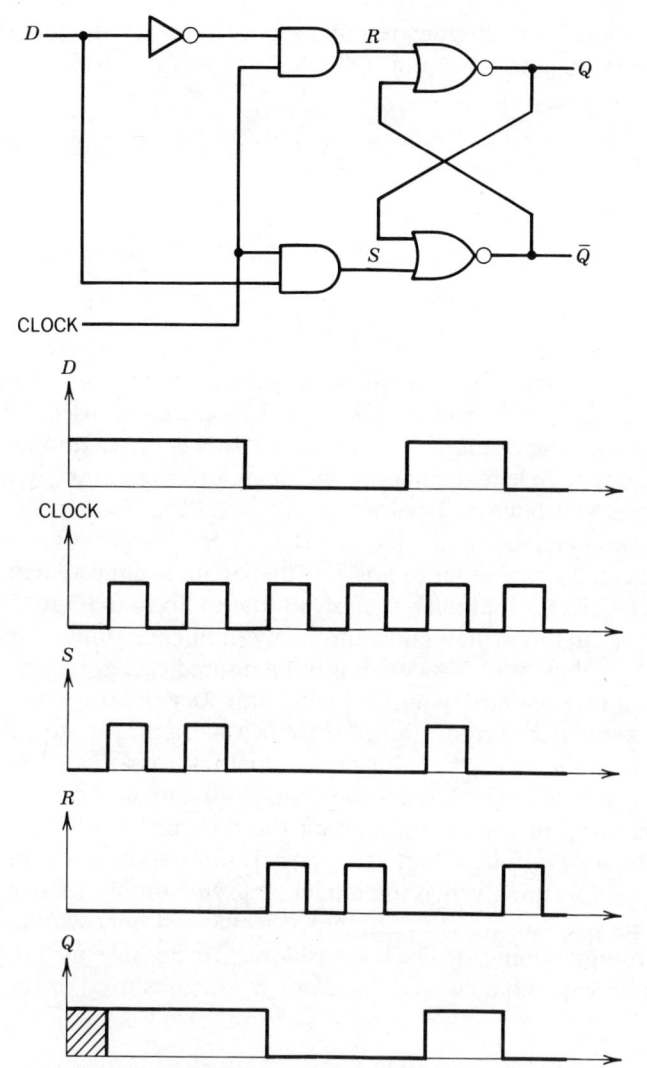

FIGURE 5.21 Operation of cross-coupled NOR circuit with CLOCK control.

pulse occurs, a pulse appears either for signal S (when D is equal to Logic-1) or for signal R (when D is equal to Logic-0). Hence, the stored bit is either SET ($D = 1$) or RESET ($D = 0$) according to the value of the D input at the time of the positive edge of the clock pulse. The time-dependent signal functions shown in Figure 5.21 demonstrate this behavior clearly.

Apparently, we have constructed a D-type flip-flop that behaves according to its specifications: the stored bit is SET when D is equal to 1, and it is reset when D is equal to 0. But on closer examination it turns out that this flip-flop cannot be used in a general synchronous sequential system. We can

144 INTRODUCTION TO SEQUENTIAL SYSTEMS

demonstrate this by attempting to build a synchronous sequential circuit that continuously "toggles" its output. The STE for this circuit is

$$Q(t_{n+1}) = Q'(t_n)$$

The transition equation for the D-type flip-flop is

$$Q(t_{n+1}) = D(t_n)$$

and the two preceding equations define the D input of the flip-flop as the complement of its output:

$$D = Q'$$

for all times. The proposed toggling sequential circuit and its STD are shown in Figure 5.22, where cross-coupled NAND gates are used. Through an INVERTER gate the output of the flip-flop (Q) is connected to its input (D). The circuit has no independent input and, therefore, the arrows in the assumed synchronous STD have no labels.

Unfortunately, the circuit does not behave according to its proposed STD. When the clock signal is set to Logic-1, the circuit is almost identical to that shown in Figure 5.10. Similar analysis to that in Section 5.3 reveals that the output of the circuit oscillates uncontrollably (influenced only by internal gate delays). The behavior of this circuit is quite unpredictable, and in fact, it may be used as a random binary number generator. Depending on the exact time instance of the 1-to-0 transition of the CLOCK input, the stored output bit could be in either state when the clock signal returns to its Logic-0 value.

The reason why this circuit does not work correctly is that the output signal is allowed to change throughout the time period of the clock pulse, and the changed output values are "sensed" continuously by the input circuitry during this time. When the model of a synchronous sequential system was defined, the sensing of the output was allowed only within small time periods around significant clock transitions. To be able to build practical synchronous sequential circuits, the same restrictions must apply to the cir-

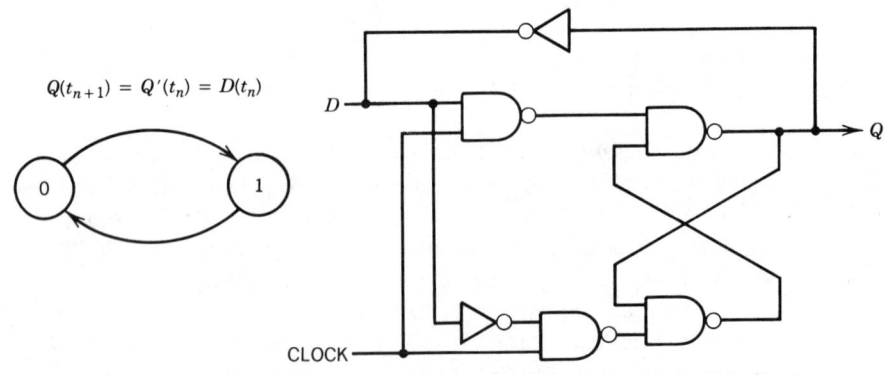

$Q(t_{n+1}) = Q'(t_n) = D(t_n)$

FIGURE 5.22 CLOCK-controlled cross-coupled NAND gates with additional feedback connection.

cuits. There are two standard methods by which the sensing of the input values can be restricted to these small time intervals around the significant edges of the SYSTEM CLOCK signal, and these will be discussed in the following two sections. The type of flip-flops shown in the preceding analysis, which change their outputs throughout the time period of the clock pulse, are referred to as **dc-controlled** circuits. The dc-controlled flip-flops are called **latches**.

Master-Slave D-Type Flip-Flop

Two dc-controlled latches may be used to construct a proper synchronous sequential memory called a Master-Slave D-Type flip-flop. The logical arrangement of the two latches and the actual circuit constructed from cross-coupled NAND gates are shown in Figures 5.23 and 5.24, respectively. The operation of a master–slave flip-flop can be understood from the diagram in Figure 5.24. The clock control signal is shown as $\overline{\text{CLOCK}}$, which indicates that the significant time transition occurs during the negative edge of the clock pulse (the 1-to-0 transition). The first latch is referred to as the "master" latch because it determines the output of the other, the "slave" latch. Let us begin by assuming that the control signal $\overline{\text{CLOCK}}$ is equal to Logic-1. As we have demonstrated before, whereas $\overline{\text{CLOCK}}$ is equal to 1, the output Q_M of the master latch follows the value of the input signal D. But since the clock control signal to the slave latch is equal to the complement of $\overline{\text{CLOCK}}$, or CLOCK, whose value is equal to 0, the overall output of the flip-flop, Q_S, does not follow its inputs $\overline{S}_S$ and $\overline{R}_S$. The slave latch is blocked and $\overline{S}_S = \overline{R}_S = 1$.

After the 1-to-0 transition of the CLOCK signal the master latch is disabled from its D input, and changes in the D signal no longer appear at the latch outputs Q_M and $\overline{Q}_M$. However, a small time delay later (which depends on the internal time delay of the inverter in the CLOCK signal line), the slave latch is enabled, and the stored output values of the master latch are transferred to the slave latch; that is, Q_S becomes equal to Q_M and $\overline{Q}_S$ to $\overline{Q}_M$. During the time when the $\overline{\text{CLOCK}}$ signal is equal to Logic-0, the master latch is

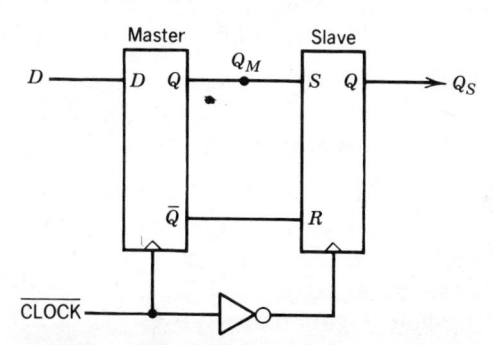

FIGURE 5.23 Master–slave connections for two asynchronous SR flip-flops.

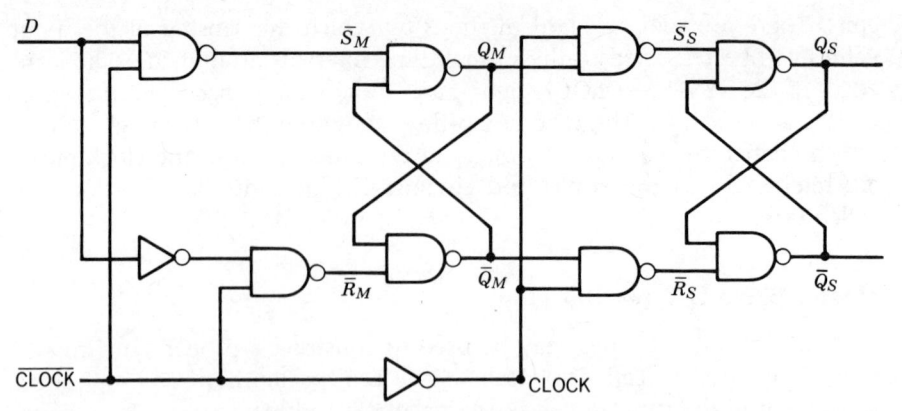

FIGURE 5.24 Circuit diagram of master–slave D-type flip-flop using NAND and INVERTER gates.

disabled, and variations in the D input value are blocked. Hence, output signal changes occur only for a very small time period after the negative transition of the $\overline{\text{CLOCK}}$ signal, and the stored output value is controlled by the value of the input signal in a small time period immediately before the negative edge of the $\overline{\text{CLOCK}}$ signal. The operation of this master-slave flip-flop is consistent with our synchronous sequential model and has been proved extremely reliable in practice.

Note that the correct operation of the master-slave arrangement depends on the internal time delays associated with practical devices used to construct the circuits. Let us suppose that the D-type master latch has large internal delays associated with it, whereas the slave RS latch does not. If the output Q_S is connected to input D through an inverter (just, as before, when we attempted to build a toggling circuit), then the internal delay in the D-type master latch may allow the latch to be enabled for a longer period of time, and both latches could be enabled at the same time. This would cause the same type of oscillations as those we showed for dc-controlled latches. On the other hand, if the circuit is constructed as shown in Figure 5.24 and if the NAND gates all have similar gate delays, then it is expected that both master and slave latches will have similar time delays, and the circuit will perform satisfactorily.

Edge-Triggered D-Type Flip-Flop

A different approach to constructing proper synchronous sequential flip-flops is to arrange logic gates in such a manner that their internal gate delays restrict the "active" time period of the flip-flop to a small region around the edge of the clock pulse. It may be surprising, but the circuit of a well known edge-triggered D-type flip-flop (Figure 5.25) proves that it is possible to design asynchronous circuits that satisfy the operating requirements of synchronous sequential circuits. The complete asynchronous analysis of this circuit is rather

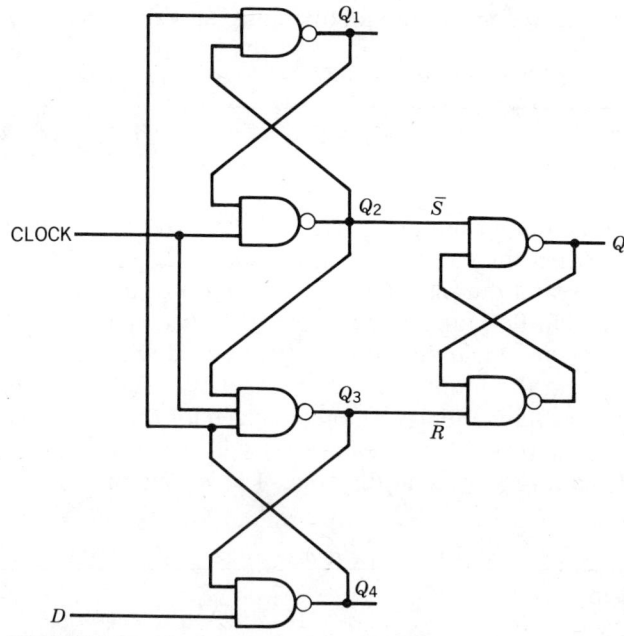

FIGURE 5.25 Edge-triggered D-type flip-flop.

tedious, and we will not attempt it here. It is possible and easier to analyze the circuit's operation on the basis of the already familiar characteristics of cross-coupled NAND circuits.

The output of the circuit (Q) is the output of two cross-coupled NAND gates with inputs designated as $\overline{S}$ and $\overline{R}$. We know that if the two binary values for these two inputs ($\overline{SR}$) are 11, then the flip-flop remains in its "passive" state, with the stored bit being either a Logic-1 or a Logic-0. In order to SET the flip-flop, the logic values 01 are required for inputs $\overline{S}$ and $\overline{R}$, whereas input values 10 RESET the flip-flop. When the CLOCK input is equal to Logic-0, then both Q_2 and Q_3 (or $\overline{S}$ and $\overline{R}$) signals are at the Logic-1 level, and the cross-coupled NAND gates are in their passive state, as required. While the CLOCK signal is at the Logic-0 level, changes in the D input effect only Q_4 and Q_1.

Examine now the 0-to-1 transition of the clock pulse and its effect on the rest of the circuit. First, assume that D is equal to 0. In this case the signal values can be traced easily (the output of a NAND gate is equal to 1 if any one of its inputs is equal to 0) to give:

	CLOCK	D	Q_1	Q_2 ($\overline{S}$)	Q_3 ($\overline{R}$)	Q_4
Before transition	0	0	0	1	1	1
After transition	1	0	0	1	0	1

which are the correct conditions for the RESET operation. When the input signal D is equal to Logic-1, the following signal values occur:

	CLOCK	D	Q_1	Q_2 ($\overline{S}$)	Q_3 ($\overline{R}$)	Q_4
Before transition	0	1	1	1	1	0
After transition	1	1	1	0	1	0

which cause the output flip-flop to be SET. We have shown that during the time periods when the CLOCK signal is equal to Logic-0, the output of the circuit is stable, and during the 0-to-1 transition of the CLOCK signal, the correct D-type flip-flop functions are executed. It remains to be shown that the output remains stable during the time periods when the CLOCK signal is equal to 1 and the input signal D changes. We start with the first case, when the signal values after transition are given for $D = 0$. The output values for the four gates (Q_1–Q_4) are determined with the D output changing and the CLOCK input remaining at the Logic-1 level. We find:

	CLOCK	D	Q_1	Q_2 ($\overline{S}$)	Q_3 ($\overline{R}$)	Q_4
After transition	1	0	0	1	0	1
After D changes	1	1	0	1	0	1

We can see that the effect of the D input is blocked, and there is no change in the $\overline{S}$ or $\overline{R}$ signal values.

Finally, when the D input value is equal to 1 after the transition, we get the following signal changes for a 1-to-0 transition in the D input signal:

	CLOCK	D	Q_1	Q_2 ($\overline{S}$)	Q_3 ($\overline{R}$)	Q_4
After transition	1	1	1	0	1	0
After D changes	1	0	1	0	1	1

This again shows no change in the $\overline{S}$ and $\overline{R}$ signal values and, hence, no change in the value of the output (Q).

We have already shown that the edge-triggered circuit operates correctly when the CLOCK signal is equal to Logic-0, when it is equal to Logic-1, and when it makes a 0-to-1 transition. At the negative transition of the CLOCK signal the signals $\overline{S}$ and $\overline{R}$ return to their Logic-1 value, which does not affect the output of the cross-coupled NAND gates. This concludes the analysis that proves the correct operation of this D-type flip-flop.

Useful Elementary Flip-Flop Types

Two methods were shown with which practical synchronous D-type flip-flop circuits could be constructed. It is also possible to add logic gates to basic D-type flip-flops or to use different arrangements of gates and construct other flip-flop types that work on the same master-slave, or edge-triggered, prin-

5.4 PRACTICAL FLIP-FLOP CIRCUITS   149

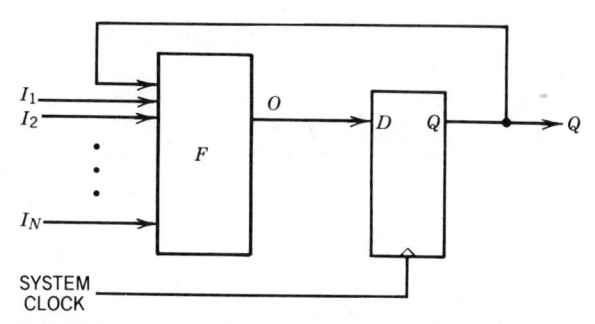

FIGURE 5.26 General 1-bit memory using D-type flip-flop.

ciples but have different output–input characteristics. These will become useful in our later studies of synchronous sequential systems.

In Section 5.2 the model of a general synchronous 1-bit memory was introduced. Its block diagram and functional specifications are shown in Figure 5.13. The D-type flip-flop and a combinational circuit may be used to construct this general 1-bit memory model, as shown in Figure 5.26. The STE of the memory is given in terms of the stored bit (Q):

$$Q(t_{n+1}) = D(t_n) = O(t_n) = F[I_1, I_2, \ldots, I_N, Q] \quad \text{(all at } t_n)$$

where the combinational circuit function $F(\ )$ has $N + 1$ variables. From this equation the STKM and the STT can be constructed. The STT specifies both the functional specifications and the STDs for the memory.

There are four types of 1-bit memories with one or two inputs that are

S R	Function
0 0	None
0 1	RESET
1 0	SET
1 1	Not allowed

Block symbol   Functional table   Transition K-map

$$Q(t_{n+1}) = S + \overline{R} \cdot Q(t_n)$$

$$S \cdot R \neq 1$$

Transition diagram   Transition equation

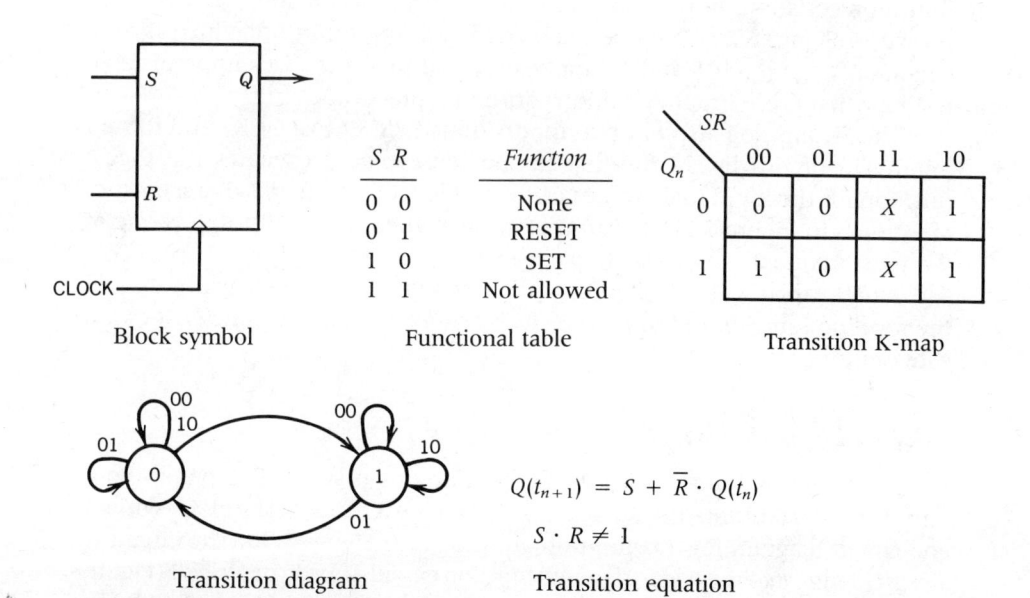

FIGURE 5.27 The SR flip-flop.

The JK flip-flop

J	K	Function
0	0	None
0	1	RESET
1	0	SET
1	1	TOGGLE

Functional table

Transition K-map:

Q_n \ JK	00	01	11	10
0	0	0	1	1
1	1	0	0	1

$$Q(t_{n+1}) = J \cdot \overline{Q}(t_n) + \overline{K} \cdot Q(t_n)$$

Transition equation

FIGURE 5.28 The JK flip-flop.

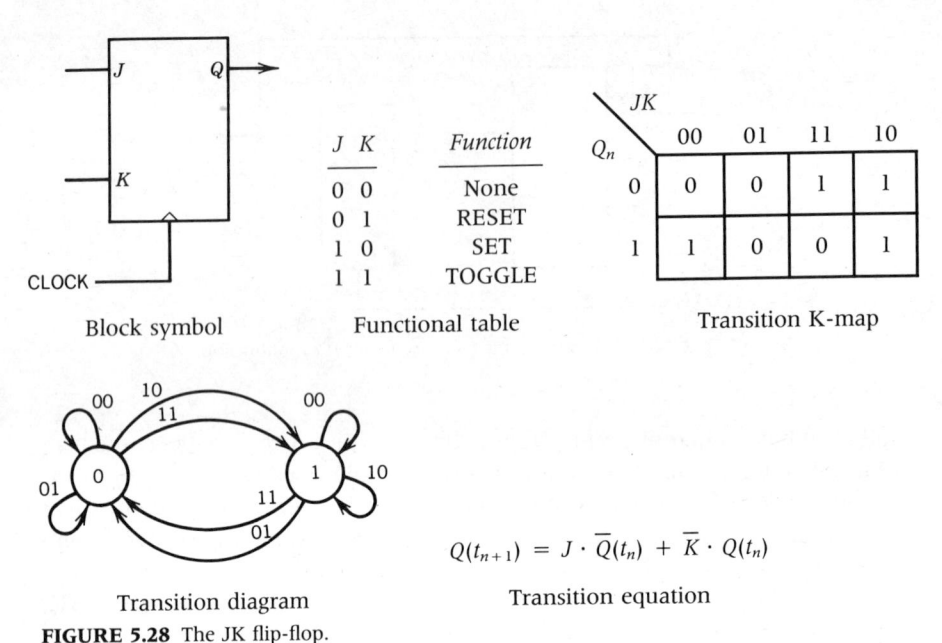

frequently used in practical circuits. The D-type flip-flop is, of course, one of them. Its characteristics are shown in Figures 5.14 and 5.15. The other three types are listed in what follows.

The set–reset (SR) flip-flop has already been discussed. Its block diagram, transition equation, functional specifications, transition K-Map, and TSD are shown in Figure 5.27. The operation of the SR flip-flop is undefined for the condition when both S and R inputs are equal to 1 and gives unpredictable results when these input conditions arise in a practical circuit.

The JK flip-flop also has two inputs (plus SYSTEM CLOCK) and it is an improvement on the SR flip-flop in the sense that it executes the toggle function for the input logic values of 11. Its block diagram and characteristics are shown in Figure 5.28. Because of its versatility, the JK flip-flop is one of the most frequently used flip-flops in practice.

Finally, the toggle flip-flop has only one input that can enable or disable the toggling function of the output. Its block diagram and characteristics are shown in Figure 5.29.

TTL 7474-Type Digital IC Device

We conclude this chapter with the description of a practical D-type flip-flop, one of the two flip-flops included in the 7474-type IC device. The block symbol and circuit diagram for this flip-flop are shown in Figure 5.30. The circuit is almost identical to the edge-triggered flip-flop circuit already described (Figure 5.25). Two additional inputs are shown that are called PRESET and CLEAR.

5.4 PRACTICAL FLIP-FLOP CIRCUITS

Block symbol

T	Function
0	None
1	TOGGLE

Functional table

Transition K-map

$Q_n \backslash T$	0	1
0	0	1
1	1	0

Transition diagram

$$Q(t_{n+1}) = T \cdot \overline{Q}(t_n) + \overline{T} \cdot Q(t_n)$$

Transition equation

FIGURE 5.29 The T flip-flop.

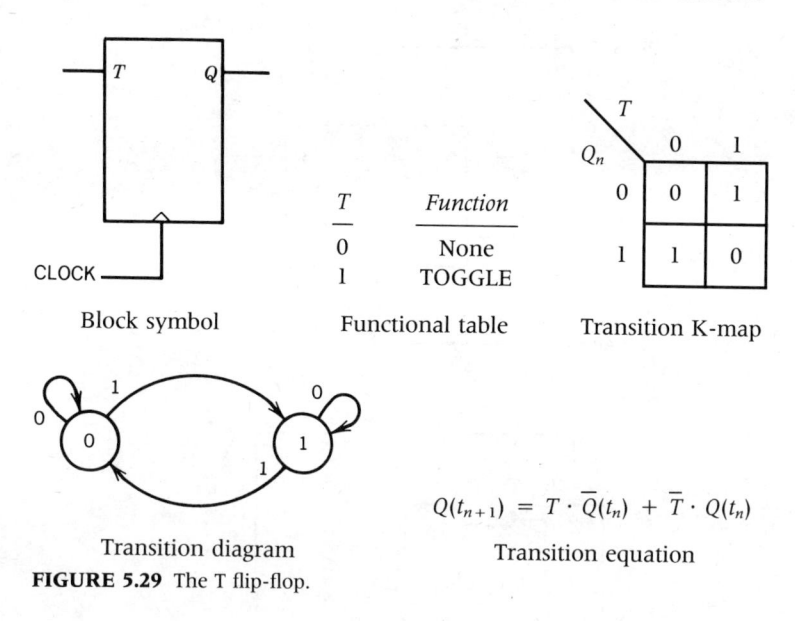

When these inputs are activated, the flip-flop is dc controlled. This means that the actions of the PRESET and CLEAR inputs override the normal operation of the edge-triggered D-type flip-flop. As long as either one of the PRESET or CLEAR inputs is active, clock transitions are ignored. Viewed from the actions of these two dc inputs, the flip-flop behaves as an asynchronous SET–RESET latch. The two small circles at the PRESET and CLEAR input terminals indicate that these inputs are active when they are at the Logic-0 level.

We have already seen the need for the initialization of sequential systems. These two additional dc inputs are intended for the initialization of the D-type flip-flop. When the state of the system is stored in this type of flip-flop, then either the PRESET or the CLEAR input of each flip-flop is connected to an input signal called $\overline{\text{MASTER RESET}}$. With a Logic-0 value of the $\overline{\text{MASTER RESET}}$ input, all the flip-flops in the system can be set to their required initial values. (Whether a flip-flop is SET or RESET by the active $\overline{\text{MASTER RESET}}$ signal depends on whether its PRESET or CLEAR input is connected to the $\overline{\text{MASTER RESET}}$ input.) This IC device will be used extensively in Experiment 5.

At the start of this introductory chapter on sequential circuits we analyzed asynchronous binary circuits with feedback connections. The material has been concluded by a list of useful synchronous flip-flop types and the description of a practical IC device. Throughout this chapter we developed the model of a general synchronous sequential system and showed that it is possible to build such systems in practice. Once the basic principles of synchronous sequential systems are understood, systems based on these principles can be designed and constructed without analyzing their actual time-depen-

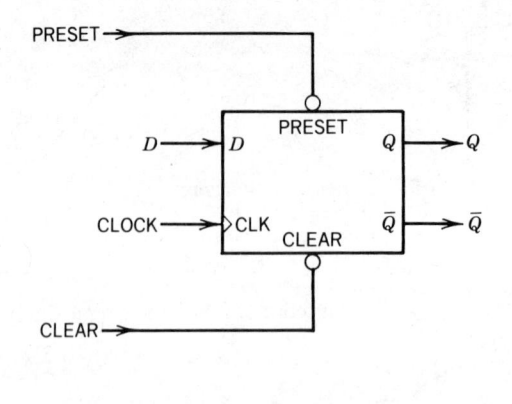

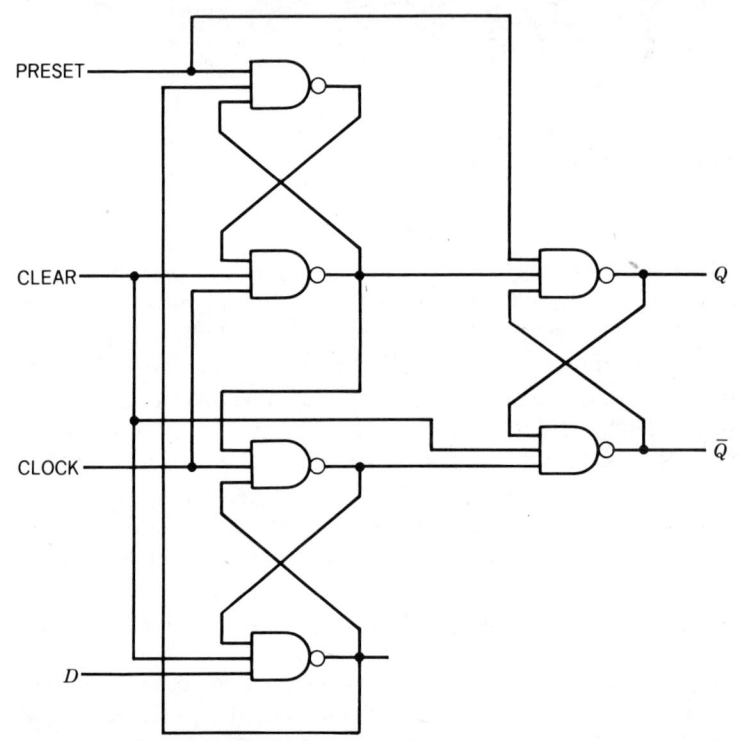

FIGURE 5.30 Block diagram and circuit of practical edge-triggered D-type flip-flop with asynchronous PRESET and CLEAR control inputs.

dent behavior. A synchronous system (or finite-state machine) can be viewed as a collection of stored bits, called the **state** of the system, which exhibits **state transitions** at controlled time instances. The transitions are completely defined by the state of the system and the values of the independent input variables. In the following chapters this synchronous sequential model is applied to the solution of various practical problems.

EXPERIMENT 5
Time-Dependent and Sequential Circuits

E5.1 Summary and Objectives

As an introduction to the construction and testing of sequential circuits, both asynchronous and synchronous circuits are included in this experiment. The type 7474 IC device is used because it can fulfill both roles of a synchronous flip-flop and of an asynchronous latch. The main objectives of this experiment are to demonstrate in practice the principles discussed in Chapter 5 and to gain experience in tracing and measuring time-dependent digital signals.

The number of completed experiments depends on available time and equipment.

E5.2 Switch Bounce Eliminator Circuit

Connect the Q' output of one of the flip-flops of a type 7474 IC device to its D input. Convince yourself that according to the transition equation of this circuit, the flip-flop should toggle its output for each 0-to-1 transition of the clock input and should not change its state for the 1-to-0 transitions. Note that no inputs are left unconnected (floating). This is essential for the reliable operation of sequential circuits. Connect a simple switch to this circuit as shown in Figure E5.1a and demonstrate that the circuit does not work properly. Explain your results.

Construct the switch debouncer circuit shown in Figure E5.1b. In this circuit another D-type flip-flop of the 7474 type is used as a dc-controlled SET–RESET flip-flop. The asynchronous CLEAR and PRESET inputs of this flip-flop take the roles of the R and S inputs of the SR flip-flop. Demonstrate that when the debounced clock signal is used as the CLOCK input, the toggling flip-flop changes its output for each 0-to-1 transition.

E5.3 Asynchronous Time-Dependent Digital Signals

Add two more toggling circuits (constructed from two additional D-type flip-flops) and two NAND gates to produce the circuit shown in Figure E5.2. Notice that the SYSTEM CLOCK signal is connected only to the first flip-flop; flip-flop outputs are used for the clock inputs of the other two flip-flops. This type of circuit (discussed in the next chapter) produces asynchronous outputs because of internal gate delays. Assume that all flip-flops are RESET initially. Complete the timing diagram in Figure E5.3 by carefully tracing the binary values for signals Q_1, Q_2, Q_3, O_1, and O_2 as functions of time for eight clock pulses. Test your circuit by comparing the actual signal outputs to your expected values. Remember that a possible change in the output of a flip-flop occurs only when its CLOCK input exhibits a 0-to-1 transition.

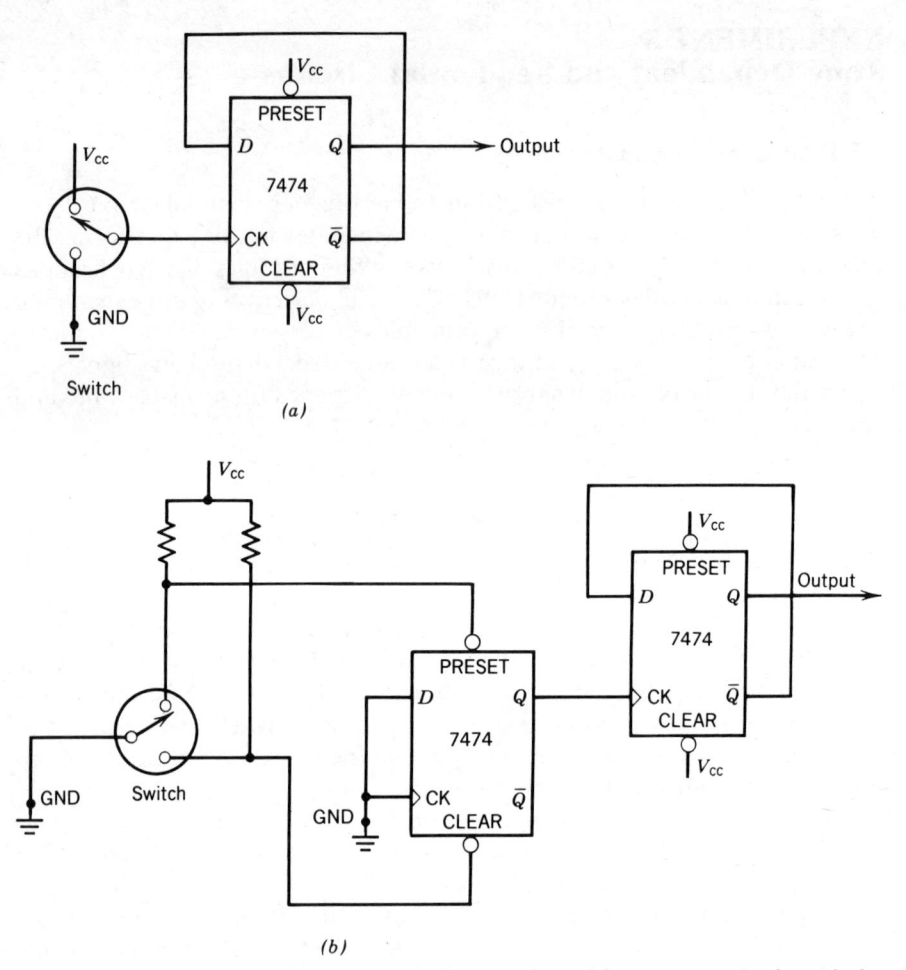

FIGURE E5.1 Using switch for CLOCK signal: (a) without debouncing circuit; (b) with debouncing circuit.

E5.4 Detection of Unwanted Signal Transitions (Glitches)

One problem with asynchronous outputs is that signals do not all change at specified clock transition times, and when these signals are connected to inputs of other circuits, unexpected results may occur. For example, if the two inputs to either NAND gate in Figure E5.2 change from values 01 to values 10, then the output of the NAND gate should remain at the Logic-1 level at all times.

However, if the inputs do not change at the same time (one may have been delayed by internal gate delays), momentarily, the output could become Logic-0. Carefully trace the time-dependent output of the NAND gate shown in Figure E5.4 whose two input values change from 01 to 10. The expected output when the two input signals change simultaneously are shown. Trace

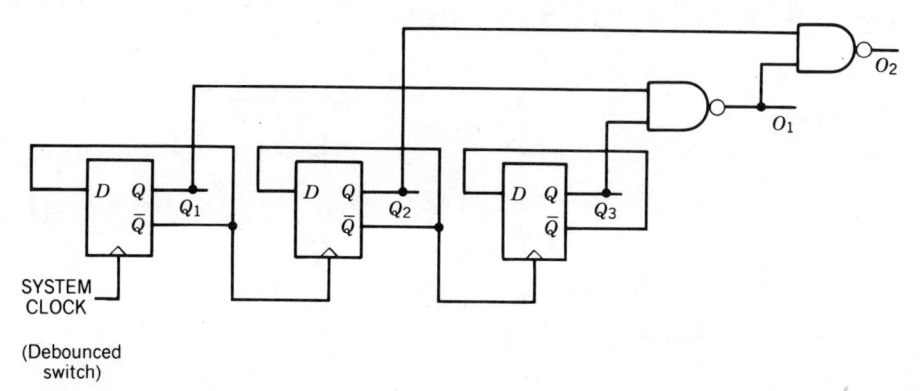

FIGURE E5.2 A synchronous sequential circuit.

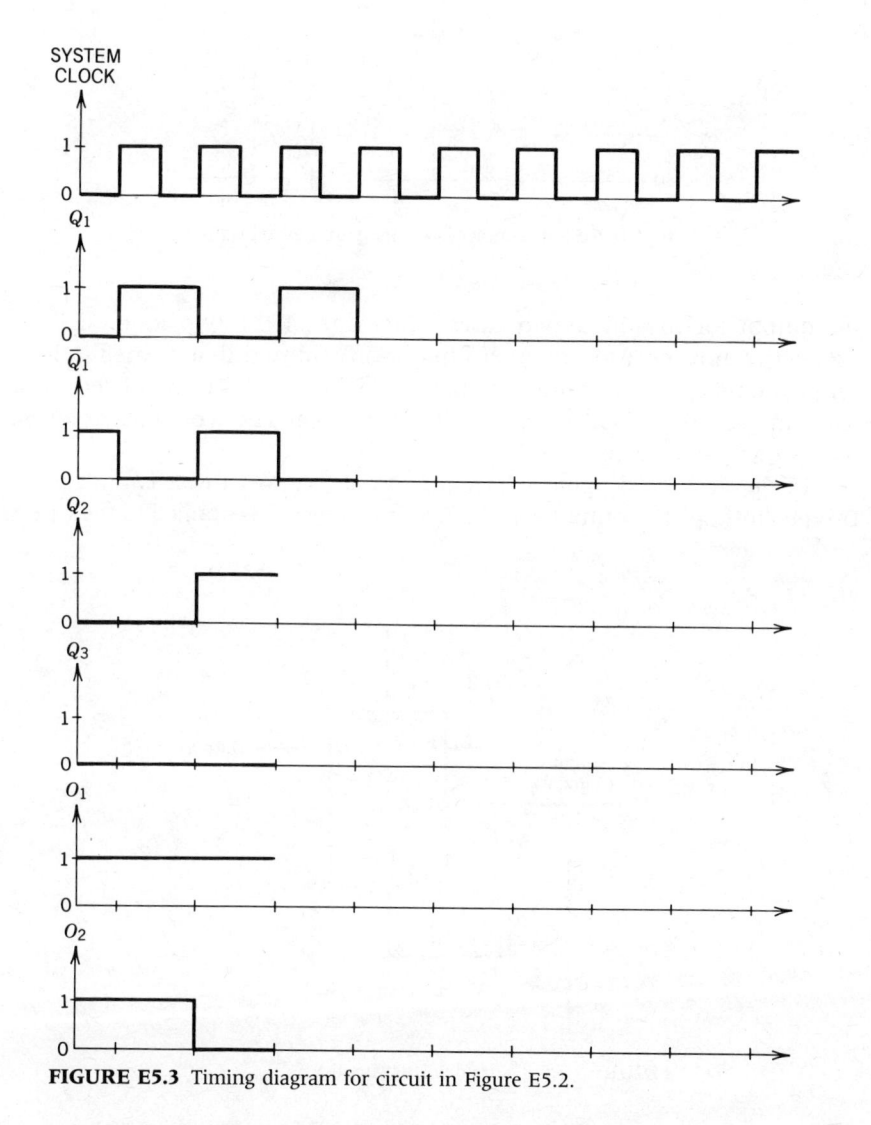

FIGURE E5.3 Timing diagram for circuit in Figure E5.2.

155

156 INTRODUCTION TO SEQUENTIAL SYSTEMS

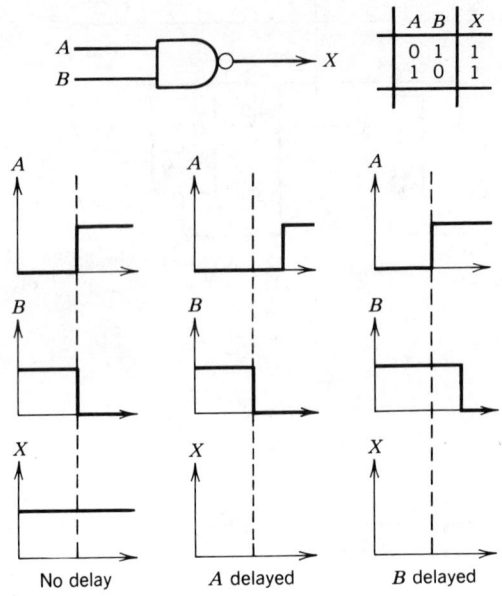

A	B	X
0	1	1
1	0	1

FIGURE E5.4 Timing diagrams with delayed signals.

the output for the other two cases when one of the two input signals is delayed with respect to the other input. You will find that a small pulse, or **glitch,** appears in the output of the circuit. From your completed timing diagram identify possible clock transitions when glitches could occur for output O_1 or O_2 for the circuit in Figure E5.2.

In Figure E5.5 a "glitch detector" circuit is shown that is built from a D-type flip-flop. The output is RESET when a Logic-0 is applied to the $\overline{\text{CLEAR}}$

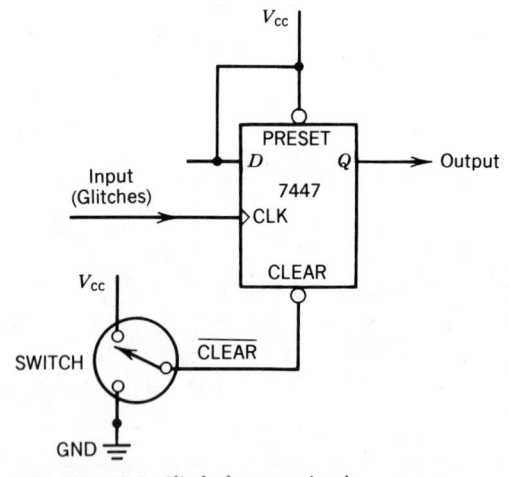

FIGURE E5.5 Glitch detector circuit.

control signal. With the $\overline{\text{CLEAR}}$ signal at the Logic-1 level, any 0-to-1 transition of the input (which is connected to the CLOCK input of the D-type flip-flop) sets the output of the flip-flop to the Logic-1 value. This indicates the occurrence of at least one 0-to-1 transition in the input signal value. Use this circuit to prove that glitches do appear where they should and do not appear where they should not for the two output signals O_1 and O_2 and the circuit shown in Figure E5.2. Glitches may occur when both inputs to a NAND gate change simultaneously (values change from 01 to 10 or from 10 to 01).

E5.5 A Synchronous Sequential System

The circuit in Figure E5.6 is constructed from two D-type flip-flops. It is a synchronous circuit because the clocks of both flip-flops are controlled by SYSTEM CLOCK. Determine the STT and the STD for this circuit by the following method. Assume that the values of the stored bits of the two flip-flops are one of the possible four combinations of 1's and 0's. Determine the D input values for the two flip-flops from the assumed output values. The "new" output values (i.e., the output values of the flip-flops after the next significant CLOCK transition) are equal to the "current" D input values. The new output values specify the state transition for the assumed current state. Repeat this procedure for the three other assumed states and complete the STT and the STD.

Initialize the circuit for states 00 ($Q_1 = 0$, $Q_2 = 0$), and by operating the SYSTEM CLOCK signal, show that the state transitions occur as predicted by the STD. Do the same for initial state 11. Finally, show that no glitches appear

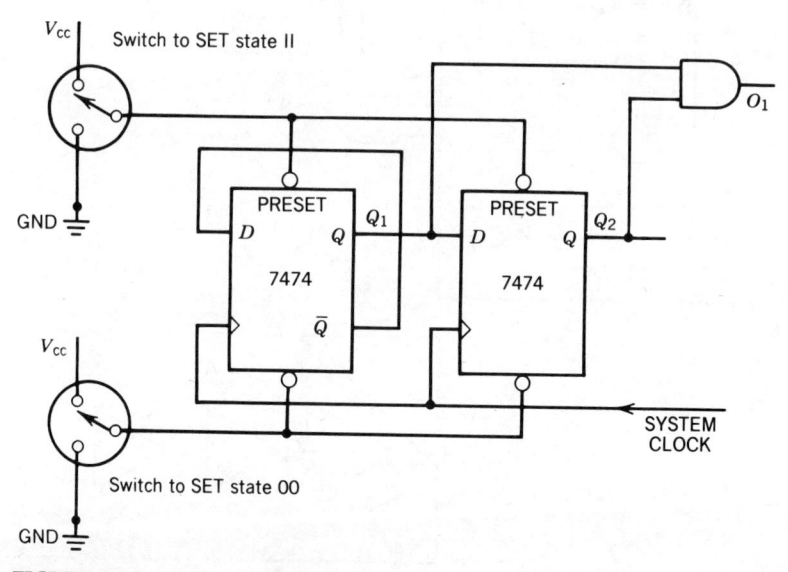

FIGURE E5.6 Synchronous sequential circuit.

for output O_1, even though, the inputs to the NAND gates change their values from 01 to 10.

E5.6 A Faulty Master-Slave Flip-Flop

Someone has suggested that in Figure 5.24 the master–slave flip-flop can be changed and an uninverted CLOCK input used if the INVERTER gate in the CLOCK line is turned around. This is shown in Figure E5.7, where the flip-flop is used for the construction of a toggling circuit with its inverted output $\overline{Q}$ connected to its input D. There is a danger in this circuit because some delay is introduced by the INVERTER gate in the CLOCK line, which may influence the proper operation of the flip-flop. Carefully analyze this circuit. Show that if there are only small delays then it should be a correctly operating master-slave flip-flop.

Demonstrate that gate delays could influence the operation of this toggling circuit by introducing pairs of INVERTER gates into the CLOCK line (which provide equivalent logic functions but increased gate delays), as indicated in Figure E5.7. Observe when the circuit stops operating as a proper

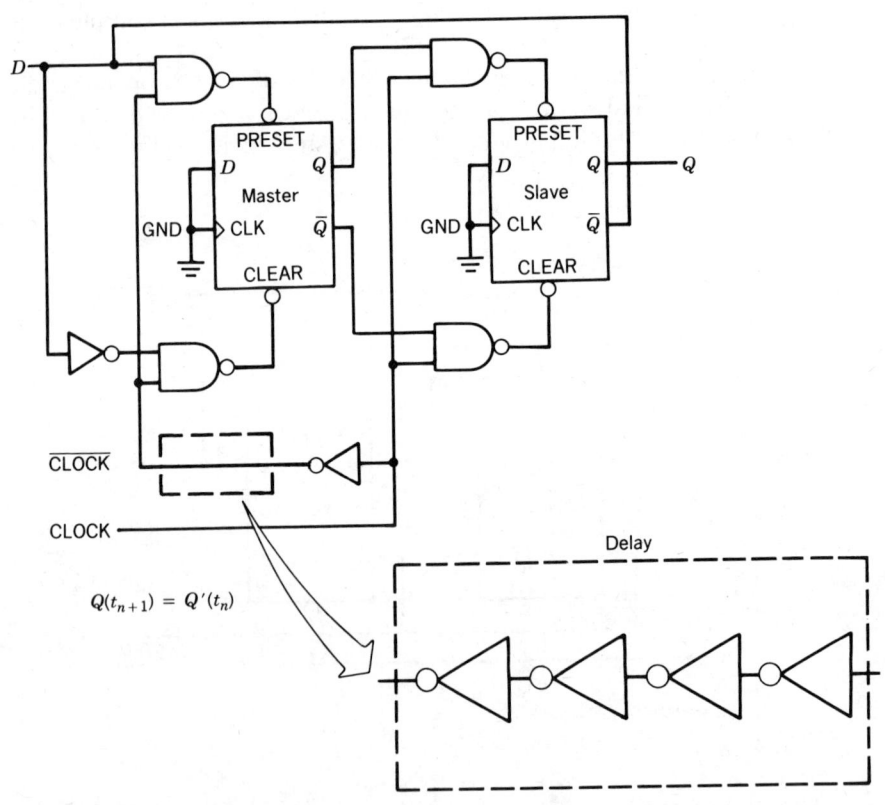

FIGURE E5.7 Faulty master–slave flip-flop.

toggling circuit. Justify the circuit's behavior by analyzing its detailed, time-dependent operation.

E5.7 Display of Time-Varying Digital Signals

An important diagnostic technique for time-varying digital signals is their display as functions of time on the cathode-ray tube (CRT) of an oscilloscope or digital analyzer instrument. There are two types of displays. Most oscilloscopes show signals in real time, with the display updated at least 30 times per second for a flicker-free display. Digital analyzers (or storage oscilloscopes) have the capability to "capture" a digital signal trace, store it, and then display this single trace (voltage as a function of time) using different time scales or voltage scales. In both cases the exact time the trace starts is important.

Oscilloscopes and digital analyzers are used by supplying a continuously changing SYSTEM CLOCK signal to the circuits and the signal outputs are connected to the measuring instruments. Normally, the SYSTEM CLOCK signal is the output of a square-wave generator with equal time periods for which the CLOCK signal is in the Logic-1 or Logic-0 state.

The specification for the start of the trace is called **triggering.** Any time-varying digital signal may be used for triggering the start of the trace, and either 0-to-1 or 1-to-0 transitions may be selected. Figure E5.8 demonstrates a common problem with the display of general time-varying digital signals. A digital waveform (X) is shown that is periodic for time interval T_p but for which $t_1, t_2, t_3,$ and t_4 are not equal. The SYSTEM CLOCK signal is also shown. Trace 1 shows the type of oscilloscope trace we should expect to see if the triggering occurs at any positive edge of the SYSTEM CLOCK signal. This confusing signal is displayed because when triggering occurs, the tracing of the signal starts from left to right. After a full trace is shown, another trace of the signal is displayed, but the trace may start at any positive edge of the SYSTEM CLOCK signal, and the repeated traces will not be the same.

Trace 2 shows the oscilloscope trace we should expect when triggering occurs at the positive transitions of the X signal. The trace is still confused because there are two positive transitions for signal X that produce different traces. The digital analyzer is capable of storing and displaying a single trace, in which case no such confusion of overlapped signal traces occur. An oscilloscope trace can be reliably displayed if the triggering occurs only once in the period of the waveform. A new digital signal must be generated that has one positive transition in time period T_p, and this signal must be used for the triggering of the oscilloscope. A proper triggering signal is shown in Figure E5.8.

Produce a proper triggering signal for output O_2 for the circuit you built in Section E5.3. Show the different traces of the signals by displaying outputs O_1 and O_2 on an oscilloscope. Use different triggering signals (SYSTEM CLOCK, O_1, O_2) and change the time scale of the display. Show that the displayed traces are confusing; they vary and are difficult to interpret. Finally, show

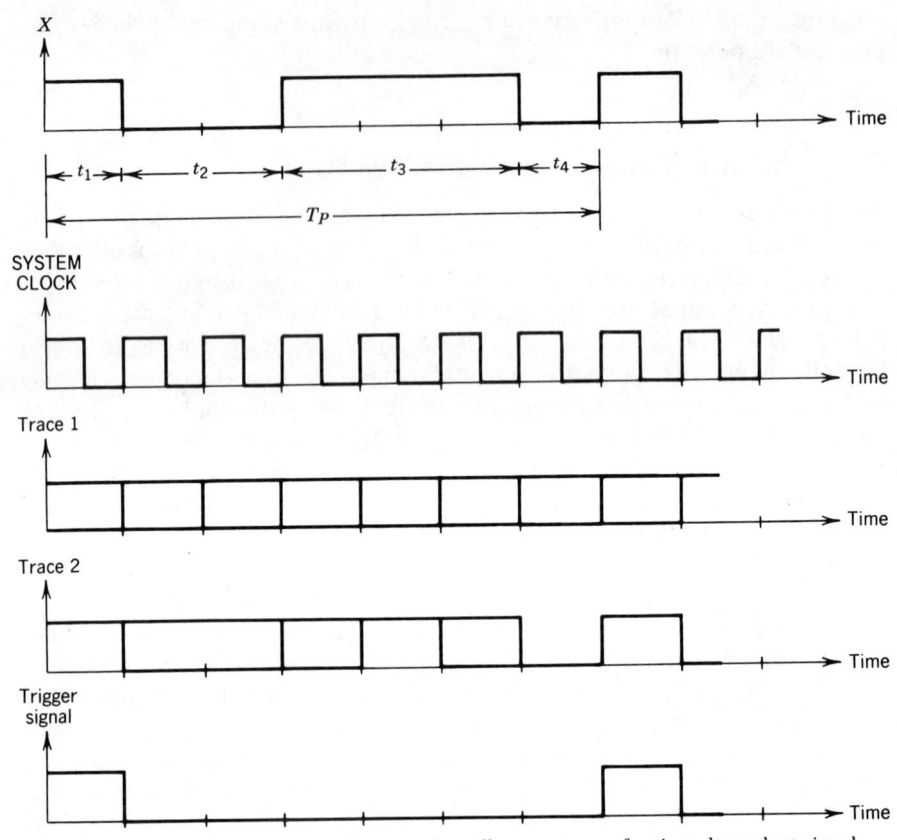

FIGURE E5.8 Timing diagrams and expected oscilloscope traces for time-dependent signal.

that if the proper triggering signal is used, then the output traces are stable and correct at any selected time scale.

References

1. Moore, E. F. *Sequential Machines: Selected Papers.* Reading, Mass.: Addison-Wesley, 1964.
2. Mealy, G. H. "A Method for Synthesizing Sequential Circuits" *Bell System Tech. J.*, **34,** 1045–1080 (Sept. 1955).

Drill Problems

5.1. Analyze the gates with the feedback connection shown in Figure D5.1 and determine their output tables.

5.2. Analyze the circuit shown in Figure D5.2. How could you use this circuit to detect a 0-to-1 transition of a signal?

EXPERIMENT 5   **161**

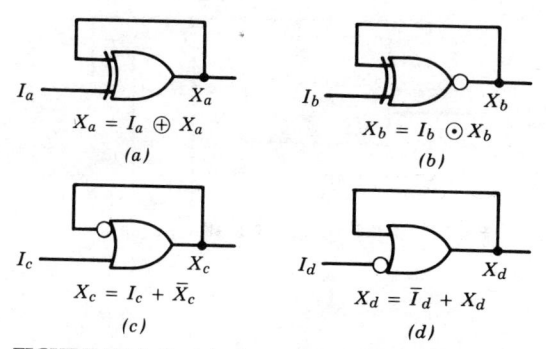

I_a
X_a
$X_a = I_a \oplus X_a$
(a)

I_b
X_b
$X_b = I_b \odot X_b$
(b)

I_c
X_c
$X_c = I_c + \bar{X}_c$
(c)

I_d
X_d
$X_d = \bar{I}_d + X_d$
(d)

FIGURE D5.1 Four logic gates with feedback.

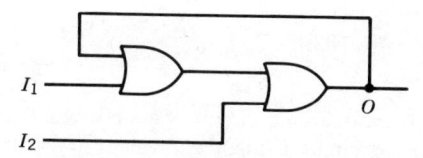

I_1
I_2
O

FIGURE D5.2 Simple circuit with feedback.

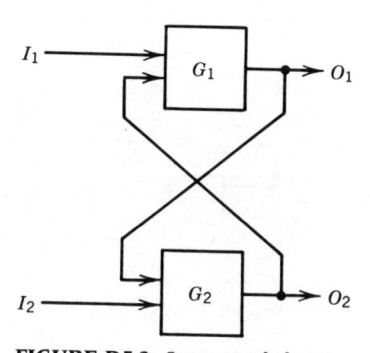

I_1 → G_1 → O_1

I_2 → G_2 → O_2

FIGURE D5.3 Cross-coupled gates.

T → Combinational circuit → D Q

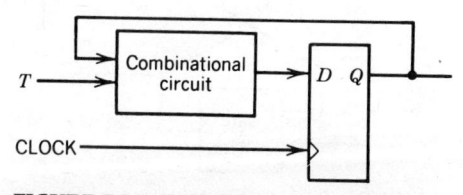

CLOCK

FIGURE D5.4 Design of a T-type flip-flop.

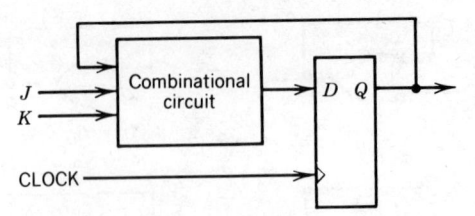

FIGURE D5.5 Design of a J-K type flip-flop.

5.3. In Figure D5.3 the schematic diagram of two two-input cross-coupled gates are shown. Determine the output table of the circuit, and if it can be used as a memory, characterize how it may be set or reset. Use the following gate types:
   (a) G_1 = AND, G_2 = AND.
   (b) G_1 = OR, G_2 = AND.
   (c) G_1 = NAND, G_2 = NOR.
   (d) G_1 = AND, G_2 = XOR.

5.4. In Figure D5.4 a combinational circuit is added to a D-type flip-flop, and the overall behavior of the circuit is made equivalent to a T flip-flop. (The output of the flip-flop is unchanged if $T = 0$ and it toggles if $T = 1$.) Determine and draw the circuit diagram of the combinational circuit.

5.5. Similar to Problem D5.4, the overall behavior of the circuit shown in Figure D5.5 is equivalent to a *JK* flip-flop. Determine and draw the circuit diagram for the indicated combinational circuit.

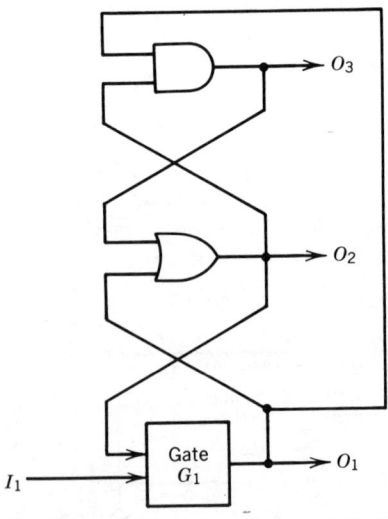

FIGURE D5.6 Circuit with several feedback connections.

5.6. Replace all NAND gates in Figure 5.25 with NOR gates. Analyze the circuit and show that it is a properly operating edge-triggered flip-flop. Characterize its operation.

5.7. Determine the state transition diagrams and the output table of the circuit shown in Figure D5.6 with the following gate substitutions:
  (a) $G_1 = $ AND.
  (b) $G_1 = $ NOR.
  (c) $G_1 = $ EQU.

CHAPTER 6

Design of Counters

6.1 INTRODUCTION

As the first practical example of sequential circuits, we will study the design of digital counters. A counter is specified by a "counting sequence" that determines the "next" state of the counter for each one of its "current" states. A simple digital counter requires no independent input variables and, usually, the output variables are the stored bits of the system; consequently, the counter requires no output circuits either. For each significant transition of SYSTEM CLOCK, the counter executes a specific state transition. Eventually, the counting sequence becomes periodic. The two most common counters are the simple binary up counter, for which each state is numerically one larger than its preceding state, and the binary down counter, for which the up counting sequence is reversed. Counters with arbitrary counting sequences are also useful for controlling digital computers.

It will be shown that the design of a counter is divided into two parts: the selection of the memory elements (flip-flop types) and the design of combinational circuits. The combinational circuits will be designed using K-maps. Both STDs and STTs are useful for the design of counters. The design of counters with other than D-type flip-flops will need a new form of sequential circuit description: the flip-flop **excitation table,** which will be discussed in Section 6.3.

If there are N stored bits, the system has 2^N states, but a counter with a specific count sequence may not use all the possible states. For the correct operation of a counter with unused system states, either the initialization of the counter is required (so that it starts in one of its used states) or the counter has to be "self-starting," which means that no matter which unused state it starts from, it must drop into its required counting sequence after a number of initial clock transitions. The design problems associated with unused states will be demonstrated.

We will also demonstrate the design of counters that generate a number of different counting sequences. Similar to variable Boolean function generators (Experiments 3 and 4), **control** inputs are added to the counter, and for each combination of 1's and 0's of the control inputs, a different count sequence is generated. Counters with variable counting sequences are frequently used as important system elements in digital computers and other complex digital data-processing systems.

At first, all the design examples will be demonstrated by using the simplest type of memory elements, D-type flip-flops. After a few examples of counter design, the use of the other three flip-flop types will be presented. We will show general design methods by which SR, JK, and T flip-flops can be utilized in the construction of practical counter circuits. A short description of asynchronous (ripple) counters and practical counter devices will conclude this chapter. Having studied the examples presented in this chapter, the student should be adequately prepared for Experiment 6, the design of a five-state up–down counter.

6.2 DESIGN OF SYNCHRONOUS COUNTERS WITH D-TYPE FLIP-FLOPS

A simple counter is defined by a counting sequence that requires no independent inputs. Each state of the counter is defined by the combination of 1's and 0's of its stored bits. The simplest way of characterizing states is to assign them the positive binary value of the stored bits. For example, the states of a 3-bit counter with stored bits Q_2, Q_1, and Q_0, are numbered from 0 (binary 000) to 7 (binary 111). The ordinary counting sequence for a 3-bit binary counter is given by

$$0 \to 1 \to 2 \to 3 \to 4 \to 5 \to 6 \to 7 \to 0$$

This sequence is referred to as counting "up," whereas the sequence for counting "down" is given by

$$0 \to 7 \to 6 \to 5 \to 4 \to 3 \to 2 \to 1 \to 0$$

Note that the counting sequence must be specified as a periodic sequence of states (the last state must be equal to the first state). The STD demonstrates the operation of the counter graphically, and the STT specifies the counter's operation in terms of 1's and 0's of its stored bits. These are shown in Figure 6.1 for the up sequence.

If only D-type flip-flops are used for storing bits Q_0–Q_2, the design of the counter becomes straightforward. In the STT each bit Q_i is given for time instance t_{n+1} [$Q_i(t_{n+1})$] as a function of the stored bit values at time t_n. Since for each D-type flip-flop:

$$Q_i(t_{n+1}) = D_i(t_n)$$

the transition tables may also be interpreted as functions for D_i (the inputs to the flip-flops) in terms of Q_i (their stored output bits). Hence, the STT in Figure 6.1 defines three Boolean combinational functions, each with inputs Q_0, Q_1, and Q_2, and one output connected to the D input of one of the three flip-flops. The design of the counter is reduced to the design of a three-input, three-output combinational circuit.

The three transition K-maps are shown in Figure 6.2. Since only D-type flip-flops will be used, these maps are also the K-maps for the combinational

DESIGN OF COUNTERS

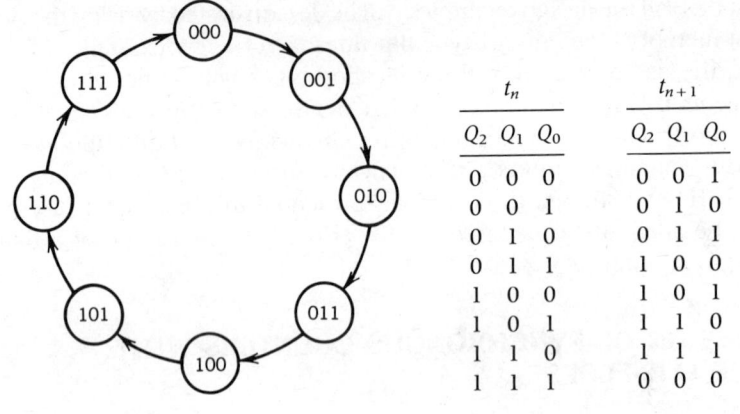

State transition diagram

t_n			t_{n+1}		
Q_2	Q_1	Q_0	Q_2	Q_1	Q_0
0	0	0	0	0	1
0	0	1	0	1	0
0	1	0	0	1	1
0	1	1	1	0	0
1	0	0	1	0	1
1	0	1	1	1	0
1	1	0	1	1	1
1	1	1	0	0	0

State transition table

FIGURE 6.1 The STD and STT for 3-bit binary up counter.

circuits from which the following design equations are deduced:

$$Q_2(t_{n+1}) = D_2 = Q_2' \cdot Q_1 \cdot Q_0 + Q_2 \cdot Q_1' + Q_2 \cdot Q_0'$$

$$Q_1(t_{n+1}) = D_1 = Q_1' \cdot Q_0 + Q_1 \cdot Q_0'$$

$$Q_0(t_{n+1}) = D_0 = Q_0'$$

Minimization for 0's produces similar results. In this form the equations require a two- and a three-input OR gate, one three-input AND gate, four

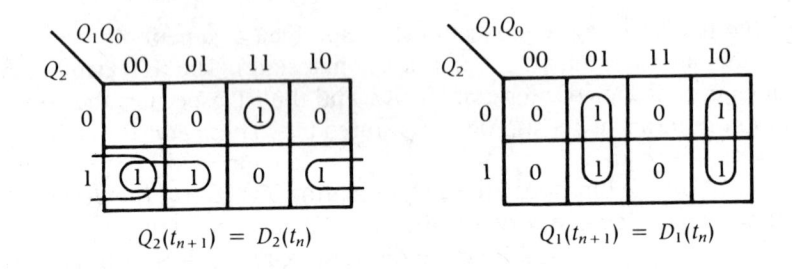

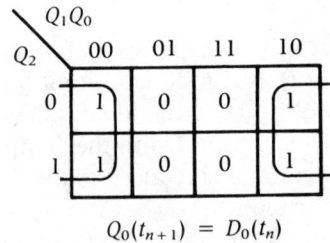

$Q_2(t_{n+1}) = D_2(t_n)$

$Q_1(t_{n+1}) = D_1(t_n)$

$Q_0(t_{n+1}) = D_0(t_n)$

State transition K-maps

FIGURE 6.2 State transition K-maps for 3-bit binary up counter.

6.2 DESIGN OF SYNCHRONOUS COUNTERS WITH D-TYPE FLIP-FLOPS

two-input AND gates, and three INVERTER gates. (Frequently, in practice, the complement of the stored bits is also available and INVERTER gates are not required.) It is possible to factor the first equation and use the Boolean equation for the XOR function ($A \oplus B = A' \cdot B + B' \cdot A$) to reduce the circuits considerably:

$$D_2 = Q_2' \cdot Q_1 \cdot Q_0 + Q_2 \cdot (Q_1' + Q_0')$$
$$= Q_2' \cdot (Q_1 \cdot Q_0) + Q_2 \cdot (Q_1 \cdot Q_0)'$$
$$= Q_2 \oplus (Q_1 \cdot Q_0)$$
$$D_1 = Q_1 \oplus Q_0$$
$$D_0 = Q_0'$$

The completed logic circuits for the 3-bit binary up counter using these reduced equations are shown in Figure 6.3. You may have noticed that the input for flip-flop Q_0 is equal to its inverted output, which produces a toggling circuit. Since the SYSTEM CLOCK signal controls all the sequential elements (flip-flops) of the circuit, the system in Figure 6.3 is synchronous and is called a synchronous counter.

Using similar K-map minimization techniques, a general counter that follows the sequence of eight states in any order may be designed. Similar to the procedures described earlier, both the STT and the K-maps can be constructed from the given counting sequence. Once the K-maps are known, the minimization techniques can be applied. The design procedures are not much more difficult when the number of used counter states are smaller than the number of possible states. This will be demonstrated by the next example, the design of a five-state counter.

It occurs frequently in sequential circuit and digital computer design that a counter is required that follows a predetermined sequence of states. Let us consider the following required sequence:

$$1 \rightarrow 6 \rightarrow 2 \rightarrow 3 \rightarrow 4 \rightarrow 1$$

This counting sequence contains five states. States 0, 5, and 7 do not appear in the counting sequence and are called **unused** states. The minimum number

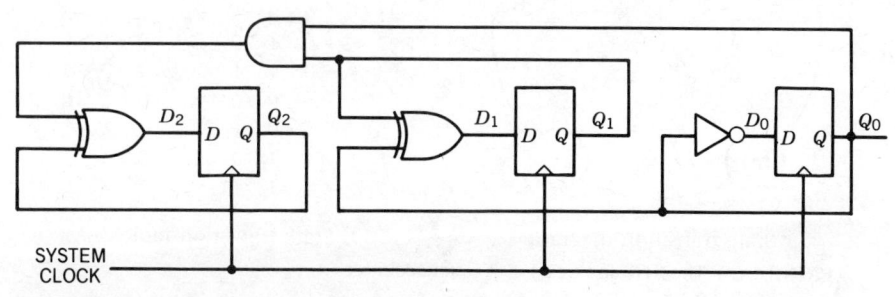

FIGURE 6.3 Circuit diagram for 3-bit binary up counter.

168   DESIGN OF COUNTERS

of stored bits that can produce five distinct states is still 3; therefore, all eight states exist (by using the PRESET or SET inputs of practical flip-flops, the stored bit values could be set to any combination of 1's and 0's and hence to any one of the eight possible states), but if the counter operates correctly, the unused states should never appear. This means that the state transitions that occur from the unused states could be ignored. The bits of the "new" output states, which are expressed as the D inputs to the flip-flops, become "don't care" entries in the K-map.

The STD and the STT for this counter are shown in Figure 6.4. The next states from the unused states are marked with X's in the table, indicating don't care entries. Three K-maps are constructed from this STT, and they are shown in Figure 6.5. A selection of groups of 1's are also shown that produce the following design equations:

$$Q_2(t_{n+1}) = D_2 = Q_0$$

$$Q_1(t_{n+1}) = D_1 = Q_2' \cdot Q_1' + Q_1 \cdot Q_0'$$

$$Q_0(t_{n+1}) = D_0 = Q_2' \cdot Q_0' + Q_2 \cdot Q_1'$$

Substitution of the bits of the five used states into values Q_2-Q_0 at t_n proves that the counting sequence satisfies its specifications. Hence, if the counter is started in any one of its used states, it will operate correctly. Let us examine how the counter behaves according to the three design equations when it is started in one of its unused states. All don't care terms included in groups of 1's in the K-maps (Figure 6.5) generate 1's for the output. We can construct the transition tables for the unused states either from the K-maps (by assuming 1's for the outputs of included don't care terms) or from the design equations (by substituting the bits for the unused states into Q_2-Q_0 at t_n). The results are:

	t_n	t_{n+1}
State	$Q_2\ Q_1\ Q_0$	$Q_2\ Q_1\ Q_0$
0	0 0 0	X X X
1	0 0 1	1 1 0
2	0 1 0	0 1 1
3	0 1 1	1 0 0
4	1 0 0	0 0 1
5	1 0 1	X X X
6	1 1 0	0 1 0
7	1 1 1	X X X

State transition diagram   State transition table

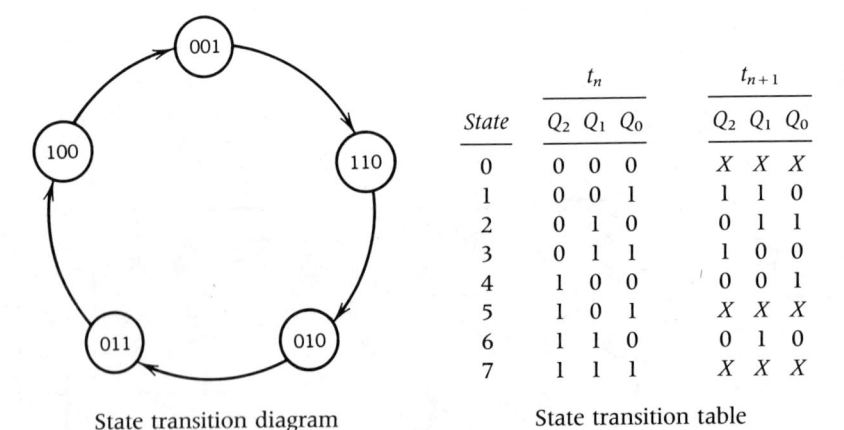

FIGURE 6.4 The STD and STT for five-state counter.

6.2 DESIGN OF SYNCHRONOUS COUNTERS WITH D-TYPE FLIP-FLOPS

Unused state	Bits at t_n, $Q_2Q_1Q_0$	Bits at t_{n+1}, $Q_2Q_1Q_0$	Next state
0	000	011	3
5	101	101	5
7	111	100	4

Even though the stored bit combinations that represent the unused states should never occur at the outputs of the flip-flops, in a practical circuit the counter may find itself in one of its unused states. When power is turned on, the initial state of the counter is undetermined. Unless the flip-flops of the counter are initialized by specially designed circuitry, the counter can start from any one of its eight states. If the state transitions from the unused states are such that the counter would find its proper counting sequence from any initial state, then the counter is called **self-starting.** For the preceding example the proper counting sequence is followed by the counter if the starting state is state 0 (000) or state 7 (111). However, if the counter starts from state 5 (101), then the counter remains in state 5. We can prove this by the substitution of the stored bit values into the design equations. With $Q_2Q_1Q_0$ equal to 101 (current state is state 5), we get

$$D_2 = Q_0 = 1$$
$$D_1 = Q_2' \cdot Q_1' + Q_1 \cdot Q_0' = 0 \cdot 1 + 0 \cdot 0 = 0$$
$$D_0 = Q_2' \cdot Q_0' + Q_2 \cdot Q_1' = 0 \cdot 0 + 1 \cdot 1 = 1$$

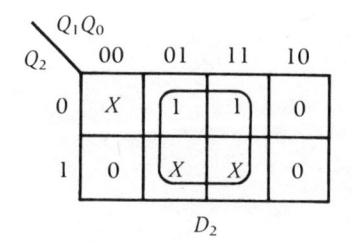

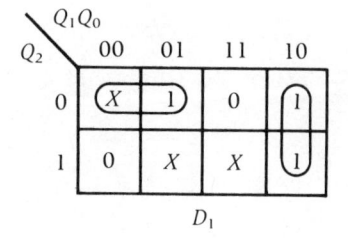

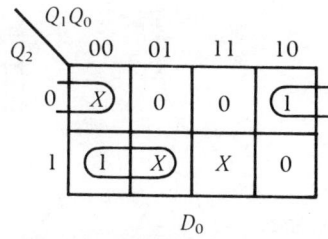

FIGURE 6.5 The state transition K-maps for five-state counter (not self-starting).

170   DESIGN OF COUNTERS

which for inputs $D_2D_1D_0$ generates the next state with bits 101; therefore, the counter will never leave state 5.

If the requirement is for a self-starting counter, and the minimized design equations do not generate a self-starting solution, then the design equations must be changed. In this case the selection of the don't care outputs must be changed in such a way that no matter from which state the counter starts, eventually, it will drop into its proper counting sequence. In the preceding example only state 5 caused a nonstarting problem. If the selected groups of 1's in Figure 6.5 are chosen differently, the state transition from state 5 may also change, and the nonstarting problem is removed. For this example the groups of 1's in the third K-map can be changed without adding extra terms to the design equation for D_0. The changed K-map for input D_0 is shown in Figure 6.6. The new design equation for D_0 is given by

$$D_0 = Q_2' \cdot Q_0' + Q_1' \cdot Q_0' = Q_0' \cdot (Q_1' + Q_2')$$

which produces a Logic-0 value for state 5, that is, for $Q_2Q_1Q_0 = 101$. This changes the next state of state 5 to state 4 (100). By comparing the changed K-map with the original one, we can see that the change influences only the state transition from state 5 and, thus, makes the counter self-starting. The completed STD for the changed circuits is shown in Figure 6.6.

Another method for finding a self-starting counter design is to minimize the K-maps for 0's that may produce different don't care outputs. Minimization for 0's for the same K-maps are shown in Figure 6.7. The design equations for these groups of 0's are

$$D_2 = Q_0$$

$$D_1 = (Q_1 + Q_0) \cdot (Q_1' + Q_0') = Q_1 \cdot Q_0' + Q_1' \cdot Q_0 = Q_1 \oplus Q_0$$

$$D_0 = Q_0' \cdot (Q_2' + Q_1')$$

State transition diagram

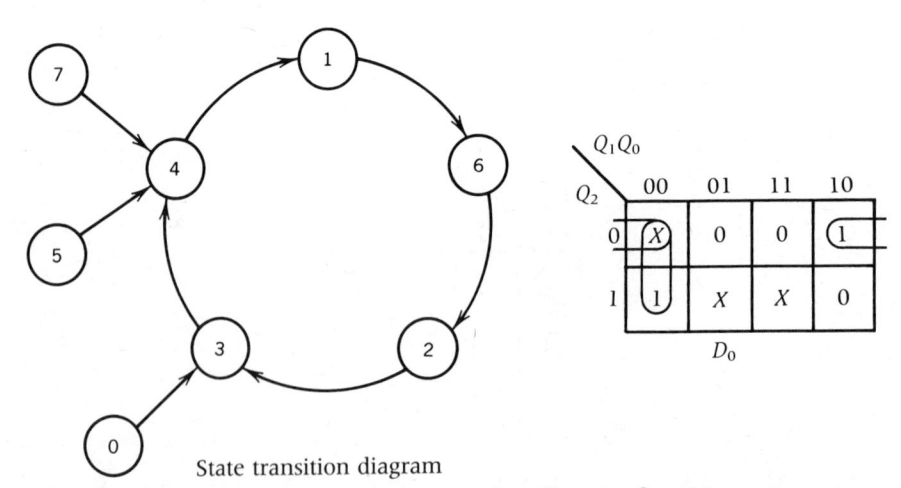

FIGURE 6.6 The STD and state transition K-map for self-starting five-state counter.

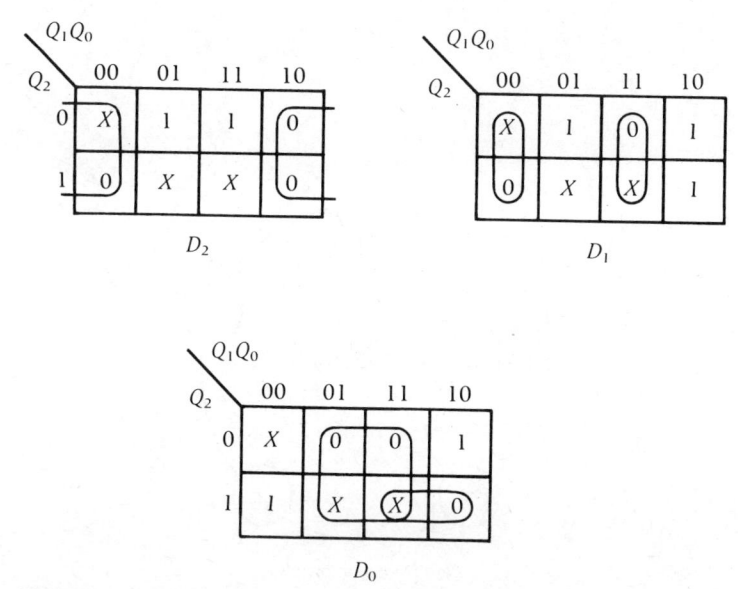

6.2 DESIGN OF SYNCHRONOUS COUNTERS WITH D-TYPE FLIP-FLOPS

FIGURE 6.7 Minimization for state transition K-maps of five-state counter using 0's.

which produce the following state transitions for the unused state:

Unused state	Bits at t_n, $Q_2Q_1Q_0$	Bits at t_{n+1}, $Q_2Q_1Q_0$	Next state
0	000	001	1
5	101	110	6
7	111	100	4

where all outputs included in groups of 0's in Figure 6.7 produce 0 bits and all others produce 1's. The STD for the counter using these design equations is shown in Figure 6.8. We can see that it is different from the circuits shown in Figure 6.6.

Counters with Controlled Counting Sequences

We have studied variable Boolean function generator circuits (Experiments 3 and 4) that produced various Boolean functions according to the settings of a number of control variables. Similarly, the behavior of sequential circuits may also be controlled by a number of input variables. There are counters, for example, that generate different counting sequences, depending on the value(s) of input variable(s). As an example of a controlled counter circuit, we will design a two-bit up-down binary counter that can operate either as a binary up counter or as a binary down counter. The counting sequences

172 DESIGN OF COUNTERS

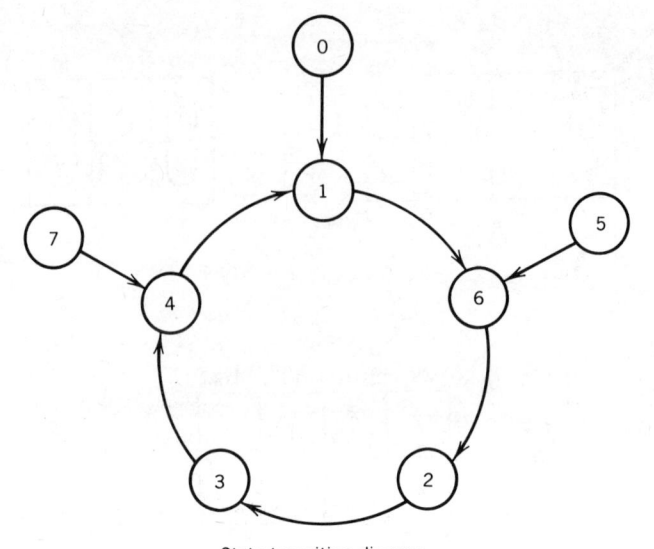

State transition diagram

FIGURE 6.8 The STD for five-state counter when 0's are used for minimization.

for this counter are

$$0 \to 1 \to 2 \to 3 \to 0 \quad \text{(up)}$$

$$0 \to 3 \to 2 \to 1 \to 0 \quad \text{(down)}$$

and since there are only two sequences to choose from, only one additional input variable is required. The block diagram and STDs for this counter are shown in Figure 6.9. When the control input, shown as C, is set to Logic-1, the circuit behaves as an ordinary binary up counter. When the control input is set to Logic-0, then the counting sequence becomes the reverse of the normal binary sequence.

The design of this counter follows the same design procedures as those

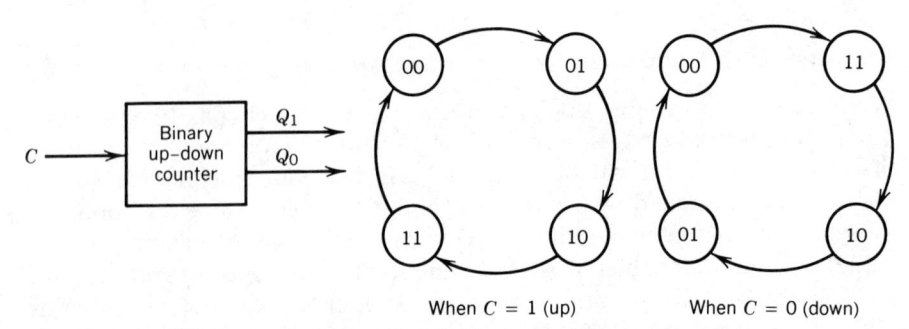

When $C = 1$ (up)　　　　When $C = 0$ (down)

FIGURE 6.9 Block diagram and STDs for 2-bit binary up-down counter.

6.2 DESIGN OF SYNCHRONOUS COUNTERS WITH D-TYPE FLIP-FLOPS   173

used before. The only difference for this controlled counter is that the control variable C appears as an additional, independent input in the STT and, hence, in the K-maps. For this example the K-maps contain three variables: input C and the two stored bits Q_1 and Q_0. These are shown in Figure 6.10. The pattern of 1's in the K-map for D_1 is equal to the Exclusive-OR function of three variables:

$$Q_1(t_{n+1}) = D_1 = C \oplus Q_1 \oplus Q_0 = C \oplus (Q_1 \oplus Q_0)$$

whereas D_0 is simply equal to Q_0' (a toggling circuit for the least significant bit). The completed counter circuit is shown in Figure 6.11. Since all four states for the two stored bits appear in both counting sequences, no unused states are present.

As the last example for counter design with D-type flip-flops, a variable five-state counter will be used. The specifications for the five-state counter in Figure 6.4 are modified by the addition of a control variable (C). When C is equal to 0, the counting sequence, shown in Figure 6.4, is generated. When C is equal to 1, the counter behaves as a Modulo-5 binary counter. A Modulo-5 counter has five states, and it counts up normally from state 0 to state 4. Since the last used state is state 4, at the next CLOCK transition it changes its state from state 4 to state 0. The combined specifications of this counter are shown below:

When $C = 0$, the sequence is $1 \to 6 \to 2 \to 3 \to 4 \to 1$.
When $C = 1$, the sequence is $0 \to 1 \to 2 \to 3 \to 4 \to 0$.

State transition table

t_n			t_{n+1}		
C	Q_1	Q_0	Q_1	Q_0	
0	0	0	0	1	
0	0	1	1	0	Up
0	1	0	1	1	
0	1	1	0	0	
1	0	0	1	1	
1	0	1	0	0	Down
1	1	0	0	1	
1	1	1	1	0	

State K-maps

D_1:

C \ Q_1Q_0	00	01	11	10
0	0	1	0	1
1	1	0	1	0

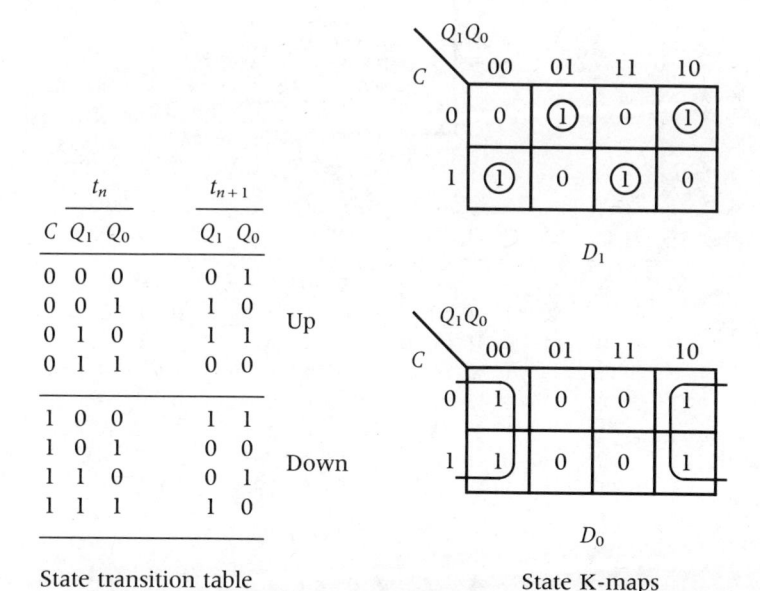

D_0:

C \ Q_1Q_0	00	01	11	10
0	1	0	0	1
1	1	0	0	1

FIGURE 6.10 The STT and state transition K-maps for 2-bit binary up-down counter.

174 DESIGN OF COUNTERS

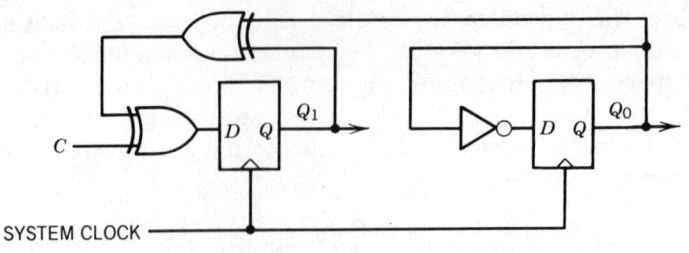

FIGURE 6.11 Circuit diagram for 2-bit binary up-down counter.

The block diagram and the STT for this five-state controlled counter are shown in Figure 6.12. Note that the two counting sequences of this counter have different don't care states, but these states do not have to be considered during the initial stages of the design. When C is equal to Logic-1, the don't care states are states 5, 6, and 7. When C is equal to Logic-0, as before, the don't care states are states 0, 5, and 7. The three K-maps for this controlled five-state counter are shown in Figure 6.13. Minimization for groups of 0's shown in Figure 6.13 produces the following design equations:

$$D_2 = Q_0 \cdot (C' + Q_1)$$
$$D_1 = Q_1 \cdot Q_0 + Q_1' \cdot Q_0' = Q_1 \odot Q_0$$
$$D_0 = Q_0' \cdot (C' + Q_2') \cdot (Q_2' + Q_1') = Q_0' \cdot (Q_2' + C' \cdot Q_1')$$

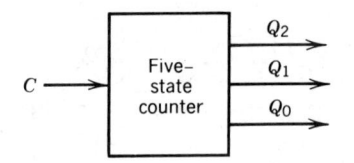

t_n			t_{n+1}			t_n			t_{n+1}		
Q_2	Q_1	Q_0	Q_2	Q_1	Q_0	Q_2	Q_1	Q_0	Q_2	Q_0	Q_1
0	0	0	X	X	X	0	0	0	0	0	1
0	0	1	1	1	0	0	0	1	0	1	0
0	1	0	0	1	1	0	1	0	0	1	1
0	1	1	1	0	0	0	1	1	1	0	0
1	0	0	0	0	1	1	0	0	0	0	0
1	0	1	X	X	X	1	0	1	X	X	X
1	1	0	0	1	0	1	1	0	X	X	X
1	1	1	X	X	X	1	1	1	X	X	X

$$C = 0 \qquad\qquad C = 1$$

$$1 \to 6 \to 2 \to 3 \to 4 \to 1 \qquad 0 \to 1 \to 2 \to 3 \to 4 \to 0$$

FIGURE 6.12 Block diagram and STT for controlled five-state counter.

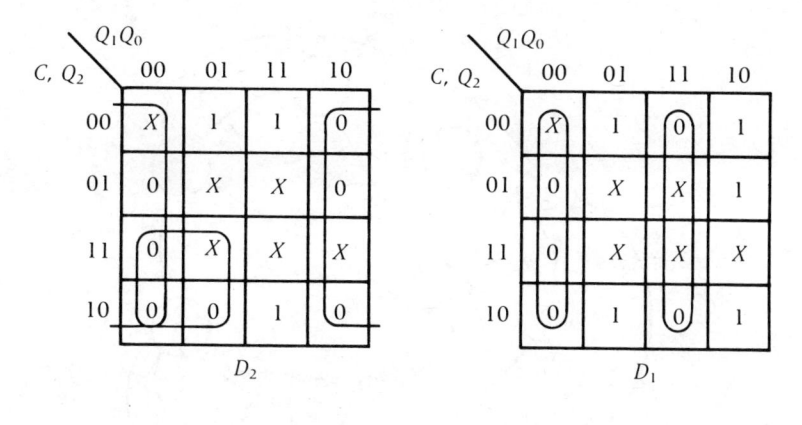

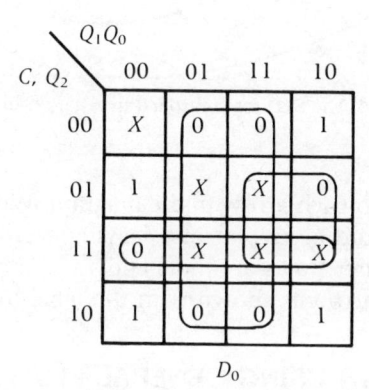

6.2 DESIGN OF SYNCHRONOUS COUNTERS WITH D-TYPE FLIP-FLOPS

FIGURE 6.13 State transition K-maps for controlled five-state counter.

Once the design equations are known, the state transitions for the unused states can be evaluated. (As before, don't care terms included in groups of 0's produce 0 results; others produce 1's.) The transition states for the unused states are:

C	Unused state	Bits at t_n, $Q_2Q_1Q_0$	Bits at t_{n+1}, $Q_2Q_1Q_0$	Next state
0	0	000	001	1
0	5	101	110	6
0	7	111	100	4
1	5	101	010	2
1	6	110	010	2
1	7	111	100	4

The counter is proved to be self-starting for both counting sequences. The complete STD for this circuit is shown in Figure 6.14. Since this is a controlled counter, it includes one independent input variable whose value should ap-

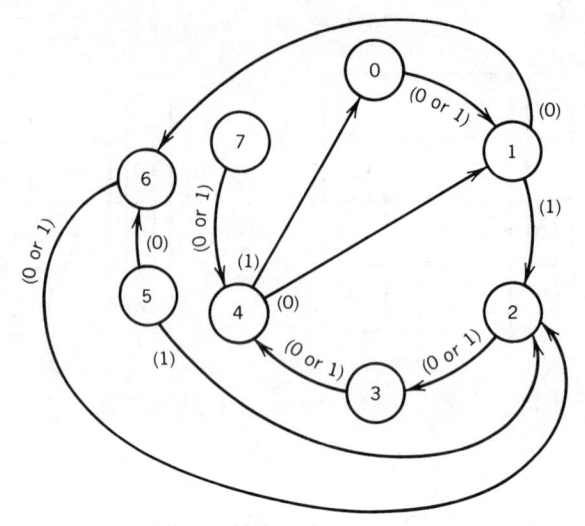

FIGURE 6.14 The STD for controlled five-state counter.

pear as a label attached to each arrow in the diagram. With one independent input variable there should be two arrows leaving each state, but since the next states of some counter states are independent of the control variable C, in these cases only one arrow is shown with the label: 0 or 1.

6.3 COUNTER DESIGN USING GENERAL FLIP-FLOP TYPES

In the previous section we demonstrated the design of counters using D-type flip-flops. The described design methods are not restricted to the design of counters only. They may be used for any general sequential circuit (not only counters), as will be demonstrated in the following chapters. These methods were based on the fact that the next state of the D-type flip-flop, $Q(t_{n+1})$, is equal to the value of the flip-flop's input D at time t_n. Hence, the K-maps specifying the next-state outputs of D-type flip-flops can be interpreted as combinational circuits whose outputs are applied directly to the inputs of the flip-flops. The same methods do not apply to general 1-bit memories, such as the SR, JK, and T flip-flops introduced at the end of Chapter 5.

The methods with which general flip-flop types can be used in the design of sequential circuits are based on the so-called **excitation tables** for flip-flops and **output transition tables** for the system. The output transition table lists the transition of each stored bit as a function of the "current state" of the system and can be derived directly from the STT. For example, when a 3-bit binary counter executes the state transition from state 0 (000) to state 1 (001), the output transition table lists the transitions $0 \rightarrow 0$ for bit Q_2, $0 \rightarrow 0$ for bit Q_1, and $0 \rightarrow 1$ for bit Q_0.

It is also possible to construct an excitation table for each flip-flop type that defines the required flip-flop input(s) for a required flip-flop output transition. There are four possible transitions ($0 \to 0$, $0 \to 1$, $1 \to 0$, and $1 \to 1$); therefore, the flip-flop excitation table has four entries. Since the output transition table defines which bit transitions are required for the flip-flop outputs, and the flip-flop excitation table specifies the required inputs for a given bit transition, the two tables together determine the required input values for each starting state of the system. The use of these tables will be demonstrated in the following sections.

Three-Bit Binary Counter Design with SR Flip-Flops

As the first example of this general design method, the three-bit binary counter will be designed using SR (set–reset) flip-flops. The state transition equations and STT for the SR flip-flop are shown in Figure 5.27. The output of the SR flip-flop is unchanged for input values $SR = 00$; it is SET (the output becomes 1 after the CLOCK transition) for input values $SR = 10$, and it is RESET (the output becomes 0) for input values $SR = 01$. The operation of the SR flip-flop is undefined when both $S = 1$ and $R = 1$, and these input values must be avoided in practice.

Let us calculate the excitation table for the SR flip-flop. First, the $0 \to 0$ transition of the output bit is considered. If the current stored bit of the SR flip-flop is equal to 0 and its next stored bit value is also 0, there are two possible input bit combinations that are consistent with this transition. First, if $S = 0$ and $R = 0$, then the output of the flip-flop remains unchanged (the stored bit is equal to 0 both before and after the significant CLOCK transition). Second, if $S = 0$ and $R = 1$, the flip-flop is RESET, so that its output becomes 0 independently from its output value before the CLOCK transition. Both these cases provide a 0-to-0 bit transition; therefore, in order to cause this transition, it is sufficient to set the S input to Logic-0 and the R input can be ignored. This can be indicated by the following entry in the excitation table:

Bit transition $Q(t_n) \to Q(t_{n+1})$	Required inputs S R
$0 \to 0$	0 X

where the R input is an unspecified, or in other words, a don't care, input. The bit transition of $0 \to 1$ can be produced only by the SET operation; hence, the input values $S = 1$ and $R = 0$ must be applied. Similar analysis for the other two bit transitions completes the specifications for the excitation table of the SR flip-flop, which is shown in Figure 6.15.

Entries in the STT are used as the first step in the design of the counter. These are listed in the leftmost two columns in Table 6.1. The entries are copied directly from the specified counting sequence or from the STT of the system. In the next three columns the required output bit transitions are entered. These entries are copied directly from the state transitions and may be skipped by an experienced designer who can translate state transitions

178 DESIGN OF COUNTERS

Transition $Q(t_n) \rightarrow Q(t_{n+1})$	Required $S\ R$
$0 \rightarrow 0$	$0\ X$
$0 \rightarrow 1$	$1\ 0$
$1 \rightarrow 0$	$0\ 1$
$1 \rightarrow 1$	$X\ 0$

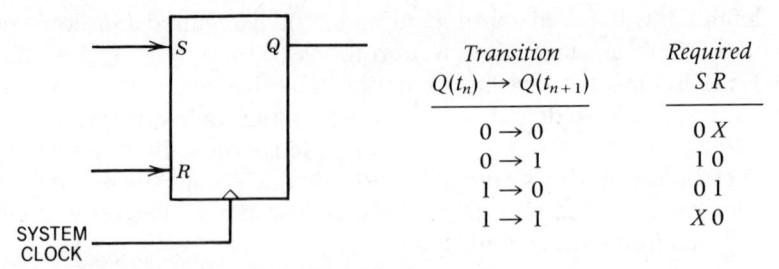

FIGURE 6.15 Excitation table for SR flip-flop.

directly to required input values. The output bit transitions, along with the flip-flop excitation table, determine the required input values, which are listed in the last six columns of Table 6.1.

Once the required flip-flop input values are known, six three-variable K-maps can be constructed. These K-maps express the required input functions for the three SR flip-flops of the 3-bit counter. The K-maps with the indicated groups of 1's specify the required Boolean functions for the inputs of the SR flip-flops. The K-maps and the resulting design equations are shown in Figure 6.16. The completed circuit constructed according to these design equations is shown in Figure 6.17. We already know that the circuit for Q_0 is a simple toggling circuit. It is interesting to note that if the inverted output is not available, then, similar to the D-type flip-flop, the SR flip-flop requires also one inverter for this circuit.

For one combination of input values ($SR = 11$) the operation of the SR flip-flop is not defined, and if these values appear in practice, the output of the flip-flop becomes unpredictable after the significant CLOCK transition. Since according to the reliable fundamental proposition of Murphy, "If anything can go wrong, it will," a system with a possible unpredictable behavior may be very difficult to "debug." Another disadvantage of the SR flip-flop is that one-quarter of its capabilities (one combination of input values out of four) is wasted. A more general flip-flop type, the JK flip-flop, eliminates this problem. Design with JK flip-flops will be the topic of the next two sections.

TABLE 6.1 State Transition Table for Binary Up Counter Using SR Flip-Flops

Current state	Next state	Output transitions Q_2	Q_1	Q_0	Required inputs $S_2 R_2$	$S_1 R_1$	$S_0 R_0$
000	001	$0 \rightarrow 0$	$0 \rightarrow 0$	$0 \rightarrow 1$	$0\ X$	$0\ X$	$1\ 0$
001	010	$0 \rightarrow 0$	$0 \rightarrow 1$	$1 \rightarrow 0$	$0\ X$	$1\ 0$	$0\ 1$
010	011	$0 \rightarrow 0$	$1 \rightarrow 1$	$0 \rightarrow 1$	$0\ X$	$X\ 0$	$1\ 0$
011	100	$0 \rightarrow 1$	$1 \rightarrow 0$	$1 \rightarrow 0$	$1\ 0$	$0\ 1$	$0\ 1$
100	101	$1 \rightarrow 1$	$0 \rightarrow 0$	$0 \rightarrow 1$	$X\ 0$	$0\ X$	$1\ 0$
101	110	$1 \rightarrow 1$	$0 \rightarrow 1$	$1 \rightarrow 0$	$X\ 0$	$1\ 0$	$0\ 1$
110	111	$1 \rightarrow 1$	$1 \rightarrow 1$	$0 \rightarrow 1$	$X\ 0$	$X\ 0$	$1\ 0$
111	000	$1 \rightarrow 0$	$1 \rightarrow 0$	$1 \rightarrow 0$	$0\ 1$	$0\ 1$	$0\ 1$

$S_2 = Q_2' \cdot Q_1 \cdot Q_0$

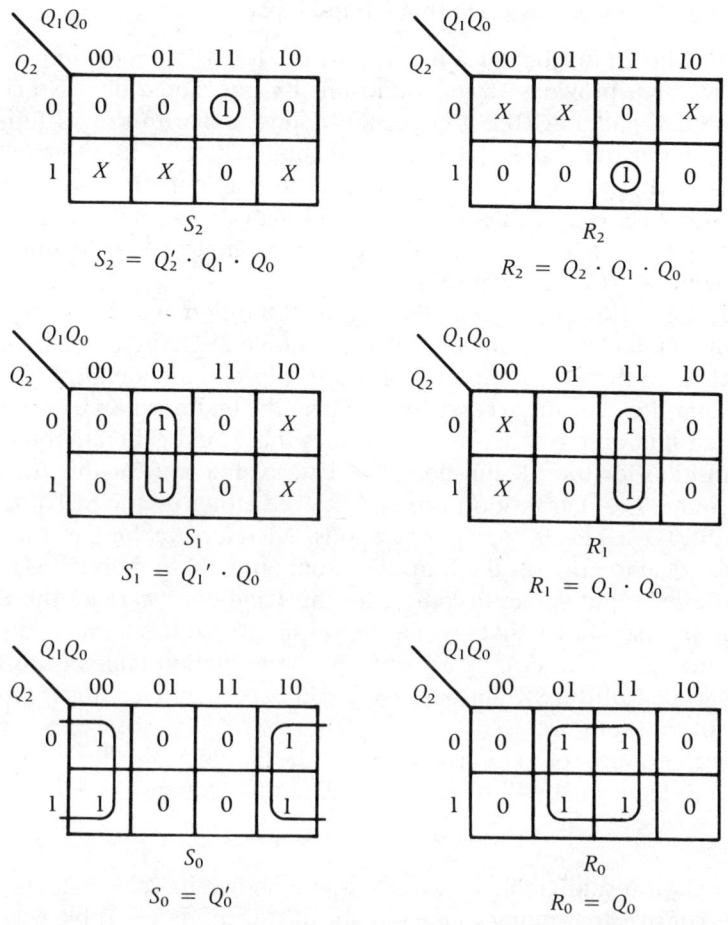

$R_2 = Q_2 \cdot Q_1 \cdot Q_0$

$S_1 = Q_1' \cdot Q_0$

$R_1 = Q_1 \cdot Q_0$

$S_0 = Q_0'$

$R_0 = Q_0$

FIGURE 6.16 State transition K-maps for 3-bit binary up counter using SR flip-flops.

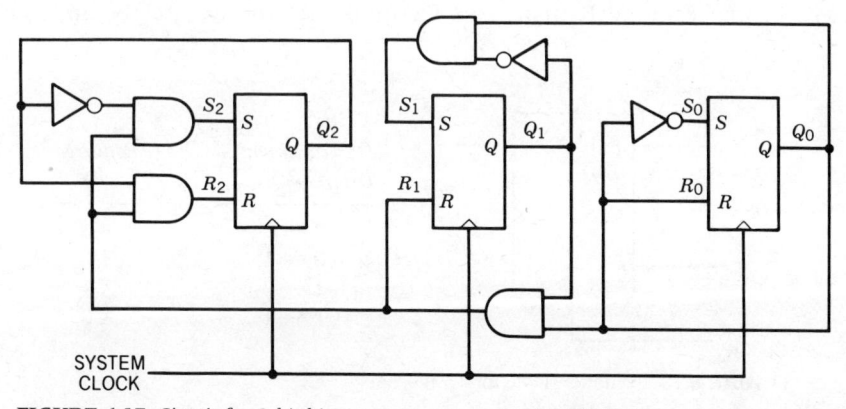

FIGURE 6.17 Circuit for 3-bit binary up counter using SR flip-flops.

179

Five-State Counter Design with JK Flip-Flops

We have shown in Chapter 5 that the JK flip-flop is the most universal flip-flop because it provides all four functions for one stored bit: "No change," SET, RESET, and TOGGLE. The JK flip-flop is similar to the SR flip-flop. It has two inputs named J and K. For the combination of input values $JK = 00$, $JK = 01$, and $JK = 10$, it operates in the same manner as the SR flip-flop, with J being equivalent to the S and K to the R inputs. For the input values of $JK = 11$, the JK flip-flop toggles (complements) its output after each significant CLOCK transition.

Since the JK flip-flop is similar to the SR flip-flop, we should expect that its excitation table will also be similar. As shown in Figure 6.18, the first and last entries in the excitation table of the JK flip-flop are identical to those of the SR flip-flop. For the second entry, where the bit transition $0 \rightarrow 1$ is listed, we could find only one set of input values that could generate the required bit transition for the SR flip-flop. This case is different for the JK flip-flop because the $0 \rightarrow 1$ transition can be achieved either by the SET ($JK = 10$) or by the TOGGLE ($JK = 11$) operations. Therefore, as long as the J input value is equal to Logic-1, the transition from output 0 to output 1 is ensured. The required input values become 1 for the J and don't care for the K input. Similar analysis shows that the $1 \rightarrow 0$ transition also has a don't care entry. Since there are more don't care terms in the excitation table of the JK flip-flop than that of the SR flip-flop, we could expect simpler circuit equations when JK flip-flops are used.

As an example of a circuit design with JK flip-flops, the five-state counter defined in Figure 6.4 will be used. The counting sequence is defined as

$$1 \rightarrow 6 \rightarrow 2 \rightarrow 3 \rightarrow 4 \rightarrow 1$$

The output transition table for this counter is shown in Table 6.2. The output table is constructed similar to the one for the SR flip-flops (Table 6.1). There are three unused states in this case for which no next states are specified. The input requirements for both J and K inputs and for all three flip-flops are don't care values in these cases. These don't care inputs are added to the ones that are specified from the excitation table. The large number of don't care terms in the resulting K-maps (see Figure 6.19) produce greatly simplified

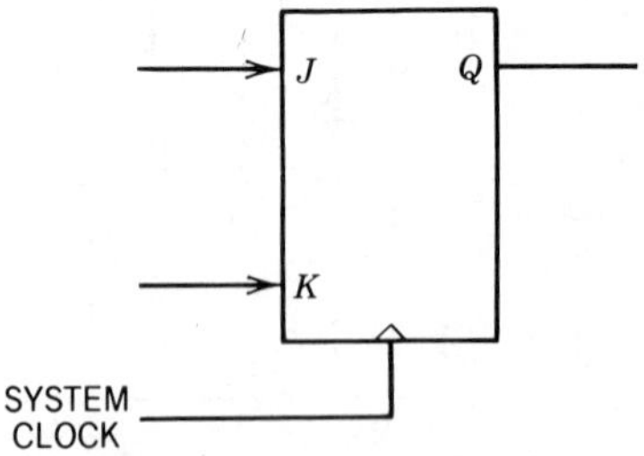

Transition $Q(t_n) \rightarrow Q(t_{n+1})$	Required JK
$0 \rightarrow 0$	$0\,X$
$0 \rightarrow 1$	$1\,X$
$1 \rightarrow 0$	$X\,1$
$1 \rightarrow 1$	$X\,0$

FIGURE 6.18 Excitation table for JK flip-flop.

TABLE 6.2 STT for Five-State Counter Using JK Flip-Flops

Current state	Next state	Output transitions Q_2	Q_1	Q_0	Required inputs J_2K_2	J_1K_1	J_0K_0
000	—		Not specified		X X	X X	X X
001	110	$0 \to 1$	$0 \to 1$	$1 \to 0$	1 X	1 X	X 1
010	011	$0 \to 0$	$1 \to 1$	$0 \to 1$	0 X	X 0	1 X
011	100	$0 \to 1$	$1 \to 0$	$1 \to 0$	1 X	X 1	X 1
100	001	$1 \to 0$	$0 \to 0$	$0 \to 1$	X 1	0 X	1 X
101	—		Not specified		X X	X X	X X
110	010	$1 \to 0$	$1 \to 1$	$0 \to 0$	X 1	X 0	0 X
111	—		Not specified		X X	X X	X X

J_2
$J_2 = Q_0$

K_2
$K_2 = 1$

J_1
$J_1 = Q_0$

K_1
$K_1 = Q_0$

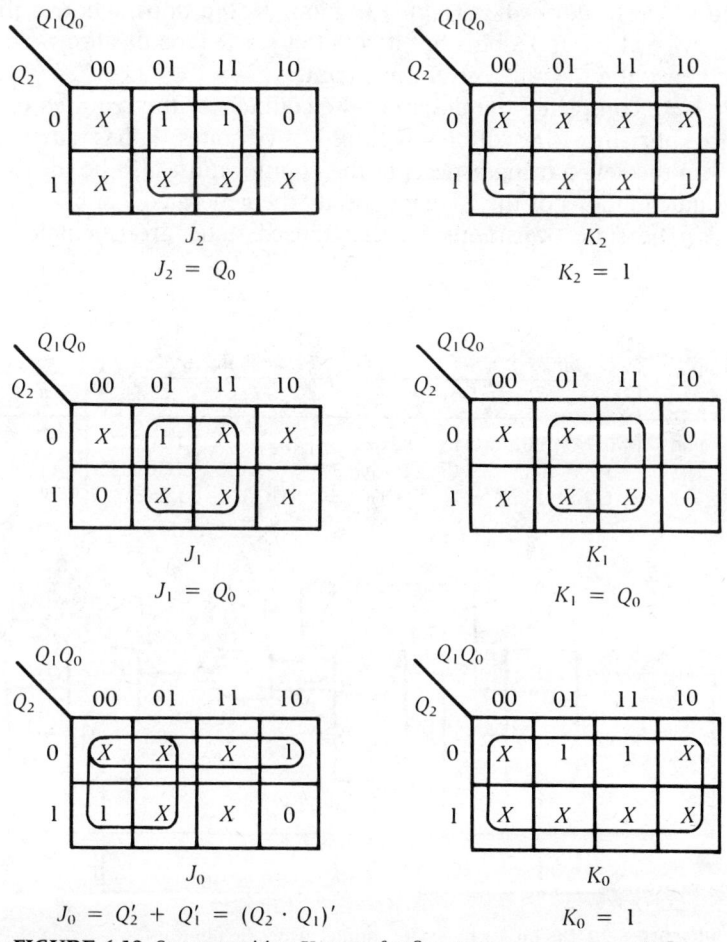

J_0
$J_0 = Q_2' + Q_1' = (Q_2 \cdot Q_1)'$

K_0
$K_0 = 1$

FIGURE 6.19 State transition K-maps for five-state counter using JK flip-flops.

182 DESIGN OF COUNTERS

design equations and, consequently, simplified logic circuits. In two cases (K_0 and K_2) the flip-flop inputs become constant (Logic-1), and in three other cases (J_1, K_1, and J_2) the inputs are equal to one of the flip-flop outputs. All these flip-flop inputs require no additional circuitry. The completed circuit diagram is shown in Figure 6.20, where only one two-input NAND gate is required to produce the specified counting sequence. These circuit equations are significantly simpler than those shown in Section 6.1, when the same circuit was designed by using D-type flip-flops. This example demonstrates the possible large savings in circuit components when JK flip-flops are used in place of D-type flip-flops. It is not certain that fewer circuit components will provide an overall simpler circuit because the JK flip-flop has one more input than the D-type flip-flop, and therefore, twice as many input functions are needed. In many cases, however, the functions are trivial and require no logic gates at all. Since the D-type flip-flop has only one input, more D-type flip-flops can be packed into one IC device than JK flip-flops. For example, the type 74376 IC device has 16 pins and four JK flip-flops, whereas the type 74378 device also has 16 pins but it contains six D-type flip-flops. This may also influence the overall cost of the circuit.

We have completed the design of the counter for the required counting sequence, but since the counter uses only five states, it has three unused states, and the self-starting aspects of the counter must also be investigated. The design equations or the K-maps specify the generated J and K inputs for all states. The state transitions for the unused states are calculated in the following manner:

State	At t_n, $Q_2Q_1Q_0$	Inputs at t_n J_2K_2	J_1K_1	J_0K_0	Flip-flop functions FF$_2$	FF$_1$	FF$_0$	State	At t_{n+1}, $Q_2Q_1Q_0$
0	000	01	00	11	RESET	None	TOGGLE	1	001
5	101	11	11	11	TOGGLE	TOGGLE	TOGGLE	2	010
7	111	11	11	01	TOGGLE	TOGGLE	RESET	0	000

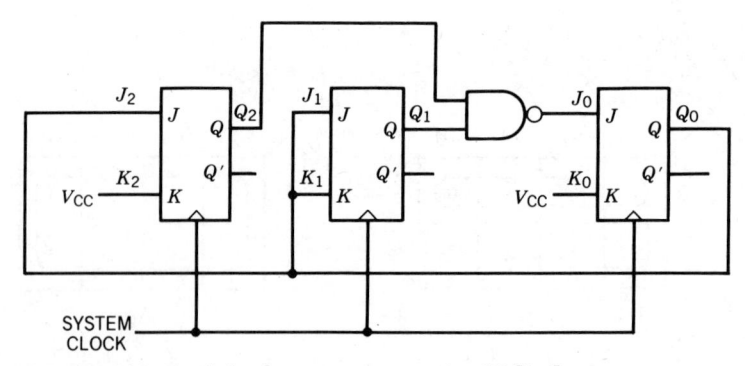

FIGURE 6.20 Circuit for five-state counter using JK flip-flops.

The initial state specifies the initial bit values, which are shown in the first two columns. The six inputs to the three flip-flops (J_2–K_0) can be determined directly from the K-maps, since, the circled don't care terms produce 1's in this case (they were minimized for 1's). The flip-flop inputs specify the flip-flop functions shown in the next three columns. The flip-flop functions and the flip-flop outputs at time t_n determine the stored bit values after the CLOCK transition (at time t_{n+1}). As shown, these design equations produce a self-starting counter. Even though the next state for the unused state 7 is equal to state 0, itself an unused state, the counter is self-starting because the next state for state 0 is equal to state 1, which is one of the states of the required counting sequence. A similar table constructed for the states of the counting sequence may be used to check the design equations. A complete check of the design process may be achieved if the flip-flop input values are determined directly from the logic circuit diagram. In this case the circuits are very simple, and the following table is constructed:

State	At t_n, $Q_2Q_1Q_0$	J_2K_2	J_1K_1	J_0K_0	FF$_2$	FF$_1$	FF$_0$	State	At t_{n+1}, $Q_2Q_1Q_0$
1	001	11	11	11	TOGGLE	TOGGLE	TOGGLE	6	110
2	010	01	00	11	RESET	None	TOGGLE	3	011
3	011	11	11	11	TOGGLE	TOGGLE	TOGGLE	4	100
4	100	01	00	11	RESET	None	TOGGLE	1	001
6	110	01	00	01	RESET	None	RESET	2	010

The state transitions in the table show that the counter follows the required counting sequence (1–6–2–3–4–1).

JK Flip-Flop Design Using Partitioned K-Maps

There is an alternate design procedure for JK flip-flops that may result in simpler circuits than the general sequential circuit design procedure that uses excitation tables. This alternate procedure can be used only for JK flip-flops, but it demonstrates an interesting application of a general principle applicable to all Boolean functions. Since we will apply the method for the design of the controlled five-state counter shown in the previous section, we will discuss this method for three stored bits (Q_2, Q_1, and Q_0) and one input variable (C). The extension of this method to other cases should be reasonably straightforward.

The STT for the controlled five-state counter defines the three four-variable K-maps shown in Figure 6.13. Let us consider the first map, which defines the Boolean function for $Q_2(t_{n+1})$. The Boolean function for $Q_2(t_{n+1})$ may be defined in general as

$$Q_2(t_{n+1}) = F_2(C, Q_2, Q_1, Q_0) \quad \text{(at } t = t_n\text{)}$$

It is possible to partition the Boolean function $F_2(\ )$ into two new functions $F_{21}(\ )$ and $F_{22}(\ )$ that have only three variables. Using the values of bit Q_2 and its complement Q_2', we get

$$F_2(C, Q_2, Q_1, Q_0) = Q_2' \cdot F_{21}(C, Q_1, Q_0) + Q_2 \cdot F_{22}(C, Q_1, Q_0)$$

The partitioning of $F_2(\ )$ is always possible since its Boolean expression can be transformed into a logical sum of logical product terms for which the terms containing the variable Q_2' yield the function $F_{21}(\ )$, and the terms that contain variable Q_2 produce function $F_{22}(\ )$. The functions $F_{21}(\ )$ and $F_{22}(\ )$ do not contain the variable Q_2; hence, they are functions of only three variables in this case.

Now, let us recall the transition equation of the JK flip-flop given for the stored bit Q_2 as

$$Q_2(t_{n+1}) = J_2 \cdot Q_2'(t_n) + K_2' \cdot Q_2(t_n)$$

Comparing the JK flip-flop transition equation with the partioned function $F_2(\ )$, we can see that it is possible to design sequential circuits with JK flip-flops when the Boolean functions for the J and K inputs of a flip-flop are not functions of the output of the same flip-flop. If we partition the $F_2(\ )$ function into functions $F_{21}(\ )$ and $F_{22}(\ )$, we find

$$J_2 = F_{21} \quad \text{and} \quad K_2 = F_{22}'$$

Partitioning of function $F_2(\ )$ may be done algebraically, but it could be performed much more easily from the K-map of the function. We find that if we set the bit Q_2 in the preceding partitioned equation equal to 0, then the function $F_2(\ )$ becomes equal to $F_{21}(\ )$, whereas setting Q_2 to 1 turns $F_2(\ )$ into $F_{22}(\ )$:

$$F_{21}(\ ) = F_2(Q_2 = 0) \quad \text{and} \quad F_{22}(\ ) = F_2(Q_2 = 1)$$

Consequently, we can partition the four-variable K-map for $Q_2(t_{n+1})$ into two three-variable K-maps, one including all those regions for which Q_2 is equal to 0 (this is the map for function J_2) and the other for which Q_2 is equal to 1 (this K-map defines the Boolean function for K_2'). Figure 6.21 shows the original K-map for $Q_2(t_{n+1})$ and the two partitioned maps. The map for J_2 is constructed from the first and last rows of the original map (since for these two rows Q_2 is equal to 0), whereas the map for K_2' is given by the second and third rows (for which $Q_2 = 1$ in the four-variable map). The design equations for J_2 and K_2 are also shown in Figure 6.21.

A similar analysis for bits Q_1 and Q_0 yields four more three-variable K-maps. As shown in Figure 6.22, the partitioned maps for J_1 and K_1' are constructed by using the first two or the last two columns of the original K-maps, respectively. The minimized design equations are also shown in Figure 6.22. Figure 6.23 shows the results for J_0 and K_0. The three-variable

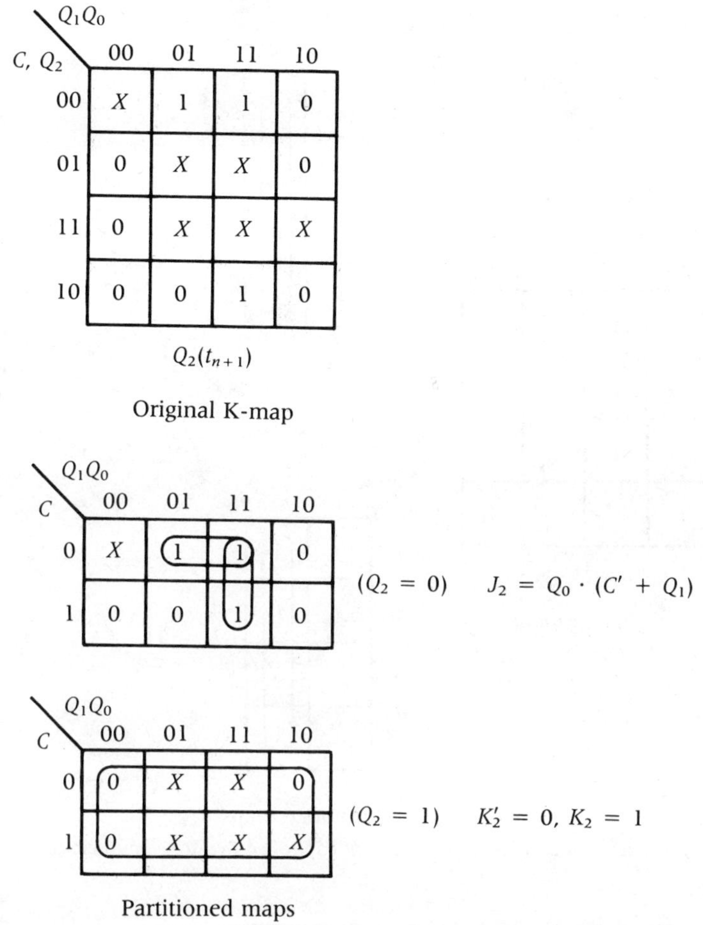

6.3 COUNTER DESIGN USING GENERAL FLIP-FLOP TYPES

Original K-map (for $Q_2(t_{n+1})$ with rows C, Q_2 and columns $Q_1 Q_0$):

$C,Q_2 \backslash Q_1 Q_0$	00	01	11	10
00	X	1	1	0
01	0	X	X	0
11	0	X	X	X
10	0	0	1	0

Partitioned maps:

($Q_2 = 0$) $J_2 = Q_0 \cdot (C' + Q_1)$

($Q_2 = 1$) $K_2' = 0,\ K_2 = 1$

FIGURE 6.21 Partitioned K-maps for output bit Q_2 of controlled five-state counter.

K-map for J_0 is constructed from the first and last columns of the four-variable map (these are the columns for which $Q_0 = 0$), whereas the second and third columns yield the K-map for K_0'. As shown in Figures 6.20–6.22, the design equations are relatively simple since only two functions (J_2 and J_0) require additional circuitry. They are summarized as follows:

$$J_2 = Q_0 \cdot (C' + Q_1) \qquad J_1 = Q_0 \qquad J_0 = Q_2' + C' \cdot Q_1'$$
$$K_2 = 1 \qquad K_1 = Q_0 \qquad K_0 = 1$$

Compared to the design equations for the D-type flip-flop at the end of Section 6.2, the equations for the JK flip-flop are much more simple. We still have to examine whether the counter is self-starting. The output transition table

186 DESIGN OF COUNTERS

Original map

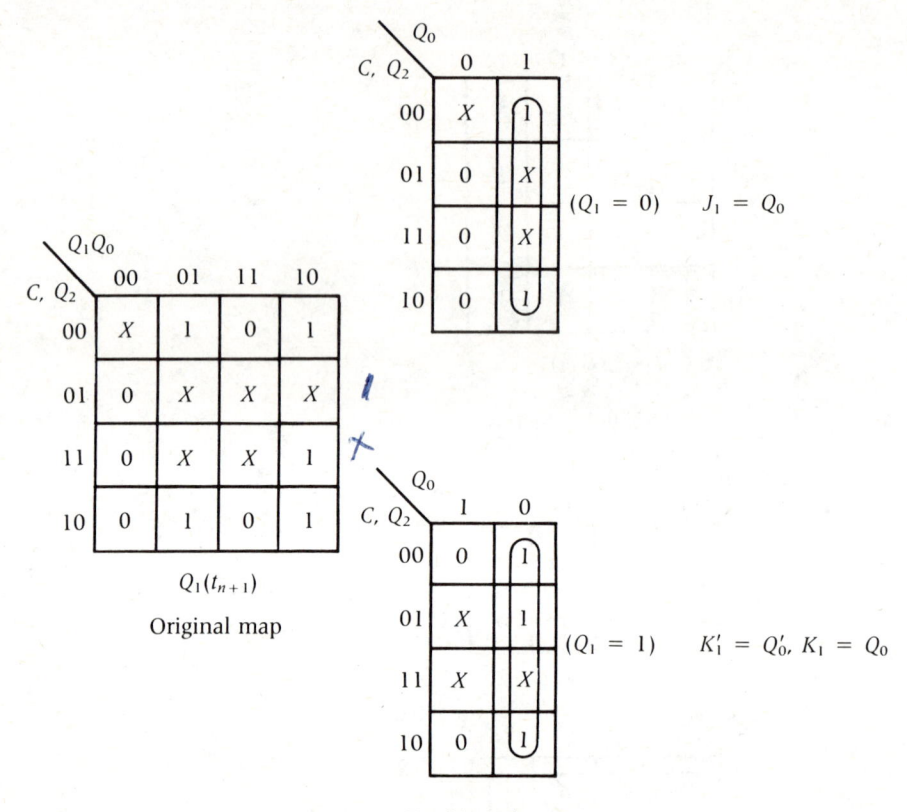

$(Q_1 = 0)$ $J_1 = Q_0$

$(Q_1 = 1)$ $K_1' = Q_0', K_1 = Q_0$

Partitioned maps

FIGURE 6.22 Partitioned K-maps for bit Q_1 of controlled five-state counter.

for these design equations and for the unused states follows:

	At t_n		Inputs at t_n			Flip-Flop Functions			At t_{n+1},	
C	$Q_2Q_1Q_0$	J_2K_2	J_1K_1	J_0K_0	FF_2	FF_1	FF_0	$Q_2Q_1Q_0$	Transition
0	000	01	00	11	RESET	None	TOGGLE	001	0 → 1
0	101	11	11	11	TOGGLE	TOGGLE	TOGGLE	010	5 → 2
0	111	11	11	01	TOGGLE	TOGGLE	RESET	000	7 → 0
1	101	01	11	01	RESET	TOGGLE	RESET	010	5 → 2
1	110	01	00	01	RESET	None	RESET	010	6 → 2
1	111	11	11	01	TOGGLE	TOGGLE	RESET	000	7 → 0

In this case the counter is self-starting for both 0 and 1 values of the control input C.

This concludes the demonstration of the partitioned K-map method. To complete the use of general flip-flops for counter design, design with the T flip-flop will be shown in the next section.

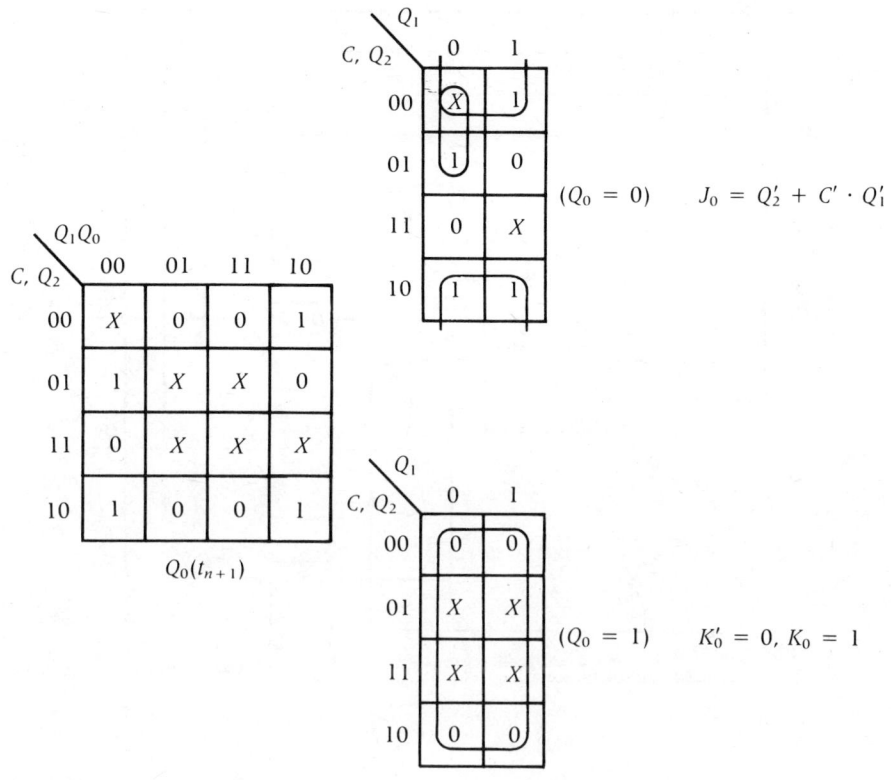

6.3 COUNTER DESIGN USING GENERAL FLIP-FLOP TYPES 187

FIGURE 6.23 Partitioned K-maps for bit Q_0 of controlled five-state counter.

Down Counter for Excess-3 Decimal Digits Using T Flip-Flops

In Chapter 4 the Excess-3 representation of decimal digits was introduced. Four bits are required to represent a decimal digit, and in Section 4.8 we designed a decimal decrementer circuit that had four inputs (D_0–D_3) and four outputs (R_0–R_3). This combinational circuit produces as output the Excess-3 representation of a decimal digit that is one smaller than the same representation of the decimal digit entered as input to the circuit. A down counter for the same representation of decimal digits sequences through the digits in reverse order, that is, the counting sequence is given by

$$0 \to 9 \to 8 \to 7 \to 6 \to 5 \to 4 \to 3 \to 2 \to 1 \to 0$$

Four D-type flip-flops and the digital decrementer circuit (shown in Figure 4.29) may be used to construct the down counter, as shown in Figure 6.24, where the inputs to the decrementer block has been renamed. They are shown as I_3–I_0 because of the possible confusion between the inputs for the decrementer block and the D inputs for the flip-flops. Since the outputs of the decrementer circuit generate the digit with one smaller value than the digit

188 DESIGN OF COUNTERS

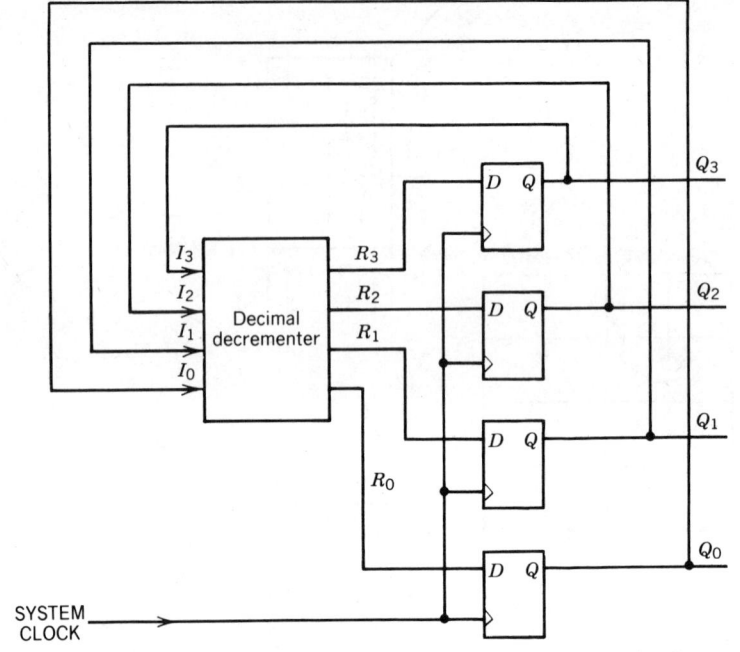

FIGURE 6.24 Circuit of decimal decrementer circuit using D-type flip-flops.

represented by the inputs, the D-type flip-flops produce the correct counting sequence for the down counter as long as the counter is started from one of its used states.

As a demonstration, the same circuit will be designed using T (toggle) flip-flops. The T flip-flops were introduced in Chapter 5. The T flip-flop has one input (T). When T is equal to 0, the state of the flip-flop remains constant (no action). Setting T equal to 1 produces a toggling circuit [$Q(t_{n+1}) = Q'(t_n)$], and we get the transition equation

$$Q(t_{n+1}) = T \cdot Q'(t_n) + T' \cdot Q(t_n)$$

The block symbol and the excitation table for the T flip-flop are shown in Figure 6.25. As we should expect for a flip-flop with only one input, there are no don't care terms. The required input bit is 1 when the stored bit toggles and 0 when it remains unchanged. This excitation table is applied to the output transition table for the decimal down counter, and the results are shown in Table 6.3. There are six unused states that produce don't care entries. Notice that when digit 0 (binary 0011 in Excess-3 representation) is decremented, digit 9 (binary 1100) is produced. This is consistent with our earlier definition of the decrementer circuit in Section 4.8. From the observed bit transitions the required flip-flop inputs are determined for each used state of the counter. In this case it is easy to determine the required input values directly from the characteristics of the T flip-flop. Whenever the stored bit

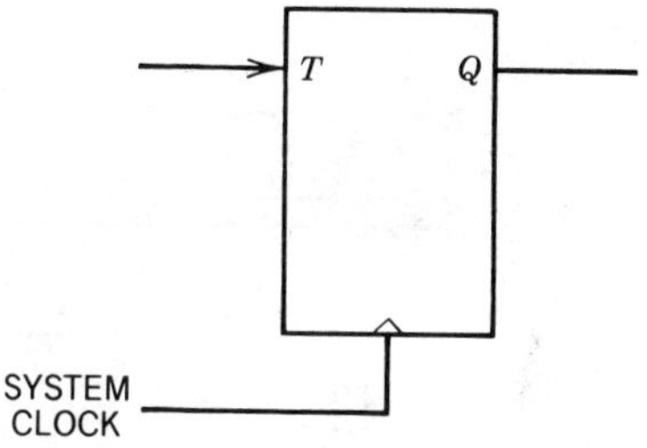

6.3 COUNTER DESIGN USING GENERAL FLIP-FLOP TYPES 189

Transition $Q(t_n) \rightarrow Q(t_{n+1})$	Required T
$0 \rightarrow 0$	0
$0 \rightarrow 1$	1
$1 \rightarrow 0$	1
$1 \rightarrow 1$	0

FIGURE 6.25 Excitation table for T flip-flop.

(Q) remains the same, a $T = 0$ input is required. If the bit value changes (toggles), then T must be set equal to 1.

The four four-variable K-maps are constructed from the required T inputs (the last column in Table 6.3), and these are shown in Figure 6.26. The indicated groups of 1's produce the following design equations:

$$T_3 = Q_2' \cdot Q_1' \cdot Q_0' + Q_3' \cdot Q_2' = Q_2' \cdot (Q_3' + Q_1' \cdot Q_0')$$

$$T_2 = Q_1' \cdot Q_0' + Q_3' \cdot Q_2'$$

$$T_1 = Q_0' + Q_3' \cdot Q_2'$$

$$T_0 = 1$$

Similar to the decimal decrementer in Chapter 4, the counter for the least significant bit also requires a simple toggling circuit. In this case, when T flip-flops are used, a toggling circuit needs the constant 1 value for T_0 and requires

TABLE 6.3 STT for Decimal Decrementer Using T Flip-Flops

Digit, Excess-3	$Q(t_n)$, $Q_3Q_2Q_1Q_0$	$Q(t_{n+1})$, $Q_3Q_2Q_1Q_0$	Required inputs, $T_3T_2T_1T_0$
X	0000	Not used	XXXX
X	0001	Not used	XXXX
X	0010	Not used	XXXX
0	0011	1100	1111
1	0100	0011	0111
2	0101	0100	0001
3	0110	0101	0011
4	0111	0110	0001
5	1000	0111	1111
6	1001	1000	0001
7	1010	1001	0011
8	1011	1010	0001
9	1100	1011	0111
X	1101	Not used	XXXX
X	1110	Not used	XXXX
X	1111	Not used	XXXX

190 DESIGN OF COUNTERS

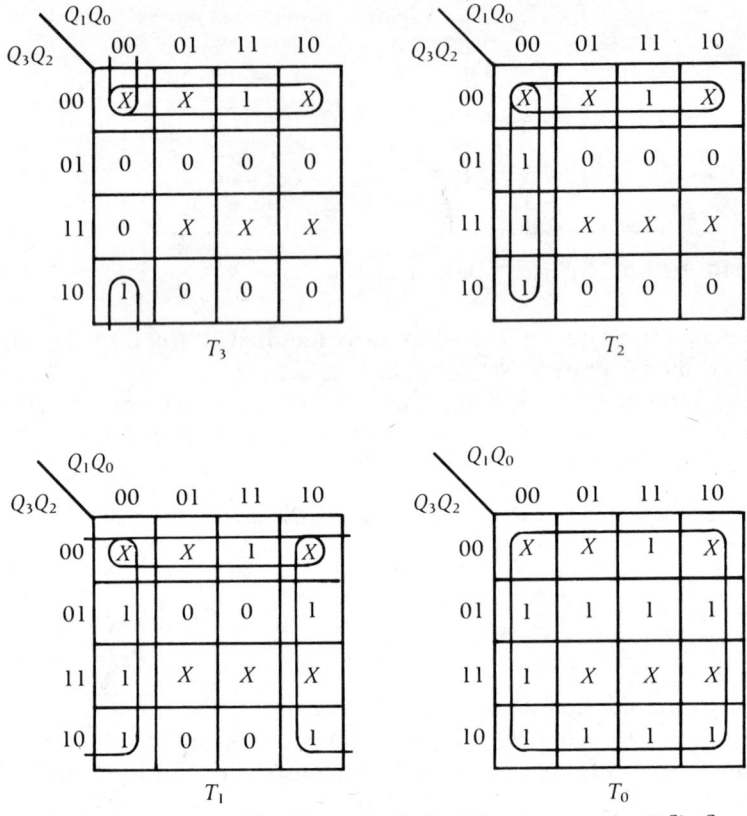

FIGURE 6.26 State transition K-map for decimal decrementer using T flip-flops.

no additional logic gates. By comparing these results with the design equations for the decimal decrementer in Section 4.8, we can see that the T flip-flops yield considerably simplified design equations. The terms of $Q_1' \cdot Q_0'$ and $Q_3' \cdot Q_2'$ are shared between three expressions that provide further savings. This example demonstrates the possible savings in circuitry when other than D-type flip-flops are utilized.

The analysis is not complete without examination of the unused states that should provide a self-starting operation for the counter. From the encircled groups of 1's in the K-maps (or the design equations) the following transitions are indicated:

State $Q_3Q_2Q_1Q_0$	Inputs $T_3T_2T_1T_0$	FF_3	Flip-Flop Actions FF_2	FF_1	FF_0	Next Bits $Q_3Q_2Q_1Q_0$	Digit
0000	1111	TOGGLE	TOGGLE	TOGGLE	TOGGLE	1111	X
0001	1111	TOGGLE	TOGGLE	TOGGLE	TOGGLE	1110	X
0010	1111	TOGGLE	TOGGLE	TOGGLE	TOGGLE	1101	X
1101	0001	None	None	None	TOGGLE	1100	9
1110	0011	None	None	TOGGLE	TOGGLE	1101	X
1111	0001	None	None	None	TOGGLE	1110	X

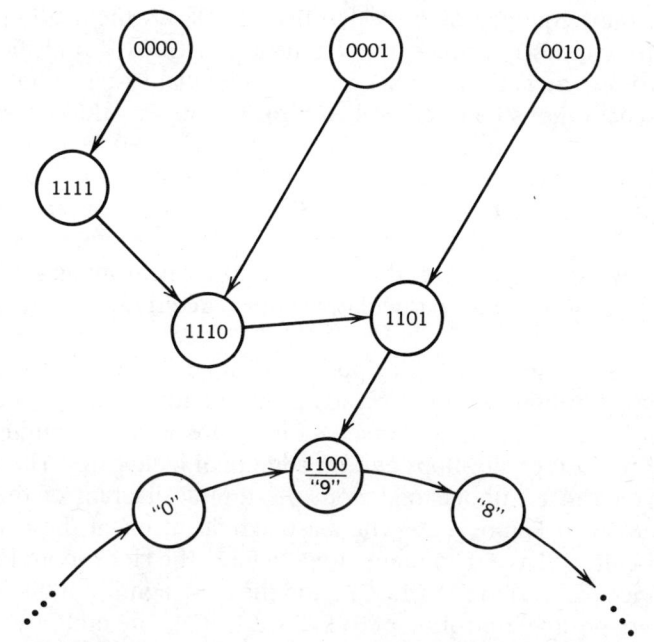

FIGURE 6.27 State transitions for unused states of decimal decrementer circuit.

Since most of the next states for the unused states are themselves unused states, it is easier to interpret the results by drawing a STD for the unused state. This is shown in Figure 6.27. The STD shows clearly that no matter which unused state is the starting state of the counter, ultimately, the counter will be in binary state 1100, which is the correct representation for the decimal digit 9. The STD also shows that in the worst case, when the counter starts from binary state 0000, four significant state transitions are required before the counter leaves its unused states.

6.4 ASYNCHRONOUS (RIPPLE) COUNTERS

For all our circuits so far the SYSTEM CLOCK input signal was connected to the CLOCK input of all the flip-flops in the system. This arrangement provides synchronous sequential systems and, as long as all the practical flip-flop devices used in the circuits have similar internal gate delay times, the output signal transitions occur almost simultaneously. Of course, as we discussed in Chapter 5, it is not practical to assume that two signals change exactly at the same time. In asynchronous systems the order in which signals change may influence the operation of the circuits. The practical problems associated with the respective order of signal changes are called **hazards.** In synchronous sequential systems the effect of hazards is minimized as long as signals are "sensed" only within small time periods around significant CLOCK transitions, and the frequency of the clock periods allows the settling of signal values to their quiescent states by the time a new CLOCK transition occurs.

The safest way is to use synchronous sequential systems, but there is a class of asynchronous counters, called **ripple** counters, that provides large savings in circuit component costs and, therefore, is used frequently in practice. Ripple counters are asynchronous systems because the CLOCK inputs of flip-flops are connected to the outputs of logic circuits or outputs of other flip-flops. The significant clock transitions for different components of such counters may be delayed and, therefore, occur asynchronously. The binary up and down counters use particularly simple circuitry and are discussed in what follows.

The circuit of a binary 3-bit ripple up counter is shown in Figure 6.28. Three D-type flip-flops are used, which provide both Q and Q' (the complement of the stored bit) outputs. As shown in Figure 6.28, the counter requires only the three D-type flip-flops and no additional logic gates. The operation of the circuit can be understood from the timing diagram of the flip-flop outputs shown in Figure 6.29. The least significant bit of the counter is a toggling circuit we have seen many times before. The clock input to this flip-flop is connected to SYSTEM CLOCK, and the least significant bit changes its state at every positive transition of SYSTEM CLOCK. The other two flip-flops are also wired as toggling circuits (their inverted outputs Q' are connected directly to the D inputs). This means that each flip-flop will toggle its output at each significant transition of its CLOCK signal input. The CLOCK inputs for two flip-flops are derived from flip-flop outputs, which makes the circuit asynchronous. The inverted signal Q_0' is used as the CLOCK input of flip-flop FF_1; therefore, the significant CLOCK transition for FF_1 is the 1-to-0 transition of the Q_0 signal. The timing diagram clearly shows that for each 1-to-0 transition of the Q_0 signal, flip-flop FF_1 changes its state. Since the CLOCK signal for the FF_2 flip-flop is equal to Q_1', the same relationship holds between outputs Q_1 and Q_2. At each 0-to-1 transition of the Q_1 signal the output of flip-flop FF_2 (Q_2) changes. As shown in Figure 6.29, the effect of this simple relationship between the inverted output of one flip-flop and the CLOCK input of another produces the required binary counting sequence.

The systematic design of ripple counters, which produce arbitrary counting sequences, is based on the design of asynchronous circuits. The techniques for asynchronous circuit design are quite complex and are clearly beyond the

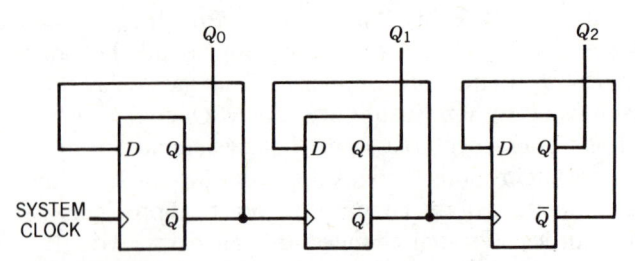

FIGURE 6.28 Circuit diagram for ripple binary up counter.

6.4 ASYNCHRONOUS (RIPPLE) COUNTERS   193

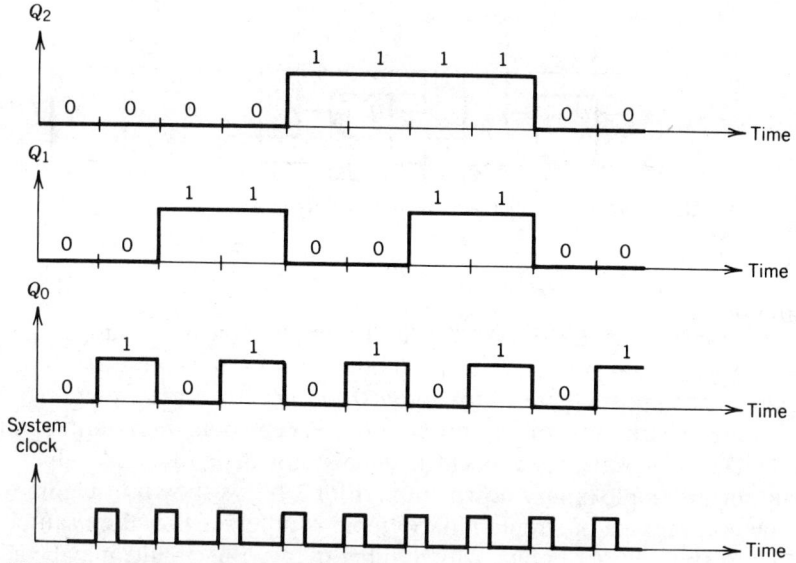

FIGURE 6.29 Timing diagrams for ripple up counter.

scope of this discussion. But as an example of another simple ripple counter, we will examine the timing diagram of the binary down counter. The counting sequence starts with binary state 111 (7), and after seven CLOCK transitions it counts down to state 000. At the next significant transition the counter changes its state to state 111. The timing diagram is shown in Figure 6.30. As in the case of the up counter, the least significant bit (Q_0) toggles at each

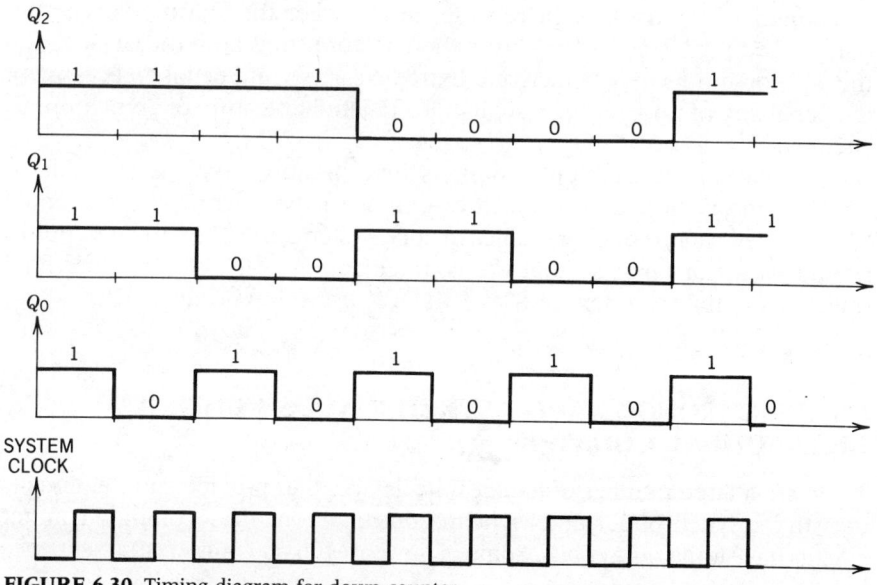

FIGURE 6.30 Timing diagram for down counter.

194 DESIGN OF COUNTERS

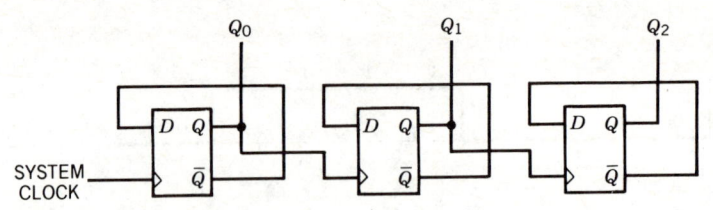

FIGURE 6.31 Circuit diagram for ripple binary down counter.

positive [negative] transition of SYSTEM CLOCK. The middle bit (Q_1) changes its state at each positive transition of bit Q_0, and the most significant bit (Q_2) changes its state at each positive transition of bit Q_1. Hence, the binary down counter has a similar structure to the up counter, except that the counter output signals Q_0 and Q_1 are used as clock signals, not their inverted values. The circuit for the ripple binary down counter for 3 bits is shown in Figure 6.31.

One advantage of simple binary ripple counters is that they can be extended to any number of bits. Any number of toggling circuits may be added to the counter. The CLOCK input for the toggling circuit with counter bit Q_i is equal to Q'_{i-1} for the up counter and Q_{i-1} for the down counter. In this way a ripple counter may be built for any number of bits. It would be considerably harder to extend a synchronous counter for, say, 8 bits.

It is difficult to design ripple counters that produce arbitrary counting sequences. In general, T or JK flip-flops can be also used. We have shown only the simplest form, when all flip-flops are changed to simple toggling circuits. In the general case both the CLOCK and the flip-flop inputs are driven by logic circuits. For these circuits the design starts with a timing diagram that demonstrates the relationships between the transition of counter outputs and their logic values. From these relationships both the CLOCK and the flip-flop inputs may be derived. Extreme care should be taken because of the possibility of uncontrolled oscillations (a familiar feature of asynchronous circuits).

We have included ripple counters here because they are available in practice, and if they are used with care in a strictly synchronous sequential system, they can be made reliable. In such systems SYSTEM CLOCK pulses should occur far apart so that the counter outputs can settle to their final states by the time the next SYSTEM CLOCK transition occurs.

6.5 SYNCHRONOUS AND ASYNCHRONOUS CONTROL OF PRACTICAL COUNTERS

There are a large number of practical IC devices that provide counter circuits of both the synchronous and asynchronous kind. The most common counter circuits have four counter bits. Some of these are binary counters that sequence

6.5 SYNCHRONOUS AND ASYNCHRONOUS CONTROL OF PRACTICAL COUNTERS

through all possible 16 states, whereas others are decimal counters, which include only 10 states (state 0 to state 9) in their counting sequence. Some counters include an additional input connection that determines whether the counting sequence is forward (up) or backward (down). These are called up-down counters.

In addition to these features, which we have already discussed, some additional inputs and/or outputs can increase the usefulness of counters quite considerably. One of the extended uses of 4-bit counters is to use them for the construction of counters that contain more than 4 bits. In this case more than one "stage" of 4 bits must be used. Many practical counter circuits provide an **ENABLE** input and a **RIPPLE CARRY** output, which allow the construction of counters with an arbitrary number of bits. The ENABLE input of the counter must be **active** in order for the counter to operate. The RIPPLE CARRY output of the counter becomes true when the counter has reached the last state of its normal sequence. In Figure 6.32 an 8-bit counter is shown. The 8 bits of the counter are numbered from B_0 (LSB) to B_7 (MSB) and are constructed from two 4-bit counters. The four lower significant bits (B_0–B_3) belong to the first stage, the 4-bit counter that is always enabled. The higher order 4 bits (B_4–B_7) are outputs of the second stage, another 4-bit counter that is enabled by the RIPPLE CARRY output of the lower stage. If the counter is a normal binary up counter, then the RIPPLE CARRY becomes 1 when the state of the 4-bit counter is equal to 1111 (for a decimal up counter the last state is equal to state 1001). The second stage of the counter starts with its first state [e.g., state 0 (0000)], and it remains in this state for the first 16 clock transitions because its ENABLE input is equal to Logic-0. After the sixteenth state, the first stage outputs are equal to 1111, the second stage outputs are still equal to 0000, and the RIPPLE CARRY output from the first stage becomes 1. After the next CLOCK transition the second stage outputs change to 0001 (since the counter is now enabled), whereas the first stage of the counter drops back to its first state (0000), and the correct binary number for 8 bits (00010000) is produced. If the counter operates as a down

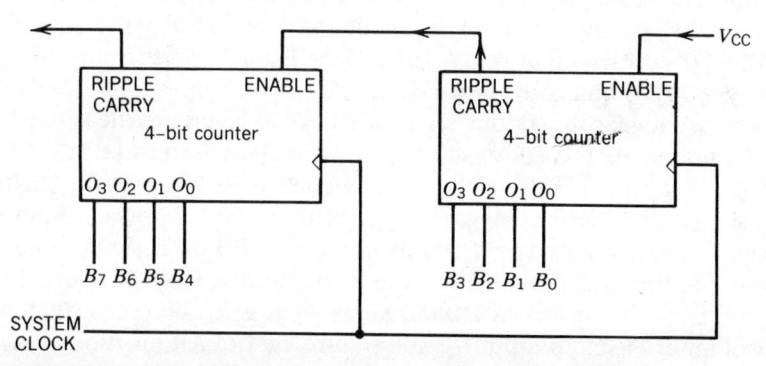

FIGURE 6.32 Construction of 8-bit counter from two 4-bit counters.

196 DESIGN OF COUNTERS

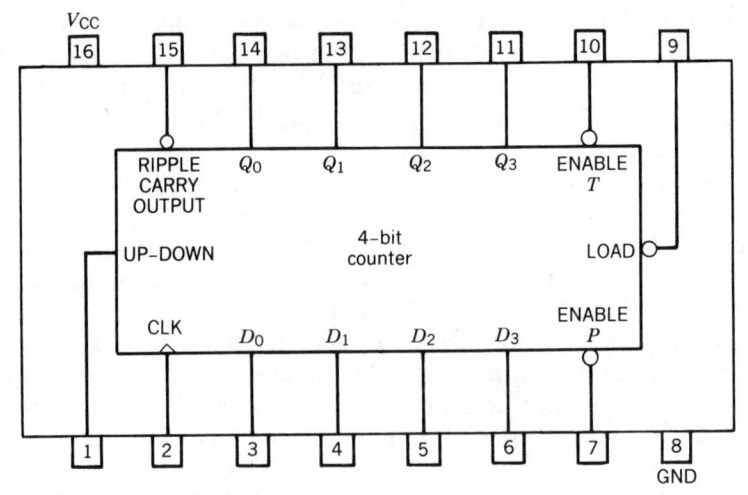

FIGURE 6.33 Practical 4-bit up-down counter with LOAD and ENABLE control inputs.

counter, then the outputs of the "last" state of the counter are equal to 0000, and this is the state for which the RIPPLE CARRY output should become active.

Another type of input and/or output controls the clearing or presetting of the counter output bits to an arbitrary combination of 1's and 0's. The simplest additional input is a **CLEAR** input that, when active, resets the output bits of the counter to 0 values. Some counters are built with asynchronous CLEAR control, which means that the counter bits are reset immediately when the CLEAR signal becomes active. Others are provided with synchronous CLEAR function, which means that if the CLEAR input is active, then the counter bits are reset to 0 values only at the next significant CLOCK transition. In a synchronous system synchronous control functions provide reliable circuits.

The **LOAD** function is provided by another control input called LOAD and four additional inputs called **DATA INPUTS.** When the LOAD control input is active, the output bits of the counter become equal to the combination of 1's and 0's presented at DATA INPUTS. In Figure 6.33 the pin assignment of the type 74169 binary up-down counter is shown. Some inputs and outputs use negative-logic conventions for which Logic-0 levels are the active levels. In order to operate this counter, both pin 7 and pin 10 must be at the Logic-0 (ground) level since this device is equipped with two separate ENABLE input signals and both use negative logic. The counter counts up if pin 1 has a Logic-1 value; otherwise, it counts down. If the LOAD signal becomes a Logic-0 (its active state), then at the next positive CLOCK transition the outputs are set to the DATA INPUT values, that is, Q_0 becomes equal to the value of input D_0, Q_1 to input D_1, and so on. The CLEAR function allows the initialization of the counter to all 0's, whereas the LOAD function is used if initialization to an arbitrary state is required.

EXPERIMENT 6
Design of Five-State Up–Down Counter

E6.1 Summary and Objectives

The main objectives of this experiment are the design, construction, and testing of a synchronous five-state up-down counter. Each experimenter is given a unique counting sequence. Two designs must be completed, one containing only D-type flip-flops, the other only JK type flip-flops. The design that produces the simplest circuits is to be wired and tested.

Since only five of the possible eight states are used, the counter has three unused states. For both the up and down cases the counter must drop into the proper counting sequence even if it starts from any one of the unused states. A RESET input should be also provided. The RESET input, when pulsed, places the counter into the 00 (unused) state.

E6.2 Description of Design Problem

A five-state general up-down counter is to be designed. The counter uses three flip-flops with outputs Q_0, Q_1, and Q_2. The block diagram of the counter is shown in Figure E6.1. The CLOCK input signal uses positive logic; hence, the state transitions occur at the positive transition of the CLOCK signal. The up counting sequence is specified in a table of possible counting sequences (Table E6.1). When the value of the control input C is equal to Logic-1, the counter follows the specified counting sequence from left to right (this is called the up sequence). When the value of C is equal to Logic-0, the counter follows the specified counting sequence from right to left (the down sequence).

In addition to the control and the clock inputs, the counter also includes a RESET input. When the RESET input is momentarily in the Logic-1 state, the outputs of all three flip-flops are set to Logic-0 irrespective of the CLOCK input (asynchronous CLEAR function). For normal counting operation the RESET input must be in the Logic-0 state.

If the counter starts from any one of its three unused states, then it must drop into its proper counting sequence for both the up and down cases. In fact, the zero state (outputs 000) is always one of the unused states. For this experiment it is immaterial how the counter reaches its normal counting sequence when it starts from any one of its unused states.

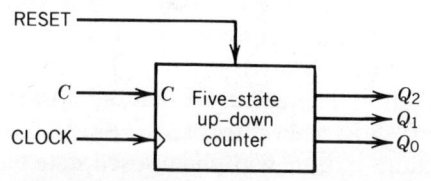

FIGURE E6.1 Block diagram of controlled five-state counter.

198 DESIGN OF COUNTERS

TABLE E6.1 Counting Sequences

10s	\multicolumn{10}{c}{Sequence number}
	0	1	2	3	4	5	6	7	8	9
		13456	13645	13654	14365	14356	14635	14653	16345	16354
10	13467	13476	13647	13674	14367	14376	14637	14673	16347	16374
20	23467	23476	23647	23674	24367	24376	24637	24673	26347	26374
30	12467	12476	12647	12674	14267	14276	14627	14672	16247	16274
40	13267	13276	13627	13672	12367	12376	12637	12673	16327	16372
50	13427	13472	13247	13274	14327	14372	14237	14273	12347	12374
60	13462	13426	13642	13624	14362	14326	14632	14623	16342	16324
70	53467	53476	53647	53674	54367	54376	54637	54673	56347	56374
80	52467	52476	52647	52674	54267	54276	54627	54672	56247	56274
90	53267	53276	53627	53672	52367	52376	52637	52673	56327	56372
100	53427	53472	53247	53274	54327	54372	54237	54273	52347	52374
110	53462	53426	53642	53624	54362	54326	54632	54623	56342	56324
120	15467	15476	15647	15674	14567	14576	14657	14675	16547	16574
130	15267	15276	15627	15672	12567	12576	12657	12675	16527	16572
140	15427	15472	15247	15274	14527	14572	14257	14275	12547	12574
150	15462	15426	15642	15624	14562	14526	14652	14625	16542	16524
160	13567	13576	13657	13675	15367	15376	15637	15673	16357	16375
170	13527	13572	13257	13275	15327	15372	15237	15273	12357	12375
180	13562	13526	13652	13625	15362	15326	15632	15623	16352	16325
190	13457	13475	13547	13574	14357	14375	14537	14573	15347	15374
200	13452	13425	13542	13524	14352	14325	14532	14523	15342	15324

E6.3 Selection of Counting Sequence

Each experimenter is given a sequence number in the range 1–209. Each number corresponds to a unique counting sequence in Table E6.1. The up counting sequences are specified in the table by the counter states from left to right. The down sequences follow the same states from right to left. For example, the first entry in the table is 13456. The two counting sequences for this entry are

Up sequence: $1 \to 3 \to 4 \to 5 \to 6 \to 1$.
Down sequence: $1 \to 6 \to 5 \to 4 \to 3 \to 1$.

E6.4 Expected Results and Testing

The expected results must be presented both in a STD and in two timing diagrams. A separate timing diagram is required for the up and down counting sequences. The STD must be shown for all eight possible states of the wired circuit. This diagram should demonstrate that the counter is self-starting. The timing diagrams must show the values of SYSTEM CLOCK and the three flip-flop outputs as functions of time with the unused state 000 as the initial state. The timing diagrams should be drawn for at least eight clock periods.

Testing can be down with ordinary switches for both RESET and C inputs, but a "debounced" switch must be used for the CLOCK input. At first, the state of the counter should be reset (set to state 000), C set to 1, and the switch connected to the CLOCK should be operated to prove that the counter follows the drawn STD for the up sequence. The same procedure should be repeated then for $C = 0$, which should provide the correct down sequence. Depending on the flip-flops used, it may or may not be possible to set the counter into an arbitrary unused state. If the flip-flops in the circuit include PRESET inputs, then it is possible to set an individual flip-flop output to Logic-1, and after the RESET function the counter may be set up for any unused state. If this is possible, then transitions from all those states that were not included in the original testing of the up and down sequences should be individually checked.

E6.5 Design Requirements

The STTs and the transition K-maps should be constructed for both D and JK flip-flops. It is up to the experimenter whether the flip-flop excitation method or the partition method is used for the design with JK flip-flops. If time permits, both methods should be tried to see whether different results may be obtained. Designs for both D and JK flip-flops must be completed, but only the one with the simpler circuit has to be wired. After selection of don't care terms, the final K-maps should be drawn and the unused states checked. If the counter does not drop automatically into the proper counting sequence when it is started from one of the unused states, then a different combination of don't care values must be chosen.

The minimum requirements for this experiment are the transition tables for the up and down counting sequences, the initial transition K-maps, the maps minimized for both 1's and 0's, and the final K-maps that show the actual values for the don't care terms. Two minimized circuits for the D-type flip-flops should be the result of the K-map minimization procedures for 1's and 0's. This entire design process should be repeated for JK flip-flops, which provides two additional minimized circuits. Out of the four circuits the simplest one should be built and the circuit tested.

E6.6 Alternate Experiments

Investigate the amount of additional circuitry that would be required to change the specifications of this counter so that it goes through all its unused states (in any order) when it starts from state 000 and before it drops into its normal counting sequence.

Design the counter using T flip-flops. Compare the minimized circuits found with those for D-type flip-flops, noting that a D-type flip-flop may be easily changed to a T flip-flop when an additional INVERTER gate and a two-input AND gate is used.

Drill Problems

6.1. There are six unique counting sequences for two flip-flops and four states (0123, 0132, 0213, 0231, 0312, and 0321). Assuming that only one type of flip-flop is used for the counter, there are three possibilities (D-type, JK, and T flip-flops). The possible combinations yield 18 different design problems. Try as many as you can.

6.2. It is also possible to mix flip-flop types. For example, it is possible to use one D-type and one JK flip-flop for a four-state counter. Mixed flip-flop types provide three more variations (one D-type and one JK, one D-type and one T, and finally, one JK and one T). These three new types provide a further 18 possible counting circuits for a four-state counter.

6.3. There are six unique counting sequences for two flip-flops and three states (012, 013, 021, 023, 031, and 032). Use either mixed or matched flip-flop types to design the counter for any one of these sequences and for the added requirement that
 (a) the counter should be self-starting or
 (b) the counter should be stuck when it is placed in its unused state.

6.4. Using three flip-flops of any type, design an up-down counter with one control input C and the following counting sequences:

$C = 1$, up, $1 \rightarrow 3 \rightarrow 5 \rightarrow 7 \rightarrow 1$.
$C = 0$, down, $0 \rightarrow 2 \rightarrow 4 \rightarrow 6 \rightarrow 0$.

Check whether the counter is self-starting and show that the same counter may be built using only two flip-flops.

6.5. Build a 2-bit counter with four possible counting sequences. The counter has two control inputs C_1 and C_2 and the following counting sequences:

$C_1 = 0, C_2 = 0; 0 \rightarrow 1 \rightarrow 2 \rightarrow 3 \rightarrow 0$.
$C_1 = 0, C_2 = 1; 0 \rightarrow 3 \rightarrow 2 \rightarrow 1 \rightarrow 0$.
$C_1 = 1, C_2 = 0; 1 \rightarrow 2 \rightarrow 3 \rightarrow 1$.
$C_1 = 1, C_2 = 1; 3 \rightarrow 2 \rightarrow 1 \rightarrow 3$.

Check whether the counter is self-starting for the last two sequences.

6.6. Design a 4-bit binary ripple up–down counter. One control input (C) determines whether the counting sequence is up ($C = 1$) or down ($C = 0$).

CHAPTER 7

Synchronous Controllers and State Assignments

7.1 INTRODUCTION

We have learned to design counters using several types of basic flip-flops. Counters belong to one class of general sequential systems. The main common feature of counters is that the counting sequence uniquely specifies the logical output levels of the flip-flops, which are used to store the states of the system. For example, if we specify that the "next" state after state 3 for a 3-bit counter is state 6, then it is understood that the counter is in state 3 when the stored bits of its three flip-flops are equal to 011 and the next state of the counter (state 6) is expressed by the three flip-flop output values of 110. This direct correspondence between stored states and flip-flop output values is not required for the specifications of other types of sequential systems. As stated in Chapter 5, any synchronous sequential system may be defined by a state transition diagram (STD) that shows both the system output values and the next states for every stored state and all combinations of input values. The stored states must be identified, but they can be identified by labels without reference to the binary values of flip-flop outputs.

Let us demonstrate this labeling of stored states by a very simple example, the JK flip-flop. The JK flip-flop has two inputs, one output, and two states. Let us label these two states with the letters A and B. The STT and the STD are shown in Figure 7.1. In place of the SET (10) and RESET (01) functions used earlier for the JK flip-flop (when the output of the flip-flop was set to 1 or 0, respectively), these more general specifications are given in terms of two states named State-A and State-B. Inputs 10 provide the "Set to State-B," whereas inputs 01 perform the "Set to State-A" functions. The other two functions ("do nothing" and TOGGLE) are the same as the ones used earlier. Ultimately, when the JK flip-flop is in a circuit, binary output values must be assigned to both state A and state B. Since there is only one flip-flop that has two states, only two choices are possible. State A may be associated with the output value of Logic-1 (in which case, state B is equivalent to output value of Logic-0), or with the output value of Logic-0 (when for state B the output is equal to Logic-1). The selection of flip-flop output values for a system state is called **state assignment**.

When there is a large number of flip-flops and, consequently, a large number of stored states, the number of different state assignments becomes

202 SYNCHRONOUS CONTROLLERS AND STATE ASSIGNMENTS

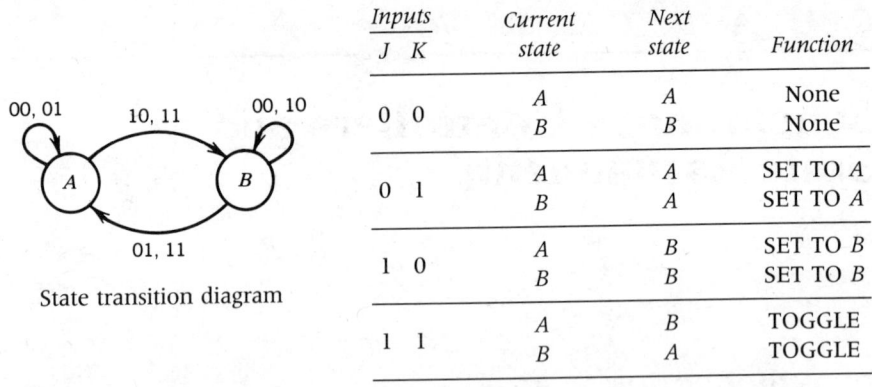

Inputs J K	Current state	Next state	Function
0 0	A	A	None
	B	B	None
0 1	A	A	SET TO A
	B	A	SET TO A
1 0	A	B	SET TO B
	B	B	SET TO B
1 1	A	B	TOGGLE
	B	A	TOGGLE

FIGURE 7.1 The STD and STT of JK flip-flop in terms of two general states A and B.

unwieldingly large. A sequential system having four flip-flops and nine states has around 10^{10} different state assignments. Even though not all of these different state assignments produce different circuits (we shall demonstrate this fact in Section 7.3), the number of different assignments for four flip-flops and nine states is still larger than 10^7, [1] which is still too large for an exhaustive search. The complexity of the final circuit depends very much on the particular state assignment used, and when the number of possible state assignments is very large, some design rules may be applied that produce reasonably simple circuits. We shall discuss these rules in Section 7.9.

State assignments and problems associated with the specification and design of general sequential state controllers will be demonstrated by examples. As with the functional representation of combinational circuits, sequential circuits may also be specified by a verbal functional specification, not a state transition table (STT) or a STD. In such cases, first, the number of stored states must be established and then the STT constructed. We will demonstrate the technique of constructing a STT from functional specifications by studying the class of bit-sequence detector circuits. As the first step in our discussions on general sequential systems, we will present sequential state controllers, which may be described in a functional manner without reference to their specific circuit implementation.

7.2 STATE CONTROLLERS

One of the most important practical applications of sequential circuits is the control of the operating characteristics of digital systems. The operation of most complex digital systems is organized into a small number of operational states. For example, the process controller of a chemical plant would use different control laws according to the different phases of the chemical process and would have entirely different operating rules in case of a plant emergency. It is customary (and obviously useful) to show the operating state of a controller on the control panel of the system. One practical way of designating operating states is to switch on different colored lights where each of these lights specifies a different operational state. This is the simplest view of such a **state controller** circuit, which may be characterized as follows: a digital system has N digital (Boolean) inputs and K recognizable distinct states indicated by K digital outputs (Figure 7.2). Each distinct state is recognized by one of the K outputs being at the Logic-1 level (at the same time all other outputs are at the Logic-0 level). The circled numbers indicate lights. At any one time only one light is on, and this light indicates the selected state of the controller. Depending on the values of the input variables, the state control laws, and possibly the past sequence of selected states, the controller determines a new state that is selected after a clock pulse is applied to the circuit.

This model of a state controller is more general than a sequential circuit because the selection of a new state may depend on past selection of states. For example, if the controller's selected state is state 2 at present, then its future operation may depend on whether the controller arrived at state 2 from state 1 or from another state (a general sequential circuit's "next state" may depend only on its inputs and "current state"). The implication for the general nature of this controller model is that the sequential circuit designed for the controller may have more states than the number of operational states of the controller.

The other important aspect of this general model is that the control laws are given in terms of recognizable states (state 1, state 2, etc.), not by the logic levels of flip-flop outputs. In fact, no flip-flops are shown at all. The sequential circuit that implements these control laws will have to include stored bits, of course, but it is not specified which particular combination of stored bit values represent which state of the controller. For example, a simple way to implement a state controller is to use a decoder at the output stage of the system, as shown in Figure 7.3. Depending on the selected state assignments, the respective indicator lights must be connected to the correct outputs of the decoder, but the selection of binary output values of the flip-flops for any controller state is entirely up to the designer. This leads to the already mentioned state assignment problem. Since the number of ways one could associate controller states with flip-flop output bits may be very large, one would like to find the particular assignment that leads to the least expensive circuit.

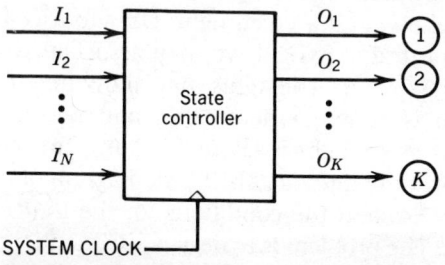

FIGURE 7.2 Block diagram of state controller with K states.

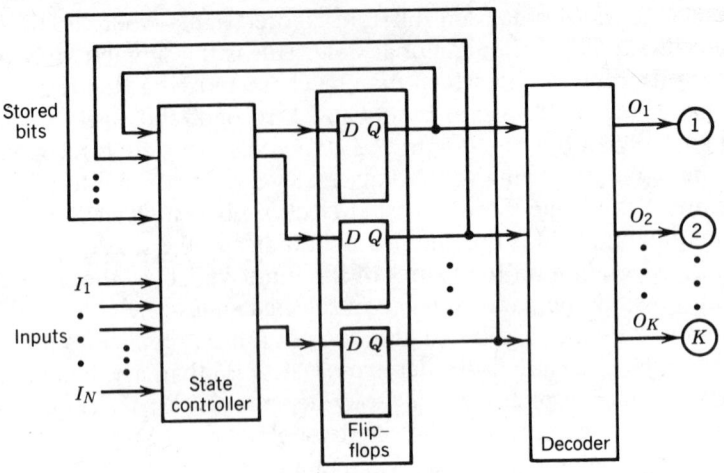

FIGURE 7.3 State controller with flip-flops for its stored bits and decoder generating its outputs.

Following this general model of a state controller, the design of a practical state controller for traffic control is discussed below in the next section. This example introduces the basic concepts of state assignment and state controller design. The methods shown may be applied to many similar state controller problems.

7.3 DESIGN OF TRAFFIC SIGNAL CONTROLLER

As a practical example, the controller for the traffic lights of a simple road crossing is considered. The problem is described in terms of two traffic lights. The first light, named LIGHT-1, controls east–west traffic. The second light, LIGHT-2, controls north–south traffic. There are two digital (binary) sensors. The first sensor, called I_1, indicates whether a car is waiting at or approaching the intersection from the east–west direction. The second sensor, called I_2, senses cars traveling in the north–south direction. The value $I_1 = 0$ means that there are no cars in the east–west direction. The value $I_1 = 1$ indicates that there is one or more cars waiting or traveling in this direction. The input sensor I_2 operates similarly in the north–south direction. Each traffic signal contains a red, a yellow, and a green light. Only four combinations of these six lights control the traffic, which we may associate with four "operating states" of the traffic crossing. The lights may allow free traffic to flow in one direction (green light is shown), whereas they stop traffic in the other direction (red light). When a green light is about to change to red, a yellow light is turned on for a short time interval. The block diagram of the traffic controller is shown in Figure 7.4, and the conditions for the four operating states are shown in Table 7.1. The problem is to design a sequential circuit that controls the traffic lights according to the sensed flow of traffic in both directions.

7.3 DESIGN OF TRAFFIC SIGNAL CONTROLLER

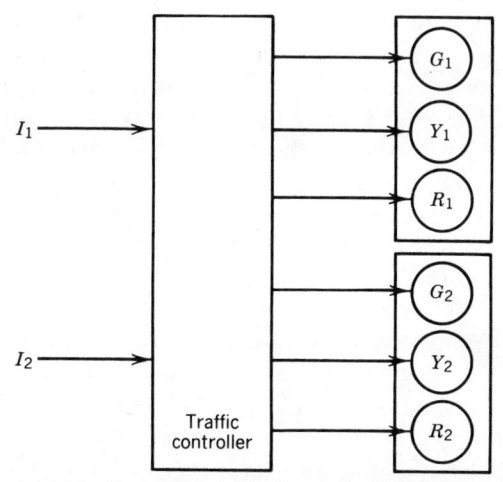

FIGURE 7.4 Block diagram of traffic controller.

The four combinations of the lights are the four possible states of this state controller. These states are designated by symbols S_1, S_2, S_3, and S_4. In terms of these four states the traffic control laws may be stated as follows: If the controller selected state S_1 (traffic flows freely in east–west direction), then it remains in state S_1 as long as the input sensors I_1 and I_2 show values of 00 (no traffic) or 10 (traffic in east–west direction); if the inputs show values of 10 (traffic in north–south direction) or 11 (traffic in both directions), then state S_1 changes to state S_2. State S_2 changes to state S_3 regardless of the input sensor values (since states S_2 and S_4 are transitory states). Similarly, if the controller selected state S_3 (traffic flows freely in the north–south direction), then it remains in state S_3 as long as the input sensor values are 00 or 01, both of which indicate no traffic in the east–west direction. If the input sensors show values 11 or 10, then the controller changes its state from state S_3 to S_4. Since S_4 is a transitory state, the controller always changes state S_4 to state S_1 regardless of the input sensor values.

TABLE 7.1 Four States of Traffic Controller

Signal 1	Signal 2	Function	State
Green	Red	Traffic flows freely east–west	S_1
Yellow	Red	Lights are changing	S_2
Red	Green	Traffic flows freely north–south	S_3
Red	Yellow	Lights are changing	S_4

Since the next state of this controller depends only on its current state and its input values, it is a proper sequential system and, therefore, its STT (and STD) may be constructed directly from the verbal description of its functional behavior. The STT and STD are shown in Figure 7.5. The state table lists the next-state values for the controller for each combination of input values and current-state values. The same information is presented graphically in the STD. We can see that the states S_2 and S_4 change into states S_3 and S_1, respectively, for all input value combinations. Depending on the input values, the system may remain in state S_1 or S_3 indefinitely, or it may change to state S_2 or S_4, as shown by the STD in Figure 7.5.

Since there are four operating states of the traffic controller, the sequential circuit designed for it must have at least four distinct states. As we shall demonstrate in the next section, it is possible to use more states for the sequential circuit than required by the controller, but of course, one may not use less. In order to store four distinct states, two flip-flops are required. The outputs of the flip-flops are designated (as usual) with symbols Q_1 and Q_0, and the four states of the sequential circuit are distinguished by the values of the two stored bits (00, 01, 10, and 11). We now face the state assignment problem. There are four distinct states indicated as S_1, S_2, S_3, and S_4, and four states are defined by the binary values of flip-flop outputs Q_1 and Q_0. Each state of the sequential circuit is now "assigned" to represent one state of the controller. A convenient way of showing state assignments is to use a Karnaugh map (K-map) (yet another use we have found for this successful graphic representation). One possible state assignment is shown in Figure 7.6, where the state assignment Karnaugh map (SAKM) shows that state S_1 is assigned binary values 00, state S_2 is assigned values 01 ($Q_1 = 0$ and $Q_0 = 1$), and so on. The question one may ask immediately is how many different ways can we select state assignments.

When the first assignment is made for S_1, there are four possibilities. These are the four possible combinations of values for the two stored bits Q_1 and Q_0 (the four empty squares in the K-map). After a state is assigned to S_1, another state has to be assigned to state S_2. There are now three possibilities

Transition state table
(next states)

Sensors		Current state			
I_1	I_2	S_1	S_2	S_3	S_4
0	0	S_1	S_3	S_3	S_1
0	1	S_2	S_3	S_3	S_1
1	0	S_1	S_3	S_4	S_1
1	1	S_2	S_3	S_4	S_1

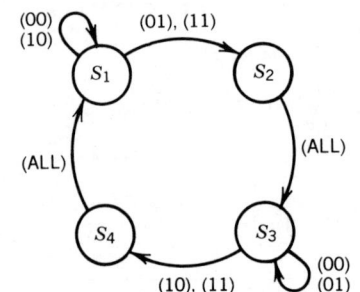

Transition state diagram

FIGURE 7.5 The STT and STD for traffic controller in terms of states S_1, S_2, S_3, and S_4.

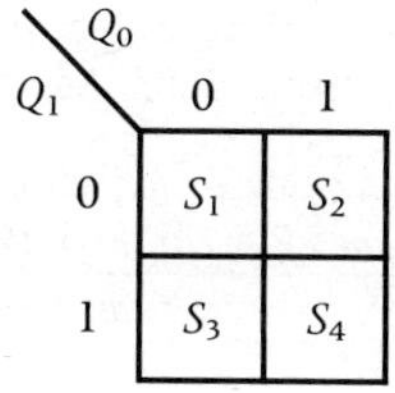

$Q_1 \backslash Q_0$	0	1
0	S_1	S_2
1	S_3	S_4

State Assignment K-Map

State	Assignment $Q_1 Q_0$
S_1	00
S_2	01
S_3	10
S_4	11

FIGURE 7.6 One particular state assignment K-map and encoding of flip-flop bit values for four states.

(since one state was already used from the four possible states). There are two possible ways to assign the third state, and we find that the total number of different state assignments for four states is equal to 4 × 3 × 2 × 1 = 24. Once a state assignment is chosen, the STT may be constructed in terms of flip-flop output values, and the sequential circuit may be designed using the same methods as those we used for designing counters in Chapter 6.

It would be possible to try all 24 different state assignments, but not all the different assignments produce different circuits; therefore, only those have to be tried that do produce different sequential circuits. It can be seen that the identities of the two flip-flops may be exchanged without producing different circuits since only the input circuits for flip-flop Q_1 and Q_0 would have to be exchanged. For example, if the roles of flip-flop Q_1 and Q_0 are exchanged for the state assignments shown in Figure 7.6, the assignments for S_1 (00) and S_4 (11) remain the same (no change occurs when the two binary values are interchanged), but the assignment for S_2 becomes 10 (it was 01) and the assignment for S_3 becomes 01. Hence, there are only 12 state assignments that produce different circuits. The other 12 may be derived from this original 12 assignments by exchanging the two binary values of the flip-flops for each state.

If the flip-flops used in the sequential circuit provide both normal (Q) and inverted (Q') outputs, then further reductions in the number of state assignments that produce unique circuits are possible. By complementing the output of one, the other, or both flip-flops, we can produce three additional variations that will produce the same circuits. Since for each variation we may exchange the roles of Q_1 and Q_0, six additional state assignments may be produced. With the original two assignments we have a total of *eight* assignments that produce the same circuits. Consequently, there are only *three* unique state assignments for four states that produce different circuits when both the normal and the complement of the flip-flop outputs are available.

In Table 7.2 we show the rules for generating the eight state assignments that produce equivalent sequential circuits. All eight assignments may be considered equivalent as far as the complexity of the final sequential circuits are concerned. Once one of the three unique assignments is chosen, any one

208 SYNCHRONOUS CONTROLLERS AND STATE ASSIGNMENTS

TABLE 7.2 Three Unique State Assignments for Four States and Rules for Generating Other Twenty-One Distinct State Assignments

Controller state	Unique assignments			Generate other assignments by substituting any one of seven alternates for original						
	(1) Q_1Q_0	(2) Q_1Q_0	(3) Q_1Q_0	Original Q_1Q_0	Substitute any column						
					1	2	3	4	5	6	7
S_1	00	00	00	00	00	10	01	01	10	11	11
S_2	01	01	11	01	10	11	11	00	00	10	01
S_3	10	11	10	10	01	00	00	11	11	01	10
S_4	11	10	01	11	11	01	10	10	01	00	00

of eight assignments may be used for our design procedures. The three unique state assignments are shown in three columns of Table 7.2.

Since there are only three different state assignments, it is practical to try all of them. We will compare the two sequential circuits designed for the first two assignments for which we will use D-type flip-flops. In order to show an example with JK flip-flops, the third assignment will be designed using JK flip-flops.

The design for D-type flip-flops and the first assignment are summarized in Figure 7.7. The state assignment defines the relationships between the four states (shown as S_1, S_2, S_3, and S_4) and the binary output values for the two flip-flops, Q_1 and Q_0. Once the binary values are known, as shown in the STT in Figure 7.7, the two K-maps can be generated, one for Q_1 and the other for Q_0. Like counters, the K-maps show the output values for the next states $[Q_i(t_{n+1})]$ with the four input values given by inputs I_1 and I_2 and outputs Q_1 and Q_0 for the current states. The minimization carried out for 1's is also shown, and the resulting two design equations define the input circuits for the two D-type flip-flops.

The same design procedure carried out for the second state assignment is shown in Figure 7.8. Note that a considerable simplification in the circuitry for D_1 occurred, but no similar savings are found in the circuits for D_0.

In Figure 7.9 the design for the third state assignment is shown, which uses JK flip-flops. If the design is tried for D-type flip-flops (a possible groupings for 1's are shown), the input circuitry for D_1 becomes the most complex of the three assignments. It is instructional to compare the three K-maps for D_1 for the three different state assignments because they clearly demonstrate that state assignments may significantly influence the complexity of the resulting sequential circuits. We use the partition method for JK flip-flops in Figure 7.9, partly to review this design method and partly to demonstrate that even the most complex K-maps yield simplified circuits for the JK flip-flops when they are compared to those that use D-type flip-flops and simpler K-maps.

From the three possible circuits, we choose the second state assignment and show the completed sequential circuits in Figure 7.10. This circuit is

State Assignment

Q_1 \ Q_0	0	1
0	S_1	S_2
1	S_3	S_4

State Transition Table

Inputs I_1 I_2	Current state	Assignment Q_1 Q_0	Next states	t_{n+1} Q_1 Q_0
0 0	S_1	0 0	S_1	0 0
0 1	S_1	0 0	S_2	0 1
1 0	S_1	0 0	S_1	0 0
1 1	S_1	0 0	S_2	0 1
0 0	S_2	0 1	S_3	1 0
0 1	S_2	0 1	S_3	1 0
1 0	S_2	0 1	S_3	1 0
1 1	S_2	0 1	S_3	1 0
0 0	S_3	1 0	S_3	1 0
0 1	S_3	1 0	S_3	1 0
1 0	S_3	1 0	S_4	1 1
1 1	S_3	1 0	S_4	1 1
0 0	S_4	1 1	S_1	0 0
0 1	S_4	1 1	S_1	0 0
1 0	S_4	1 1	S_1	0 0
1 1	S_4	1 1	S_1	0 0

Karnaugh Maps for Next-State Equations

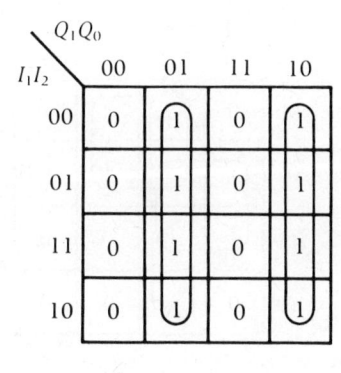

$Q_1 Q_0$

$I_1 I_2$	00	01	11	10
00	0	1	0	1
01	0	1	0	1
11	0	1	0	1
10	0	1	0	1

$Q_1(t_{n+1}) = D_1(t_n)$

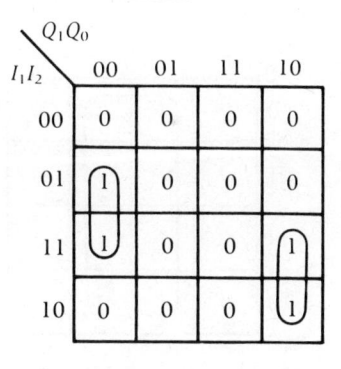

$Q_1 Q_0$

$I_1 I_2$	00	01	11	10
00	0	0	0	0
01	1	0	0	0
11	1	0	0	1
10	0	0	0	1

$Q_0(t_{n+1}) = D_0(t_n)$

Flip-flop Input Equations

$D_1 = Q_1' \cdot Q_0 + Q_1 \cdot Q_0'$
$ = Q_1 \oplus Q_0$

$D_0 = I_2 \cdot Q_1' \cdot Q_0' + I_1 \cdot Q_1 \cdot Q_0'$
$ = Q_0' \cdot (I_2 \cdot Q_1' + I_1 \cdot Q_1)$

FIGURE 7.7 The STT, K-maps, and design equations for traffic controller using first state assignment.

State Assignment

Q_1\\Q_0	0	1
0	S_1	S_2
1	S_4	S_3

State Transition Table

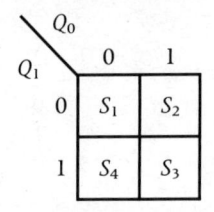

Inputs $I_1\ I_2$	Current state	Assignment $Q_1\ Q_0$	Next states	t_{n+1} $Q_1\ Q_0$
0 0	S_1	0 0	S_1	0 0
0 1	S_1	0 0	S_2	0 1
1 0	S_1	0 0	S_1	0 0
1 1	S_1	0 0	S_2	0 1
0 0	S_2	0 1	S_3	1 1
0 1	S_2	0 1	S_3	1 1
1 0	S_2	0 1	S_3	1 1
1 1	S_2	0 1	S_3	1 1
0 0	S_3	1 1	S_3	1 1
0 1	S_3	1 1	S_3	1 1
1 0	S_3	1 1	S_4	1 0
1 1	S_3	1 1	S_4	1 0
0 0	S_4	1 0	S_1	0 0
0 1	S_4	1 0	S_1	0 0
1 0	S_4	1 0	S_1	0 0
1 1	S_4	1 0	S_1	0 0

Karnaugh Maps for Next-State Equations

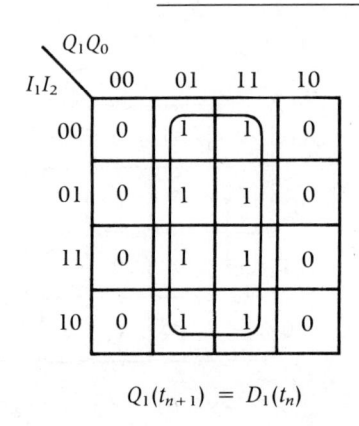

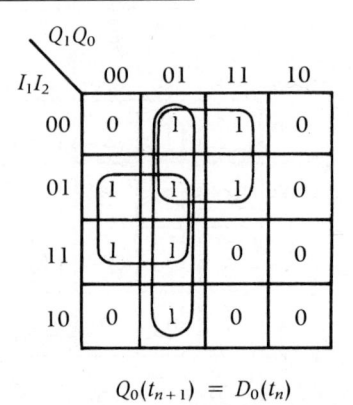

Q_1Q_0 / I_1I_2

	00	01	11	10
00	0	1	1	0
01	0	1	1	0
11	0	1	1	0
10	0	1	1	0

$Q_1(t_{n+1}) = D_1(t_n)$

Q_1Q_0 / I_1I_2

	00	01	11	10
00	0	1	1	0
01	1	1	1	0
11	1	1	0	0
10	0	1	0	0

$Q_0(t_{n+1}) = D_0(t_n)$

Flip-flop Input Equations

$D_1 = Q_0$

$D_0 = I_1' \cdot Q_0 + I_2 \cdot Q_1' + Q_1' \cdot Q_0 = Q_0 \cdot (I_1 \cdot Q_1)' + Q_1' \cdot I_2$

FIGURE 7.8 The STT, K-maps, and design equations for traffic controller using second state assignment.

State Assignment

Q_1 \ Q_0	0	1
0	S_1	S_4
1	S_3	S_2

State Transition Table

Inputs I_1 I_2	Current state	Assignment Q_1 Q_0	Next states	t_{n+1} Q_1 Q_0
0 0	S_1	0 0	S_1	0 0
0 1	S_1	0 0	S_2	1 1
1 0	S_1	0 0	S_1	0 0
1 1	S_1	0 0	S_2	1 1
0 0	S_2	1 1	S_3	1 0
0 1	S_2	1 1	S_3	1 0
1 0	S_2	1 1	S_3	1 0
1 1	S_2	1 1	S_3	1 0
0 0	S_3	1 0	S_3	1 0
0 1	S_3	1 0	S_3	1 0
1 0	S_3	1 0	S_4	0 1
1 1	S_3	1 0	S_4	0 1
0 0	S_4	0 1	S_1	0 0
0 1	S_4	0 1	S_1	0 0
1 0	S_4	0 1	S_1	0 0
1 1	S_4	0 1	S_1	0 0

Karnaugh Maps for Next-State Equations

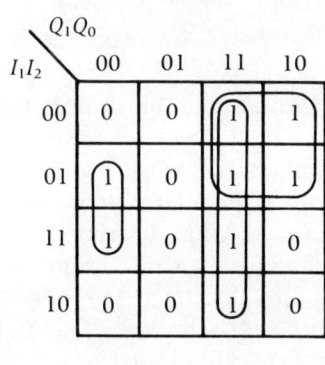

$Q_1(t_{n+1}) = D_1(t_n)$

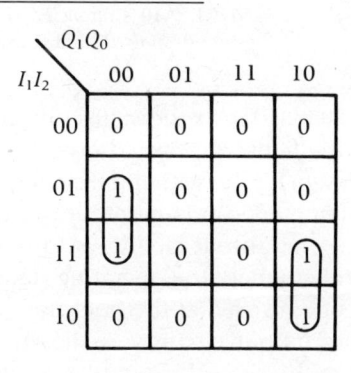

$Q_0(t_{n+1}) = D_0(t_n)$

Design for JK Flip-flops

$J_1 = I_2 \cdot Q_0'$

$K_1 = I_1 \cdot Q_0'$

$J_0 = I_2 \cdot Q_1' + I_1 \cdot Q_1$

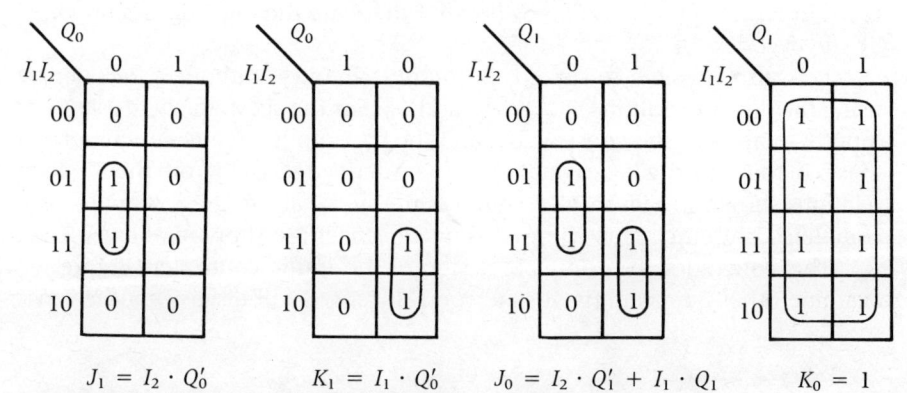

$K_0 = 1$

FIGURE 7.9 The STT, K-maps, and design equations for traffic controller using third state assignment.

212 SYNCHRONOUS CONTROLLERS AND STATE ASSIGNMENTS

Design equations:
$$\begin{cases} D_1 = Q_0 \\ D_0 = Q_0 \cdot (I_1 \cdot Q_1)' + Q_1' \cdot I_2 \end{cases}$$

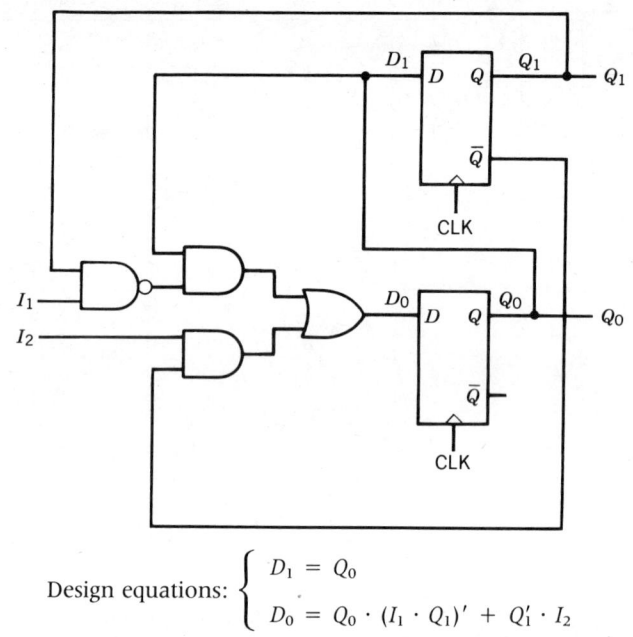

FIGURE 7.10 Completed circuits of traffic controller generating flip-flop outputs Q_0 and Q_1.

complete as far as the state transitions are concerned, but it does not show how the lights are turned on or off by the circuit. Once we have completed the design for the sequential circuit, we still have to provide the correct outputs by adding appropriate output circuitry to the sequential system. One way of adding the output circuits is to return to our original description of a state controller where the operating states are described by single output lines and a decoder is used at the final stage, which provides the system outputs. This type of output circuitry is shown in Figure 7.11. Observe that in order to operate the six traffic lights, two more OR gates are needed. These gates are necessary because it is not individual traffic lights but rather a combination of traffic lights that indicates a system state. The conditions for switching on the traffic lights are also shown in Figure 7.11. The output equations are given in terms of the states S_1–S_4 because these are the logic signals provided by the decoder.

It is also possible to design the output circuitry without a decoder by using the flip-flop outputs Q_1, Q_1', Q_0, and Q_0'. For the six traffic lights six two-input K-maps may be constructed, and the six output circuits designed from these. In this case the logic equations can be derived directly from the Boolean relations shown in Figure 7.11, and it is not necessary to draw K-maps. The completed output circuits without using a decoder are shown in Figure 7.12.

This concludes the complete design of the traffic controller. Before we examine other applications of state controllers, we will discuss another pos-

7.4 ONE-FLIP-FLOP-PER-STATE METHOD   213

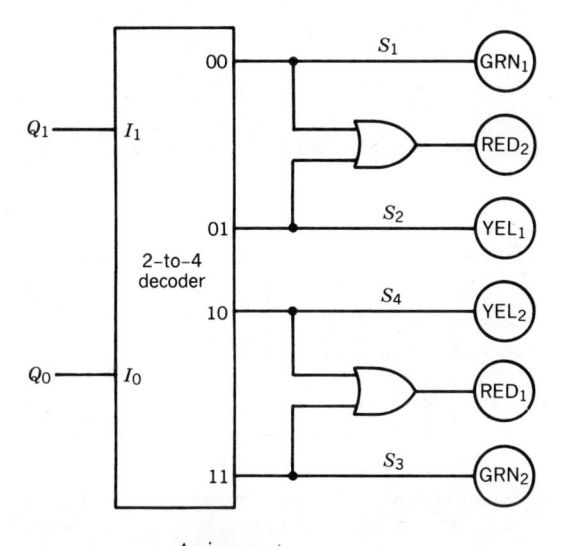

State	Assignment $Q_1\ Q_0$	Lights turned on
S_1	0 0	$G_1\ R_2$
S_2	0 1	$Y_1\ R_2$
S_3	1 1	$R_1\ G_2$
S_4	1 0	$R_1\ Y_2$

$$G_1 = S_1 \qquad G_2 = S_3$$

$$Y_1 = S_2 \qquad Y_2 = S_4$$

$$R_1 = S_3 + S_4 \qquad R_2 = S_1 + S_2$$

FIGURE 7.11 Generation of output signals (traffic light controls) from flip-flop outputs using decoder.

sible way of designing a sequential circuit for the same traffic control problem that suits well a system whose binary output variables indicate the operating states of the system.

7.4 ONE-FLIP-FLOP-PER-STATE METHOD

Until now we have been designing sequential systems with the minimum number of flip-flops that can provide the required number of system states. It is possible to find other design methods that use more than the required minimum number of flip-flops. For those systems for which a single output signal indicates the existence of an operating state, the one-flip-flop-per-state method is particularly useful. This method eliminates the decoder because the flip-flop outputs behave as if they were the outputs of a decoder; that is, at any one time only one of the outputs is equal to Logic-1, whereas the

214 SYNCHRONOUS CONTROLLERS AND STATE ASSIGNMENTS

Output design equations:

$$G_1 = Q_1' \cdot Q_0' \qquad G_2 = Q_1 \cdot Q_0$$
$$Y_1 = Q_1' \cdot Q_0 \qquad Y_2 = Q_1 \cdot Q_0'$$
$$R_1 = Q_1 \qquad R_2 = Q_1'$$

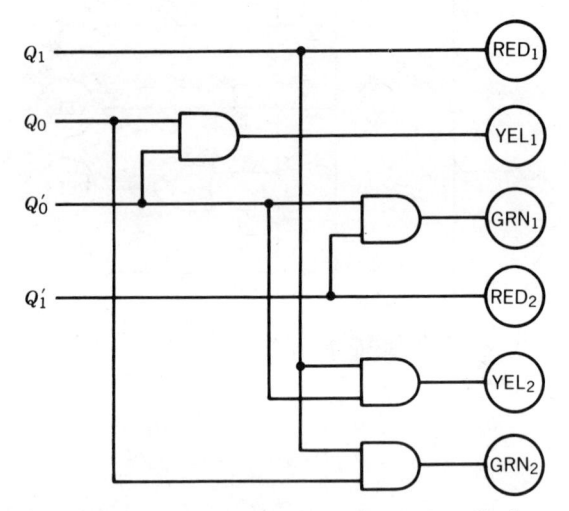

FIGURE 7.12 Generation of output signals from flip-flop outputs without using decoder.

others are equal to Logic-0. This design method uses the maximum reasonable number of flip-flops for a state controller.

As the name of this method suggests, the number of flip-flops used for the design is equal to the number of selected states. Only one flip-flop output may be in the Logic-1 state at any one time. Each flip-flop indicates the selection of one particular state. The Boolean equations for these flip-flops are determined from the next-state entries listed in the STT. No state assignments are necessary because the binary value of 1 indicates the presence of a particular operating state, and the state transition equations are expressed in terms of state variables (in this case the state variables are S_1, S_2, S_3, and S_4).

The STT of the traffic controller first shown in Figure 7.5 is duplicated in Figure 7.13. In the general case four six-variable K-maps would be required for the design of this controller since there are six variables (two input variables plus the four flip-flops representing the current states), but in this case the design equations are simple and can be expressed by the shown Boolean equations. Since the outputs of the four flip-flops indicate the selected states, they are also named S_1, S_2, S_3, and S_4. For example, if the output S_1 is in the Logic-1 state, the controller has selected state S_1. Therefore, the next-state equations for the four flip-flops are simply the next-state equations for states

7.4 ONE-FLIP-FLOP-PER-STATE METHOD 215

Transition State Table (next states)

Sensors		Current state			
I_1	I_2	S_1	S_2	S_3	S_4
0	0	S_1	S_3	S_3	S_1
0	1	S_2	S_3	S_3	S_1
1	0	S_1	S_3	S_4	S_1
1	1	S_2	S_3	S_4	S_1

Next-State Equations

$$S_1(t_{n+1}) = S_1 \cdot (I_1' \cdot I_2' + I_1 \cdot I_2') + S_4$$
$$= S_1 \cdot I_2' + S_4$$

$$S_2(t_{n+1}) = S_1 \cdot (I_1' \cdot I_2 + I_1 \cdot I_2) = S_1 \cdot I_2$$

$$S_3(t_{n+1}) = S_2 + S_3 \cdot (I_1' \cdot I_2' + I_1' \cdot I_2)$$
$$= S_2 + S_3 \cdot I_1'$$

$$S_4(t_{n+1}) = S_3 \cdot (I_1 \cdot I_2' + I_1 \cdot I_2) = S_3 \cdot I_1$$

FIGURE 7.13 Design equations for STT of traffic controller using one-flip-flop-per-state method.

S_1–S_4. The next-state equations are derived from the state table by enumerating all the conditions necessary for a particular state to be selected. For example, state S_3 becomes the next selected state if either the current state is equal to S_2 or if the current state is equal to S_3 and the input values are either 00 or 01.

Using the design equations in Figure 7.13 and D-type flip-flops, the completed traffic controller circuit is shown in Figure 7.14. This circuit is not much more complicated than the one designed by conventional sequential circuit methods (Figures 7.10 and 7.11), but of course, it has four rather than two flip-flops. And since it has more flip-flops than the minimum required number, it has a large number of unused states. In this case the number of

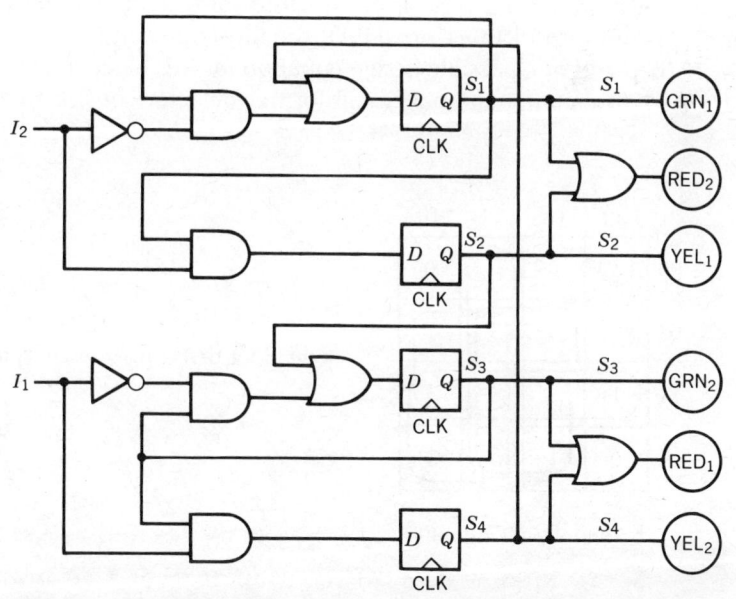

FIGURE 7.14 Completed circuit for traffic controller using one flip-flop per state.

216 SYNCHRONOUS CONTROLLERS AND STATE ASSIGNMENTS

possible states is equal to $2^4 = 16$; therefore, there are 12 unused states. Similar to counters with unused states, arrangements must be made to ensure that the sequential system will not enter into one of its unused states. One method we have seen before relies on an initialization procedure that places the system in one of its used states, say 0001 (where the four binary values refer to the outputs of the four flip-flops S_1–S_4). The initialization can occur asynchronously, using the asynchronous PRESET and CLEAR inputs of the flip-flops. Once the system is in one of its normal states, the design equations should ensure that it will not drop into any one of its unused states. (This can be checked by examining the design equations, which set only one state to Logic-1 and set all three other states to Logic-0 for all input conditions when the system is in any one of its four used states, 0001, 0010, 0100, or 1000.)

For many systems such an initialization procedure would be sufficient. However, for a system such as a traffic controller a more "fail-safe" method would be necessary since an unused state (e.g., green lights in both direction) could cause confusion and accidents on the road. In order to ensure that the system never enters into any one of its unused states, the asynchronous initialization procedure must be executed whenever an unused state is detected. Hence, we have to design and build an "unused state detector" circuit. This is a combinational circuit whose output becomes a Logic-1 whenever the system is in one of its unused states, that is, more than one output is equal to Logic-1 or all outputs are equal to Logic-0. The output of this detector circuit may be used to activate the asynchronous PRESET and CLEAR functions. This arrangement would ensure that no unused states will appear at the outputs for more than a very short time interval (the time it takes for an asynchronous input to PRESET or CLEAR the flip-flop's output).

The K-map and one possible design equation for this detector are shown in Figure 7.15. The combinational circuit for this detector consists of a four-

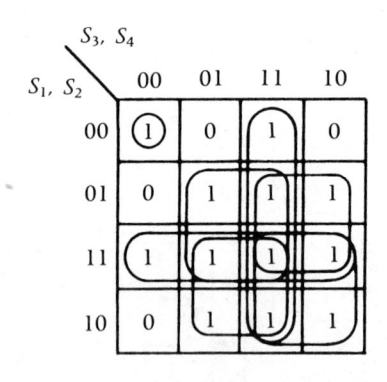

S_1, S_2 \ S_3, S_4	00	01	11	10
00	1	0	1	0
01	0	1	1	1
11	1	1	1	1
10	0	1	1	1

K-map for detecting unused state

K-map design equation:

$$\overline{S_1} \cdot \overline{S_2} \cdot \overline{S_3} \cdot \overline{S_4} + (S_1 + S_4) \cdot (S_2 + S_3) + S_2 \cdot S_3 + S_1 \cdot S_4$$

FIGURE 7.15 Karnaugh map and design equation for detecting unused state for one-flip-flop-per-state method.

input AND, a four-input OR, three two-input AND, and two two-input OR gates (seven gates altogether) and turns out to be much more complicated than the sequential controller circuit. We showed this example to demonstrate that in the real world of circuit design, circuit savings in one place may be outweighed by other requirements that must be met in order to ensure the proper operation of the circuit. It is important to point out that probably none of the methods and procedures used in this book will work all the time and provide satisfactory solutions to all practical problems. It is not the details in the design methods, but rather the approach to design problems, that should be learned from this course.

We have seen two different methods of designing sequential circuits for state controller problems. In these examples we have seen that the number of system states could be easily determined from the verbal description of the problem because the specifications were given in terms of operating states. In many of these practical problems the number of system states was equal to the number of operating states, and the system and operating states could be made equivalent. Once the required number of system states is known, the STT can be determined. Once the STT is known and the states are assigned binary flip-flop values, the K-maps and the circuits can be constructed in a straightforward manner.

The functional descriptions for general sequential (finite-state) systems are not always given in terms of states. In these general cases the number of required system states that can solve a given problem must be determined first. There are no foolproof general methods for identifying system states from general verbal descriptions of sequential systems. This selection-of-states procedure is very similar to the selection of variables in a computer program designed according to some external specifications and not given in terms of variables. Indeed, a computer program can be described as a sequential system whose variables hold its system states. We have the same freedom of choosing system states for sequential systems as choosing variables for a computer program. We will demonstrate this procedure of selecting system states for a class of sequential systems called **bit-sequence detectors** in the following sections.

7.5 BIT-SEQUENCE DETECTOR CIRCUITS

Another important practical application for sequential circuits is the detection of a given sequence of 1's and 0's while a time-varying digital signal is observed. Many practical digital systems operate "serially." This means that they require only two wires (signal and ground connections) for a signal input or output and that they receive or transmit digital information through one signal only (i.e., 1 bit per clock period). Computer terminals are a common example of serial devices. It is a frequent requirement that such serial devices "recognize" a given sequence of 1's and 0's.

The block diagram of a general bit-sequence detector circuit is shown in Figure 7.16. We assume that there is an initial quiescent state when the output

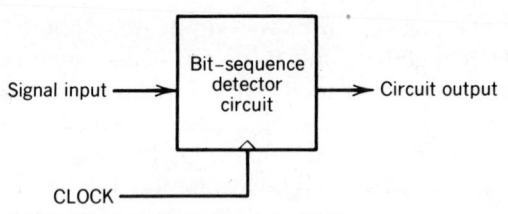

218 SYNCHRONOUS CONTROLLERS AND STATE ASSIGNMENTS

FIGURE 7.16 Block diagram of BSD.

is equal to Logic-0 and the input transmission has not yet begun. Given a sequence of N 1's and 0's (where N is equal to the number of bits in the sequence), the output of the bit-sequence detector (BSD) becomes Logic-1 after a correct sequence of bits has been transmitted through the input of the system. For demonstration purposes we choose a bit sequence of 101 ($N = 3$) and show the operation of one type of BSD by its timing diagrams in Figure 7.17. This is a synchronous system and all signals, including the input signal, change only during the negative transition of the system clock. The "sensing" of the input signal occurs during a very small time interval immediately before the negative transition of the clock signal, which is consistent with our definition of a strictly synchronous sequential system (see Chapter 5). It is important to note for the subsequent discussions about different types of detector circuits that the output of such a strictly synchronous system becomes a Logic-1 **after** the significant clock transition that has detected the correct sequence (e.g., the fourth negative clock transition in Figure 7.17).

There are two main variations in the operation of synchronous BSDs. The BSD may operate "continuously" (i.e., its output could alternate between Logic-1 and Logic-0 values depending on whether the last transmitted N bits are equal to the required sequence or not) or it could "reset" its operation after a correct sequence has been found and start from its quiescent state. In the first case the detected bit sequences may "overlap." For example, in the 9-bit sequence 011010100 (the leftmost bit arrives first and the rightmost bit arrives last), there are two sequences of the 3-bit sequence 101, the first

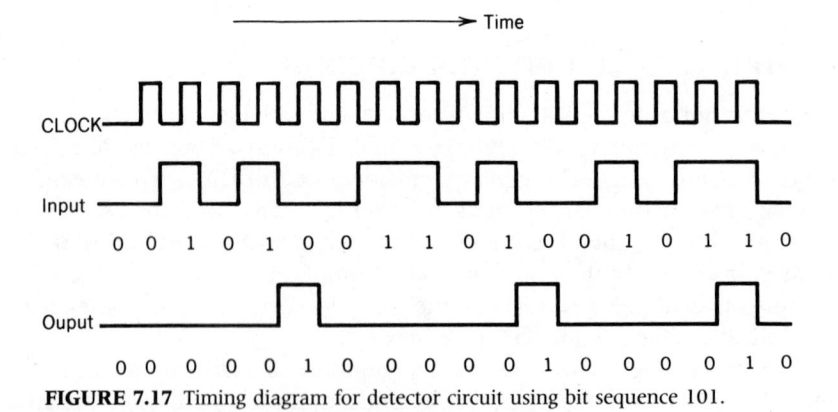

FIGURE 7.17 Timing diagram for detector circuit using bit sequence 101.

7.6 INPUT AND OUTPUT SIGNAL SYNCHRONIZATION IN SEQUENTIAL SYSTEMS   219

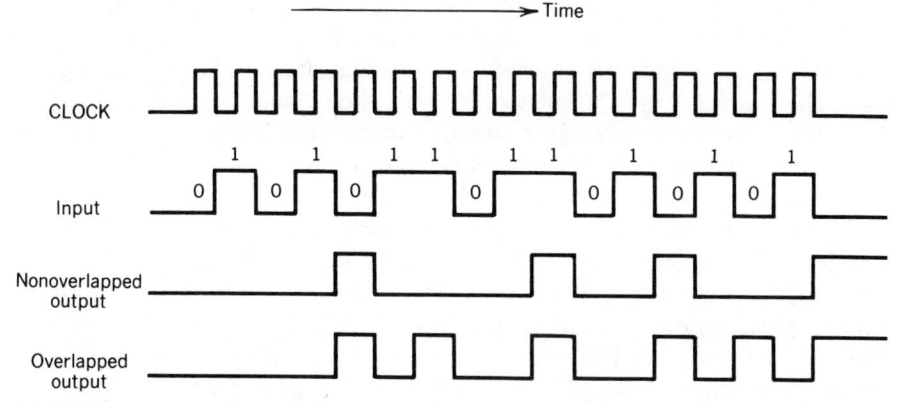

FIGURE 7.18 Timing diagrams for nonoverlapped and overlapped synchronized detectors using bit sequence 101.

starting with the third bit on the left, the second with the fifth bit. The two small bit-sequences **overlap** since the total number of bits containing the two sequences is equal to 5. Two complete sequences of 101 require 6 bits, and the bit sequence 011011010 contains two complete sequences of 101. Observe that the BSD that detects overlapped sequences would detect nonoverlapped sequences as well, whereas the nonoverlapped BSD could miss many of the overlapped sequences. In Figure 7.18 the operation of these two types of synchronous BSDs are demonstrated by their timing diagrams.

We have been emphasizing the fact that the circuits we have chosen so far were all strictly synchronous systems. In many practical circuits inputs may change at any time; hence, they are asynchronous signals. It is possible to use asynchronous input signals of BSD circuits in such a way that the output of the BSD becomes also asynchronous. Therefore, before studying the design and construction of sequence detector circuits, we will discuss the problems of synchronizing input and output signals for general sequential systems.

7.6 INPUT AND OUTPUT SIGNAL SYNCHRONIZATION IN SEQUENTIAL SYSTEMS

The synchronization for the operation of sequential circuits is provided by the SYSTEM CLOCK signal. All the flip-flops that generate the states of the sequential circuit are operated by the same CLOCK signal; consequently, their operation is synchronized. For these discussions we will assume that the significant CLOCK transitions for all flip-flops occur at the trailing edge of the clock pulse (at the 1-to-0 transition of the CLOCK signal), and the "sensing" of the input signals occurs during a very small time interval immediately preceding the significant CLOCK transitions. During these small sensing time intervals, which are called flip-flop **setup** times, all input values are assumed to be stationary.

Input signals are independent of the system and are controlled by devices (or people) whose behavior is independent of the system we are designing; therefore, it seems inconsistent to state conditions for them. Since input signals can behave in all sorts of unpredictable ways, how could we ensure that their behavior follows our rules for synchronous systems? The answer is that we cannot. On the other hand, if we need strictly synchronous input signals, then we can insert a D-type flip-flop between inputs and our system, as shown in Figure 7.19, and transform the input signal to a synchronous variable. We pay for this "buffered" synchronous input by increased circuit cost (which, for a serial system, is insignificant since it requires only one D-type flip-flop) and also by a time delay. The D-type flip-flop delays the input signal by one clock period, and this delay may be significant for a BSD circuit, which, for example, could be designed to alert a system to a dangerous condition.

Now let us examine the output of the BSD circuit. We have been using strictly synchronous systems for our examples. The output of the detector changed to Logic-1 after the significant clock transition that detected the arrival of all the required bits in the sequence. But, looking at the problem without worrying about its implementation, the system should be able to detect a correct sequence sooner than it does. Looking at the timing diagrams in Figure 7.18, after the third significant CLOCK transition the system should "know" that the past two input bits were 10 in sequence and that the **current** input bit is equal to 1. In other words, the correct bit sequence could be detected one clock period sooner. In order to make this decision earlier, the output of the BSD circuit must be made a function of the **current** input value as well as the state of the system (where the state of the system indicates the already arrived $N - 1$ bits of the sequence).

We have already mentioned in Chapter 5 that sequential systems whose outputs depend only on system states are called Moore circuits, whereas those whose outputs are functions of both system states and system inputs are called Mealy circuits. The design of BSD circuits demonstrates the difference between these two types of sequential systems very well. If speed of response is important, then the Mealy-type circuit should be selected, though, as we shall show in what follows, the asynchronous nature of these circuits could cause problems. If speed is not that important, then for better reliability, the Moore-type synchronous circuit should be used. We shall work out examples for both types, and we shall see that the design of BSD circuits demonstrates the procedures of selecting system states from the verbal description of the design problem very well.

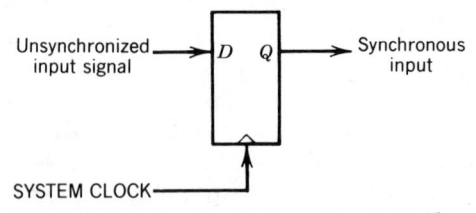

FIGURE 7.19 Synchronization of input signals.

7.6 INPUT AND OUTPUT SIGNAL SYNCHRONIZATION IN SEQUENTIAL SYSTEMS   221

For the ultimate speed of detection we have to work with an asynchronous input signal and a Mealy-type circuit. Since the output of the BSD is a function of the input signal, the output of the system becomes asynchronous as well. This means that any unexpected noise in the input signal may trigger a false indication of the presence of the sequence, since the output signal is a function of the input at all times. Let us demonstrate this with our example sequence of 101. Let us assume that the bit sequence sent to our detector is equal to 100; hence, the output is equal to 0. Now, some noise on the input line causes the input to become Logic-1 for a short time interval (we have a glitch on the line!). A Mealy-type circuit will immediately register a Logic-1 in its output. A Moore-type circuit will not "see" this temporary disturbance. Of course, if the noise occurs during the very small sensing (setup) time interval of the flip-flop, the Moore circuit will also register an error. However, such noise disturbances have usually very small time periods in practice, and with the extremely small setup times of current digital devices (on the order of nanoseconds, or 2^{-12} seconds) the probability that this may happen in a Moore-type circuit is extremely small.

The timing diagrams for Mealy-type BDS circuits and for bit sequence 101 are shown in Figure 7.20. We can see that although the output waveforms are the same as those in Figure 7.18, they are shifted to the left by one clock period. Indeed, we have two ways to make our system strictly synchronous. One way is to use a Moore-type circuit, the other is to synchronize the input signal with a flip-flop and use a Mealy-type circuit. Since we shall see that Mealy-type circuits use one less system state than Moore-type circuits, the trade-off between these two methods may be evaluated only after the circuits are designed in both ways and the results compared.

The detection for one particular bit sequence can be done in four different ways depending on whether the detection is for overlapped or for nonoverlapped sequences and whether the circuit is constructed as a Moore- or as a Mealy-type circuit. For the bit sequence 101 we shall examine all four of these cases in the following sections. Comparison of these four design ex-

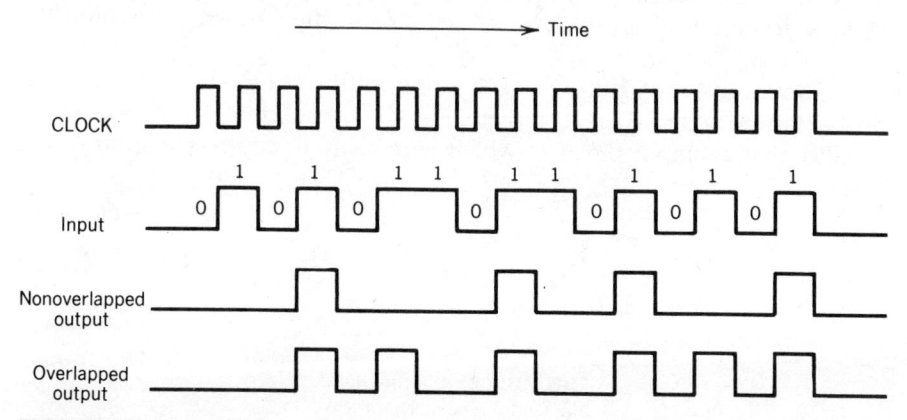

FIGURE 7.20 Timing diagrams for Mealy-type detectors using bit sequence 101.

7.7 MOORE DETECTOR CIRCUITS FOR BIT SEQUENCE 101

Both the overlapped and the nonoverlapped detection circuits will be designed in this section. The design of a BSD circuit is based on the construction of its STD rather than on its STT. Just like using flowcharts or data flow diagrams for computer programs, graphical methods seem to be easier to use for overall (top-down) design. One does not know in advance how many states are needed, but each state is selected for a given phase of the detection process. The states are numbered S_1, S_2, and so on. The first state, S_1, is always the "starting" state (which we called earlier the "quiescent" state and is often called the "reset" state). When the detector is in the starting state, the detection process has no information about the preceding input bits, and the output of the BDS is equal to 0. We shall see later that the system may return to its starting state (S_1) under some input conditions. Some systems may employ a RESET input signal that, when active, would force the system into its starting state. This is the reason why this first state is often called the reset state.

We start with the design of a Moore-type circuit (the output is a function only of the system state). This means that there will be only one state with output 1 (the state which indicates that a correct sequence of bits was found), whereas all other states will generate Logic-0 for the BSD output. We begin with state S_1, the starting state. The starting state with output 0 indicates that as far as the system knows, a detectable bit sequence has not started at the input. Therefore, as long as the input bit has the opposite value to that of the first bit of the correct sequence (Logic-1 is expected in this case), the system stays in its starting state. This is shown in Figure 7.21, which demonstrates the first step in the design process. The value 0 in the lower half of the circle in Figure 7.21 indicates the output of the circuit. Since outputs of Moore circuits depend only on the state of the circuit, they are shown within the state circles.

If an input bit of 1 arrives at the input of the circuit, then this bit may be the starting bit of a correct bit sequence. The second state, S_2, of the detector signifies that an input bit of 1, which is possibly the correct first bit of the

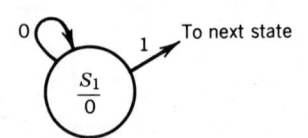

FIGURE 7.21 Starting state for Moore-type detector circuit with bit sequence 101.

7.7 MOORE DETECTOR CIRCUITS FOR BIT SEQUENCE 101   223

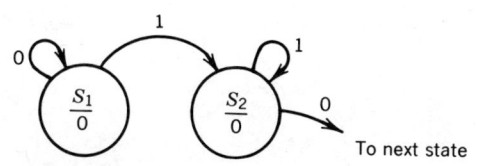

FIGURE 7.22 First two states for Moore-type detector circuit.

sequence, has already arrived. If the next bit is also a Logic-1, then this bit cannot be the second bit of the correct sequence, but it may signify the start of the correct sequence; therefore, as shown in Figure 7.22, input bit 1 returns the system to state S_2 (and not to the starting state S_1). If the next bit is a Logic-0, then the first 2 bits of the correct bit sequence may have arrived. In this case the system proceeds to a new state, S_3. The state diagram for the first two states of this Moore-type detector is shown in Figure 7.22.

Following similar reasoning, we can now construct all the required states of the system until the "detection" state (the state with output of Logic-1) is found. The state following state S_2 is state S_3. In this state the 2-bit sequence 10 has already been detected. If the detector is in state S_3 and the next input bit is equal to 1, then the correct bit sequence 101 has arrived, and the detector drops into its final state, S_4, which is the "detection" state (note that state S_4 has output of 1). In case the last bit is equal to 0, the bit sequence received so far is equal to 100, which is neither a correct sequence nor could it be the beginning of a new sequence; therefore, the system reverts to its starting state, which indicates that the beginning of a correct sequence has not yet started. These four states and their transitions are shown in Figure 7.23.

So far we have not considered whether the detection is for overlapped or for nonoverlapped sequences because we have followed the operation of this BSD only once from its starting state and, therefore, we considered the occurrence of only the first correct bit sequence, and overlapped sequences could not have yet occurred. In order to complete the state diagram, the state transitions from state S_4 have to be added to the diagram in Figure 7.23.

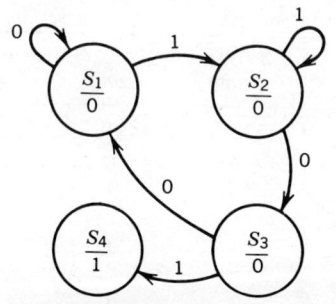

FIGURE 7.23 First four states for Moore-type circuit.

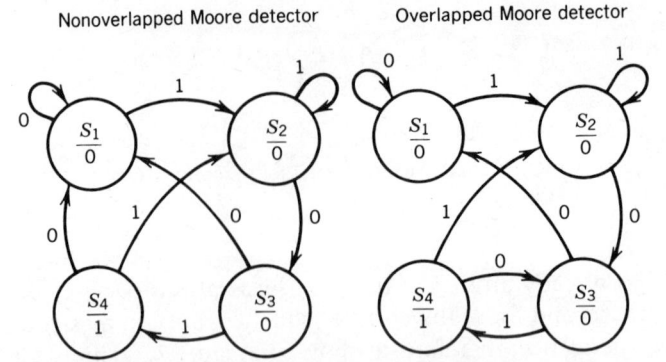

FIGURE 7.24 Completed STDs for both overlapped and nonoverlapped detectors using bit sequence 101.

These transitions are different for the two different detection methods. If the nonoverlapped detection is required and the next input bit is a Logic-0 when the detector is in state S_4, then the detector must return to its starting state because a new correct sequence has not yet started. If the input bit is equal to 1, then a new sequence may be starting again, and the system should return to state S_2.

If the overlapped detection method is used, then the correct interpretation for input bit 0 (when the system is in state S_4) is that the last received 4-bit sequence is equal to 1010, which could indicate the first 2 bits of a new overlapped sequence (bits 10). Therefore, the system should return to state S_3, which is used to detect this partial sequence. If the next input bit is equal to Logic-1 when the detector is in state S_4, then the received 4-bit sequence is equal to 1011, which indicates the possible starting 1 bit for a new bit sequence (this is true, incidentally, for both the overlapped and the nonov-

TABLE 7.3 State Transition Table for Overlapped and Nonoverlapped Moore-Type Detector Circuits for Bit Sequence 101

	Nonoverlapped (*Moore*)			Overlapped (*Moore*)	
Input	Current state	Next state	Input	Current state	Next state
0	S_1	S_1	0	S_1	S_1
1	S_1	S_2	1	S_1	S_2
0	S_2	S_3	0	S_2	S_3
1	S_2	S_2	1	S_2	S_2
0	S_3	S_1	0	S_3	S_1
1	S_3	S_4	1	S_3	S_4
0	S_4	S_1	0	S_4	S_3
1	S_4	S_2	1	S_4	S_2

erlapped methods). Hence, in this case the detector changes its state to S_2. The completed state diagrams for both nonoverlapped and overlapped sequence detectors are shown in Figure 7.24.

From the STD the STT can be constructed in the usual manner, as shown in Table 7.3. This concludes the selection of states and the construction of the STT for the Moore-type circuits. After a suitable assignment of binary values for the outputs of two flip-flops, the circuits may be designed similar to state controllers.

7.8 MEALY DETECTOR CIRCUITS FOR BIT SEQUENCE 101

Since for Mealy-type circuits the output signals are functions of both system states and the inputs, the STD and the STTs have somewhat different appearance from those for the Moore circuits. Since the outputs are functions of both states and inputs, the outputs cannot be associated with the states only, and they cannot be placed within the circles that represent the different states in the state diagram. The output values are associated with the transition arrows since each arrow originates from a known state and is specified for a given set of input values. Otherwise, the STD for the two types of systems does look similar. In the STT output values must be given for all states and all variations of input values. The simplest way of indicating output values is to associate them with the current conditions since these occur for all possible input values in the STT.

The STD for Mealy-type circuits is constructed similar to those of the Moore type. We start, as before, with the starting state, state S_1. Whenever transition arrows are shown, they are labeled with both input and output values and the labels are indicated as input/output (see Figure 7.25). The design of the STD from state S_1 to S_3 for the Moore- and Mealy-type circuits follow the same procedure (see Figures 7.21 and 7.22). The correct conditions for the detection of the 101 bit sequence require that the sequence 10 should have already been detected (the system is in state S_3) and that the current

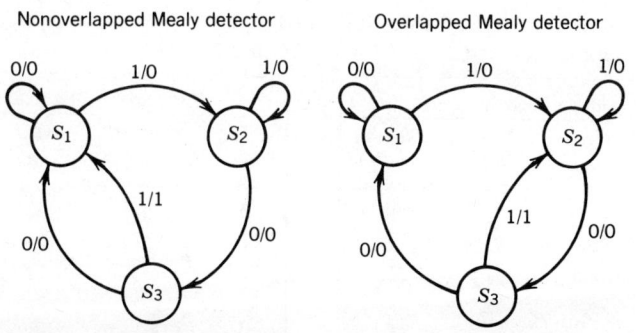

FIGURE 7.25 Completed STDs for Mealy-type detector circuits for bit sequence 101.

226 SYNCHRONOUS CONTROLLERS AND STATE ASSIGNMENTS

input bit is equal to 1. Hence, only three states are needed, and the transition arrow that leaves state S_3 and is labeled with input bit 1 should show an output of 1. This may be considered the detection transition. The completed STD for the nonoverlapped Mealy-type circuit is shown in Figure 7.25.

In order to construct the overlapped Mealy-type detector, only the last state (S_3) of the STD has to be changed. This is similar to the procedure we used for Moore-type circuits, and the STD for the completed Mealy-type overlapped detector circuit is also shown in Figure 7.25. We can see that only one of the transition arrows that leave state S_3 is different when we compare the nonoverlapped and the overlapped state diagrams.

Since we are dealing with Mealy-type circuits, the STTs (Table 7.4) show both the outputs and the next states of the system for all current states and all current input conditions. The STTs may be easily constructed from the STDs in Figure 7.25. The outputs are labeled Current output because they depend on the current values of the inputs. This means that the output may be changed by changing the input values of the system even though it remains in the same system state and no CLOCK pulse is applied to the circuit. The next state occurs after the significant transition of the clock pulse is applied to the circuit, and it is also a function of the current input. It is important to remember these timing considerations during the testing and debugging of a Mealy-type circuit since what are current input, current output, current state, and next state are easily confused.

To complete this section, we will design one of these four detector circuits using D-type flip-flops. We are going to select the overlapped sequence detector, which uses a Mealy-type circuit (STT in Table 7.4). The sequential circuits that generate the next states are designed independently from the output circuit that generates the system output. Both circuits require only combinational circuit design, and K-maps will be used. First, let us design the sequential part of the circuit.

The STT in terms of the three general states (S_1, S_2, and S_3) is shown on the right in Table 7.4. The first step is to choose the number of flip-flops used and a set of state assignments. Since there are three system states, two flip-

TABLE 7.4 State Transition Tables for Overlapped and Nonoverlapped Mealy-Type Detector Circuits for Bit Sequence 101

Nonoverlapped Mealy detector				Overlapped Mealy detector			
Input	Current state	Next state	Current output	Input	Current state	Next state	Current output
0	S_1	S_1	0	0	S_1	S_1	0
1	S_1	S_2	0	1	S_1	S_2	0
0	S_2	S_3	0	0	S_2	S_3	0
1	S_2	S_2	0	1	S_2	S_2	0
0	S_3	S_1	0	0	S_3	S_1	0
1	S_3	S_1	1	1	S_3	S_2	1

7.8 MEALY DETECTOR CIRCUITS FOR BIT SEQUENCE 101 227

	Q_0	
Q_1	0	1
0	S_1	S_3
1	S_2	X

Input	Current state	Assignment Q_1Q_0	Next state	Assignment Q_1Q_0	Current output
0	S_1	00	S_1	00	0
1	S_1	00	S_2	10	0
0	S_2	10	S_3	01	0
1	S_2	10	S_2	10	0
0	S_3	01	S_1	00	0
1	S_3	01	S_2	10	1
0	S_4	11	Any	XX	X
1	S_4	11	Any	XX	X

FIGURE 7.26 State assignments and STT for Mealy-type circuit detecting bit sequence 101.

flops will be used, which we will designate as Q_1 and Q_0. In Figure 7.26 a set of state assignments and the resulting STT are shown. Since there are only three used states, there is one unused state, and the output of the system is shown as a don't care value for this unused state. For the correct operation of this detector, it must be ensured that the system never drops into its unused state, since this could produce a Logic-1 output, which, obviously, is an error. The same procedures we used for counters can be used here to make certain that the sequential system always drops into one of its correct states even if it starts from its unused state (which could happen when the power is turned on). If we use this procedure, we could assign output 0 for the unused state as well, in which case there would never be an erroneous output; however, if the system is allowed to start from its unused state, it may miss the detection of an incoming bit sequence. As we will show later, we will select asynchronous CLEAR functions and will add one gate to ensure that the system never stays in its unused state; therefore, the don't care selection of the output is a reasonable choice (see Figure 7.29).

The two K-maps for the sequential circuit, shown in Figure 7.27, are constructed in the usual manner from the STT in Figure 7.26. The minimized circuit equations are also shown. The output function can also be constructed from the same STT, and this is shown in Figure 7.28. The output equation is also very simple, and the completed circuit is shown in Figure 7.29. Since the unused state was chosen to have binary values of 11, a NAND gate is

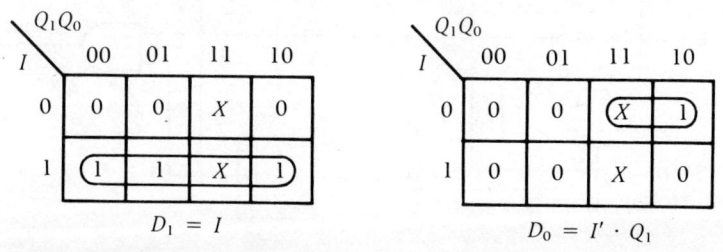

I \ Q_1Q_0	00	01	11	10
0	0	0	X	0
1	1	1	X	1

$D_1 = I$

I \ Q_1Q_0	00	01	11	10
0	0	0	X	1
1	0	0	X	0

$D_0 = I' \cdot Q_1$

FIGURE 7.27 Karnaugh maps for Mealy-type circuit and state assignments shown in Figure 7.26.

228 SYNCHRONOUS CONTROLLERS AND STATE ASSIGNMENTS

I \ Q_1Q_0	00	01	11	10
0	0	0	X	0
1	0	(1	X)	0

Output = $I \cdot Q_0$

FIGURE 7.28 Karnaugh map for output of detector.

used for the asynchronous CLEAR function that becomes active when both flip-flop output values become 1. As shown in Figure 7.29, a NAND gate instead of an AND gate is used for the detection of state 11 because for most practical flip-flops the active CLEAR function requires a ground signal.

7.9 STATE ASSIGNMENT SELECTION

There are many problems associated with the selection of state assignments for sequential systems, which are described by general verbal functional specifications. So far we have covered only a few of these. It is possible to construct a STT (or STD) that has redundant states, which means that the next states and the system output values are the same for two states when identical input conditions are applied. In this case the two states are indistinguishable from each other, and one of the states may be eliminated. The detection and elimination of redundant states have enjoyed a large amount of interest and research, but it is beyond the scope of this introductory text. If the system specifications are well stated and the system designer is careful in selecting system states only when absolutely necessary, the appearance of redundant states can be avoided.

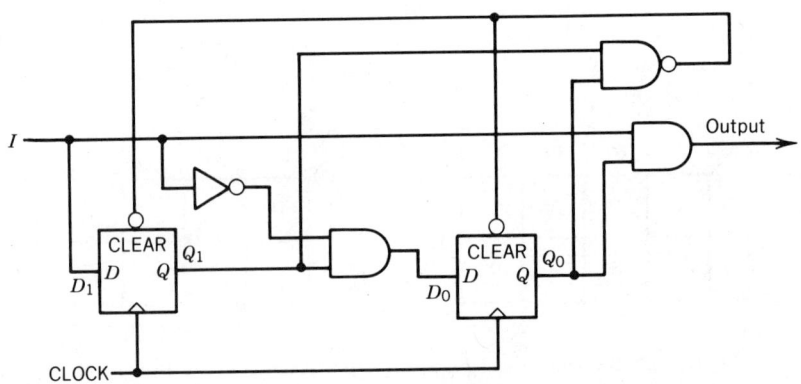

FIGURE 7.29 Completed circuit for Mealy-type detector circuit for bit sequence 101.

Another well-researched area of sequential circuit design is the production of rules for state assignments that produce minimum sequential circuits. This area has had much less success in dealing with general systems, and few, if any, methods can ensure optimized results. Fortunately, with digital hardware prices decreasing, the need for finding the least expensive circuit is rapidly losing its importance. The minimization problem is complicated by the fact that there are two sets of circuits that one must consider. The design of the sequential circuits and the output circuits may be carried out independently of each other, but the state assignments will affect the complexity of both. Since state assignments are used in Experiment 7, we will discuss a few known rules that should provide some help in the selection of state assignments that produce simplified sequential circuits. We will take a pragmatic approach and will describe these rules and their use without much explanation of how they were derived. The basis of any of these rules is to try to cluster 1's or 0's in the final K-maps so that simplifications become possible. It appears that there are a few simple rules that were found to connect the arrangement of states in the SAKM and the arrangement of 1's and 0's in the K-maps for the sequential and/or the output circuits. The most successful of these rules are described in what follows.

As we have seen before, the state assignments are made in a K-map that includes only the stored bits of the system (the input variables are not shown). The rules point out which states should be placed in adjacent rectangular regions, where *adjacent* is used in the same sense as *neighboring* regions were used for the minimization of 1's or 0's in the K-map. When one finds that two states should be adjacent, if possible, the state assignments should be made in such a way that the two states form a group of two neighboring regions. When three or four states should be placed within adjacent regions, they should, for example, form groups of four adjacent K-map regions. The rules that suggest that two states should be placed in adjacent regions in the SAKM are derived from the STT. We shall use only the two most important rules quoted in the literature:

Rule 1 All those states whose next states are the same for the same input values should be given adjacent state assignments.

Rule 2 The next states of a state produced by applying adjacent input conditions should be given adjacent state assignments.

The selection method is used by setting up a "scoring" system. Considering the states in pairs, whenever two states satisfy one of these rules, an extra point is given to that pair of states. Separate scores are kept for the two rules because rule 1 is more important and should be given more weight than rule 2. This means that we should try to satisfy the adjacency conditions that arise from applying rule 1 first. Let us demonstrate this method first by the traffic controller problem, which has only four system states and, therefore, three unique state assignments.

230 SYNCHRONOUS CONTROLLERS AND STATE ASSIGNMENTS

The STT for the traffic controller is shown in Figure 7.5. Applying rule 1, we find that in the first row of the STT (input values 00) the pair of states S_1 and S_4 produce the same state (state S_1) as well as the pair of states S_2 and S_3 (state S_3). In the second row both S_2 and S_3 produce S_3 as the next state. In row 3 the two states S_1 and S_4 produce the next state, S_1, and, finally, in the fourth row no two states produce the same next state for the same input conditions. Hence, the number of times rule 1 could be applied is equal to 2 for the two states S_1 and S_4 and is equal also to 2 for the two states S_2 and S_3. From applying rule 1, we should conclude that state S_1 should be adjacent to state S_4 and state S_2 to state S_3. There are two assignments that satisfy these criteria, namely, the assignments shown in Figures 7.8 and 7.9. For the assignment shown in Figure 7.7 the two states S_1 and S_4 are not in adjacent regions. Therefore, it is eliminated from the set of preferred state assignments.

In order to make a choice between the two assignments that are consistent with the results derived from rule 1, we apply rule 2. Comparing the first and second rows in the STT of Figure 7.5 (adjacent input conditions 00 and 01), state S_1 produces states S_1 and S_2; therefore, these two states should be adjacent to each other. Note that the other three states produce the same next states for the different input conditions; therefore, no useful adjacency conditions may be derived from these entries. Taking the first and third row (another adjacent input conditions, i.e., 00 and 10), the pair of states S_3 and S_4 will be designated as desirable adjacent states. Similar procedures for rows 2 and 4 (input conditions 01 and 11) and rows 3 and 4 (input conditions 10 and 11) provide a total score of 2 for adjacent states S_1 and S_2 and also 2 for states S_3 and S_4. Hence, we should select the state assignment that most closely satisfies the required conditions that states S_1 and S_2 and states S_3 and S_4 should be adjacent. Clearly, the assignment shown in Figure 7.8 is the only one that satisfies these conditions as well as the conditions we derived when we applied rule 1. We find that the state assignments chosen by these rules (shown in Figure 7.8) produce the simplest K-maps among the three possible state assignments, and we can conclude that the rules worked well in this case.

To conclude this chapter, we will work through a design problem that involves a BSD for 6 bits, and we will apply the adjacency rules for its STT.

Design Example: Design Mealy Circuit That Detects Nonoverlapped Bit Sequence 100110

The STD and the STT for this BSD circuit are shown in Figure 7.30. The STD is derived similarly to the earlier examples shown for bit sequence 101. In this case six states are required. The results derived from the application of the two rules for state assignment selection are shown in Table 7.5. We shall assume here that rule 1 is much more important than rule 2. Therefore, the first priority is to satisfy the adjacency of states S_1 and S_6 (score of 2 for rule 1). The next group that should be satisfied has scores of 1 for rule 1 and

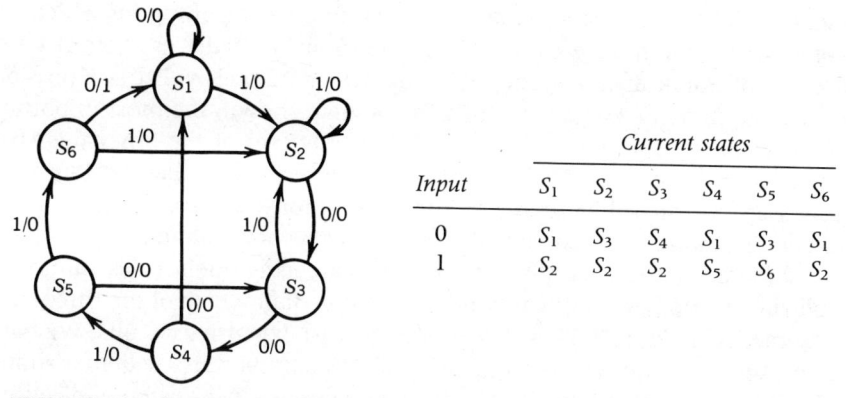

| | Current states | | | | | |
Input	S_1	S_2	S_3	S_4	S_5	S_6
0	S_1	S_3	S_4	S_1	S_3	S_1
1	S_2	S_2	S_2	S_5	S_6	S_2

FIGURE 7.30 The STD and STT for Mealy-type nonoverlapped detector recognizing bit sequence 100110.

TABLE 7.5 Number of Times Rule 1 and Rule 2 Can Be Applied for Adjacent States of Mealy-Type Detector Circuit Recognizing Nonoverlapped Bit Sequence 100110

| Adjacent states | Scores | |
	Rule 1	Rule 2
S_1, S_2	1	2
S_1, S_3	1	0
S_1, S_4	1	0
S_1, S_5	0	1
S_1, S_6	2	0
S_2, S_3	1	1
S_2, S_4	0	1
S_2, S_5	1	0
S_2, S_6	1	0
S_3, S_4	0	0
S_3, S_5	0	0
S_3, S_6	1	1
S_4, S_5	0	0
S_4, S_6	1	0
S_5, S_6	0	0

232 SYNCHRONOUS CONTROLLERS AND STATE ASSIGNMENTS

a score of 2 for rule 2 (state S_1 and S_2); then there is the group for which the scores are 1 for both rule 1 and rule 2, and so on. The adjacent pair of states in order of importance are shown in Figure 7.31 along with three possible sets of assignments. For two of the three K-map shown, the most important four adjacency conditions are satisfied, but only two of the other five conditions can be included. In the third K-map only three of the first four conditions are satisfied, but there are three conditions that cause adjacent assignments among the next seven suggested adjacency conditions.

These examples clearly show that the state assignment rules cannot be applied in a strictly scientific manner. It is not certain which of the three state assignments in Figure 7.31 will generate the least expensive circuits. We have shown this design example in great detail, exactly because it demonstrates that very frequently, in practice, design rules may give inconsistent indications (e.g., the first four adjacency conditions in Figure 7.31 are inconsistent with the condition that S_1 and S_3 should be adjacent). In similar situations the designer must have some experience and develop his or her own heuristic rules, in other words, become an "expert."

	Scores	
Adjacent states	Rule 1	Rule 2
S_1, S_6	2	0
S_1, S_2	1	2
S_2, S_3	1	1
S_3, S_6		
S_1, S_3		
S_1, S_4		
S_2, S_5	1	0
S_2, S_6		
S_4, S_6		
S_1, S_5	0	1
S_2, S_4		
S_3, S_4		
S_3, S_5	0	0
S_4, S_5		
S_5, S_6		

1.

$Q_2 \backslash Q_1Q_0$	00	01	11	10
0	S_1	S_2	S_5	S_4
1	S_6	S_3	X	X

2.

$Q_2 \backslash Q_1Q_0$	00	01	11	10
0	S_2	S_1	S_6	S_3
1	S_5	X	S_4	X

3.

$Q_2 \backslash Q_1Q_0$	00	01	11	10
0	X	S_1	S_2	S_5
1	S_3	S_6	S_4	X

FIGURE 7.31 State assignment selection for STT in Figure 7.30.

A few more words about output circuit minimization. So far we have ignored the fact the the complexity of the output circuits is also influenced by state assignments. One may design the complete sequential circuit on the basis of the preceding two rules and determine both the sequential and the output circuits from a specific state assignment. If the output circuits represent a very small portion of the total circuit costs, then one may accept the results as satisfactory. On the other hand, if the output circuits represent a large portion of the total circuit costs, then it may be worthwhile to change the state assignments and lower the circuit costs for the output circuits. There is a rule for state assignments based on the output of the system:

Output Rule All those states that under adjacent input conditions provide identical output should be given adjacent state assignments.

In case of Moore-type circuits the output is not a function of the inputs. Therefore, the output rule applies to all states that provide the same output.

EXPERIMENT 7
Serially Controlled Combination Lock

E7.1 Summary and Objectives

The design of a serially controlled combination lock for this experiment is based on a bit-sequence-recognizing sequential circuit. The design includes alternate state assignments and minimization for D-type or JK flip-flops. The lock is opened if a programmed sequence of 4 bits is applied to the signal input of the circuit. Alternatively, the circuit turns on an alarm light if an erroneous sequence of 0's and 1's arrives at the signal input. The main objective of this experiment is to solve a practical design problem from a verbal description of its external specifications to the design, construction, and testing of its circuits.

E7.2 Operation of Combination Lock

The combination lock is a synchronous sequential circuit with three inputs and two outputs (including the CLOCK input). Its block diagram is shown in Figure E7.1. The lock operates the door of a safe. When the lock signal L is at the Logic-1 level, the safe is locked. The safe is opened when a given

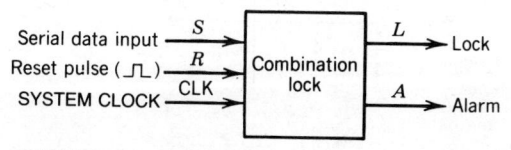

FIGURE E7.1 Block diagram of combination lock.

sequence of 4 bits are sent through the serial input line, S. The safe is opened by lowering the L output to its Logic-0 level, and it stays open as long as the L output is equal to 0.

The operation of the safe is synchronized by two pulse (push-button-type) inputs. One of these is the synchronous SYSTEM CLOCK input. The other pulse type input is the RESET input, which, when operated, places the combination lock system into its starting (reset, or idle) state. The lock remains in its idle state as long as the serial input signal S remains at the Logic-0 level even if CLOCK pulses are generated by operating the CLOCK push button. After the input signal S is set to 1 and the CLOCK button is pushed once, the system is ready to receive the first bit of its required combination bit sequence. For the next four CLOCK pulses the correct input bits must be entered for the S input, after which the lock opens ($L = 0$), and it remains open until input S is set to 0 and the CLOCK button is pushed, or the RESET button is operated.

If any one of the four bits is different from the one required by the given bit sequence, the system drops into its "alarm" state, raises its ALARM output to the Logic-1 level, and keeps the safe locked ($L = 1$). Irrespective of the values of input S or the CLOCK input, the system remains in its alarm state until it is cleared by a pulse input at its RESET input line. The RESET input line overrides all other signals and, when operated, it places the system in its idle state.

E7.3 Bit Sequence and Standard State Transition Diagram

Each student is given a "private" bit sequence. One method is to use the expression SEQ = 1 + n mod 14, where n is the student's number and is an integer between 1 and the number of students in the class, and then the integer SEQ is expressed as an unsigned binary number that yields the correct bit sequence. For example, if $n = 33$, then

$$\text{SEQ} = 1 + (33 \bmod 14) = 5 = 0101$$

This method avoids the two sequences 0000 and 1111. The sequence of bits are labeled from B_0 to B_3, as shown in Figure E7.2.

The suggested standard STD uses seven states, is of the Moore type, and is shown in Figure E7.3. The two outputs shown are for the L (LOCK) and the A (ALARM) outputs of the system. The STD is shown for general bit values B_0–B_3. As shown, the RESET pulse transfers the system into its idle state independently of the state of the system. Complete the STD by placing

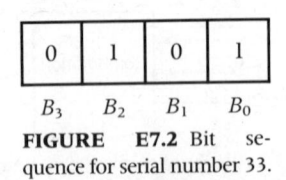

0	1	0	1
B_3	B_2	B_1	B_0

FIGURE E7.2 Bit sequence for serial number 33.

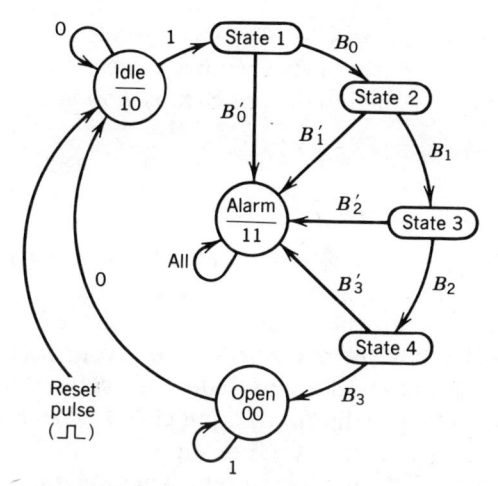

FIGURE E7.3 The STD for digital lock.

the correct incoming bit values for your combination lock system and construct the STT from your STD.

E7.4 Design Requirements

The minimum requirement is that you work through at least two versions of your combination lock system. One of these should be the standard circuit whose STD is given in Figure E7.3 and that uses three JK flip-flops. Use the assignment selection method described in Section 7.9 to choose your state assignments. After the selection of the state assignments, construct the STT in terms of the flip-flop output values Q_2, Q_1, and Q_0. Either generate three four-variable K-maps and use the partition method to find the six three-variable K-maps for the input circuits of the three flip-flops or use flip-flop excitation tables to generate six four-variable K-maps for these circuits. Determine the minimized design equations, sharing common circuits, whenever possible. Draw the circuit diagrams and, if you wish, use gate transformations to fit the circuit into as few IC packages as possible. Finally, design the two output circuits, checking whether the "unused" state could open the safe or switch on the alarm.

For the second version of your design, you may select another state assignment and use D-type or T-type flip-flops, or you may wish to be adventurous and try another form of STD. You may use a Mealy-type circuit or the one-flip-flop-per-state method. If you use a Mealy-type circuit, you must use an extra flip-flop to synchronize the input line or use the extra flip-flop to hold the safe open when it is in its open state independent of the value of its S input. Be careful to satisfy the stated operating characteristics of the combination lock, whichever system you use. Your system must operate similar to the standard STD in Figure E7.3. Work this design through and generate your circuit diagrams for this second version of the combination lock. You may choose either one of your designed systems for wiring, but before you start the construction of your circuits, you must check your design first.

E7.5 Checking and Testing Your Circuits

Since the design and construction of this experiment will be time consuming, it is essential for you to become absolutely certain that your design equations are correct before you construct your circuit. You have to check only one of your two designed circuits, the one you decided to construct. The best method is to use an independent checking procedure for your system. From the six (or three if you used D-type flip-flops) input circuit diagrams generate six (or three) Boolean expressions. For each of the seven used states evaluate the flip-flop input values, and from the current states and input conditions evaluate the next states of the system (in terms of 1's and 0's, of course). From the results of these calculations you can construct the STD of your system. Compare it to your original diagram.

The advantage of working through your system in terms of 1's and 0's is that if your circuits do not work, you can compare the measured signal values in your circuits to your calculated signal values. Use three (or as many as the number of flip-flops you used) lights to show the current state of the system. Start by "resetting" your circuits (by pushing the reset button), and follow through the sequence by setting first the required input bit value for S and pressing the clock button once. If your system drops into an unexpected state, then restart the procedure from the idle state and stop just one state before the error occurred. Check the inputs to your flip-flops, which, according to your checking procedure, should yield your next correct state. Most probably, some of the flip-flop input values will be in error (because of missing or miswired connections), and you are on your way to locating your problem.

E7.6 Alternate Experiments

It is also possible to use a partial one-flip-flop-per-state method. For example, you may use three flip-flops to indicate that the system is in the idle, open, or alarm state. Hence, when the outputs for these three flip-flops are 100, the system is in the idle state; when they are 010, it is in the open state; and when they are equal to 001, it is in the alarm state. When the outputs of these three flip-flops are equal to 000, the system is in one of its four other states (state 1 to state 4). You may use two flip-flops to represent the four extra states. For these four states only three unique assignments are possible, so minimization can be tried for all three of them.

Another way to change slightly this design problem is to use two signal inputs. Hence, the bit sequence changes to number sequence because the inputs could be from 0 (inputs 00) to 3 (inputs 11). In this case there are 4^4 = 256 different number sequences (beginning with 0000 and ending with

3333). It is true that the K-maps become five-variable maps in this case, but if you use the partition method for JK flip-flops, only four-variable maps will be needed.

Finally, you may add a 3-to-8 decoder to the output stage of this controller and design your system as a state controller. In this case you do not need output circuits at all and may find some of the other circuits simplified as well.

References

1. McCluskey, E. J., and Ungar S. H. A note on the number of internal variable assignments for sequential switching circuits. *IRE Trans. Electr. Comp.*, **EC8,** 439–440 (1959).

Drill Problems

7.1. Determine the K-maps and the minimized circuit equations for the three state assignments shown in Figure 7.31. Determine the K-maps and the minimized circuit equations also for a simple assignment of $S_1 = 000$, $S_2 = 001$, $S_3 = 010$, and so on. Did the rules select the best set of state assignments?

For all of the design problems below the systems can be designed as a state controller circuit (with an added decoder), a Moore- or a Mealy-type controller, or a circuit that uses the one-flip-flop-per-state method. Hence, each design problem may be solved in four different ways.

7.2. A bit-sequence generator circuit has one input, one output, and two operating modes. In mode 1 its output produces Logic-0 bits; in mode 2 it produces alternating 1's and 0's (sequence 01010101 . . .). It remains in one mode as long as the input remains at the Logic-0 level. For each Logic-1 input bit it switches its mode. Design a sequential circuit for this system.

7.3. Design a bit-sequence generator circuit with three operating modes, one output, and two input variables. In mode 1 it outputs 0's, in mode 3 it outputs 1's, and in mode 2 the outputs alternate between 1's and 0's. The two-input bits control the operation of this system in the following manner:

Input bits	Action
00	Remain in current mode
01	Switch to mode 1
10	Switch to mode 2
11	Switch to mode 3

7.4. Design a controller for a toll collector machine. The machine accepts nickels and dimes, and the correct toll is 15 cents. The machine has two outputs. One is a light that is either green or red, the other is an alarm that rings a bell if a car passes a red light. The light becomes green if the inserted coins total or exceed 15 cents. There are two input lines (plus SYSTEM CLOCK input). The following meanings are associated with the input bits:

Input bits	Meaning
00	Reset the system (turn alarm off)
01	A nickel was deposited
10	A dime was deposited
11	The car has passed the light

7.5. Any BSD may be used to detect the complement of its bit sequence by the addition of one inverter. The circuits for detecting bit sequence 101 were already designed. With the addition of an inverter the same circuits may be used to detect the sequence 010. Hence, there are only four unique 3-bit sequences; the other three are 111, 110, and 100. Design any number of BSDs for Mealy- and Moore-type circuits and for the overlapped and nonoverlapped cases for these additional 3-bit sequences, and among the ones you have designed, find the bit sequence that requires the detector with the simplest circuits.

7.6. Design a sequential circuit that detects the occurrence of at least three 1's arriving at the input. There are three versions of this detector:
  (a) (Simple) The detector starts in its initial state (output equal to 0), and its output remains 0 until three 1's arrive at the input. The three 1's do not have to arrive during consecutive clock periods (in this mode 0's are ignored). Once the output becomes 1, it remains 1 as long as the input bits are equal to 1. When the first 0 is received, the detector drops back into its initial state.
  (b) (More difficult) This detector works similarly to the simple one, but its output remains 1 until it resets itself by receiving three consecutive 0 input bits (whenever the input sequence 000 is detected whereas its output is equal to 1).
  (c) (Most difficult) This detector works similarly to version (b), but the detector resets itself (drops into its initial state) at any time three successive 0's are received at the input (regardless of the output value of the decoder).

7.7. Design a pattern recognizer with one output and one input that recognizes an input sequence of three equal bits (sequence 000 or 111).

7.8. Design a sequence comperator circuit that has two inputs and one output and whose output is equal to 1 whenever the last three received pairs of input bits are equal. This detector detects the condition that the last two 3-bit sequences received at the two inputs are the same.

CHAPTER 8

Number Representation and Computer Arithmetic

8.1 INTRODUCTION

We have been studying binary or logic variables as examples of general two-valued variables. The digital or logic design problems we have discussed did not include numerical variables; the numbers 0 and 1 always meant Logic-0 and Logic-1 values. Logic variables may be used for logic operations such as AND, OR, XOR, and so on, or as outputs to logic processors that may determine solutions to complex problems with yes or no answers. One such processor was used for the solution of the digital lock problem (Experiment 7), which decided whether a correct sequence of 1's and 0's had been received by the system.

One of the most important applications of digital computers is numerical computation. For numerical computations input and output data must be in the form of commonly used numbers. Internally, most digital computers can store and process only binary variables; therefore, a correspondence must be made between commonly used numbers and their representation inside the computer. We wish to show here only the most elementary concepts of digital computer number representations and arithmetic. These are the binary representations for both positive and negative integers and the binary operations for the addition and the subtraction of these numbers. We will also limit our discussion to number representations and arithmetic operations that use the same number of bits for each number. The limitation is appropriate to most current digital computers, which are organized around a fixed number of bits for most of their arithmetic operations.

With the introduction of **registers** and **register transfer** operations we will start to draw parallels between logical circuits and computer digital hardware which we will continue to expand in the succeeding chapters. We will assume that the student is familiar with at least one higher level computer language such as BASIC, FORTRAN, or Pascal and will start to obtain some insight into the inner workings of computers. In particular, we will investigate how some of the simple computer program statements are executed by a digital computer using its own "private" machine language.

Experiment 8, a serial arithmetic unit, will be the first experiment described in terms of a collection of digital subelements, an essential requirement for the description of complex systems. This experiment will demonstrate how

8.2 REGISTERS AND REGISTER TRANSFER OPERATIONS

combinational and sequential systems are controlled and connected together to form a single, more complex system that performs a given task.

The concept of a register is fundamental to the understanding of the operation of digital computers and all complex digital systems. In fact, if you look at the hardware description of any modern digital computer (from the microcomputers to the supercomputers), you will find that the detailed operation of the computer is described in terms of its registers. In general, a register is defined as an ordered number of stored bits. The number of bits in a register is called its **length.** For easier identification, the register is given a name that represents all the bits that belong to the register and that is often used in place of the actual bit values stored in the register. For a register containing N bits, the individual bits are identified by indices of integer numbers. The least significant bit (LSB) has index 0, the more significant bits have indices 1, 2, 3, and so on, whereas the index of the most significant bit (MSB) is equal to $N - 1$. A register is often represented graphically as a long rectangle with the bits of the register identified by $NAME_{index}$. A 6-bit register called R is shown in Figure 8.1. The bit pattern stored for this register is equal to 101101, which is called the **contents** of register R. Often the contents of a register is indicated by an equal sign, which is shown in this case as $R = 101101$. This means that R_0 (LSB) $= 1$, $R_1 = 0, \ldots, R_5$ (MSB) $= 1$ (see Figure 8.1).

All the bits of a register form a unit. Logical operations on the bits of a register may be indicated by the logical operation on the name of the register. For example, the symbol R', or $\overline{R}$, indicates the complement of the bits of register R. Hence, the contents of the complement register, or R', is equal to 010010, or, alternatively, we can write that $R' = 010010$.

Logical operations are also possible between two registers of equal length. For example, if the contents of another 6-bit register, say X, are binary 110100 ($X = 110100$), then the symbols $A + X$ (A OR X) indicate the ORing of the bits of the two registers (R_0 is ORed with X_0, R_1 with X_1, etc.). The result is then equal to the binary number $A + X = 111101$. Similarly, the symbols $A \cdot X$ indicate the logical AND function with the result $A \cdot X = 100100$.

Another fundamental concept for digital systems is the register transfer

Register R

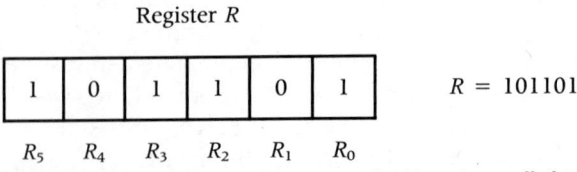

1	0	1	1	0	1
R_5	R_4	R_3	R_2	R_1	R_0

$R = 101101$

FIGURE 8.1 Graphical representation of 6-bit register called R and its contents.

8.2 REGISTERS AND REGISTER TRANSFER OPERATIONS

operation. Register transfer is indicated by the left arrow symbol, $\leftarrow$. The symbols $X \leftarrow R$ mean that the contents of register R are transferred to register X without changing the contents of the R register. After the operation $X \leftarrow R$ is executed, the contents of register X will be equal to the contents of register R. Although the logical operations on a register can be executed by a combinational circuit (e.g., the $X \cdot R$ operation can be implemented by N two-input AND gates), the register transfer operation is a "sequential" operation; it requires a significant CLOCK transition after which the results of the operation are stored in the register on the left side of the register transfer operation symbol (no clock signal is needed for the register on the right). The symbolic indication of a register transfer operation and its hardware implementation are shown in Figure 8.2, where N D-type flip-flops are used for storing the contents of each register. Obviously, there is similarity between register transfer operations and assignment statements such as $A = B$ in BASIC and FORTRAN and $A := B$ in Pascal. In fact, the variable names in higher level computer languages may be thought of as names of registers and the assignment statements as register transfer operations. As in BASIC, FORTRAN, or Pascal, operators and register transfer operations may be combined. The results of the two register transfer operations $A \leftarrow (X + R)$ and $B \leftarrow (X \cdot R)$ are shown in Figure 8.3, and the hardware implementation of one of these register transfer operations is shown in Figure 8.4, where only one CLOCK signal is shown for register A. This is the usual way of indicating that the CLOCK signal operates on all the storage elements of the register.

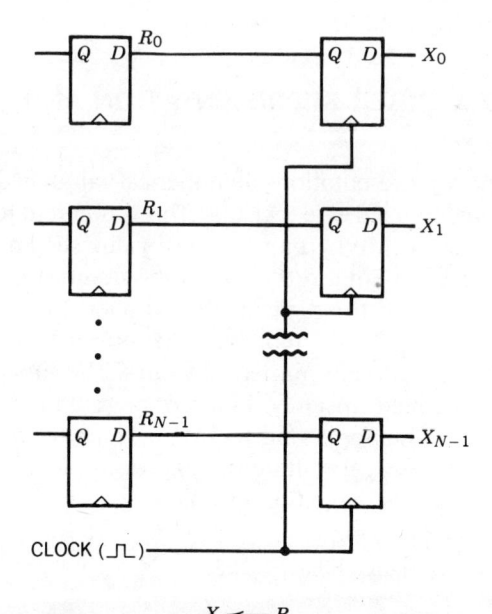

FIGURE 8.2 Register transfer operation and its hardware realization using D-type flip-flops.

|1|0|1|1|0|1| |1|1|0|1|0|0|

R   X

|1|1|1|1|0|1| |1|0|0|1|0|0|

A   B

$A \leftarrow R + X$ $B \leftarrow R \cdot X$

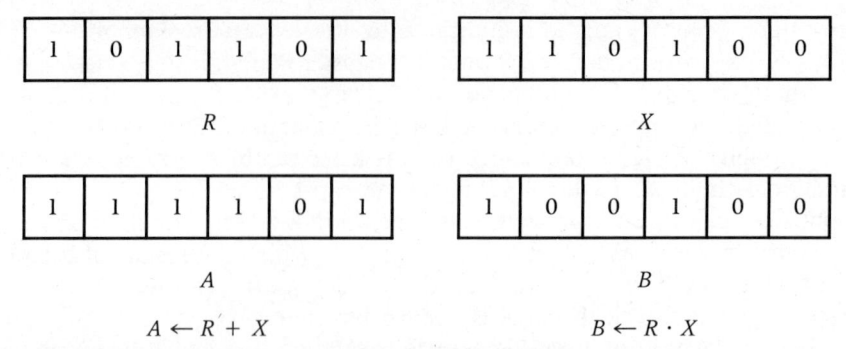

FIGURE 8.3 Combining binary (logical) and register transfer operations.

In order to discuss arithmetic operations on the contents of registers, a correspondence between the stored bits in the register and the numerical value represented by the contents must be established. It is important to realize that at the most basic level all information is given in terms of binary variables or bits and that the same bits could be interpreted in various ways, giving a different meaning to the contents of a register. In most modern digital computers registers can store any types of data. Hence, the same bits may represent unsigned (positive) or signed integers, floating-point numbers, printable characters, or even computer instruction codes. It is the digital processor, the sequential and combinational elements of the computer, that must "know" which register contains which type of data and handle the various types of information accordingly. We proceed by showing three different ways of representing numerical data in a register.

8.3 UNSIGNED NUMBER REPRESENTATION AND ARITHMETIC

We discuss the binary representations of numerical values in digital computers in terms of bits stored in fixed-size registers. This is different from the ordinary representation of numbers. When a numerical value such as 92,578 or 12 is written down, there is no limit to the size of the number as long as the page is large enough to hold all the digits of the number. In a fixed-size register with N bits there are only 2^N different combinations of 1's and 0's; therefore, there could be only 2^N different numerical values. The simplest number representation is the so-called unsigned binary representation, which we have already introduced in Chapter 4, and which provides positive integers in the range of from 0 to $2^N - 1$. Recalling the expression for evaluating the numerical value of N bits, we have the formula

$$\text{Unsigned binary value} = \sum_{i=0}^{i=N-1} R_i \cdot 2^i$$

where the bits of the number, R_i, are interpreted as numbers 1 and 0 (of course, Logic-1 is equivalent to value 1 and Logic-0 to 0), and the numerical sum determines the value of the number.

8.3 UNSIGNED NUMBER REPRESENTATION AND ARITHMETIC

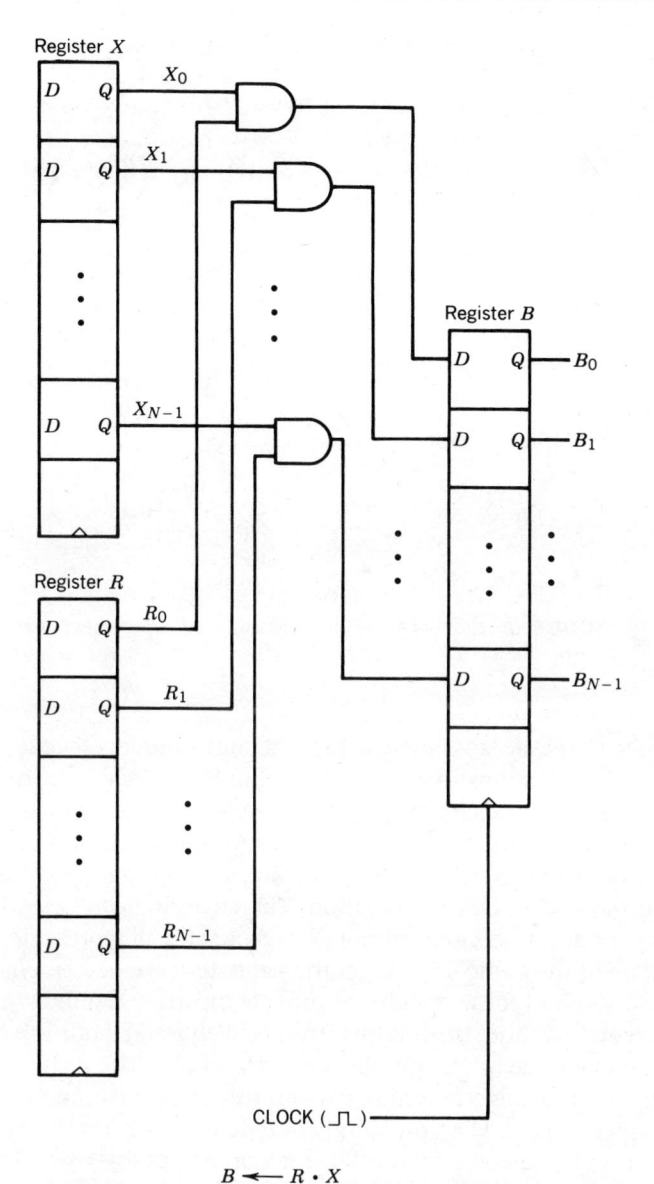

$$B \leftarrow R \cdot X$$

FIGURE 8.4 Hardware realization of logical AND operation with register transfer.

We will describe the numerical representations for binary numbers using 4-bit numbers as examples. The unsigned binary representation for the 16 possible values of 4-bit numbers is shown in Figure 8.5. The numbers are placed around a circle. The 16 binary combinations of 4 bits are shown on the outside, whereas the numerical values attributed to these bit values are shown on the inside of the circle. This circle is equivalent to the number line, which is used traditionally for natural numbers. For fixed-size numbers the

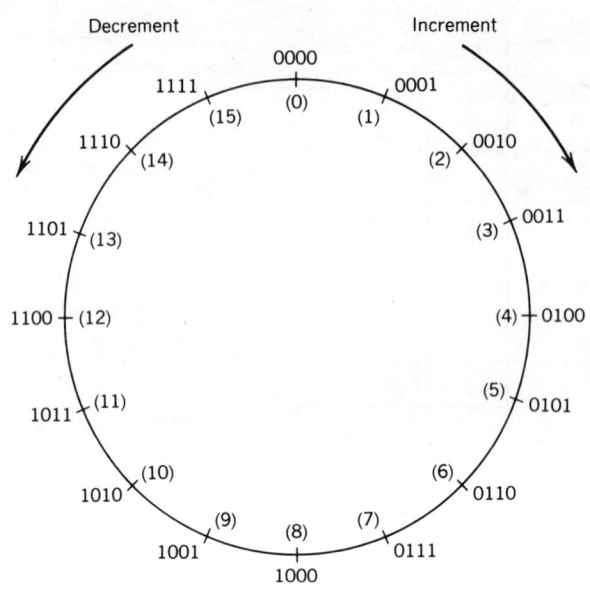

FIGURE 8.5 Number circle for unsigned binary representation of 4-bit numbers.

line becomes a circle because there are only a finite number of values expressed by different combinations of 1's and 0's around the circle.

Moving around the circle represents simple counting, which is the basis of all arithmetic operations. Moving in a clockwise direction is equivalent to the incrementing operation. Similarly, moving in a counterclockwise direction is equal to the decrementing operation. The circle indicates clearly that incrementing (or decrementing) cannot be executed for all representable values. If we increment the value 15, we get the value 0; if we decrement the value 0, we get 15; both of these operations provide incorrect results. We call these "errors" **overflow** and **underflow,** respectively, and both are the direct result of the fixed size of our number representations. The result of an arithmetic operation is incorrect because of overflow when the numerically correct result is larger than the largest number represented. If the binary number 1111 (value 15) is incremented, then the correct result (16) is beyond the capacity of 4 bits, and overflow occurs. It is incidental that the result becomes the binary 0000 because this depends on the representation of the numbers we have chosen. The important information is that overflow occurred. Similarly, if the number 0000 (value 0) is decremented, then the numerical result should be equal to -1, which is also beyond the capabilities of our number representation. In this case an underflow occurred. For correct operation of a digital computer, its arithmetic processor must be able to detect both overflow and underflow in addition to its ability to perform arithmetic functions.

The numerical operation of **addition** for unsigned binary numbers is executed similar to ordinary number addition with the bit values interpreted

8.3 UNSIGNED NUMBER REPRESENTATION AND ARITHMETIC

as binary 1's (which we will designate sometimes as $+1$) and 0's ($+0$) instead of their logical equivalents. The rules for adding two binary numbers are shown in Figure 8.6, where the word *plus* is used for the addition operator instead of the normal plus sign so as not to confuse it with the logical OR operation. We can see that it is possible to generate a carry when two binary 1's are added. Similar to ordinary decimal numbers, the addition of two binary numbers proceeds from the LSBs toward the more significant bits with the carries added to successive bits of the two addends. The carry generated by the LSBs, CY_0, is added to the sum of the next higher significant 2 bits, that is, to A_1 and B_1. All but the least significant columns require the addition of three bits. Two bits are from the addends, whereas the third bit is the carry bit brought over from the less significant column. The arithmetic sum of these 3 bits is equal to the bit of the result, and a new carry bit may be generated that is used in the higher order column. The general case of adding two 4-bit numbers is shown with an example:

Carries (added) $\rightarrow$ CY_2 CY_1 CY_0	1 1 0	Values
Bits of A $\rightarrow$ A_3 A_2 A_1 A_0	0 1 1 1	(7)
Bits of B $\rightarrow$ B_3 B_2 B_1 B_0	0 1 1 0 plus	(6)
Bits of A plus B $\rightarrow$ R_3 R_2 R_1 R_0	1 1 0 1	(13)
Carries (generated) CY_3 CY_2 CY_1 CY_0	0 1 1 0	

In this case the carry from the most significant column, CY_3, is equal to 0; therefore, no overflow occurred, and the resulting bit values correctly represent the numerical value for A plus B ($6 + 7 = 13$).

Now we would like to build logic circuits that perform the addition of unsigned binary numbers. Let us consider the addition of 2 bits first. By interpreting the binary value of $+1$ as Logic-1 and the binary value of $+0$ as Logic-0, the truth tables, K-maps, and Boolean equations for the binary addition of 2 bits are calculated as shown in Figure 8.7. The SUM bit is equal to the XOR and the CARRY bit to the AND function of the two addend bits.

In the least significant column only 2 bits are added, but in the higher significant columns there are 3 bits. The rules for adding 3 bits may be derived from the rules for 2 bits, shown in Figure 8.6 by adding either $+1$ or 0 to the sum of 2 bits. These rules are shown in two truth tables, K-maps, and

	Sum	Carry
0 plus 0	0	0
0 plus 1	1	0
1 plus 0	1	0
1 plus 1	0	1

FIGURE 8.6 Rules for adding 2 binary bits.

246 NUMBER REPRESENTATION AND COMPUTER ARITHMETIC

Addition of 2 bits

AB	Sum	Carry
00	0	0
01	1	0
10	1	0
11	0	1

Sum $= A \oplus B$

Carry $= A \cdot B$

Sum K-map:

	B=0	B=1
A=0	0	1
A=1	1	0

Carry K-map:

	B=0	B=1
A=0	0	0
A=1	0	1

FIGURE 8.7 Truth tables and K-maps for addition of 2 bits.

Boolean expressions in Figure 8.8. The two-input, two-output combinational circuit that performs the addition of 2 bits and provides the SUM and CARRY outputs is called a **half adder** (shown on the left of Figure 8.9). Clearly, a half-adder circuit is required to add the two LSBs of two binary numbers. The three-input, two-output logic circuit that adds 3 bits is called a **full adder.** For the addition of the higher significant columns full adders are required. The full-adder circuit may be designed according to the Boolean equations shown in Figure 8.8, but we use a different method for constructing this circuit because it results in a simpler circuit. Since the addition of 3 bits is the same as the addition of 2 bits plus the addition of the sum to another bit, we can use two half adders to build a full adder. This is shown on the right of Figure 8.9. The resulting SUM is simply equal to the "sum" of the SUM

Addition of 3 bits

A B C	Sum	Carry
0 0 0	0	0
0 0 1	1	0
0 1 0	1	0
0 1 1	0	1
1 0 0	1	0
1 0 1	0	1
1 1 0	0	1
1 1 1	1	1

Sum $= A \oplus B \oplus C$

Carry $= A \cdot B + B \cdot C + A \cdot C$

SUM K-map:

A \ BC	00	01	11	10
0	0	1	0	1
1	1	0	1	0

CARRY K-map:

A \ BC	00	01	11	10
0	0	0	1	0
1	0	1	1	1

FIGURE 8.8 Truth tables and K-maps for adding 3 bits.

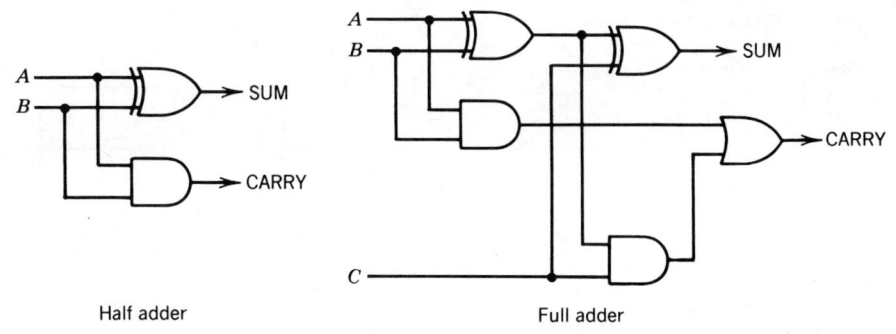

8.3 UNSIGNED NUMBER REPRESENTATION AND ARITHMETIC 247

FIGURE 8.9 Combinational circuits for the full and half adder.

of 2 bits (bits *A* and *B*) and the *C* bit. The CARRY may be generated either by the addition of the 2 bits (*A* and *B*) or by the second addition; hence, the CARRY bit of the full adder is equal to the logical ORing of the carry bits of the two half adders. This method of designing the full-adder circuit is a good example of practicing functional design, which in this case is not only simpler than the construction and minimization of K-maps, but it yields simpler circuits as well.

Before considering the required circuits for adding two numbers that both have *N* bits, we will briefly review the **subtraction** operation and its logic circuit implementations for unsigned binary numbers. Rules for binary subtraction may not be as well known as those for addition. The rules are similar to the subtraction of decimal numbers with the possibility of a **borrow** generated when two digits are subtracted. If 0 is subtracted from a binary digit (0 or 1) it remains unchanged without a borrow present. A binary 1 subtracted from 1 results in binary 0 without a borrow and finally, a 1 subtracted from 0 results in 1 with a borrow generated (BORROW = 1). These rules are summarized in Figure 8.10. We can see that the DIFFERENCE (result of subtraction of two binary digits) has the same truth table as the SUM (result of addition). The BORROW signal, however, is different from the CARRY; thus, if for unsigned binary numbers both addition and subtraction are required, then the binary adder circuitry must be expanded so that it can handle subtraction as well.

In order to subtract two 4-bit unsigned binary numbers, similarly to

	Difference	*Borrow*
0 minus 0	0	0
0 minus 1	1	1
1 minus 0	1	0
1 minus 1	0	0

FIGURE 8.10 Rules for subtracting 2 binary bits.

248 NUMBER REPRESENTATION AND COMPUTER ARITHMETIC

Subtraction of 2 bits

AB	Difference	Borrow
00	0	0
01	1	1
10	1	0
11	0	0

Difference = $A \oplus B$

Borrow = $A' \cdot B$

K-map for Difference:

	B=0	B=1
A=0	0	1
A=1	1	0

K-map for Borrow:

	B=0	B=1
A=0	0	1
A=1	0	0

FIGURE 8.11 Truth tables and K-maps for subtraction of 2 bits.

addition, we subtract two digits for the LSB but may require to subtract the borrow from the difference of the digits for the higher order columns. This is demonstrated with a numerical example below:

Borrows (subtracted) → BW_2 BW_2 BW_0	1 1 0	Values
Bits of A → A_3 A_2 A_1 A_0	1 1 0 1	(13)
Bits of B → B_3 B_2 B_1 B_0	0 1 1 0 minus	(6)
Bits of A minus B → R_3 R_2 R_1 R_0	0 1 1 1	(7)
Borrows (generated) BW_3 BW_2 BW_1 BW_0	0 1 1 0	

From the rules stated in Figure 8.10 the truth tables, K-maps, and design equations for the **half subtractor** can be determined as shown in Figure

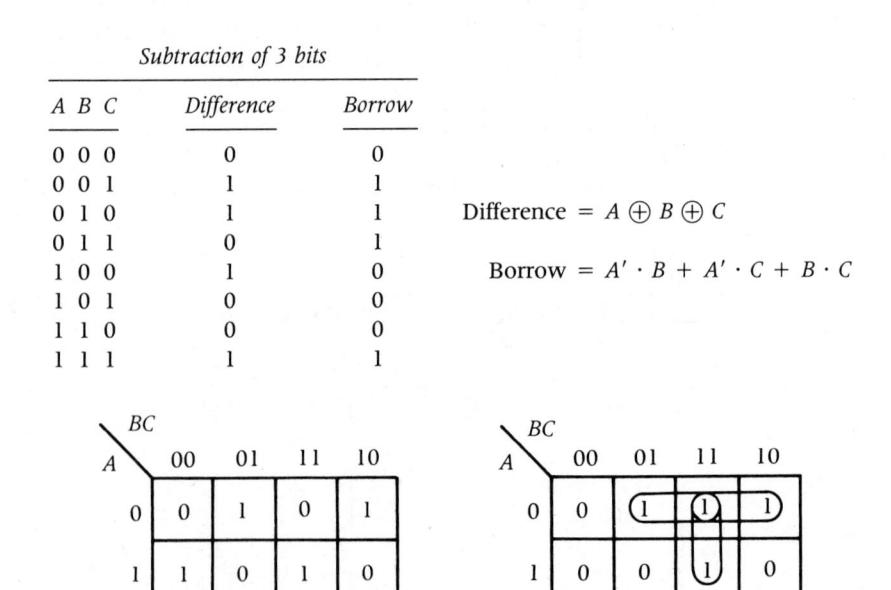

Subtraction of 3 bits

A B C	Difference	Borrow
0 0 0	0	0
0 0 1	1	1
0 1 0	1	1
0 1 1	0	1
1 0 0	1	0
1 0 1	0	0
1 1 0	0	0
1 1 1	1	1

Difference = $A \oplus B \oplus C$

Borrow = $A' \cdot B + A' \cdot C + B \cdot C$

DIFFERENCE K-map:

A\BC	00	01	11	10
0	0	1	0	1
1	1	0	1	0

BORROW K-map:

A\BC	00	01	11	10
0	0	1	1	1
1	0	0	1	0

FIGURE 8.12 Truth tables and K-maps for subtracting 3 bits.

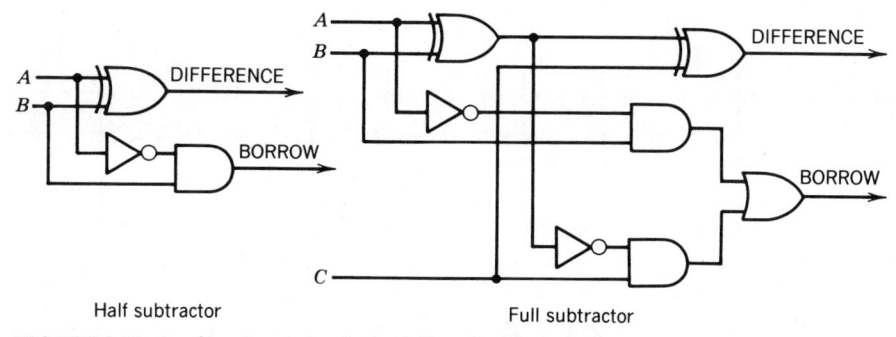

FIGURE 8.13 Combinational circuits for full and half subtractor.

8.11. Similar to the adder, we also need a subtractor for 3 bits in order to perform a complete subtraction of two binary numbers. The 3-bit subtractor, or **full-subtractor,** circuit has three binary input signals (*A*, *B*, and *C*) and the difference (*A* minus *B*) minus *C* is generated by it. The rules for the 3-bit subtraction can be derived by applying the rules of subtraction for 2 bits first and than to the difference of the result and the third bit (*C*). The results are shown in the truth tables in Figure 8.12, where both the K-maps and the design equations are also included. The circuits for the half and full subtractors are shown in Figure 8.13, where the full subtractor was designed by using two half-subtractor circuits, similar to the full adder in Figure 8.9.

It is true that subtractor circuits are seldom used in practice because, as will be shown in Section 8.5 (where the 2's complement representation of negative numbers will be introduced), it is easier to perform subtraction by adding the negative of a number than by providing a separate subtracting circuit. However, in a system that handles positive numbers only, and the "negation" operation (generate the negative representation of a positive number) does not exist, subtraction is still necessary. We may notice that the adder and the subtractor have equivalent circuits for their SUM outputs, but the BORROW circuit is different from the CARRY circuit and two additional inverters are needed for the subtractor.

8.4 PARALLEL ADDITION AND SUBTRACTION

Two *N*-bit binary numbers may be added (or subtracted) with one half adder and $N - 1$ full-adder circuits (we use subtractor circuits if subtraction is required). This type of operation is called **parallel** because all the bits of both operands are available at the same time, and the operation can occur in one step. Also, there is a parallel structure of hardware, that is, *N* similar adder (or subtractor) circuits are working in parallel. In the next section we will study **serial** operations when the bits of the numbers are made available at the rate of 1 bit per clock pulse, and when an operation lasts for *N* clock periods. We gain time in parallel operations and have to pay the price with increased hardware costs. The circuit connections for the parallel addition of two 4-bit numbers are shown in Figure 8.14. The connections are exactly the

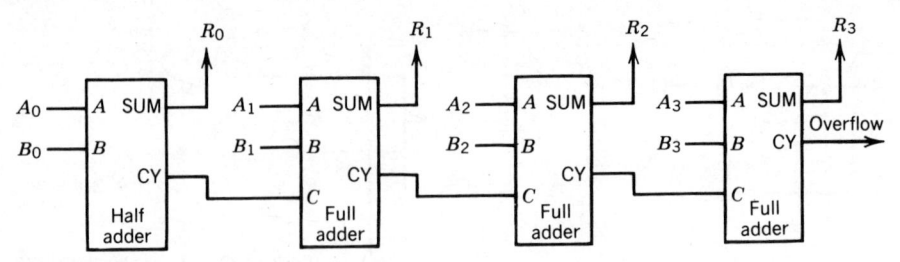

FIGURE 8.14 Parallel addition of two 4-bit numbers.

same for a 4-bit subtractor, for which half- and full-subtractor circuits replace the adder circuits in Figure 8.14. We can see that the carries (borrows) generated for 2 bits are used as the third input bit for the adders (subtractors) that operate on the next higher order 2 bits. The CARRY (BORROW) output from the MSBs indicates overflow (underflow) for unsigned (positive) binary numbers. If this bit is equal to Logic-1, then overflow (underflow) occurred. The presence of overflow (or underflow) is not equivalent to the carry (borrow) value from the MSB column when both positive and negative numbers are used. This simple equivalence works only for unsigned binary numbers. The problem of detecting overflow (underflow) for signed binary numbers will be discussed later in this chapter. The parallel hardware structure is the direct equivalent of the addition (subtraction) example we showed in Section 8.3. The four output bits of the adder circuits form the result register with bits R_0–R_3. The eight input bits to this full-adder circuit are from the addend A and B registers. This circuit performs the operation $R = A$ plus B (arithmetic sum of A and B). A similar circuit with subtractors would perform subtraction.

There are many different practical IC devices that perform binary addition. The most common ones are the 1-bit (type 7480), 2-bit (7482), and 4-bit (type 74283) full adders, and each of these multibit adders is contained in one IC package. These adders provide FAST CARRY outputs, which means that they contain additional circuitry that generates the CARRY (out) signal faster than the so-called RIPPLE CARRY arrangement shown in Figure 8.14. We will return to the discussion of propagated (ripple) and parallel carry generation in the next chapter, where parallel- and serial-type processes will be studied.

8.5 SERIAL ADDITION

We have seen that parallel addition or subtraction requires a repeated hardware structure and that the number of such structures is equal to the number of processed bits. For digital systems with a large number of bits per word the fully parallel processing of information could require very large and expensive hardware. When the speed of processing is not extremely important, considerable savings in hardware can be achieved by the technique of serial data processing. Fully serial data processing is based on a serial data input and output. The bits of the data operands are transmitted on so-called serial data input or output lines, which need only one digital input signal for each operand. The processing clock signal synchronizes data transmission, and successive bits of the operands are transmitted between successive clock pulses. Usually, the bits are transmitted in sequence, and as we shall see later, for serial addition or subtraction the LSB of the number has to be transmitted first.

We have seen that during the parallel addition (or subtraction) process a carry (or borrow) is generated by each full adder that is used as input to the adder for the higher significant bits of the addends. Hence, bits A_i, B_i, and C_i (C_i is the carry input for the ith column, as shown by examples in Section 8.3) generate the carry output CY_i, which is used as the carry input for the next column, that is, $C_{i+1} = CY_i$. If bits of the two numbers A_i and B_i are available serially, then during the ith clock period the carry CY_i is generated by a full adder, and if during the next clock period the more significant bits A_{i+1} and B_{i+1} are transmitted, then by delaying the saved carry bit one clock period, the correct carry input can be made available to the appropriate bits of the addends. A simple D-type flip-flop can be used for the delay of exactly one clock period. This arrangement of a full adder and a D-type flip-flop, which delays the carry signal, is shown in Figure 8.15.

The serial data addition process generates 1 bit of the result on the serial data output line at any one time. Between two clock pulses the two serial data input lines transmit bits A_i and B_i. The full adder generates bit R_i for the sum, and CY_i for the carry. The third input to the full adder is equal to CY_{i-1} (CY_{i-1} is the delayed carry bit; hence, $C_i = CY_{i-1}$, which is the correct expression for the carry input). This procedure is equivalent to the parallel addition process, except that the bits of the result are generated serially in time, that is, 1 bit per clock period. In addition to the very significant simplification of hardware (only one full adder is required), this serial adder can produce the sum of two numbers with an arbitrary number of bits. There is no limit to the generation of the sum as long as one is patient enough to wait for the arrival of all the bits of the result. Of course, if a full subtractor is substituted in place of the full adder in Figure 8.15, the serial subtraction of two numbers can be executed.

We have shown that the addition of the 2 bits of the addends A_i and B_i and the carry input C_i are correctly handled by the serial adder in Figure 8.15, but the equation $C_i = CY_{i-1}$ is valid only if i is larger than 0. During the very first clock period, when the input data bits are A_0 and B_0 (LSB), the third

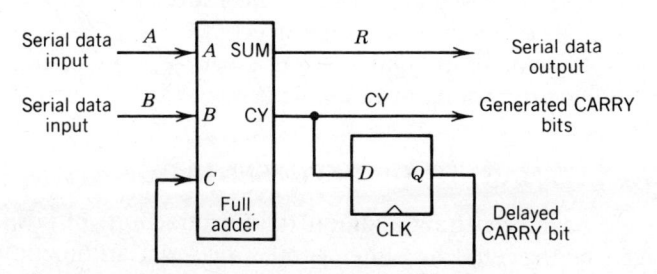

FIGURE 8.15 Serial addition of two binary numbers.

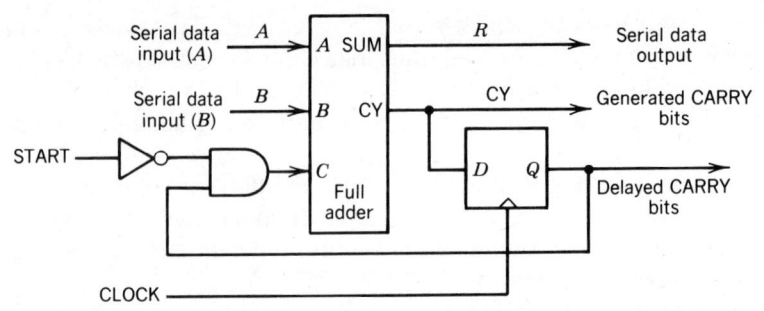

FIGURE 8.16 Circuit for serial addition of two binary numbers that includes the correct carry input for LSBs.

input to the full adder is not specified. Normally, the third input is equal to the output of the D-type flip-flop, which is equal to the input of the flip-flop one time period earlier. However, at the time the LSBs of the operands are transmitted, the addition process has just begun, and the output of the D-type flip-flop is unknown. For the correct addition of two binary numbers, when the LSBs of the numbers are added, the value of the third input to the full adder must be at the Logic-0 level. (Parallel addition dealt with this problem by using only a half adder for the LSBs.) This means that the circuit must be "warned" when the LSBs of the two numbers are added so that it could take appropriate action (e.g., set the D-type flip-flop output to 0). We call this "warning" **data synchronization**.

Since serial processing could start during any clock period and go on indefinitely, usually data synchronization is required, which indicates when data transmission starts and when it ends. This may be done by the transmission of special bit sequences or by the addition of control signals. We will use the latter form of synchronization for this example. In this case we add an additional input signal called the START signal, which is at the Logic-1 level only for the first clock period, when the LSBs of the operands are present at the two serial data input lines. This control signal can be used either to clear the D-type flip-flop or to make the third input to the full adder zero for this initial clock period. For the remaining clock periods the START signal is at the Logic-0 level, and normal serial addition proceeds. In Figure 8.16 we demonstrate one method by which the input carry for the LSBs is forced to the Logic-0 value. An AND gate is inserted before the C input of the full adder. When the START signal is at the Logic-1 level, the output of the AND gate is forced to be equal to 0, and the correct SUM and CY are produced by the full adder. The circuit in Figure 8.16 is a complete and operational serial adder, and it is worth noting how simple it is.

8.6 SIGN-PLUS-MAGNITUDE REPRESENTATION

Now that we have seen how addition (and subtraction) of positive binary numbers can be executed by either parallel or serial arithmetic processing circuits, we expand our number representations to include negative numbers

8.6 SIGN-PLUS-MAGNITUDE REPRESENTATION

as well. We know that for numbers of N bits there are only 2^N different bit combinations possible. We have used all of these to represent positive numbers in the magnitude range of from 0 to $2^N - 1$. When both negative and positive numbers are used, some of these binary bit patterns are used for positive numbers, whereas others are used for negative numbers. In most practical applications the number of positive and negative numbers are approximately equal. Hence, the range of numbers extends from approximately $-2^{N-1} + 1$ to $+2^{N-1} - 1$ with zero being in the middle of the range of available numbers.

Negative numbers are included in all ordinary arithmetic operations, and in the ordinary use of numbers negative numbers are defined by the **sign-plus-magnitude** representation. This means that a negative number, such as -25, has a negative sign ($-$) to indicate that it is negative and that its absolute value (magnitude) is equal to 25. If the negative of a number is taken (this operation is often called **negation**), its magnitude does not change, only its sign changes: The negative of -25 is equal to $+25$, or 25. It would be natural to look for a similar number representation among binary numbers, and the earliest computers did use sign-plus-magnitude representation for negative numbers. Since the sign of the number is a binary variable (it can be either positive or negative), 1 bit of the binary number can be designated as the **sign bit.** Normal practice is to use either the MSB or the LSB for the sign bit and, either the 0 or the 1 value can be used for the representation of either the positive or the negative sign. We will use the MSB as the sign bit and the 0 value to represent positive numbers because this representation has the advantage that the positive numbers (in the range of from $+0$ to $+2^{N-1} - 1$) have exactly the same representation as that of the unsigned binary numbers discussed in Section 8.3.

The number circle for the sign-plus-magnitude representation of 4-bit numbers is shown in Figure 8.17. As we mentioned before, the MSB is used as the sign bit, with values of 0 for positive and 1 for negative numbers. The remaining lower order 3 bits are used for the magnitude of the number that has a range of from 0 (000) to 7 (111); hence, the entire range of numbers is from -7 to $+7$.

The sign-plus-magnitude representation has quite a few unpleasant features, two of which are shown clearly on the number circle. The first anomaly is that there are two representations for the zero value. There is a "positive" zero (0000) and a "negative" zero (1000), which are represented by two different bit patterns but for which it is very difficult to find any kind of meaningful interpretation in ordinary arithmetic calculations. Computers are often asked to determine whether a number is equal to 0, and in case of the sign-plus-magnitude representation the number would have to be compared to 2-bit patterns before the answer to this test could be found. The other anomaly is that incrementing or decrementing is not a simple "walking around the circle" operation. If we want to increment a number on the number circle in Figure 8.17, we have to step in the clockwise direction for positive numbers (as before) and in the counterclockwise direction for negative numbers.

There are problems also with the simple addition and subtraction operations with which we are familiar from ordinary arithmetic. In case of

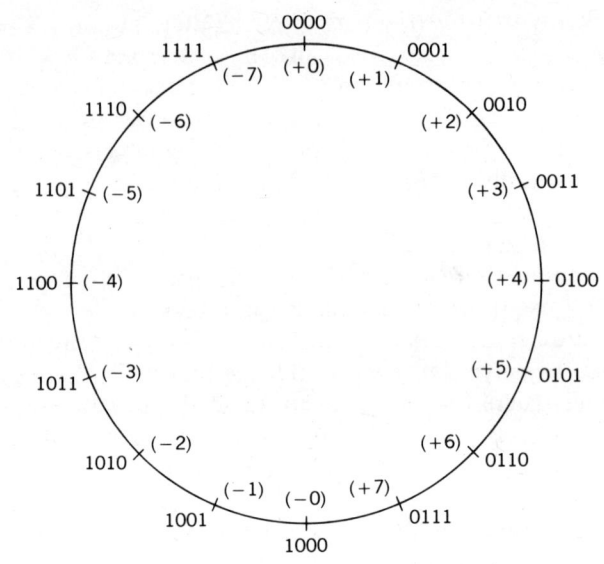

254 NUMBER REPRESENTATION AND COMPUTER ARITHMETIC

FIGURE 8.17 Number circle for sign-plus-magnitude representation of 4-bit numbers.

addition (or subtraction) the signs of the two numbers have to be examined in order to determine whether addition or subtraction is required. Equal signs require addition, whereas opposite signs require subtraction of the two magnitudes of the numbers. A further complication arises when subtraction is required, since the smaller magnitude must be subtracted from the larger one, which would require the execution of a COMPARE instruction by the computer before subtraction could take place. Even after the addition or subtraction of the magnitude parts of the numbers have been executed, the computer would have further work to do because it would have to determine the value of the sign bit of the result. To do serial processing on sign-plus-magnitude numbers is even more difficult. In order to know whether addition or subtraction is required, the two sign bits of the addends must be transmitted first. However, the sign bit of the result cannot be determined until the magnitudes of both numbers are tested; hence, the sign bit of the result can be transmitted only as the last bit of the serial data. This shows, in fact, that purely serial processing of the sign-plus-magnitude representation of numbers is not possible. All these factors show that there are serious difficulties with the binary processing of sign-plus-magnitude representation of numbers. Since a general arithmetic unit that can add, subtract, increment, and decrement signed numbers must be a part of the basic arithmetic circuitry of even the simplest of digital computers, it would be very complicated and expensive to build such a general arithmetic unit for the sign-plus-magnitude representation. Fortunately, a much simpler representation, the **2's complement** representation for negative numbers, eliminates most of the complications mentioned. This is another instance when computer designers were "lucky" to discover a different method of building computers that made arithmetic addition and subtraction simple and orderly and avoided all the complications mentioned. There is one operation that is simple for the sign-plus-magnitude representation: **negation.** This requires only the complementing of the sign bit. But fortunately, as we will see in the next section, the negation operation for the 2's complement representation is also a simple operation.

8.7 TWO'S COMPLEMENT REPRESENTATION

There are several ways to derive the 2's complement representation for signed binary numbers. We will use the number circle to show that a "natural" negative-number representation may be generated from the number circle by the simple counting (incrementing or decrementing) operations. Let us consider again the number circle (Figure 8.18), where the binary codes are shown in their usual order. Let us define incrementing and decrementing operations (as before) by the operation of movement around the circle in the clockwise or counterclockwise directions, respectively. Hence, starting at 0 (0000) and incrementing by 1, we get +1 (0001), and decrementing by 1, we get −1, or the binary code 1111. This is quite different from the sign-plus-magnitude representation (1001). Continuing with the incrementing and decrementing operations, the codes will meet at the bottom of the circle where the code 1000 must be assigned to either +8 or −8. We select −8 for this code, and in fact, we have generated the 2's complement representation for signed 4-bit numbers, which are in the range of from −8 to +7.

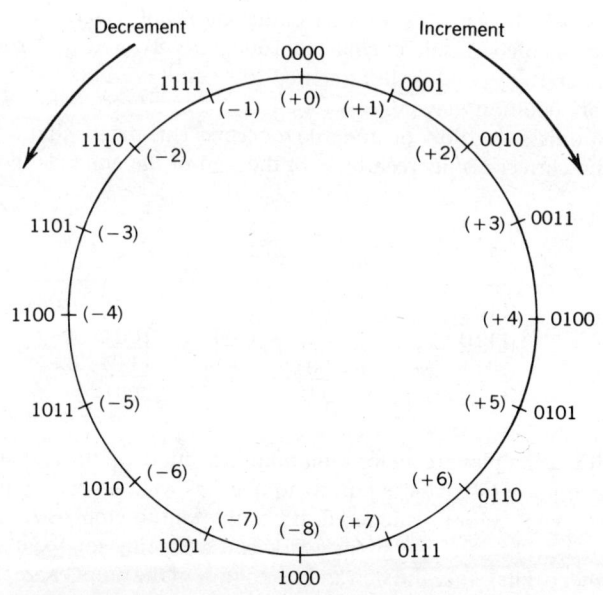

FIGURE 8.18 Number circle for 2's complement representation of 4-bit signed numbers.

The following are some of the advantages of the 2's complement representation:

1. The MSB of positive numbers is always 0; the MSB of negative numbers is always 1. Hence, the MSB may be considered as the sign bit, and there is an easy test for determining whether a number is positive or negative.
2. The representation of positive numbers is equivalent to their unsigned binary representation.
3. There is only one zero element (0000).
4. Simple binary incrementing (and decrementing) works correctly for both positive and negative numbers unless overflow (or underflow) occurs.
    We already know that incrementing/decrementing works for positive numbers since they are equivalent to their unsigned binary representation. Incrementing/decrementing also works for negative numbers because we defined their bit patterns according to these two simple operations. The following two examples show that the incrementing and decrementing operations work correctly for negative numbers where simple binary addition and subtraction are used:

Increment	Decrement
1101 (−3)	1101 (−3)
+1	−1
1110 (−2)	1100 (−4)

In these calculations we have ignored the carry generated by the most significant column of bits. As will be shown, the carry generated by the MSBs of 2's complement numbers have a different meaning from that of the unsigned binary representation. Notice also that the incrementing of bit pattern 1111 (−1) results in pattern 0000 (0), which is the correct numerical result. Overflow occurs if 0111 (+7) is incremented (result is equal to 1000 = −8), and underflow occurs if binary number 1000 (−8) is decremented (the result is 111 = +7).

5. Simple binary addition may be used for all signed numbers, and the correct result is produced unless overflow or underflow occurs. This means that a binary adder produces the correct results regardless of the sign of the addends. Four examples are shown:

Binary addition (without overflow)

0011 (+3)	1101 (−3)	0111 (+7)	0101 (+5)	Addends
0011 (+3)	1110 (−2)	1100 (−4)	1001 (−7)	
0110 (+6)	1011 (−5)	0011 (+3)	1110 (−2)	Result
0011	1100	1100	0001	Carries

The preceding examples are all for 4-bit numbers, but the same rules may be easily demonstrated for any fixed-size binary numbers as well. As before, the carry generated in the MSB column is ignored. It is important to emphasize again that the demonstrated arithmetic operations work correctly only for fixed-size numbers (fixed number of bits) since the sign and magnitude of the number are not separated. This restriction is not a disadvantage for ordinary digital computers since most computers are designed around fixed-size data registers.

There are some operations that become more complicated when the 2's complement rather than the sign-plus-magnitude representation is used. The two most important among these are the detection of overflow and the negation operation, which are discussed in what follows.

It can be shown that overflow does not occur when the two addends are of different polarity. We will demonstrate this for 4-bit numbers. Obviously, the largest number that can be produced by adding two numbers with different sign bits is equal to $+6$, which is the result of the binary addition of $+7$ (0111) and -1 (1111). The binary addition of these two numbers yields 0110, which is equal to $+6$. The most negative number produced by the addition of two numbers of different polarities is equal to -8 when we add -8 (1000) to $+0$ (0000), which obviously gives the correct result. Consequently, the result of the binary addition of two numbers with different sign bits is always within the range of expressible binary numbers (-8 to $+7$ for 4-bit numbers).

Overflow or underflow may occur when two numbers with the same sign bits are added together. There are two usual methods for detecting overflow during the binary addition of two 2's complement numbers. The first method examines the sign bit of the result (SUM). When the sign bits of the addends are the same and the sign bit of the result is different from those of the addends, either an overflow or an underflow had to occur (the sign bits indicate the obviously erroneous result that the addition of two positive or negative numbers yields a negative or positive result). Examples of overflow and underflow for 4-bit numbers and the rules for the detection of these conditions are shown in Figure 8.19.

The other method of overflow/underflow detection involves the examination of the carries in the MSB and in the next to the MSB columns. If the

```
0111   (+7)     1101   (-3)    Addends
0101   (+5)     1010   (-6)
1100   (-4)     0111   (+7)    (Incorrect) Result
0111            1000           Carries
```

Overflow $= A_3 \cdot B_3 \cdot R_3' + A_3' \cdot B_3' \cdot R_3$

Binary addition (with overflow/underflow)

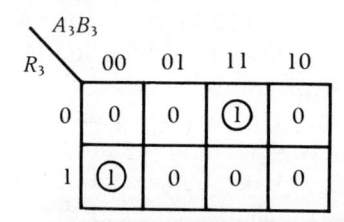

R_3 \ $A_3 B_3$	00	01	11	10
0	0	0	①	0
1	①	0	0	0

Overflow/underflow

FIGURE 8.19 Karnaugh map for detection of overflow/underflow for 2's complement representation of signed numbers.

two carries are the same, then the results of the addition are correct. If they are different, then an overflow or underflow occurs. This test leads to a simpler circuit, and its Boolean equation can be expressed for the addition of two N-bit 2's complement numbers as

$$\text{Overflow/underflow} = CY_{N-1} \oplus CY_{N-2}$$

where a simple XOR operation is used.

The negation operation for 2's complement numbers can be derived from the concept of a "negative" number. If we add a number to its negative equivalent, then the result must be equal to zero. For example, $+5$ is represented by the binary pattern 0101 and -5 by 1011. If we add these two numbers (simple binary addition is used), we get 0000, and a carry value of 1 is generated from the MSB column which can be ignored. As a result, we can calculate the negative of a 4-bit number, also called its 2's complement, by considering it as an unsigned binary bit pattern and subtracting it from the bit pattern 10000, which as an unsigned binary value is equal to $2^4 = 16$. This becomes a general rule, and the 2's complement of a N-bit binary number can be calculated by considering it as an unsigned binary number of N bits and by subtracting its bit pattern from the unsigned binary number of 2^N. Examples of 4-bit numbers added to their 2's complement are shown:

Binary addition of numbers to their 2's complement

0011	(+3)	1100	(−4)	0000	(0)	Number
1101	(−3)	0100	(+4)	0000	(0)	2's complement
0000	(0)	0000	(0)	0000	(0)	Sum
1111		1100		0000		Generated carry bits

The 2's complement operation is equivalent to negation, that is, the calculation of the negative of a number. Now it seems that subtraction must be added to the capabilities of our computer if we want it to negate a 2's complement number. But luck is with us again: There is another way to calculate the 2's complement of a number that requires addition only. We can see that as far as unsigned binary values are concerned, the bit pattern of N 1's (1111 for 4-bit numbers) is only 1 less than the unsigned binary value of 2^N. (If we increment N 1's, we get N 0's with a carry generated from the MSB column.) Now, if we subtract our original 2's complement number from all 1's (unsigned binary value $2^N - 1$) and add $+1$ to the result, we can calculate the 2's complement of our number. The subtraction of a number from all 1's is equivalent to its **1's complement,** which is simply its logical complement (every bit value complemented). So the negation operation may be executed by the binary (logical) complement of the number and the increment operation. The binary complement operation is called the 1's complement because each bit is individually complemented to 1; that is, if an N-bit number is added to its 1's complement, the result is equal to N bits of 1's. This is shown for the same three 4-bit numbers used to show the 2's complement operation:

8.7 TWO'S COMPLEMENT REPRESENTATION

Binary addition of numbers to their 1's complement

0011	(+3)	1100	(−4)	0000	(0)	Number
1100	(−4)	0011	(+3)	1111	(−1)	1's complement
1111	(−1)	1111	(−1)	1111	(−1)	Sum
0000		0000		0000		Generated carry bits

The value +1 for an N-bit binary number is equal to $N - 1$ 0 bits and one 1 bit in the least significant column. The 2's complement of a binary number is generated in two steps. First, the number is complemented bit by bit (1's complement); then the binary value +1 is added to the number. These two operations are shown for the three 4-bit numbers used as examples:

2's complement by incrementing 1's complement

0011	(+3)	1100	(−4)	0000	(0)	Number
1100	(−4)	0011	(+3)	1111	(−1)	1's complement
0001	(+1)	0001	(+1)	0001	(+1)	Value of +1
1101	(−3)	0100	(+4)	0000	(0)	2's complement
1100	(−4)	0011	(+3)	1111	(−1)	Carry bits

No subtraction is required to calculate the negative (2's complement) of a binary number because it can be carried out by the logical NOT (inversion) and the increment operations. Since subtraction of two binary numbers may be executed by the addition of the first number to the 2's complement of the second number, one may execute both addition and subtraction with a binary adder circuit. Algebraically, we can write

$$A - B = A \text{ plus } (-B) = A \text{ plus } (\text{2's complement of } B)$$
$$= A \text{ plus } \bar{B} \text{ plus } +1$$

where the word *plus* is used for the arithmetic addition operation (since the plus operator is used for logical OR), the number +1 is shown with the positive sign to indicate the numerical value of 1 (since the number 1 is often used for the Logic-1 value), and the term $\bar{B}$ indicates the complement of the binary number B bit by bit.

It is easy to modify the parallel adder circuits used in Figure 8.14 so that it can perform parallel subtraction. The modified circuits are shown in Figure 8.20. INVERTER gates complement the bits of number B, which are applied to the four full-adder circuits. In place of the half adder, a full adder is used for the LSBs of the addend, and the third input bit (C_0) is set to Logic-1. This additional input of 1 bit adds the value +1 to the two addends since a binary value of 1 is added to the LSBs of the addends (this is so since the full adder adds 3 bits together). The binary adder performs the algebraic operation A plus $\bar{B}$ plus 1, which, as we have seen, is numerically equivalent to the difference $A - B$ (A minus B).

There is yet another easy method by which the 2's complement of a binary number may be calculated. Proceeding from right to left (from LSB to

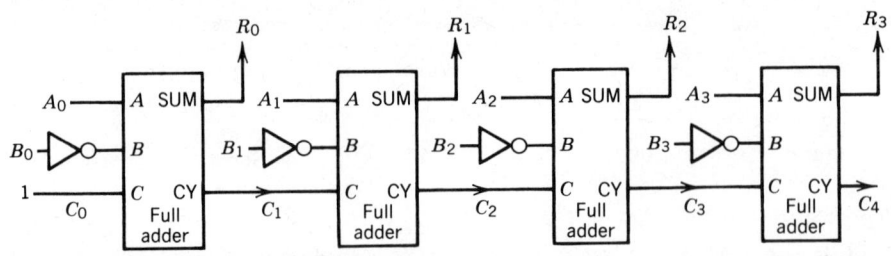

FIGURE 8.20 Circuit for parallel subtraction of two 4-bit numbers expressed in 2's complement representation.

MSB), bits that are equal to 0 are left unchanged. When the first 1 bit is encountered, it is changed to 0 (complemented), and to the left of this bit position all bits are complemented as well. This method is particularly well suited to the generation of the 2's complement of a binary number which is transmitted serially with its LSB arriving first.

8.8 OCTAL AND HEXADECIMAL REPRESENTATION OF BINARY NUMBERS

Practical digital computers use 16, 32, or even 64 bits to represent numerical data that may be of integer or of floating-point type. Most modern digital computers use 2's complement representation for negative integers and include binary addition, complementing (1's complement), negation (2's complement), incrementing, decrementing, and logical operations (AND, OR, XOR, etc.) in their basic data-processing repertoire. Most detect overflow and underflow after an arithmetic operation and also provide access to the carry bit generated from the MSBs of the addends. The range of values for 16-bit 2's complement integers is from -2^{15} to $2^{15} - 1$, that is, from $-32,768$ to $32,767$. The range for 32-bit integers is from approximately -2×10^6 to 2×10^6.

It would be extremely inconvenient to handle numbers in their binary form when a large number of bits is used, since every time we want to write down the number we would have to use a large number of 1's and 0's. A more practical representation is the **octal** representation. In the octal representation (number system based on 8) the bits are separated into groups of 3 bits from the least significant to the most significant bits. Each group of 3 bits forms an octal digit from octal digit 0 (bits 000) to digit 7 (bits 111). Since for many practical computers the number of bits is not divisible by 3, the most significant octal digit of a binary number may have only 1 or 2 bits. The smallest practical binary unit is the **byte,** which contains 8 bits, or three octal digits, but only the two lower order ones are full octal digits, and the most significant digit contains only 2 bits. The octal (unsigned) range of a byte is from 000_8 to 377_8, where the index shows the base of the number. The most common microcomputers use either 16 or 32 bits for their basic integer numbers. In the case of 16 bits the octal representation of the number

8.8 OCTAL AND HEXADECIMAL REPRESENTATION OF BINARY NUMBERS

contains five full octal digits, and the most significant "octal digit" contains only 1 bit, so that the unsigned octal representation has the range of 000000_{16} to 177777_{16}. Similarly, 32-bit octal numbers contain 11 octal digits, with the most significant digit containing 2 bits.

For a long time the machine languages of computers were expressed in octal representation, which became one of the most frequently used number systems among computer professionals. With the proliferation of 8- and 16-bit microprocessors designers felt that the **hexadecimal** representation (number base of 16) is more convenient since this provides full 4-bit digits for the popular standards of 8-, 16-, or 32-bit binary numbers. For the hexadecimal representation the binary numbers are divided into 4-bit groups, each representing one hexadecimal digit that may have the binary value of 0 (0000) to 15 (1111). Since a digit should be represented only by one symbol, the letters A–F are used for numerical values of 10 to 15. The binary, octal, and hexadecimal representations of a 16-bit number is shown:

Binary	Octal	Hexadecimal
1111100111000110	174706	F9C6

As long as binary numbers are used for unsigned integers or special codes, either the octal or the hexadecimal representation is convenient, but when the binary numbers are used for signed numerical values, the hexadecimal representation becomes more difficult to use. We will use only the octal representation for numerical calculations. Of course, one can always expand either the octal or the hexadecimal representation to binary bits and do the calculations on the basis of binary numbers, but if these conversions have to be made frequently, the usefulness of octal or hexadecimal representations becomes rather limited. The addition of two 16-bit binary numbers, their octal equivalents, and 2's complement decimal values are shown:

Binary		Octal	Value (2's complement)	
0011001110010111		031627	+13,207	(decimal)
1111100101000110	plus	174506	−1,722	(decimal)
0010110011011101	sum	026335	+11,485	(decimal)
1111001100000110	carries	110101		

Both the binary addition and the conversion to numerical values may be done by the binary representation, as in the preceding section. However, the same operations may also be done directly in the octal representation, in which case all the conversions between binary and octal representations become unnecessary. The binary addition of two octal digits may be performed by the two tables shown in Figure 8.21. One shows the sum and the other the carry generated from the addition of two octal digits. These tables can be derived simply from the binary representation of octal digits. Using these two tables, the two octal numbers may be added directly, with the generated

	Sum									Carry							
	0	1	2	3	4	5	6	7		0	1	2	3	4	5	6	7
0	0	1	2	3	4	5	6	7	0	0	0	0	0	0	0	0	0
1	1	2	3	4	5	6	7	0	1	0	0	0	0	0	0	0	1
2	2	3	4	5	6	7	0	1	2	0	0	0	0	0	0	1	1
3	3	4	5	6	7	0	1	2	3	0	0	0	0	0	1	1	1
4	4	5	6	7	0	1	2	3	4	0	0	0	0	1	1	1	1
5	5	6	7	0	1	2	3	4	5	0	0	0	1	1	1	1	1
6	6	7	0	1	2	3	4	5	6	0	0	1	1	1	1	1	1
7	7	0	1	2	3	4	5	6	7	0	1	1	1	1	1	1	1

FIGURE 8.21 Sum and carry tables for octal digits.

carries shown in Figure 8.21. The only problem left is to evaluate the decimal equivalent value of an octal number. The expression for the unsigned value can be derived from the fact that an unsigned octal number is the number represented in base 8; hence, its decimal value is given by the expression

$$\text{Value of unsigned octal number} = \sum_{i=0}^{i=N-1} D_i \cdot 8^i$$

where the octal digits have values 0–7 and are numbered from right to left beginning with index 0. For the first number in our example we get

$$31627_8 = 3 \times 4096 + 1 \times 512 + 6 \times 64 + 2 \times 8 + 7 \times 1 = 13{,}207 \text{ (decimal)}$$

The value of a positive number is equal to its unsigned decimal value; the sign of a 16-bit octal number can be determined easily from its most significant bit, which is 0 for a positive number and 1 for a negative number. To determine the value of a negative octal number requires the **negation** operation first and then the same unsigned value calculation as before. The negation is equivalent to the 2's complement of the number, as we have seen before, but it is just as easy to determine the negative of a number in its octal representation. We use the same rule as that for finding the 2's complement, which uses the complementing (1's complement or logical complement) and the incrementing operations to yield the 2's complement. In octal representation the 1's, or logical, complement becomes the "7's" complement, which for digit 0 yields digit 7, for digit 1 gives digit 6, and so on, (the digit which, when added to the original digit, produces the sum of 7). This is a simple operation, and the 7's (logical) complement and the negative of several 16-bit octal numbers are:

031627	174506	002635	000001	000000	(Octal number)
146151	003271	175142	177776	177777	(Logical complement)
+1	+1	+1	+1	+1	(Increment)
146152	003272	175143	177777	000000	(Negative of number)

Note that the MSB digit is complemented to 1, not 7 since it contains only one bit.

We have now demonstrated how one may use the convenient octal notation for all the required basic arithmetic operations and for the conversion of an octal number to its decimal equivalent. For example, the decimal equivalent of the octal number 174506 is equal to the negative of its 2's complement, or to the negative of the octal number 3272. This is equivalent to the decimal value of $-(3 \times 512 + 2 \times 64 + 7 \times 8 + 2 \times 1) = -1722$.

We have not yet shown how the binary or octal representation of a number may be determined from its decimal value. This is called decimal-to-binary conversion and is performed most conveniently in octal representation, especially when a large number of bits is involved. The method can be used only for unsigned numbers and is based on repeated division operations by the number 8 (the base of the octal number system). The remainder after each division determines the octal digits of the number in reverse order (least significant digit first). The first division by 8 is performed on the number itself (the remainder determines D_0, the least significant octal digit). The second division is performed on the quotient of the first division, and the remainder of this second division is equal to D_1. Each subsequent quotient is divided by 8, and each subsequent remainder is equal to a more significant digit of the octal number. The division operations are performed until the quotient is equal to zero. The remaining more significant bits of the binary number are set to zero. The described decimal-to-octal conversion procedure is demonstrated in Table 8.1 for the decimal number 13,207, which yields the octal result 031627. When a negative decimal number is to be converted to its octal equivalent, we use the same procedure for its magnitude and then determine the 2's complement of the number according to the methods demonstrated in the earlier part of this section.

8.9 GENERAL DATA PROCESSOR (CPU)

We have shown that the arithmetic operations for signed numbers (with the 2's complement representation for negative numbers) require only a binary adder and a logical complementer circuit. With the adder and complementer we can generate the negative of a number, and subtraction can be performed by adding a binary number to the negative of the other operand. The subtractor circuit shown in Figure 8.20 may be generalized for N bits, for which N full-adder circuits are required. In this section we would like to incorporate addition, subtraction, negation, and some logic functions into a general N-bit

TABLE 8.1 Decimal-to-Octal Conversion of Decimal Number 13,207

$\dfrac{13{,}207}{8} = 1650$	$\dfrac{1650}{8} = 206$	$\dfrac{206}{8} = 25$	$\dfrac{25}{8} = 3$	$\dfrac{3}{8} = 0$
Remainder: 7	2	6	1	3

264 NUMBER REPRESENTATION AND COMPUTER ARITHMETIC

data processor circuit that would be similar to the controllable function generator we built for Experiments 3 and 4. Such a data processor often lies at the heart of the computer's central processing unit (CPU). It will be very instructive to see that we can build with only a few circuit elements a sophisticated data processor that performs a large number of useful functions. This general arithmetic processor will also demonstrate how the already familiar circuit elements may be connected together so that they produce a new, much more useful controllable function generator.

The model of an N-bit general data-processing circuit is shown in Figure 8.22. Only one line is drawn for the N digital lines of inputs A and B and for the output RESULT. A small crossed line with the number of bits above it indicates that one line represents a "bundle" of N binary signals. The processor also has k control lines whose values select one of a possible 2^k different functions. We would expect to find useful arithmetic functions, such as addition (A plus B), subtraction (A minus B or B minus A), incrementing (A plus 1 or B plus 1), decrementing (A minus 1 or B minus 1), negation (minus A or minus B), and logical functions (A AND B, A OR B, NOT A, or NOT B). We will show now that many of these useful functions may be generated by N full-adder circuits if we insert controllable function generator circuits called input processor circuits between the inputs of the data processor and the full adders.

The input processor circuit for both input A and input B is the same and is shown in Figure 8.23. It has N data inputs, N data outputs, and two control inputs. The structure of the input processor circuit is simply N equivalent circuits of the two gates shown on the right side of Figure 8.23. This simple circuit can produce all four possible functions of 1 bit (we have already demonstrated this in Chapter 3; see Figure 3.4). The C_1 and C_2 inputs to these N circuits are all connected together so that the same function is applied to all N input bits simultaneously. There are two control inputs for input A, which are called C_1 and C_2, and a similar circuit for input B provides two more independent control inputs, which we called C_3 and C_4. The outputs of the $2N$ XOR gates are connected to the inputs of N full adders, which generate the binary sum of two N-bit binary numbers in the usual manner. The third input bit of each adder (except the adder for the LSB) is connected to the CARRY output bit of the respective less significant full adder. This arrangement

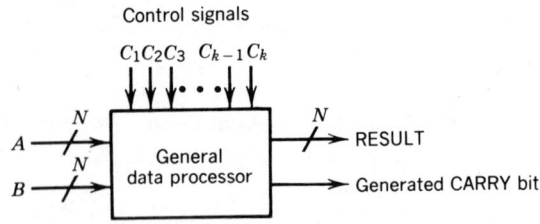

FIGURE 8.22 Block diagram of general data processor for two N-bit numbers.

8.9 GENERAL DATA PROCESSOR (CPU)

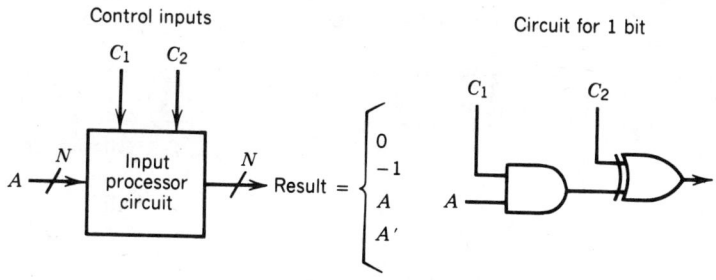

FIGURE 8.23 Input processor and its hardware realization.

provides one more independent binary input variable, namely, the third input of the full adder used to add the LSBs (A_0 and B_0). We use this additional input line also as a control line and call it C_5. The overall structure of this N-bit parallel data-processing circuit is shown in Figure 8.24.

As shown in Figure 8.23, depending on the values set for control inputs C_1 and C_2, the input processor circuit for input A generates four possible functions; hence, it produces the following four functions of A: A (no change), A' (logical complement), 0 (all bits are equal to zero, which is equivalent to a numerical value of zero), and -1 (all bits are 1, which is equivalent to a numerical value of -1 in 2's complement representation). The same functions are generated for input B by the two control inputs C_3 and C_4. Finally, since the control input C_5 determines the value of the third input bit to the adder of the LSB, it can add the numerical value of $+1$ to the sum (because $+1$ in 2's complement representation is equal to a 1 in the least significant column and all other bits are set to 0). We can now see that the different values of the five control lines produce different functions of the input numbers A and B. These functions of A and B are applied to the binary adder with the additional possibility of incrementing the resulting sum by 1. Five control lines can produce $2^5 = 32$ possible different combinations of these functions, and from the input functions the overall output function generated by the SUM outputs of the N full-adder circuits can be determined, as shown in Table 8.2.

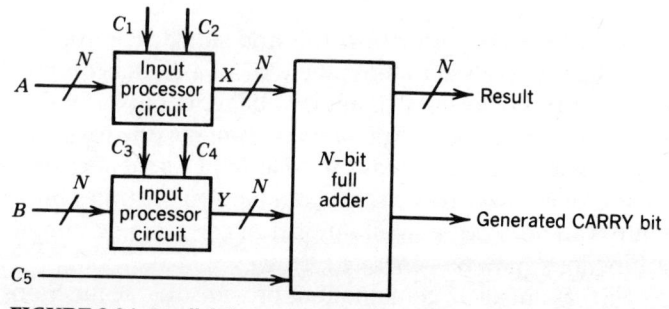

FIGURE 8.24 Parallel data processing circuit using input processors and full adders.

TABLE 8.2 Thirty-two Functions Generated by Parallel Data Processor Shown in Figure 8.24

$C_1C_2C_3C_4C_5$	X	Y	Result	Remarks
00000	+0	+0	+0	All bits are 0
00001	+0	+0	+1	Numerical value +1
00010	+0	−1	−1	All bits are 1
00011	+0	−1	+0	Generated carry = 1
00100	+0	B	B	—
00101	+0	B	B plus 1	Increment operation
00110	+0	B'	B'	1's complement
00111	+0	B'	−B	2's complement
01000	−1	+0	−1	All bits are 1
01001	−1	+0	+0	Generated carry = 1
01010	−1	−1	−2	Numerical value = −2
01011	−1	−1	−1	Generated carry = 1
01100	−1	B	B minus 1	Decrement operation
01101	−1	B	B	—
01110	−1	B'	−B minus 2	Also, B' minus 1
01111	−1	B'	B'	1's complement
10000	A	+0	A	—
10001	A	+0	A plus 1	Increment
10010	A	−1	A minus 1	Decrement
10011	A	−1	A	Generated carry = 1
10100	A	B	A plus B	Addition
10101	A	B	A plus B plus 1	—
10110	A	B'	A minus B minus 1	Also, A plus B'
10111	A	B'	A minus B	Subtraction
11000	A'	+0	A'	1's complement
11001	A'	+0	−A	2's complement
11010	A'	−1	−A minus 2	Also, A' minus 1
11011	A'	−1	A'	1's complement
11100	A'	B	B minus A minus 1	Also, B plus A'
11101	A'	B	B minus A	Subtraction
11110	A'	B'	−A minus B minus 2	Also, A' plus B'
11111	A'	B'	−A minus B minus 1	—

In Chapter 10 we will show how this and similar controllable function generator circuits can be used to build the CPU of a digital computer. In fact, setting the five control bits of our circuit may be thought of as the 5-bit binary code of a computer hardware instruction. For example, the binary code 10100 produces the sum *A* plus *B*, whereas the code 10111 generates the difference *A* minus *B* (see Table 8.2). This circuit example demonstrates how powerful digital circuitry can be. With a small number of gates a large number of data-processing functions may be generated. However, if we examine Table 8.2, we can also see that not all 32 combinations provide unique functions. Among the functions provided, there are common arithmetic functions such as sum, difference, increment, and decrement, but there are functions that would not seem to be very useful (e.g., minus A minus B minus 2). This demonstrates a very important general feature of digital hardware. Sometimes simple, regular hardware structures provide odd programming (software) features. When hardware was very expensive, computer designers paid little attention to the software aspects of a digital computer, and it was often the programmer's task to make (clever?) use of strange features of a particular hardware device. As hardware costs decreased and software costs exploded, more and more computer designers insisted on regular (and meaningful) software features even if that incurred additional hardware costs. With the present state of very low hardware and extremely high software costs, one would think that odd features in hardware devices have completely disappeared. This is not so as a result of a relatively new but obviously most important digital hardware design area: VLSI hardware design. One of the important objectives of VLSI hardware designers is to pack into a very small area of silicon a very large number of digital devices. This objective can often be met by the use of regular hardware structures that then could provide unusual "software" facilities. This dichotomy of hardware/software trade-off will always be an important aspect of digital engineering design.

Returning to the data-processing circuit we have designed, it is relatively complete as far as basic arithmetic processing functions are concerned, but it produces very few logic functions. There are commercially available IC devices that provide a complete set of both arithmetic and logic functions. The type 74181 TTL IC device is a good example of this class of arithmetic-logic function generators, and its functional capabilities are shown in Table 8.3. This device

TABLE 8.3 Logic and Arithmetic Functions of TTL Type 74181 IC Device

$S_3S_2S_1S_0$	$S_4 = 0$	$S_4 = 1$	
		$S_5 = 0$	$S_5 = 1$
0000	A'	A	A plus 1
0001	$(A + B)'$	$A + B$	$(A + B)$ plus 1
0010	$A' \cdot B$	$A + B'$	$(A + B')$ plus 1
0011	0	-1	0
0100	$(A \cdot B)'$	A plus $A \cdot B'$	A plus $A \cdot B'$ plus 1
0101	B'	$(A + B)$ plus $A \cdot B'$	$(A + B)$ plus $A \cdot B'$ plus 1
0110	$A \oplus B$	A minus B minus 1	A minus B
0111	$A \cdot B'$	$A \cdot B'$ minus 1	$A \cdot B'$
1000	$A' + B$	A plus $A \cdot B$	A plus $A \cdot B$ plus 1
1001	$A \odot B$	A plus B	A plus B plus 1
1010	B	$(A + B')$ plus $A \cdot B$	$(A + B')$ plus $A \cdot B$ plus 1
1011	$A \cdot B$	$A \cdot B$ minus 1	$A \cdot B$
1100	1	A plus A	A plus A plus 1
1101	$A + B'$	$(A + B)$ plus A	$(A + B)$ plus A plus 1
1110	$A + B$	$(A + B')$ plus A	$(A + B')$ plus A plus 1
1111	A	A minus 1	A

has six control inputs, two of which we renamed S_4 and S_5. Four of these control functions (S_0–S_3) provide 16 different combinations that generate 16 logic functions when input S_4 is equal to Logic-0. For the logic functions input S_5 is ignored. When input S_4 is equal to Logic-1, the combinations of the five other control inputs provide 32 arithmetic functions. Even though this is a commercially produced hardware device, we can see that many of the functions are not unique. Also, unusual functions are included in the table. For example, one would not expect the function $(A + B')$ plus $A \cdot B$ to be very useful in most applications. This device has less than 48 unique functions out of a possible 64 (six control bits are used), which may cause inefficiencies in the utilization of digital signals where this device is used. All these features again demonstrate the fact that digital devices are often designed with digital hardware economy in mind, which may not provide the best possible set of facilities for the ultimate user of the device.

The general data processor as a controllable function generator has been discussed in terms of a parallel processor. It is also possible to construct an equivalent serial processor that generates the same 32 arithmetic functions. The required hardware is greatly reduced since, as has been shown in the previous sections, serial processing requires only one full adder for an arbitrary number of bits (see Figure 8.16). Because only one adder is required, only two input-processing circuits are needed. Special circuitry must be used to handle the LSB because it is processed differently from the higher order bits. As shown in Figure 8.16, the presence of the LSBs is indicated by the input signal called START, which has a Logic-1 value only during the clock period when the LSBs are present at the input. Since it is required that during the LSBs the independent control input (C_5) is to be applied to the third input of the adder (C), whereas the output of the D-type flip-flop should be used for the other bits, a 2-to-1 multiplexer is the simplest device that can provide

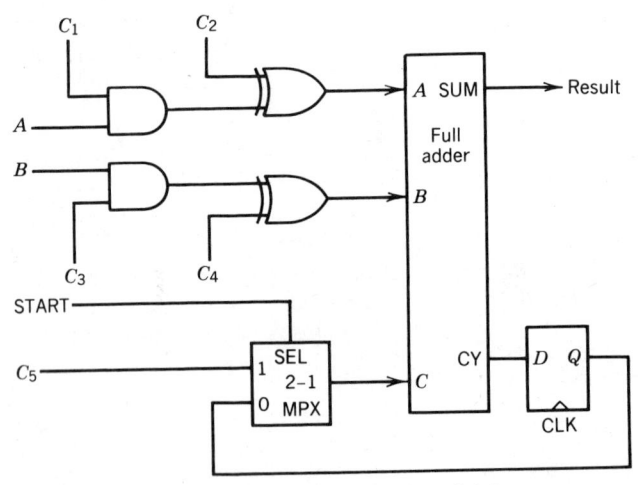

FIGURE 8.25 Detailed circuit diagram for parallel data processor.

this function. The completed general serial data processor is shown in Figure 8.25. The full adder can be built from 5 basic digital gates, the multiplexer from 4. Thus, the whole processor can be built from 13 digital gates and 1 flip-flop. It can provide all the basic arithmetic functions of two serially transmitted binary numbers when the 2's complement representation is used for negative numbers. This example again demonstrates that it is possible to execute significant data processing with a very small amount of digital hardware.

EXPERIMENT 8
Serial Data Processor

E8.1 Summary and Objectives

The serial representation and processing of numerical data are very important digital techniques. Serial processing is used mainly for compact, low-cost digital processors such as small pocket calculators. For this experiment a three-function serial processing unit is designed. The two data operands are transmitted through two digital input lines with the LSBs appearing first and the MSBs arriving last. The processor has two additional digital input lines that are used both for the synchronization of the serially transmitted input data and for the selection of the required arithmetic operation. Two's complement representation is used for negative numbers. The main objectives of this experiment are the examination of serially represented numerical data, the introduction of important concepts for serial data processing, and the design of a complete serial data processor that is constructed from combinational and sequential digital circuits. The structure of the processor is specified in terms of functional blocks. The design process must start with the clear understanding of the functional behavior of each block and their interconnections, which provide the required functions for the serial processor.

E8.2 Serial Representation of Data

The bits of two binary numbers are transmitted serially by using two digital input signals called A and B. During the first time interval (t_0) the LSBs of both A and B (A_0, B_0) are present at these input signals. After one clock pulse (i.e., at time interval t_1) the bits A_1 and B_1 appear at the inputs. For successive time intervals the more significant bits of the two binary numbers are transmitted. Once the data transmission starts, it proceeds continuously for as many clock periods as the number of bits contained in binary numbers A and B.

Two additional digital input signals, C_1 and C_2, are used for both synchronizing the serial data transmission and selecting one of the three possible data-processing functions. Transmission of the input data is synchronized by the logic levels of these two control inputs. When both control input signals are at the Logic-0 level, no data bits are transmitted. This is called the idle state. During the time interval when the logic states of either or both control inputs change to the Logic-1 level, data transmission starts. Data transmission starts with the LSBs A_0 and B_0, proceeds with higher significant bits, and ends with the MSBs A_{N-1} and B_{N-1} for N-bit numbers. The three possible binary combinations for data transmission (01, 10, and 11) determine which arithmetic or logic function is to be generated by the processor. The values of the control inputs remain constant during the entire data transmission period, after which the control inputs return to their idle, or 00, levels.

The synchronization and timing for the addition of the two 4-bit binary numbers 0011 (3) and 1110 (-2) are shown in Figure E8.1. It is assumed that the control bits $C_1 C_2 = 10$ select the addition function. All signal transitions occur at the negative transition of the SYSTEM CLOCK. In Figure E8.1 both the serial and the parallel additions of the two 4-bit binary numbers are shown. Serial addition is demonstrated by the timing diagrams for the four input signals and two output signals of the serial processor. The data input

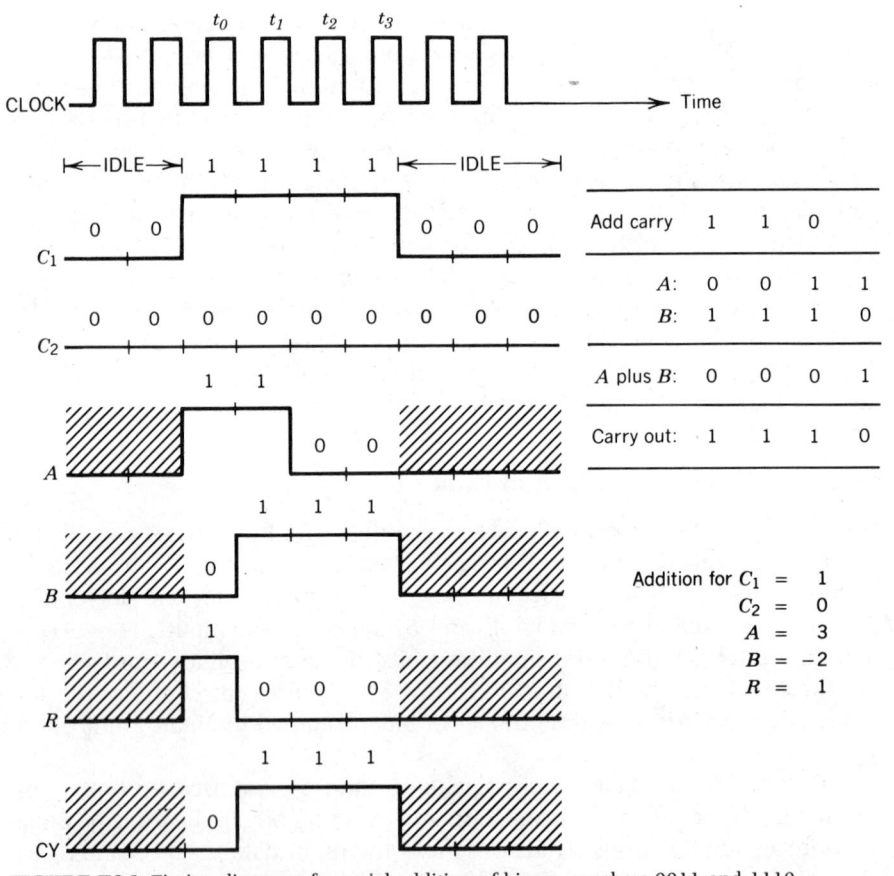

FIGURE E8.1 Timing diagrams for serial addition of binary numbers 0011 and 1110.

and output values are not specified for the idle periods (indicated by hashed areas), and these signals have don't care values during these time intervals, when both control inputs have Logic-0 values. During the addition process the control inputs have constant values: $C_1 = 1$ and $C_2 = 0$. The carry output generated by the addition process is also shown. The data bits of the resulting sum are synchronized with the input data bits. When the LSBs of the input numbers are received, the LSB of the resultant sum is generated through the output signal R. The same is true for other, more significant bits of the results. In the idle state all data inputs and outputs are ignored.

The parallel addition is shown to the right of the timing diagram. The carry added to each column of 2 bits is equal to the carry generated by the less significant column of bits to the right. In both parallel and serial cases the result of adding binary 0011 (value equal to 3) to binary 1110 (2's complement representation of -2) is equal to binary 0001 (which is equal to the value 1). Notice, however, that the add carry value for the LSBs must be 0 for a correct sum.

E8.3 Serial Processor Block Diagrams and Operation

The overall block diagram of the serial processor is shown in Figure E8.2. The more detailed internal block diagram is shown in Figure E8.3. The internal block diagram is constructed from functional blocks, which were discussed in Chapter 8. There are five major blocks, four of which should already be familiar from their descriptions in Chapter 8. Two of these, the full adder and the D-type flip-flop, require simple circuits. The input processor was discussed in detail in Chapter 8. Its circuit depends on the arithmetic or logic functions specified for the three nonzero settings of the two control input signals. Finally, the carry generator circuit, which requires the START signal to operate, was also discussed in Chapter 8. Its function is to provide the correct carry input bit to the full adder during the one clock period when the START signal is at the Logic-1 level, indicating that the LSBs of the numbers A and B are being received at the data inputs. At all other times, when the START signal value is equal to 0, the output of the carry generator (which is connected to

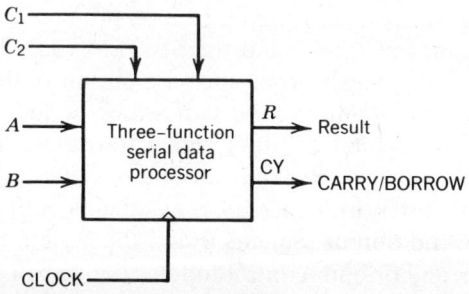

FIGURE E8.2 Block diagram of serial data processor.

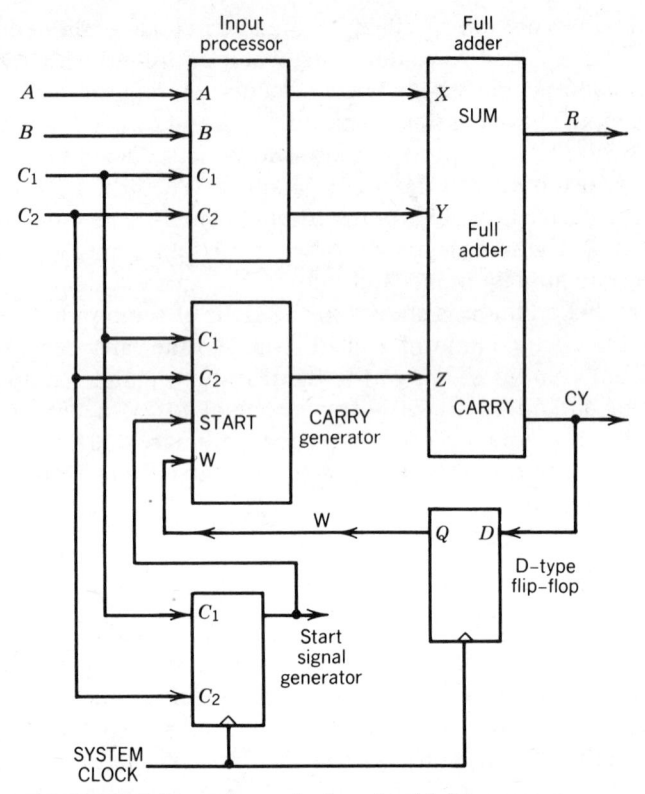

272 NUMBER REPRESENTATION AND COMPUTER ARITHMETIC

FIGURE E8.3 Internal organization of serial data processor.

the third input bit of the full adder) is equal to the output of the D-type flip-flop, and hence it is equal to the CARRY output of the full adder delayed by one clock period. This will provide, as we have learned, the correct results for serial addition.

The fifth block is called the start signal generator, and it is a new functional block. It is required because the start of the data transmission is indicated by the transition(s) of the two control inputs C_1 and C_2 and not by a simple START signal. The start signal generator block produces the correct START signal from the values of C_1, C_2, and the SYSTEM CLOCK signal. This is a sequential circuit of the Mealy type, since the output of the circuit (START) must change (from 0 to 1) during the same clock period when one or both of the control inputs exhibit a 0-to-1 signal transition. The START signal should remain at the Logic-1 level only for one clock period. Its STD is shown in Figure E8.4. As shown in Figure E8.4, as long as both control inputs (C_1 and C_2) remain 0, the output remains 0 and the system remains in its idle state. When either one or both control inputs change to Logic-1, the output changes to 1 for one clock period, after which the system changes its state to the compute state. It remains in the compute state as long as the control

Inputs: C_1 and C_2
Output: START

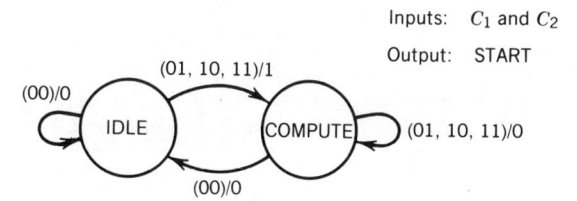

FIGURE E8.4 The STD of start signal generator circuit.

inputs do not change back to values 00. In the compute state the output of the start signal generator circuit remains at Logic-0. Finally, when the control inputs return to their initial 00 values, the system returns to its idle state.

The five functional blocks are connected together to produce the correct operation of the serial processor. When the two control inputs (C_1 and C_2) indicate that data transmission is starting, the START signal becomes Logic-1, the control signals are applied to the input processor, and they select the correct functions for inputs A and B. The same control inputs select the correct Z input to the full adder for this first clock period. During subsequent clock periods the START signal becomes 0, and the output of the D-type flip-flop is applied to the Z input of the full adder. For each clock period, when the input bits are received, the R output shows the respective bit of the result, and the CY output represents the generated carry/borrow value for the selected arithmetic-logic operation.

The input processor and the carry generator circuits are combinational circuits. They can be defined only when the three required processing functions also have been defined and binary values for C_1 and C_2 (which correspond to the selection of the processing functions) are specified. The designer is allowed to assign each specific combination of control input values to a specific processing function, similar to Experiment 4 (which was a parallel processing version of a variable function generator circuit). For example, if the control input values 10 select addition, then the input processor must generate outputs so that $X = A$ and $Y = B$ for the entire COMPUTE period, and during the first clock period (when the START signal is equal to 1) the carry generator must provide 0 for the Z input of the full adder.

E8.4 Design Requirements and Testing

Three processing functions are selected from Table E8.1, choosing one function from each group shown. The selection of specific control input values is not specified, and the most convenient assignment for the design of the input processor and/or the carry generator circuits should be made with the restriction that the control input values of 00 are reserved for the idle state. Design the required two combinational and one sequential circuits, and try to be economical with the number of gates and/or ICs. Try to minimize the overall cost as much as possible, but keep the basic functional structure (i.e., the five functional blocks) separate. This structure should help you in testing

TABLE E8.1 Selection of Three Processing Functions (One per column)

Function 1	Function 2	Function 3
A plus B A plus 1 B plus 1 A minus 1 B minus 1	A minus B B minus A minus A minus B	A B $\overline{A}$ $\overline{B}$

and debugging your circuits during their construction and final integration. Test each block thoroughly before you interconnect them for the final circuit.

After having wired the entire circuit, connect a properly debounced switch to the SYSTEM CLOCK input, four switches to the serial data inputs and two control inputs, and two lights to outputs R and CY and test the circuit using all three required processing functions. A sample of data values is shown in Table E8.2. Complete the table with your expected results and demonstrate that your circuit is working properly. If you have difficulties, connect test lights to the adder inputs X, Y, and Z, the flip-flop output Q, and the START signal and follow the operation of the processor one functional block at a time until you find your error.

E8.5 Additional Circuits and Alternate Experiment

The most useful addition to this serial processor is an overflow/underflow detector circuit. It should work in such a way that after the control inputs return to their idle state a new output signal called ERROR should be set to Logic-1 if an overflow/underflow occurred (and set to 0 if it did not occur).

TABLE E8.2 Test Values for Serial Data Processor

A Binary	Value	B Binary	Value	Function	Result Binary	Value	CY (Binary)
0101	+5	0110	+6	$F_1 =$ $F_2 =$ $F_3 =$			
1111	−1	0000	0	$F_1 =$ $F_2 =$ $F_3 =$			
1010	−6	1010	−6	$F_1 =$ $F_2 =$ $F_3 =$			

This signal should remain constant as long as the control inputs indicate the idle state. Hence, the ERROR signal always shows the overflow/underflow status of the last complete data operation executed by the processor.

As an alternate experiment, use a commercially available arithmetic-logic function generator device to build a seven-function serial data processor that provides the most common functions: addition (A plus B), subtraction (A minus B), incrementing (A plus 1), decrementing, (A minus 1), logical AND ($A \cdot B$), logical OR ($A + B$), and XOR ($A \oplus B$). Use three control inputs and the 000 state for the idle state. Include an overflow/underflow detection circuit as explained in the preceding.

Drill Problems

8.1. In Table D8.1 three numerical representations are shown (unsigned binary, sign plus magnitude, and 2's complement) which are expressed in three different number bases (binary, octal, and hexadecimal) of the same 6-bit binary bit pattern. For each example only one form of the number is shown. Fill in the table by calculating all the other forms.

8.2. Construct the DIFFERENCE and BORROW tables for octal digits and the SUM and CARRY tables for hexadecimal digits (similar to Figure 8.21).

8.3. An older model 4-bit digital computer uses the logical complement for the representation of negative numbers (e.g., -5 is represented by 1010). Draw the

TABLE D8.1 Conversion Between Binary, Octal, and Hexadecimal Representations of Binary Numbers

	Value	Binary	Octal	Hexadecimal
Unsigned binary	50,000			
		101011001011 0110		
			053724	
				A6E5
Sign plus magnitude	−32,000			
		1110101110001101		
			100642	
				8FF3
2's complement	−32,000			
		1110101110001101		
			100642	
				8FF3

number circle and discuss the advantages and disadvantages of this representation. Establish the rules for adding two signed numbers including overflow/underflow detection by using addition and incrementing operations only.

8.4. The binary full-adder and full-subtractor circuits are very similar. With the addition of two XOR gates to the full-adder design an adder/subtractor that has four binary inputs. The additional input (called S) is a control input. When $S = 0$, the circuit behaves as a full adder. When $S = 1$, the circuit behaves as a full subtractor.

8.5. Using the adder/subtractor circuit of Problem 8.4, design a full adder for the sign-plus-magnitude representation of 4-bit signed numbers. First assume that the magnitude of input A is always larger than or equal to the magnitude of B. Then, remove this restriction on the respective magnitudes of A and B by adding a preprocessor circuit block to the adder/subtractor, which has two inputs (A and B) and two outputs (X and Y) and which provides the correct output to the adder/subtractor. (*Hint:* Use multiplexers and a magnitude comperator circuit.)

8.6. Design a serial increment/decrement processor for the 2's complement representation of signed numbers. The processor has three inputs: the serial data input, the control input (which selects either incrementing or decrementing), and the START input (which indicates the transmission of the LSB). It has two outputs. One is the serial data output, the other is the overflow/underflow indicator.

8.7. Design a similar serial processor to that of Problem 8.6 that generates either the SUM or the XOR function of two unsigned binary numbers. Try to use a minimum number of gates.

CHAPTER 9

Serial–Parallel Processing and Conversion

9.1 INTRODUCTION

In Chapter 8 we introduced the concepts of registers and register transfer operations. Registers are parallel storage devices because all the bits stored in them are available at the same time. If we want to transfer data from one register to another (executing the $B \leftarrow A$ register transfer operation, for example), the simplest method is to use **parallel** data transmission when the output of each binary storage cell of one register (register A) is connected to the input of another (register B). After a clock pulse is applied to the CLOCK inputs of all the storage elements of the "destination" register (register B), parallel transfer takes place, as shown in Figure 9.1.

It is also possible to execute the same data transfer operation **serially**. In this case, as shown in Figure 9.2, two additional data-processing circuits are required. The **parallel-to-serial converter** converts the contents of a register into a serial bit stream. If we are interested in data transmission only, then the order in which the bits are sent does not matter. On the other hand, as we have seen in Chapter 8, if arithmetic processing is required, then the LSB must be transmitted first.

At the receiving end a **serial-to-parallel converter** is required to assemble the register contents from the received bit stream. When all the bits have been transmitted, a parallel data transfer is executed between the output of the data converter and the destination register. The clock pulse that executes this parallel register transfer operation occurs only once for each complete data transfer and is shown in Figure 9.2 as the DONE clock pulse. The word *done* indicates that the serial data transmission has been completed.

We will show in this chapter that serial-to-parallel and parallel-to-serial data conversions can be done most easily by using **shift registers**. Serial data transmission is necessary when the transmission is over long distances. For example, the user terminals of a large time-sharing computer network are connected through serial network connections to the host computer. Serial transmission (and data processing) is also used to lower hardware costs when the high speed of processing is not an important system requirement. We will demonstrate some other applications for shift registers and will show that bit-sequence detectors may be easily designed when a shift register is used. From the examination of the structure of shift registers, we will proceed to discuss

278 SERIAL–PARALLEL PROCESSING AND CONVERSION

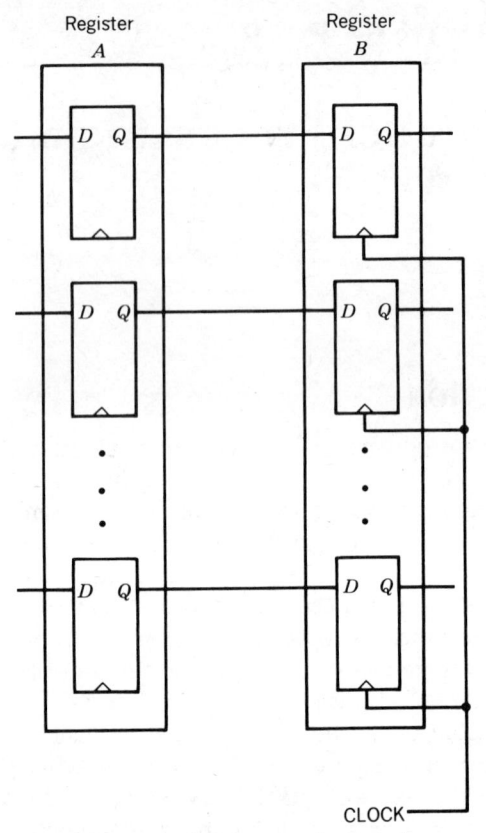

FIGURE 9.1 Parallel data transfer $B \leftarrow A$.

shift operations that can be used for both logical and arithmetic data processing. Shift operations are always included in the basic machine instruction repertoire of digital computers.

In Figures 9.1 and 9.2 we show parallel and serial transfer operations for which either all bits are present at the same time or only 1 bit is transmitted for each clock pulse. In practical digital processors mixed parallel–serial systems occur more frequently. This means that there are a number of bits that are processed in parallel, but the number of bits of the final result is larger than the number of bits processed at any one time. Serial multiplication of two binary numbers is a good example, and we will discuss this in detail later. Data processing for higher accuracies requires multiple data words, which is another example of mixed parallel–serial processing functions. Finally, we will discuss carry propagation in parallel binary adders, which will be used as a demonstration of yet another aspect of the trade-offs that can be made between expensive but fast parallel processors and inexpensive but slow serial-processing circuits.

9.2 SHIFT REGISTERS AND PARALLEL–SERIAL DATA CONVERSION

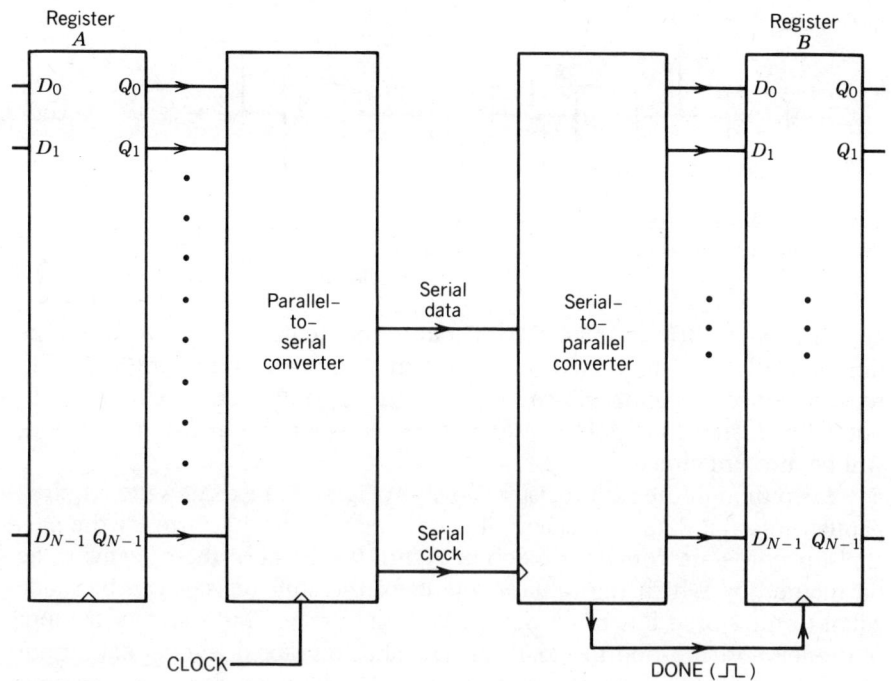

FIGURE 9.2 Serial data transfer $B \leftarrow A$.

Experiment 9 shows an interesting application of a mixed serial–parallel processor and demonstrates the operation and design of a state controller that controls these two types of processing.

9.2 SHIFT REGISTERS AND PARALLEL–SERIAL DATA CONVERSION

When data is stored, data registers must be used containing at least as many stored bits as the number of bits used for the data. Hence, storing information is inherently a parallel function (all stored bits must be present at the same time). In order to serially process or transfer data, first the data must be converted to its serial form; that is, it must be presented to the processor or transmitted down a serial line 1 bit at a time. This process is called parallel-to-serial data conversion, as shown in Figure 9.2. A **shift register,** whose 4-bit version is shown in Figure 9.3, can perform parallel-to-serial conversion. The shift register in Figure 9.3 is constructed from four D-type flip-flops. If initially the outputs D_0–D_3 are equal to the data bits to be transmitted serially, then during the initial clock period (before the first CLOCK pulse appears), the output of the shift register is equal to D_0. After the first clock pulse, its output is equal to D_1, then D_2, and so forth. We can see that the binary data that was originally stored in the shift register is "shifted out," and the bits

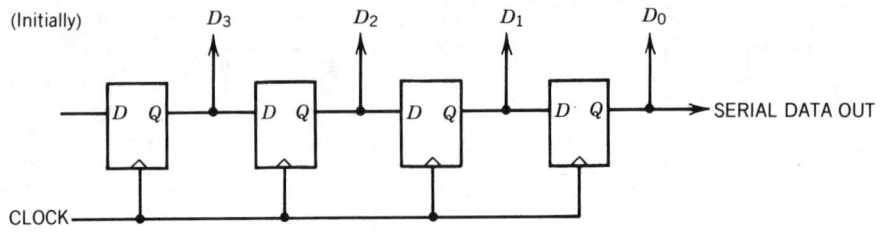

FIGURE 9.3 Four-bit shift register used for parallel-to-serial data conversion.

appear at the SERIAL DATA OUT signal in their correct order. At the same time that the bits are shifted out, other bit values are shifted into the shift register. Since in Figure 9.3 the input of the flip-flop at the extreme left is not defined, after four clock pulses, the contents of the four D-type flip-flops will be also undefined.

Even though the shift register shown in Figure 9.3 executes the required parallel-to-serial data conversion, it cannot be used in this form for the parallel-to-serial data converter block in Figure 9.2 because there seems to be no method by which the initial contents of the shift register can be made equal to the stored bits of register A. We may use the independent D input of the leftmost flip-flop to "load up" the shift register if a serial data input line is available. This is shown in Figure 9.4, where an identical 4-bit shift register is used to execute a serial-to-parallel data conversion function. Initially the SERIAL DATA IN line transmits bit D_0. After the first clock pulse the LSB (D_0) is shifted to the output of the first flip-flop on the left and the SERIAL DATA IN line transmits data bit D_1. After three more clock pulses the correct data bits from D_0 to D_3 are loaded into the shift register, as shown in Figure 9.4. If these flip-flop output lines are connected to the input lines of register B, then the done pulse (see Figure 9.2) can be used to transfer the contents of this shift register to register B. In this case the shift register can perform the complete serial-to-parallel conversion process and may be used in place of the serial-to-parallel converter block in Figure 9.2.

The "parallel loading" of the first shift register has not yet been solved. Multiplexers come to our rescue, as shown in Figure 9.5. We can view the parallel loading and the shifting processes of a shift register as two modes of its operation. In the parallel loading mode the inputs to the flip-flops are

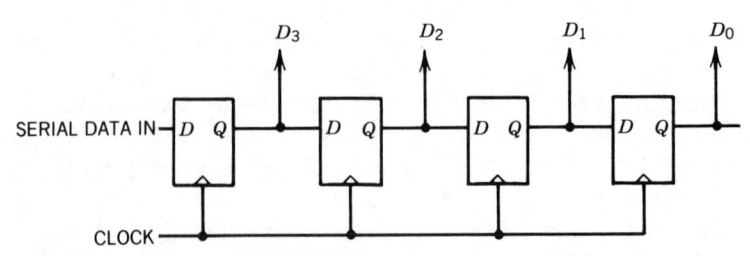

FIGURE 9.4 Four-bit shift register used for serial-to-parallel data conversion.

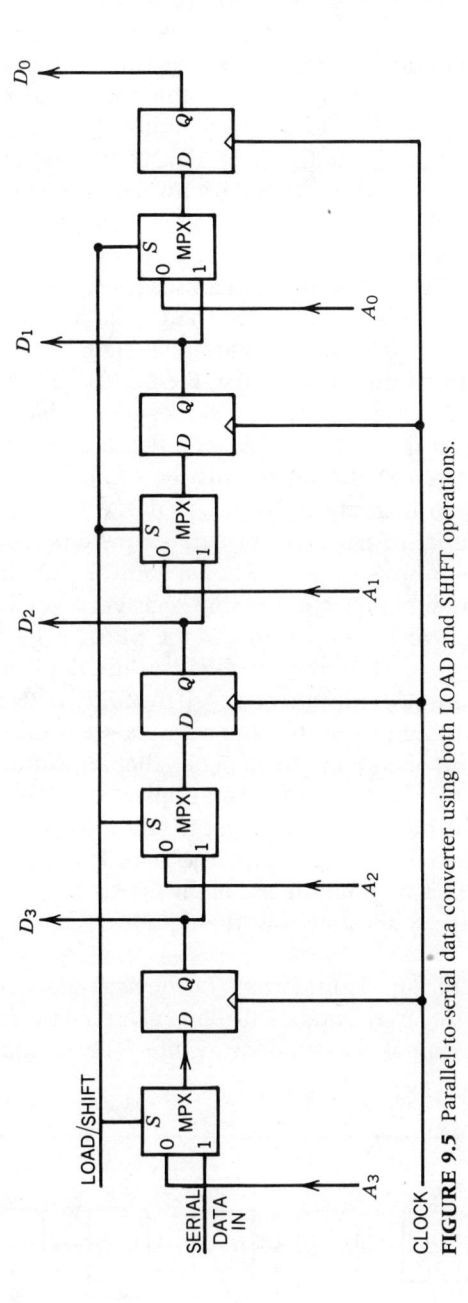

FIGURE 9.5 Parallel-to-serial data converter using both LOAD and SHIFT operations.

connected to the outputs of register A. In the shifting mode the inputs of the shift register's flip-flops are connected to the outputs of its other flip-flops, except for the leftmost flip-flop, whose input becomes the independent SERIAL DATA input of the shift register. One control input line may be used to select the mode of operation, and since for the two modes different inputs are selected for the flip-flops, four two-to-one multiplexers can provide the required selection of the inputs. The completed parallel-to-serial converter circuit is shown in Figure 9.5, and we can see now that the parallel-to-serial conversion process is more complicated than originally shown in Figure 9.2. It was not obvious at first that an additional signal (LOAD/SHIFT) is also required by the conversion process. First, the LOAD/SHIFT signal must be set to 0 (select the LOAD function) and a clock pulse applied to the converter. This loads the correct data bits into the shift register. For the four succeeding clock pulses the LOAD/SHIFT signal must be set to 1, and the data bits are shifted out serially in their correct order. Hence, a shift register can be used to perform both serial-to-parallel and parallel-to-serial conversions, but for most practical applications some additional control circuitry is required.

Another important application of shift registers involves the recirculation of the stored data. We have seen that if the SERIAL IN signal of the shift register is unused, then after four clock pulses the data, stored originally in the shift register, is lost. On the other hand, if the SERIAL IN input is connected to the SERIAL OUT output of the shift register, then during the four clock pulses of the shifting operation, the data bits that appear at the SERIAL OUT signal output will be shifted into the flip-flops of the shift register, and after four clock pulses the original contents of the flip-flops are restored (see Figure 9.6). If this arrangement is used, then the same data may be converted to serial data many times without the use of the LOAD operation. This operation is called data **rotate,** indicating that the data is being **recirculated** within the shift register.

Recirculation of data is important for devices that require continuous, periodic access to a large number of data values. For example, the most common display terminal, the cathode-ray tube (CRT), requires the refreshing

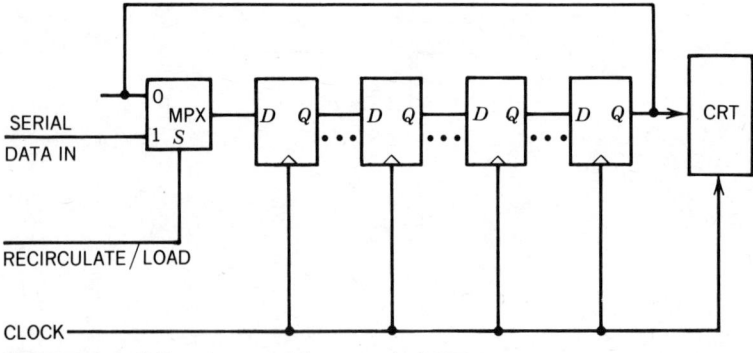

FIGURE 9.6 Shift register used for control of CRT.

9.2 SHIFT REGISTERS AND PARALLEL–SERIAL DATA CONVERSION

of the screen at least once every 30 seconds for a flicker-free display. A shift register can be used as an inexpensive memory device that contains the on–off information for each picture dot (called picture element, or pixel) that can be displayed on the screen. The operation of such a CRT device is shown in Figure 9.6. The cathode ray is deflected to a given point on the screen, and according to the bit value stored in the shift register, it is either turned on (bright spot) or turned off (left dark). A clock pulse is applied, the ray is deflected to a new position, and a new data value at the SERIAL OUT output of the shift register is used for turning the beam on or off. As long as the same data values are recirculated, the screen is refreshed with the same information, and the displayed image remains static. The contents of the shift register shown in Figure 9.6 can also be changed through the 2-to-1 multiplexer connected to the first flip-flop of the shift register. When the control line RECIRCULATE/LOAD is set to Logic-1, the data value from the signal SERIAL DATA INPUT is shifted into the shift register. This stores a new binary value for a single pixel on the screen, and if the RECIRCULATE/LOAD signal is set equal to Logic-1 only for one clock pulse, only one pixel data bit is changed. Of course, some sophisticated control circuitry is required to change the RECIRCULATE/LOAD signal value at the appropriate time when the cathode ray is deflected to the required position on the screen. Nevertheless, the shift register as a memory element is an extremely convenient and inexpensive storage device, especially when LSI or VLSI technology is used, since it requires only three signal connections and can handle a very large number of data bits. The price we have to pay is the slowness of its operation. It seems that for the updating of one pixel on the screen we have to wait until the dot is at the right position, which may be as long as the refreshing of the entire screen, that is, $\frac{1}{30}$ second. A medium-resolution display screen contains at least 100,000 pixels. At this speed it would take around 3000 seconds (almost an hour!) to change an entire screen. Obviously, this would be too slow for any practical display device. This is a typical situation where sophisticated control circuitry and the mixing of parallel and serial processes can improve the speed of operation (with additional hardware costs, of course). Nevertheless, shift registers play an important role as memory elements of display devices.

Another very simple application of shift registers is their use in bit-sequence detector circuits. In Chapter 7 we showed a general method by which Mealy- and More-type bit-sequence detectors may be designed. A bit-sequence detector circuit may also be constructed from a shift register and a D-type flip-flop, as shown in Figure 9.7. At the beginning of the detector's operation the correct bit sequence is stored in the shift register, and both D-type flip-flops are cleared. When the incoming bit sequence is transmitted through the SERIAL INPUT DATA line, the TRANSMIT signal becomes Logic-1 and the CLOCK signal follows the SYSTEM CLOCK signal. Each incoming bit is compared to the correct bit stored in the shift register by the XOR gate. As long as the incoming bit and the bit stored in the rightmost flip-flop of the shift register are equal (00 or 11), the output of the XOR gate

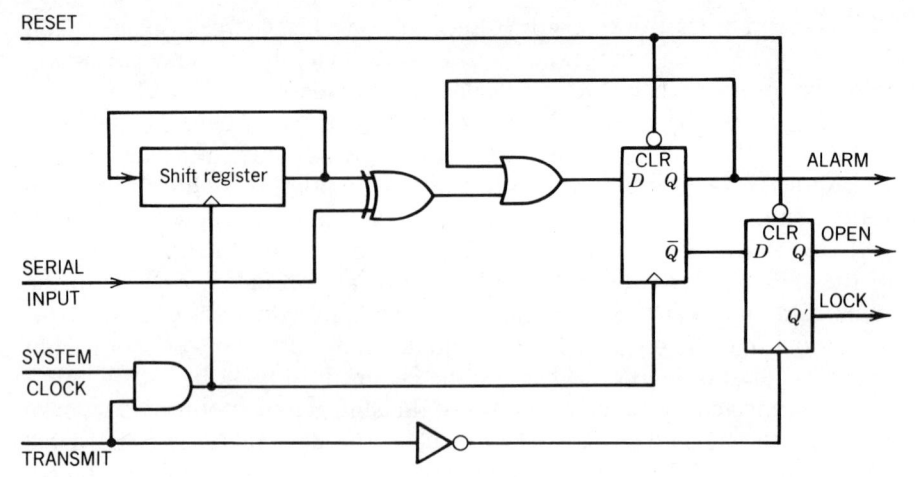

284 SERIAL–PARALLEL PROCESSING AND CONVERSION

FIGURE 9.7 Shift register used for the digital lock circuit.

is equal to 0, and the D-type flip-flop remains cleared. If at any time the bits are unequal, the output of the XOR gate becomes 1, the first D-type flip-flop is set, and the ALARM signal, which is the output of this flip-flop, is turned on. Once the ALARM signal is equal to Logic-1, the OR gate at the input of this flip-flop forces this flip-flop to remain set independently from the bits arriving at the SERIAL DATA input line.

The inverted output of the first flip-flop indicates whether the incoming bit sequence has been correct or not. The TRANSMIT signal returns to 0 when the transmission of the incoming bit sequence has been completed. At this time (at the trailing or negative edge of the TRANSMIT signal), the second flip-flop is set according to the complement of the ALARM signal. If the ALARM signal is still equal to Logic-0 at the end of the bit sequence, then the OPEN signal is set to 1 and the LOCK signal is set to 0. This opens the safe as required by the functional specifications of the digital lock circuit. For this design example both the OPEN and ALARM signals are set to Logic-0 (initialized) by an additional signal called RESET, which normally is at the Logic-0 level but should be set temporarily to 1 before the digital lock is operated again.

As shown in Figure 9.7, the data stored in the shift register is recirculated at the same time when the stored bit sequence is compared to the transmitted bit sequence. We assume that the SYSTEM CLOCK signal is repeated periodically and continuously. The clock signal that operates the shift register is equal to the logical AND of the SYSTEM CLOCK and the TRANSMIT signals. Since the TRANSMIT signal is equal to 1 only during the time the serial bits are transmitted, the number of clock pulses applied to the shift register is equal to the number of bits in the bit sequence. This ensures that the stored bit sequence is correctly recirculated and the stored bits are appropriately aligned at the end of the SERIAL DATA transmission; therefore, they can be reused for successive operations of the digital lock.

We have demonstrated that shift registers are useful as serial-to-parallel converters and as storage devices and could be used in bit-sequence detector circuits. Naturally, there are many other practical design problems that can be simplified by the use of shift registers. In the following section we will examine shift registers in their own right as storage registers that have the ability to perform "shift" operations.

9.3 MULTIPLE PRECISION, SHIFT OPERATIONS, AND MULTIPLICATION

In the last chapter we introduced the 2's complement as a convenient representation of negative integers for the two basic arithmetic operations, addition and subtraction. We have also said that most modern digital computers are organized around fixed-size data stored in registers that contain the same number of bits. For example, many microcomputers have 16-bit parallel adders and 16-bit data registers. The basic integer data size of a computer is often called its **word** size, and we will use the term *word* to designate a binary bit pattern with this specified number of bits.

The range of 16-bit signed integers is from $-32,768$ to $+32,767$, which may be sufficient for many applications but would be very restrictive for a general data processor. Since the hardware is not capable of dealing with larger integers, some software scheme must be provided that can handle integers with higher accuracy containing multiple words. The addition of integers consisting of more than one data word proceeds similarly to serial addition. The least significant data word (all 16 bits in this case) of the addends are added together, and the generated carry bit from the MSBs (bits A_{15} and B_{15}) is stored in a flip-flop. After the results are stored (this is the least significant word of the sum), the more significant data words are added, with the stored carry bit also being added to the LSBs of the addends. A new carry is generated for the higher order data words, and so on. Hence, with an additional flip-flop (which is often called the **carry** flip-flop, or carry register), addition may be executed for integers of any accuracy. The 16-bit parallel adder may be used repeatedly to add two integers of any size. The hardware must provide in this case a special operation that adds the contents of the carry flip-flop to the LSBs of the addends, which, as we have shown in Chapter 8, requires a full adder for the LSBs. As we will see shortly, the carry flip-flop finds far more numerous uses than the addition of integers with multiple precision. The carry bit becomes an integral element of general shifting operations.

We have seen how a shift register works for parallel–serial conversion. A shift register may be also used to **shift** a binary number by 1 bit. Shifting may occur to the **right** (toward the LSB) or to the **left** (toward the MSB). Hence, a general (may also be called a universal) shift register as a variable function generator can execute three functions: parallel load, right shift, and left shift. Since we need two control inputs to specify three functions, the fourth function may be designated as a "no-change" operation (NOP). The block diagram and function table of a commercially available universal 4-bit shift register (type 74194) is shown in Figure 9.8. Depending on the binary values of the two control inputs S_0 and S_1, the shift register executes one of the four possible operations upon the rising edge of the clock pulse (positive clock input is shown). The three operations that change the stored 4 bits (parallel outputs) of the shift register may be defined as follows:

SHIFT RIGHT ($S_1, S_0 = 01$):

$$D_i \leftarrow D_{i+1} \quad \text{for } i = 0, 1, 2$$

$$D_3 \leftarrow \text{SERIAL INPUT (RIGHT)}$$

SHIFT LEFT ($S_1, S_0 = 10$):

$$D_i \leftarrow D_{i-1} \quad \text{for } i = 1, 2, 3$$

$$D_0 \leftarrow \text{SERIAL INPUT (LEFT)}$$

LOAD ($S_1, S_0 = 11$):

$$D_i \leftarrow I_i \quad \text{for } i = 0, 1, 2, 3$$

In Figure 9.9 we show one possible way of constructing this type of universal shift register. This is a good demonstration of the usefulness of multiplexers. The structure of the shift register shown in Figure 9.9 is similar to that in Figure 9.5, but in this case four 4-to-1 multiplexers are used. The shift register is drawn for an arbitrary number of bits, where the flip-flops for the LSB (Q_0), the MSB (Q_{n-1}), and any other bit (Q_i) are shown. The multiplexer inputs are marked with the values for the two select inputs S_1 and

S_1	S_0	Operation
0	0	No action
0	1	SHIFT RIGHT
1	0	SHIFT LEFT
1	1	LOAD

Block diagram   Functional table

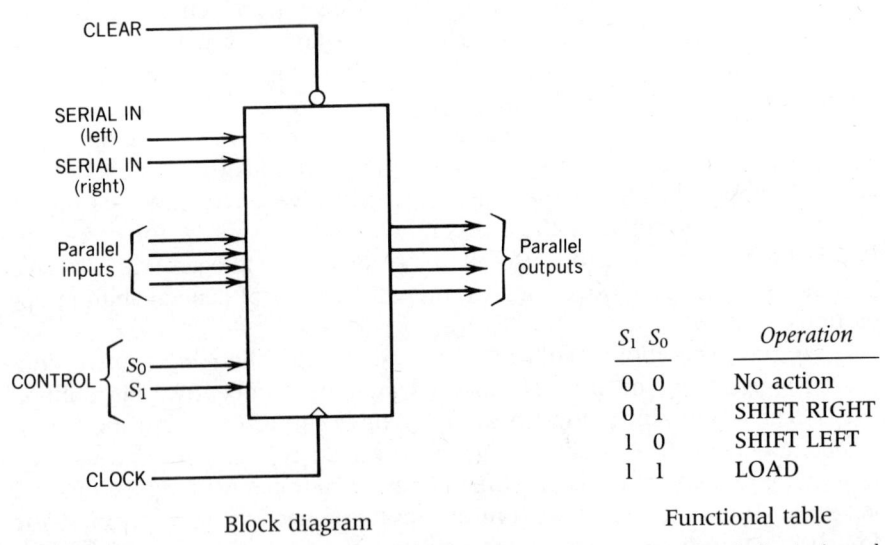

FIGURE 9.8 Block diagram and functional specifications for TTL type 74194 IC device (universal shift register).

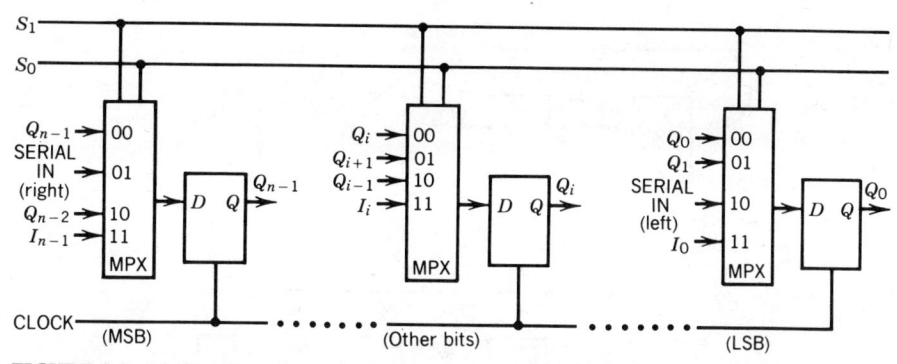

9.3 MULTIPLE PRECISION, SHIFT OPERATIONS, AND MULTIPLICATION

FIGURE 9.9 Construction of universal shift register from D-type flip-flops and multiplexers.

S_0. When the select inputs have values 00, the inputs marked with 00 are selected; when the select inputs take on values 10, the multiplexer inputs marked with 10 are selected, and so on. The serial input shown for the MSB is for RIGHT SHIFT, and the serial input shown for the LSB is for the LEFT SHIFT operation.

The shifting of a binary number adds a new family of data-processing operations to the repertoire of a digital computer. When the binary data is interpreted only as a number of bits, the shifting of the register contents is called a **logical shift** operation. Different shift operations are possible depending on where the shifted-out bit is stored and from where the shifted-in bit is derived. It is common practice to place the bit that is shifted out of the register (the MSB for left and the LSB for right shifts) into the carry flip-flop. On the other hand, there are many possibilities for the selection of the serial data input bit. If the serial input data bit is equal to the bit stored in the carry flip-flop before the shift operation, then it becomes a bit-rotate operation with the carry flip-flop attached to the register, making it a register of $N + 1$ bits. This is shown in Figure 9.10, where the bit-rotate operations are schematically illustrated. If the carry bit is not involved in the shift/rotate operation then only N bits are processed.

Since the processing units of most computers have several registers but only one carry flip-flop, the carry flip-flop can be used to shift an arbitrary number of bits of one register into another. For example, if there are two registers called A and B, then a left-shift operation for A with the carry flip-flop and then another left shift for B will shift the MSB of the A register into the LSB of the B register. Repeated shifts will transfer the bits of the A register to the B register. In this manner shift operations become essential "bit-manipulating" operations. Many computers provide several types of logical shift operations. Two types are shown in Figure 9.10. In addition to the "shift with carry" operation described above, the serial input data can be set equal to the appropriate LSB or MSB of the shifted register. In this case the operation becomes a simple bit-rotate operation. Large computers with extensive instruction repertoire may even provide shift/rotate operations with an arbitrary number of bits. Of course, this extension is not essential, but it could decrease the execution time of some often used bit-manipulating instructions.

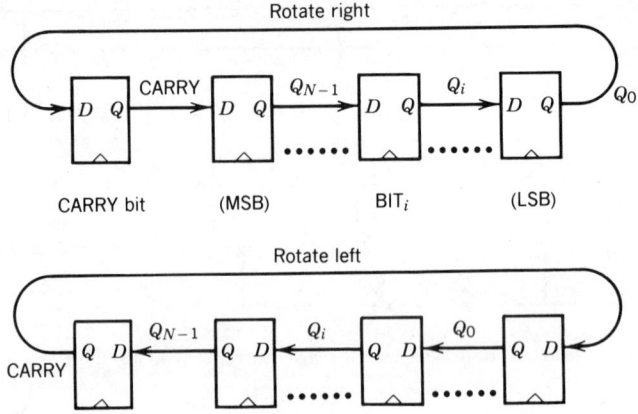

FIGURE 9.10 Schematic representation for rotation operations that include carry bit.

If the data in a register is interpreted as a signed (2's complement) number, then shifting the register is called an **arithmetic shift** operation, which effectively multiplies or divides the number by a power of 2. If the register contents are shifted left by 1 bit and a 0 is shifted into the LSB, then this operation is numerically equivalent to multiplying the signed number by 2 (as long as overflow/underflow does not occur). Notice that the same operation works for positive and negative numbers, since the 4-bit positive number 0001 (+1) shifted left 1 bit gives us 0010 (+2), whereas the negative number 1111 (−1) yields 1110 (−2). The test for overflow/underflow is identical to the one for addition; that is, if the sign bits of the numbers before and after the shift operation are different, then overflow/underflow has occurred. If an arithmetic right shift of 1 bit is used, then the value of the number should be divided by 2. This is achieved by shifting in the same value as the sign bit of the number (0 for positive and 1 for negative numbers). For example, the 4-bit positive number 0110 (+6) becomes 0011 (+3), whereas the negative number 1100 (−4) becomes 1110 (−2). There is one inconsistent condition when the negative number 1111 (−1) is shifted right arithmetically by 1 bit, since the number remains the same (1111 = −1) instead of becoming 0, which is the numerically correct result.

To summarize, shifting operations involve the shifting in of 0's or 1's from either the right (LSB) or the left (MSB) side of the word. Logical shifts usually require the shifting in of 0's. Arithmetic shifts require the shifting in of bits which result in the multiplication or division of the number by 2. Arithmetic shifts require shifting in 0's from the right (LSB) and the bit which is equal to the MSB of the number from the left.

Rotation operations circulate the bits of a register. This means that the bit shifted in at one is equal to the bit shifted out at the other side. When a number is rotated as many bits as it contains, the value of the number does not change.

9.3 MULTIPLE PRECISION, SHIFT OPERATIONS, AND MULTIPLICATION

The role of the CARRY in shift and rotate operations may be varied and sometimes confusing. The simplest is when for rotate operations the CARRY is attached to the word and, in effect, the register is extended by one bit. In this case, it does not matter whether we assume that the CARRY is attached at the right or left side of the register. For shift operations (especially for arithmetic shifts), the shift-in bit is usually not affected by the value of the CARRY bit. However, the value of the CARRY after the execution of the shift operation is often affected by the bit which is shifted out of the register. Some processors store the shifted out bit in the CARRY flip-flop, others complement the CARRY value if the shifted out bit is equal to 1. As we can see, there are many variations for handling CARRY values. For each practical processor the machine instruction manual must be consulted. The assumption that the manufacturer provided us a "logical" way of dealing with the CARRY flip-flop will be most likely wrong.

Multiplication of integer numbers can be derived from addition and shift operations. Let us consider only unsigned numbers. Multiplication is inherently a multiple precision operation since the correct result of the product of two n-bit binary numbers requires at least $2n$ bits. Instead of solving the more complicated problem of multiplying two n-bit numbers using n-bit registers (in which case the product must be stored in two registers), we shall assume that the magnitudes of the two multiplicands are less than $2^{n/2}$ (only the lowest $n/2$ bits have to be considered), and the registers are all n bits long. We have three n-bit registers called A, B, and C. The algorithm for multiplying the contents of A and B and storing the product in C is as follows:

Set the contents of C to 0 (all bits equal to 0)
FOR $n/2$ times DO
BEGIN
  IF the LSB of A is equal to 1 THEN
  BEGIN
    Add the contents of B to the contents of C and leave results in C.
  END
  Shift A right by 1 bit.
  Shift B left arithmetically by 1 bit.
END

For this algorithm we need a test of the LSB of a register, a facility supplied only by very few computers. However, if the contents of register A are shifted right, and the LSB is shifted into the carry flip-flop, then the test can be made on the stored bit value of the carry flip-flop. All computers supply the facilities to test the contents of the carry flip-flop, and thus, the algorithm can be easily implemented by a few simple instructions.

Here we have used the multiplication of unsigned integers as an example that shows how the hardware structure of a computer may affect its software capabilities. The addition of the carry flip-flop (in itself a simple hardware addition) makes multiple precision operations and bit manipulations possible. Combining the testing of 1 bit (the carry bit) and shifting operations which involve the carry bit allow the testing of any 1 bit of any register. In this case the addition of one flip-flop and minimal control circuitry (discussed in detail in the next two chapters) have increased the data processor's software capabilities significantly.

We could continue by examining how signed integers are multiplied, how integer division may be implemented, or how one could manipulate floating-point (real) numbers by hardware instructions. However, our main interest is in simple hardware structures, which demonstrate important design principles like effective hardware/software or speed/cost trade-offs. Although these various arithmetic operations are important to the understanding of digital computer hardware, they do not contribute significantly to the demonstration of design principles. Hence, we will not pursue this line of arithmetic data manipulations. We will examine carry propagation in binary adders, which will demonstrate parallel and serial processes within the interconnections of binary gates.

9.4 CARRY PROPAGATION

In Chapter 8 we described both parallel and serial addition. We have seen that for a fully parallel adder the number of adder circuits is equal to the number of bits. We have also stated that the results (the sum) appear in parallel, that is, all bits at the same time. This is not strictly correct if we include the effects of gate delay into our interpretation of parallel and serial processes. If we examine the structure of the 4-bit full adder in Figure 8.14, we can recognize the "serial" nature of the CARRY signals. Let us assume that we are looking for the fastest possible hardware implementation of a 4-bit adder. We know that it must be a parallel adder. The full-adder circuit is a combinational circuit whose outputs may be expressed as canonical (AND–OR or OR–AND) Boolean expressions of its inputs, which can be realized by two-level circuits of multi-input AND and OR gates. Consequently, the maximum delay of the SUM and CARRY outputs of the full-adder circuit is equal to two gate delays (this assumes that both the input signals and their complement are available at the same time).

The problem with the CARRY signals of the parallel adder is that they suffer accumulated delays as they propagate through the full-adder circuits. For example, in Figure 8.14 the OVERFLOW signal will be available only after eight gate delays, whereas the SUM bits (R_0–R_3) will have variable amount of gate delays. Bit R_0 will be available after two gate delays, but R_1 will have to wait for the arrival of the CARRY signal from the first adder and will become valid only after four gate delays. Similarly, the delays for R_2 and R_3 are six and eight gate delays, respectively. Even though we presented this circuit as a fully parallel adder, we can see that the output signals suffer unequal amounts of delay. In order to use this adder as a properly synchronized parallel device (all outputs are available at the same time), either we would have to accept that the adder is slow (has the delay equivalent to eight gate delays) or we would have to supply additional hardware to speed up

9.4 CARRY PROPAGATION

the generation of the outputs. This is very similar to the parallel–serial trade-offs we have discussed before. Until now we examined these trade-offs on the IC device level. Now we can see similar trade-offs on the gate level.

For example, for the full 4-bit adder we could derive a Boolean expression for the four sum bits (R_0–R_3) and the output carry bit in terms of the nine input signals (A_0–A_3, B_0–B_3, and CARRY input, or CY_0) and provide two-level circuits to ensure only two gate delays for all signals. Commercial fast IC devices, like the type 7483A 4-bit full adder with fast carry, generate their SUM outputs within a maximum delay time of 24 nanoseconds and their carry outputs within 16 nanoseconds. Thus, the speed problem is solved by commercial IC suppliers for 4-bit numbers. But, unfortunately, for meaningful arithmetic calculations we need at least 16 bits and could require even 32 or 64 bits. The problem with the parallel approach is that the complexity of the parallel circuitry is increasing very rapidly as the number of bits is increased. The question arises, could we devise a method by which some parts of the circuitry are in parallel (to increase speed) but some remain serial (keep costs low)? Such a method is found by the processes of carry **generation** and **propagation**.

Let us return to our 4-bit parallel adder (Figure 9.11). As shown in Figure 9.11, the carries propagate through the four full-adder circuits, but each CARRY output can be described also in terms of the two input bits (A and B) and the CARRY input of the adder. For the first adder we get

$$CY_1 = A_0 \cdot B_0 + (A_0 \oplus B_0) \cdot CY_0$$

which, as a logical expression, states that the CARRY output is equal to Logic-1 if both A_1 and B_1 are equal to 1 (a carry is generated by this adder) or if either one is equal to 1 (expressed by the XOR function) and the CARRY input is equal to 1 (the carry is propagated through the adder). In general, for the ith adder we can define a generating term (G_i) and a propagating term (P_i), where

$$G_i = A_i \cdot B_i \quad \text{and} \quad P_i = A_i \oplus B_i$$

and

$$CY_{i+1} = G_i + P_i \cdot CY_i$$

The SUM output can also be expressed by these terms, and we get

$$R_i = P_i \oplus CY_i$$

In fact, all four outputs can be generated by the same full adder, as shown in Figure 9.12. In Figure 9.12 we also indicate an estimated delay time for all the signals in terms of gate delays. We assumed that the XOR function requires two gate delays, a reasonable assumption for TTL devices since the XOR function can be realized with a two-level AND–OR circuit. We can see that both the CARRY and the SUM outputs are delayed by two gate delays with respect to the CARRY input and four gate delays with respect to the data input bits (A or B). If we connect four adders in the usual manner, then

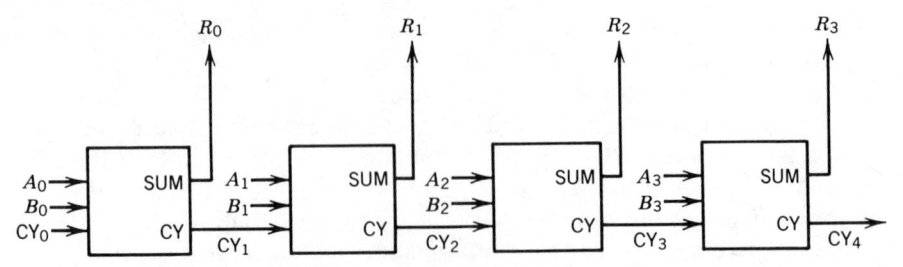

292 SERIAL–PARALLEL PROCESSING AND CONVERSION

FIGURE 9.11 Four-bit parallel adder with serially propagating carry.

the maximum delay of the last carry (CY_4) and sum (R_3) bits will be equal to 10 gate delays. All data bits are available at time $t = 0$ and the overall delay is calculated by the propogation of the carry signals.

Now let us generate all the CARRY outputs according to the expanded Boolean expressions:

$$CY_1 = G_0 + P_0 \cdot CY_0$$

$$CY_2 = G_1 + P_1 \cdot CY_1 = G_1 + P_1 \cdot G_0 + P_1 \cdot P_0 \cdot CY_0$$

$$CY_3 = G_2 + P_2 \cdot CY_2 = G_2 + P_2 \cdot G_1 + P_2 \cdot P_1 \cdot G_0 + P_2 \cdot P_1 \cdot P_0 \cdot CY_0$$

$$CY_4 = G_3 + P_3 \cdot CY_3 = G_3 + P_3 \cdot G_2 + P_3 \cdot P_2 \cdot G_1 + P_3 \cdot P_2 \cdot P_1 \cdot G_0$$
$$+ P_3 \cdot P_2 \cdot P_1 \cdot P_0 \cdot CY_0$$

We can see that the carries are all generated by two-level AND–OR expressions. This is called **look-ahead carry generation.** Since all the P_i signals are delayed by two gate delays, all four CARRY signals will be available in four gate delays. From the preceding expression for the SUM output, another XOR gate is required to generate the R_i bits, so the results will be generated within 6 gate delays instead of the 10 we calculated for the circuit using four full adders. The overall gate delays are reduced by almost half. The completed full adder using a look-ahead carry generator circuit is shown in Figure 9.13, where each of the four small block symbols indicates a full-adder circuit drawn in Figure 9.12.

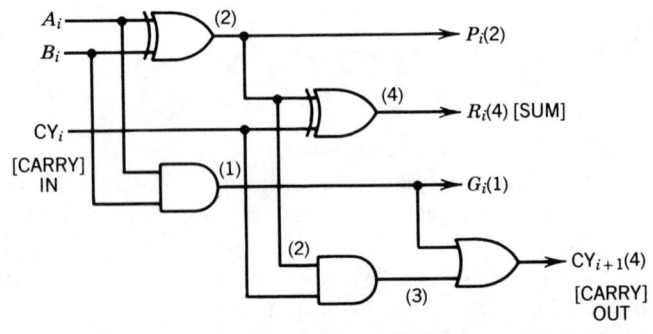

FIGURE 9.12 Full-adder circuit with generate (G) and propagate (P) signals.

9.4 CARRY PROPAGATION

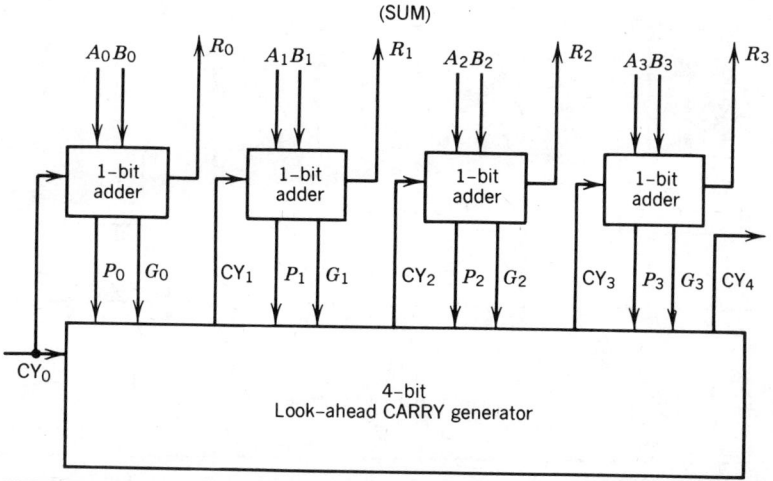

FIGURE 9.13 Use of a look-ahead carry generator circuit to increase speed of parallel 4-bit adder.

The carry generator is a two-level circuit that produces the four carry signals CY_1–CY_4 from the eight propagate and generate signals (P_i and G_i) and the LSB carry input CY_0. It is available as a practical IC device (e.g., type 74182). We have shown how we could produce a faster 4-bit adder by using such a circuit. But since 4-bit adders are readily available, the real usefulness of the look-ahead carry generator is the construction of arithmetic processors that operate on a large number of bits. The basic idea is that the look-ahead carry generator generates a carry around the 4-bit adder. Each 4-bit adder must provide a **generate** and a **propagate** output. These signals have the same functional roles as the generate and propagate signals for the simple 1-bit full-adder circuit. A generate signal for a 4-bit adder is equal to 1 if the carry from the MSB of the 4-bit adder is generated regardless of the CARRY input to the LSB of the adder. If the propagate signal is equal to 1, then the CARRY input to the LSB of the adder is propagated through the adder, and in this case the CARRY output becomes Logic-1.

One may construct a 16-bit adder using four 4-bit adders and a 4-bit look-ahead carry generator circuit. The completed circuit is shown in Figure 9.14. Each 4-bit adder is constructed from the circuits shown in Figure 9.13 and have two additional signals named P and G. These are the overall "propagate" and "generate" signals described and can be derived as follows.

The Boolean propagate output signal (P_{out}) is given by

$$P_{out} = P_0 \cdot P_1 \cdot P_2 \cdot P_3$$

In other words, if all four propagate signals are equal to 1, then the carry will be propagated through all four stages. Similarly, a "carry-generated" output may be defined as

$$G_{out} = G_3 + P_3 \cdot G_2 + P_3 \cdot P_2 \cdot G_1 + P_3 \cdot P_2 \cdot P_1 \cdot G_0$$

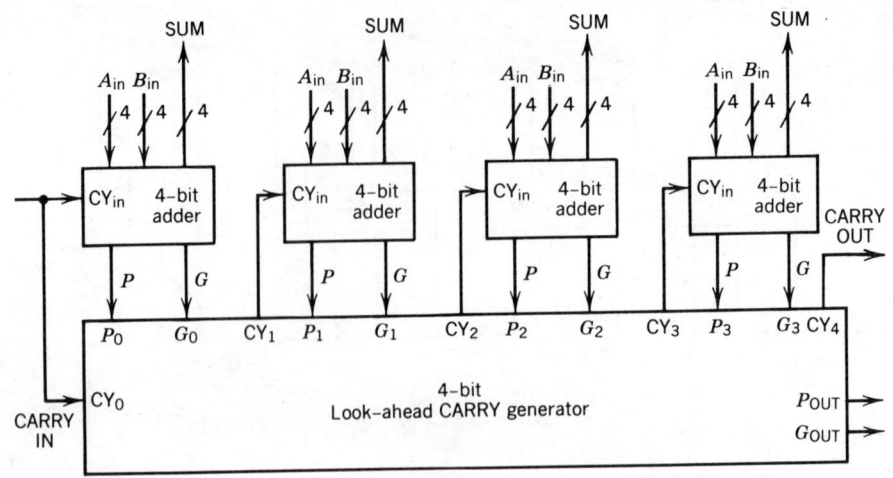

FIGURE 9.14 Use of additional carry generator circuit to increase speed of 16-bit adder.

which means that a carry out is generated if there is a carry from the last stage (stage 3) or there is carry from the second stage, which is propagated by the third stage, and so on.

The construction of this 16-bit adder from four 4-bit adders demonstrates a "recursive"-type hardware structure because the same carry generator circuit is used to speed up the 4- and 16-bit adders. Recursive-type hardware results in repeated, periodic hardware structures that are very useful for the design of efficient VLSI devices. In this case four 16-bit adders and an additional carry generator circuit can be used to build a fast 16-bit adder. The larger the structures, the more savings can be made in VLSI design if the hardware structure is of a repeated nature.

EXPERIMENT 9
Four-Bit Serial–Parallel Arithmetic Processor

E9.1 Summary and Objectives

The basic principles and techniques of serial data processing were introduced and applied in Experiment 8. Serial data-processing techniques are developed further in the following experiment by the introduction of serial-to-parallel and parallel-to-serial data conversion. The technique of using one serial data line for both data transmission and self-synchronization is also demonstrated.

The overall block diagram of this experiment is shown in Figure E9.1. There are no function selection input lines, and there is only one input line for the data bits of both input operands A and B; therefore, both the function selection and all the data information must appear on the same serial data input line. This means that at some time the incoming data bits are used for function selection, at other times for bits of number A, and yet at other times

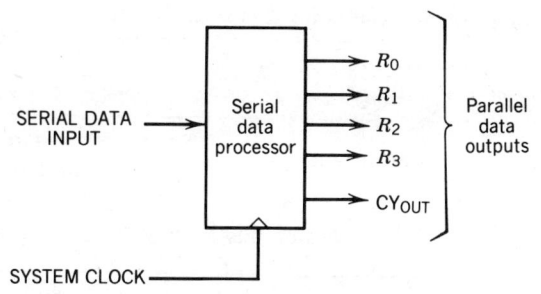

FIGURE E9.1 Block diagram of serial data processor.

for the bits of input data B. This "time multiplexing" of different data input and control information requires a much more sophisticated control circuitry than the one used for Experiment 8.

The data bits of the result are required in parallel. Universal shift registers will be used for both storing the data and converting the serial incoming data to parallel output. For the controlling circuit the use of a universal counter is recommended, which should reduce the complexity of the sequential controller for this experiment. The main objective of this experiment is to demonstrate how basic building blocks, such as shift registers, counters, multiplexers, adders, and so on, are combined to provide a reasonably sophisticated serial–parallel processing circuit.

E9.2 Serial Control, Input Data, and Expected Output Signals

The basic principles of serial data processing learned in Experiment 8 are applied in this experiment for a different type of input data transmission and for different requirements on the output signals. The overall block diagram for the serial data processor is shown in Figure E9.1. The three arithmetic or logic functions of two 4-bit numbers provided by this processor are the same as the functions used for Experiment 8. The main differences are that only one data input line is available, and the 4-bits of the result, plus the CARRY output from the MSBs, are required in parallel. These output signals are shown as R_0, R_1, R_2, R_3, and CY_{OUT}.

Since only one input line is available, a special sequence of 0's and 1's is required to indicate that data transmission is starting. It is assumed that the system starts from its **idle** state when no previous data transmission has yet occurred. As long as the serial input data bits remain at the Logic-0 level, the processor remains in its idle state. A single bit of 1 indicates that data transmission will start with the next serial input bit. From this time on 11 bits will be sent to the processor. The first 2 bits are the function selection bits, which determine the type of the required arithmetic-logic operation (the same processing functions and control bit settings may be used as for Experiment 8). The next 4 bits are the bits of one of the 4-bit operand (A), and the last 4 bits belong to the other operand (B). All data bits are sent with the

296 SERIAL–PARALLEL PROCESSING AND CONVERSION

TABLE E9.1 List of Transmitted Bits for this Experiment

Time period	Bit transmitted	Meaning	Remarks
0	0 1	Synchronization	System remains in idle state Transmission starts
1 2	C_0 C_1	Function selection bits	Only three functions are provided
3 4 5 6	A_0 A_1 A_2 A_3	First operand	LSB first
7 8 9 10	B_0 B_1 B_2 B_3	Second operand	LSB first
11 12 or 0	0 X	Synchronization Done	Output bits should show results Return to idle state

LSB first. Finally, an additional bit of 0 is sent to indicate the end of the serial transmission, which should place the processor into its idle state. The list of the transmitted bits is shown in Table E9.1.

It is assumed that as long as the system remains in its idle state, the SYSTEM CLOCK signal has no effect on the system, and the time period is indicated as T = 0. Once a Logic-1 is received, the system starts to operate in its functional (not idle) mode and transfers to time period T = 1. As shown in Table E9.1, after the two function selection bits (C_0 and C_1) have been received, the eight data bits follow. After the data bits are transmitted, the processor returns to its idle state, at which time the output bits show the results of the arithmetic-logic processing done on the received two 4-bit input data words (A and B). It is included in the functional specifications that the output bit values must remain constant as long as the processor remains in its idle state.

E9.3 Recommended Internal Organization

The state transition diagram (STD) for this processor is shown in Figure E9.2. Two alternative ways of constructing the STD are shown. We can include an extra state (done) or eliminate it from the STD because it turns out to be a redundant state according to the way the processor was specified. This is so because an additional Logic-0 bit is transmitted after the last data bit (B_3), and if the system is already in the idle state, then it will remain in the idle state when an input bit of 0 is received. This is functionally equivalent to the transition from the done state to the idle state. If the additional data bit were not specified as 0, then the done state could not have been eliminated.

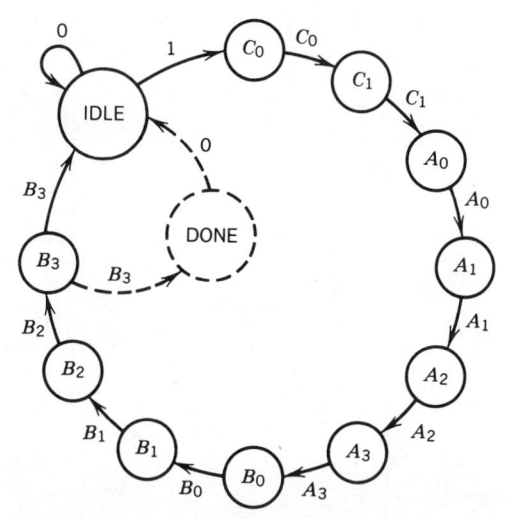

FIGURE E9.2 The STD for serial data processor.

There are many different ways this processor can be designed. We show one particular internal structure that uses only one 4-bit shift register and one full adder. One could also design the same processor using serial-to-parallel conversion for the incoming 8 data bits (which would require an 8-bit shift register), when a 4-bit parallel adder could provide the desired results. The processor shown here uses less hardware than the one with the parallel adder, and it also demonstrates the serial–parallel conversion and data-processing functions for a practical problem.

One possible internal organization of the processor is shown in Figure E9.3. There is a 2-bit shift register into which the function selection bits C_0 and C_1 are shifted at the beginning of the serial input data transmission. The stored 2 bits select the processing function. Notice that the clock that controls the 2-bit shift register is derived from the AND function of the SYSTEM CLOCK signal and a special control signal CLK_C. If CLK_C is equal to Logic-1, then the shift register will operate. If it is equal to 0, then the contents of the shift register remain constant. The CLK_C signal must be set to 1 during the two time periods when the incoming serial data input bits are C_0 and C_1. The 2 bits are stored in this shift register, after which the CLK_C signal becomes 0, and the stored selection bits will be unchanged during the rest of the processing clock periods.

The 4-bit shift register has a similar control clock signal called CLK_R. This is necessary because it is one of the requirements of the processor that its output should remain constant as long as the processor is in its idle state. For proper operation, the CLK_R signal, which controls the 4-bit shift register, has to be equal to 0 when the processor is in its idle state and must be equal to 1 when the processor is operating. The signal $CLOCK_R$ operates the 4-bit shift register. The same signal, $CLOCK_R$, is connected to the D-type flip-flop, which delays the carry output bit for the required serial addition and provides the

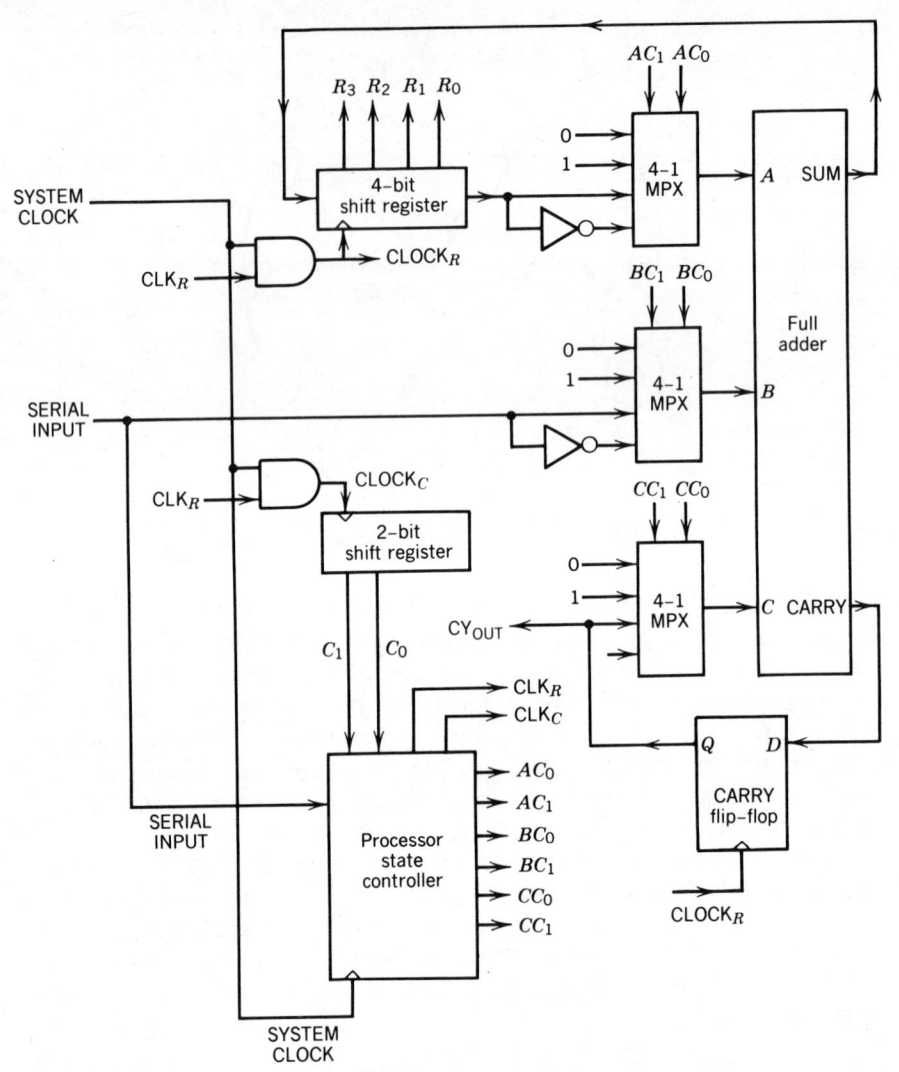

FIGURE E9.3 One possible internal organization of serial data processor.

CY_{OUT} signal after the completion of the serial processing function. When the processor is in its idle state, the contents of both the 4-bit shift register and the D-type flip-flop (CY_{OUT}) remain constant.

The 1-bit full adder receives its two data inputs from two 4-to-1 multiplexers that act as input data processors. During different phases of the input data transmission the control inputs (data selectors AC_0, AC_1, BC_1, and BC_0) are set so that inputs A and B of the full adder receive the correct input functions. Input A can be selected from values 0 and 1, the serial output of the 4-bit shift register and the complement of this output. Input B can be selected similarly with the SERIAL DATA INPUT replacing the output of the shift register.

The third input to the full adder (C) receives its input also from a 4-to-1 multiplexer. This input can be selected from values 0 and 1, and the output of the carry flip-flop (the delayed CARRY output of the adder). During the binary addition, the delayed CARRY is selected; at other times either a 0 or a 1 is selected, depending on the required function being executed by the processor. The operation of this data processor is carried out in the following manner:

1. During the idle state the outputs of the 4-bit shift register and the carry flip-flop do not change ($CLOCK_R = 0$).
2. During the first two clock periods of the data transmission, the function selection bits are shifted into the 2-bit shift register (C_0 and C_1).
3. During the next four clock periods the four input control bits are sent. The A input of the full adder is equal to 0, and its B input is equal to the incoming serial input data bits (which are equal to the bits of the first operand, A_i). The carry control bits are set so that the C input is also equal to 0. Since the other two input bits are equal to 0, the output of the full adder is equal to its B input, which is equal to the bits of the first operand. These bits are shifted into the 4-bit shift register, and after four clock periods the shift register will contain the bits of the first 4-bit operand (in the correct order!).
4. During the next clock period the SERIAL INPUT DATA line transmits bit B_0, the LSB of the second operand (see Figure E9.2). The output of the 4-bit shift register is equal to A_0, the LSB of the first operand (since this bit was shifted into the shift register during the first four clock periods). Now we are ready to execute the serial addition as discussed in Chapter 8. During this clock period the correct first carry input bit must be selected along with the correct input functions for data inputs A and B. These functions are selected by the six control signals AC_0, AC_1, BC_0, BC_1, CC_0, and CC_1.
5. For the remaining three clock periods the input functions for A and B remain the same, but the CY_{in} signal must be made equal to the delayed CARRY output of the full adder.

The operation of the processor is summarized in Table E9.2. There are five modes of operation for the complete serial processing function. During the first, or idle mode the results of the last operation are stored in the four flip-flops of the shift register and in the carry flip-flop. During the second mode of operation (two clock periods) the incoming function selection bits are stored. During the next operating mode (for four clock periods) the input bits for the first operand, A, are transmitted and stored in the 4-bit shift register.

The next mode of operation lasts only one clock period when the LSB of the second operand, B_0, is transmitted as serial input. This phase is equivalent to the start state in Experiment 8. During this clock period the C input may be adjusted to become either a 0 or a 1 by the control signals CC_0 and CC_1. The input-processing functions of the incoming bit A_0 (stored in the 4-bit shift register) and B_0 (received by the SERIAL DATA INPUT line) also have to be selected by the four control signals AC_0, AC_1, BC_0, and BC_1 according

TABLE E9.2 Operation of Data Processor with Recommended Internal Organization Shown in Figure E9.3

Time period	Phase	Select AC_0, AC_1 for input A	Select BC_0, BC_1 for input B	Select CC_0, CC_1 for input C	CLK_C	CLK_R
0	Idle	Don't care	Don't care	Don't care	0	0
1 2	Select bits C_0, C_1	Don't care	Don't care	Don't care	1	X
3 4 5 6	First operand bits A_0–A_3	0	Serial input	0	0	1
7	B_0 is received	Select according to function	Select according to function	Select 0 or 1 according to function	0	1
8 9 10	B_1–B_3 are received	Same as for Time 7	Same as for Time 7	CY_{OUT}	0	1

to the selected processing function (one of three functions, such as add, subtract, increment, decrement, etc.). The specific processing function is selected according to the binary values of the stored bits C_0 and C_1.

During the last mode of operation (three clock periods) the transmission of the serial input bits B_1–B_3 occur. During this mode of operation the same values are used for the four control signals (AC_0, AC_1, BC_0, and BC_1) as those in the last mode. But control signals CC_0 and CC_1 must be set differently so that the output of the D-type flip-flop is used for the C input of the adder. This completes the entire serial processing function.

We have described the operation of this serial–parallel data processor in terms of eight control signals. We have also shown in Table E9.2 how these control signals must be set for the correct operation of the processor. It is now the function of a sequential circuit called the processor state controller to generate the correct values for the control signals. This will be discussed in the next section.

E9.4 State Controller

Given the requirements for setting the control signals and with the recognition of five different operating modes, a general sequential circuit may be designed using the techniques with which we are already familiar. This may result in a reasonably complex circuit since there are at least 11 states (idle plus 10 data bits), as shown in Figure E9.2. There are two factors that make our recommended structure of the processor different from a general state controller. First, the STD is very simple. Once the processor leaves its idle state, it goes through all remaining states in a set sequence. This STD resembles more the STD of a simple binary counter than a general sequential circuit. Second, most control signals operate multiplexers; therefore, the state assignment problem is eliminated. Any combination of control signal values can be accepted because if the control signal value assignments change, then only the order of the input connections to the multiplexers has to be changed, and no other changes to the circuit are required.

These two factors indicate that a counter may be used for the generation of the states of this controller. In fact, we recommend that a general binary counter (e.g., types 74163, 74169, or 74191) could be used for solving most of the problems associated with the control of this processor. This example is another good demonstration of how MSI/LSI devices may find other uses than the applications they were specifically designed for.

Basically, there are two problems that have to be solved when a counter is used for this controller. First, when the counter outputs are equal to the values designated for the idle state and the SERIAL DATA INPUT is equal to 0, the counter should remain in this state and the counting disabled. When the SERIAL DATA INPUT becomes 1, the counter should start and go through 10 successive states before it drops back into the idle state. Both of these problems are solved in the circuit shown in Figure E9.4 where a counter similar to IC type 74163 is used.

The type 74163 counter has a parallel load function (similar to a general shift register) operated by the negative-logic synchronous LOAD control input. If this LOAD input is active (at 0), then during the next significant clock transition (0 to 1) the outputs of the counter are set equal to the values of the parallel data inputs I_0, I_1, I_2, and I_3. (The counter also has a synchronous

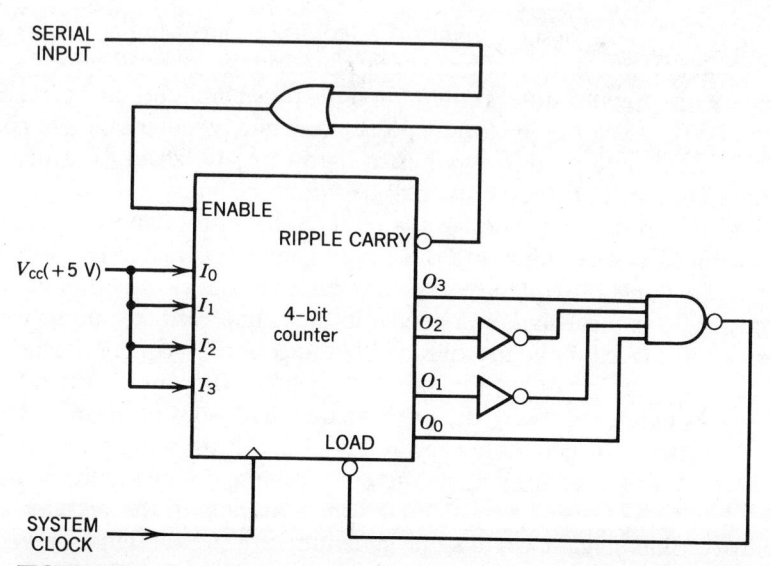

FIGURE E9.4 Circuit diagram for state controller using TTL type 74163 universal binary counter.

CLEAR control input, which, if active, causes all the output bits to become 0 at the next positive clock transition, but this feature is not used here.) It has an ENABLE input (actually, the device has two ENABLE inputs that should be connected together for this application), which must be equal to 1 for the counter to operate. It also has a negative-logic RIPPLE CARRY OUTPUT that becomes active (Logic-0) when the counter is in state 1111 (binary down counters generate an active RIPPLE CARRY OUTPUT for state 0000). A combination of these input and output facilities with some additional gates can solve both problems mentioned.

For this demonstration we have selected state 1111 as the idle state. The ENABLE input is connected to the output of a two-input OR gate, as shown in Figure E9.4. Since the RIPPLE CARRY OUTPUT is at Logic-0 for this state, as long as the SERIAL DATA INPUT is equal to 0, the ENABLE input is equal to 0, and the counter does not operate (remains in this state). As soon as a 1 appears at the SERIAL DATA INPUT, the counter is enabled and at the next positive clock transition it will change its state to state 0000. Now the RIPPLE CARRY OUTPUT becomes inactive, or Logic-1, and the counter will be enabled until it reaches state 1111 again.

If no other control circuitry were to be added to this counter, then it would go through 15 states (since this is a 4-bit binary counter) before it would reach state 1111 again. The four-input NAND gate activates the LOAD function and breaks the natural binary counting sequence. When the counter reaches state 1001, the output of the NAND gate becomes 0, and after state 1001 the counter drops into its idle state, state 1111, since the parallel data inputs are all connected to the +5-V (V_{cc}) signal. Hence, with the addition of a four-input NAND gate and a two-input OR gate, the state transitions of the controller have been realized.

There are many other schemes that could produce similar results with different bit patterns representing the idle state and the other 10 system states. If a binary up–down counter is used, then the possibilities become even more numerous. The exploration of these possibilities and whether different counting sequences could produce savings in hardware are left to the student as an important part of the experimental design procedures.

For our recommended design the required 8 control signals are all functions of the 11 system states and the binary values assigned to the 4 counter outputs (O_0–O_3). Consequently, the complexity of the circuits that generate the required 8 control signals depends on the actual counter output values. Therefore, it is worthwhile to explore different counting sequences and select the sequence that produces the simplest circuits. Of course, if a different counting sequence is chosen, then the circuits that control the counter must also be changed. In generating the required 8 control signals the use of a 3-to-8 decoder that can indicate 8 different combinations of 3 of the 4 counter output bits should also be considered. The 8 outputs of the decoder could simplify considerably the circuits that generate the 8 control signals, especially if the counting sequence is carefully chosen.

E9.5 Design Requirements and Suggestions

Obviously, the requirement for this experiment is the design and construction of a working circuit that satisfies the requirements of this serial–parallel 4-bit processor. The same three processing functions should be provided that were used for Experiment 8. It is not necessary that the same function selection bit settings be used as those of Experiment 8, although these may be the best choices for this experiment as well.

The minimum requirement is the detailed design of this processor along the lines suggested in the previous sections. The use of available MSI and LSI devices is strongly recommended. In particular, multiplexers become very useful in many parts of complex state controllers and control signal generators. Sometimes the addition of a flip-flop reduces the amount of circuitry required for the recognition of a state. For example, in this case, to uniquely decode a system state, one needs a four-input AND or NAND gate (there are four outputs of the counter) with possible inverters (because the counter does not provide inverted outputs). On the other hand, a state may also be recognized by the fact that one of its bits exhibits a particular signal transition. If a flip-flop is used to delay the bit value by one clock pulse, then a simple two-input AND gate may be used to recognize the same state through the indicated signal transition, and no inverters are needed (since most flip-flops provide inverted outputs).

The designer of this problem is encouraged to explore different structures, circuits, and design philosophy for the solution of this problem. However, any design chosen must be justified by showing that it requires less hardware than the recommended solution. This is the reason why the detailed design of the recommended processor is required. If the minimum amount of hardware that produces a solution to this problem is still beyond the capabilities of the available experimental equipment, then the simplifications suggested in the next section should be considered.

E9.6 Simplified Processors and/or Input Conditions

Since this problem may demand an unreasonable amount of hardware, simplifications of the processor requirements will be listed here. One may adopt one or a combination of simplifications to achieve a design problem that can be solved with a limited amount of hardware resources. The suggested simplifications are listed in what follows.

1. Limited Processing Capabilities

The 4-to-1 multiplexer connected to the SERIAL INPUT DATA signal may be eliminated and the required control signals reduced to 6. In this case there is only a limited number of functions that can be generated since the bits of the second operand (B_i) cannot be changed. The available functions are B,

B plus *A*, *B* minus *A*, *B* minus 1, and *B* plus 1. Actually, there are other possible functions, but they lead to more complications. Choose three or only two different functions.

2. Separate Function Control Bits

Two input signals are added to the processor that are set to the function selection bits. In this case only the eight data bits are sent through the SERIAL DATA INPUT line, and the STD of the processor is reduced to nine states. The two-input shift register is also eliminated, since the function selection bits are provided as inputs to the processor and do not have to be stored in a shift register.

3. Serial–Parallel Add and Subtract Processor

Provide a processor that serially adds or subtracts two 4-bit numbers and provides the result in a parallel form. The add–subtract function is selected by one additional input signal, and the transmitted serial data includes one synchronizing Logic-1 bit plus 8 data bits. This is the same input data sequence as the one we suggested for the simplified processor.

4. Sixteen Input Bits

If the total number of bits (including the synchronizing bits, the control bits, and the data bits) is equal to 16, then the circuits that control the counter are greatly simplified. This also allows the addition of bits transmitted either before, after, or between the input numbers, which may eliminate some of the required control circuitry. For this version of the processor you are free to specify what additional 5 bits should be transmitted serially (and in which order). These bits are in addition to the 8 bits of the input data, 2 control bits, and 1 synchronizing "start" bit.

Drill Problems

9.1. Add some additional hardware and change the type 74193 shift register so that it complements its contents (logical or 1's complement) when the function selection inputs are 00 and a clock pulse is applied.

9.2. What is the difference in the circuits of Moore- and Mealy-type bit-sequence detectors if they are built using shift registers as suggested in Section 9.2.

9.3. The bit-sequence detector in Section 9.2 detects only nonoverlapped bit sequences. Design an overlapped bit-sequence detector using a shift register. (*Hint:* use serial-to-parallel data conversion.)

9.4. Build a data processor using the least amount of hardware possible that increments or decrements a 16-bit number stored in a 16-bit shift register. (*Hint:* use recirculation of data.)

9.5. Design the algorithm of multiplying two N-bit signed (2's complement) binary numbers stored in two N-bit registers. Two other N-bit registers receive the product. A carry flip-flop is also available.

9.6. Design the algorithm for division of two unsigned binary numbers. Assume that both divisor and dividend are stored in N-bit registers and both the result and remainder are also stored in N-bit registers. Detect overflow.

9.7. Using 4-bit parallel adders (with carry generate and propagate signals) and 4-bit look-ahead carry generators, design a 64-bit adder and calculate its delay in terms of gate delays. Compare this to the delay of a 64-bit parallel adder that does not use look-ahead carry generator circuits.

CHAPTER 10

Introduction to Digital Computers

10.1 INTRODUCTION

We have been studying basic digital circuits and how they are applied to digital design problems. We are now ready to examine the structure and operation of digital computers. We will start by describing two new digital hardware building blocks, data buses and memory. With the already familiar concepts of registers and register transfer operations and the new concepts of data buses and memory, we will be able to describe the basic hardware structure of most modern digital computers. We will limit the examination of computers to their most fundamental features. In this and the next chapter two small digital computer models will be developed. These models will be based on 4- and 8-bit computers, which will serve us well as introduction to features found in all small computers and, especially, in microprocessors.

We will introduce both the hardware and the software aspects of the computer. Therefore, in addition to the examination of the basics of digital computer design, an introduction to computer machine language and assembly language will also be covered. Computer software will be limited to very simple hardware-oriented programs such as diagnostics, which demonstrate the correct operation of the computer's instruction set.

Even though the models of digital computers studied in this and the next chapters will be small and limited, they will be designed using "real" operational circuits. They will feature automatic program execution, integer arithmetic and logic functions, program sequence control (including conditional control instructions), memory register load and store instructions, effective memory address calculations, and both input and output instructions.

Digital computers incorporate all aspects of digital design. Both combinational and sequential circuits play major roles. Counters, multiplexers, demultiplexers/decoders, shift registers, sequential state controllers, and so on, are all important parts of modern digital computers. By studying the hardware structures of digital computers, we will be able to use all the different elements of digital design covered in preceding chapters.

10.2 REGISTER TRANSFERS, DATA BUS STRUCTURES, AND MEMORY SYSTEMS

We have studied the role of digital registers in data processing. If we define binary **data storage** hierarchically, then at the lowest level there is the flip-

10.2 REGISTER TRANSFERS, DATA BUS STRUCTURES, MEMORY SYSTEMS

flop, which can store 1 bit, that is, the smallest possible amount of binary data. At the next higher level there is the register that consists of a number of flip-flops and can store an ordered set of bits. The next higher level will consist of a number of registers that can store an ordered set of words. Digital computers usually contain a large number of storage registers, some of which may have specific names such as A, B, INP, SP, and so on; others may be identified as an ordered set of registers, such as R_0, R_1, and R_2; and others are treated as a block of a large number of storage registers, that is, memory.

Let us examine now two different ways that data may be transmitted within the different storage registers of a computer. We have seen in Chapters 8 and 9 the methods of both parallel and serial data transmission. In most modern digital computers data transmission between two registers (we call one the source register and the other the destination register) occurs in parallel. The most flexible method of data transmission between registers allows any register to be a source register, a destination register, or both. This most general data transmission arrangement is shown schematically in Figure 10.1, where four registers R_0, R_1, R_2, and R_3 are connected through a digital circuit called a switch. Since data transmission (register transfer) operations are sequential processes, a clock pulse is also required for the execution of the data transfer. It is important to stress that Figure 10.1 is a schematic and not a circuit diagram. Figure 10.1 illustrates that any one of the registers can act as a source register and, at the same time, can accept data and become a destination register as well.

Similar to how bits of a register can be transmitted in serial or parallel, transmission of data between a number of registers can be executed in two ways. If more than one register transfer operation can occur in parallel (at the same significant clock transition), then the switch is called a crossbar switch, which is similar to a telephone exchange since an exchange is capable of handling a large number of telephone connections in parallel. Figure 10.2 shows the most general crossbar switch for four 1-bit registers realized using

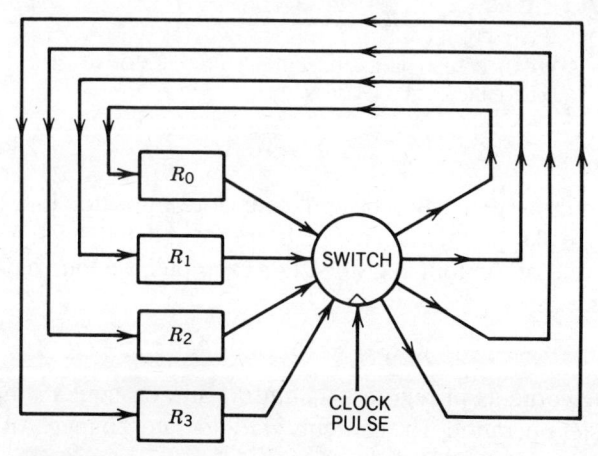

FIGURE 10.1 Schematic diagram for universal switch connecting four registers.

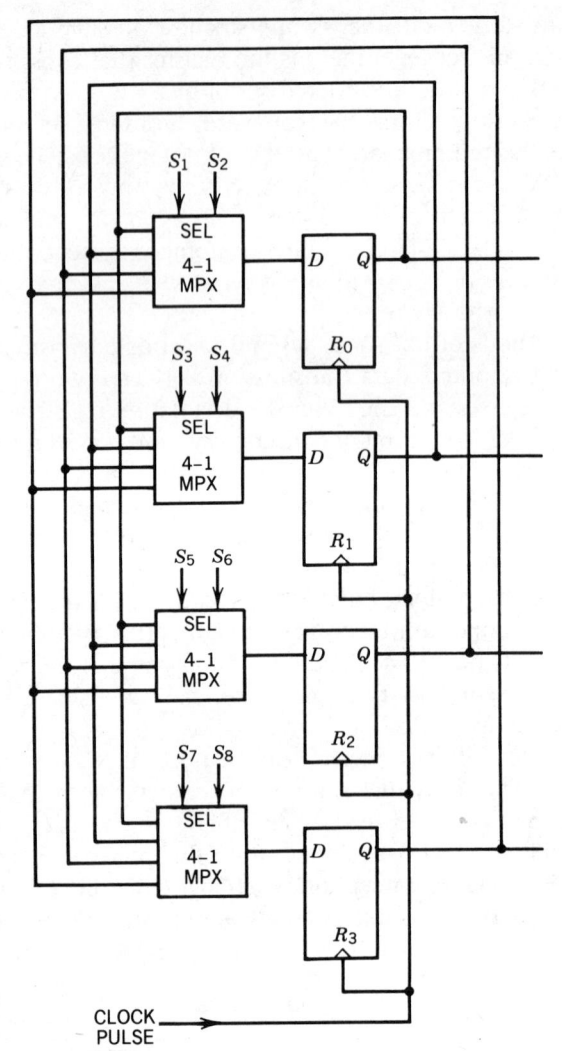

FIGURE 10.2 Hardware realization of crossbar switch between four 1-bit registers.

four 4-to-1 multiplexers. At each significant clock transition four data transfer operations take place; that is, all the registers are updated. Each source register may be any one of the four registers. For example, the four register transfer operations

$$R_0 \leftarrow R_1 \qquad R_1 \leftarrow R_0 \qquad R_2 \leftarrow R_3 \qquad R_3 \leftarrow R_3$$

exchange the contents of registers R_0 and R_1 and execute a simple $R_2 \leftarrow R_3$ register transfer operation. The contents of R_3 does not change. All four register transfer operations occur at the same time.

10.2 REGISTER TRANSFERS, DATA BUS STRUCTURES, MEMORY SYSTEMS

The same type of crossbar switch may be built for registers containing more than 1 bit. For each additional bit per register, four multiplexers must be added. Hence, for four 16-bit registers sixty-four 4-to-1 multiplexers are required. The number of selection lines of the switch remains 8 and is not a function of the number of bits stored in each register since the selection lines operate in parallel for all the multiplexers that belong to the same register.

The crossbar switch is the most general data transfer method, but it requires expensive hardware and, therefore, has not been used extensively in digital computers. A much less expensive method is to allow only one register transfer per clock pulse while allowing any one of the registers to become a source and another a destination. The hardware realization of this type of switch (called a time-multiplexed switch) for four 1-bit registers is shown in Figure 10.3. Only one multiplexer is required, which selects one of the four registers as a source register and connects its output to a digital

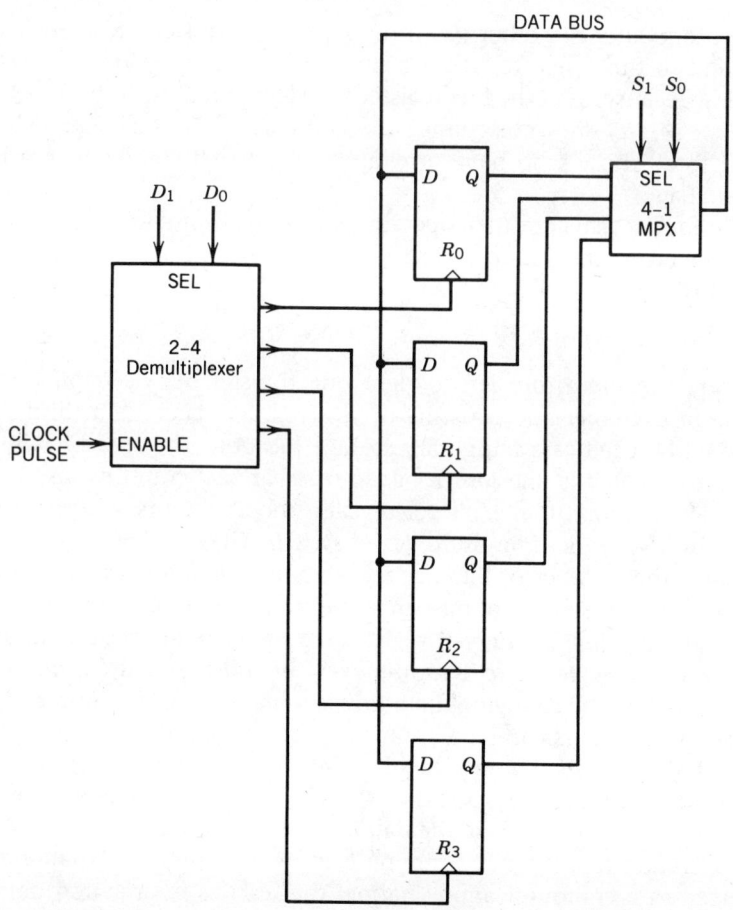

FIGURE 10.3 Hardware realization of time-multiplexed switch between four 1-bit registers.

line, labeled DATA BUS in the figure. The DATA BUS signal is connected to the D inputs of all four registers. Each register receives a separate CLOCK signal connected to a 2-to-4 demultiplexer. The input of the demultiplexer is connected to the CLOCK signal whose significant transition executes the data transfer. The destination register is selected by the two control lines of the demultiplexer (D_0 and D_1). These control lines determine the output line through which the CLOCK pulse is routed. Consequently, the 4 control lines (2 for the multiplexer and 2 for the demultiplexer) select 1 of 16 possible register transfers, which can be written as $R_j \leftarrow R_i$, where i and j can take on any value from 0 to 3. For each significant clock transition one data transfer operation can be executed. Obviously, this switch works slower than the crossbar switch. In addition, it is not as powerful as the crossbar switch because it is impossible to exchange the contents of two registers without destroying the contents of a third register. The two register transfer operations

$$R_0 \leftarrow R_1 \qquad R_1 \leftarrow R_0$$

executed one after the other (with the $R_0 \leftarrow R_1$, transfer executed first) are equivalent to the single register transfer $R_0 \leftarrow R_1$. This is not an exchange operation because, after the first register transfer operation has been executed, transfer $R_1 \leftarrow R_0$ does not change the contents of R_1 (this should be very familiar to all those who have programmed a digital computer in a higher level language).

The four register transfer operations used as examples for the crossbar switch can be executed in four clock periods by the following register transfer operations:

$$R_2 \leftarrow R_0 \qquad R_0 \leftarrow R_1 \qquad R_1 \leftarrow R_2 \qquad R_2 \leftarrow R_3$$

which are executed from left to right one transfer per clock pulse. It was possible to exchange the contents of two registers because a third register (R_2) could be used as a temporary storage location. The speed of the time-multiplexed switch is one-fourth of the crossbar switch in this case.

If there is more than 1 bit per register, then, of course, more hardware must be added to the time-multiplexed switch. The number of multiplexers is equal to the number of bits per register. But the demultiplexer circuit is sufficient for any number of bits since the updating of the registers occur in parallel (all flip-flops of a given register are updated at the same time). For 16-bit registers sixteen 4-to-1 multiplexers would be required, but we will show shortly that by changing the hardware characteristics of our IC devices, even the multiplexers can be eliminated.

We have now seen how general register-to-register data transfers can be executed using multiplexers and demultiplexers. The combination of registers and multiplexers that provide the data bus is so fundamental to digital computer operation that we will examine it in its own right. The data bus may be viewed as a communication channel that carries information only from one source at any one time but may be switched from source to source. This type of switching is called **time multiplexing,** or simply multiplexing. It is

10.2 REGISTER TRANSFERS, DATA BUS STRUCTURES, MEMORY SYSTEMS 311

used often when the provision of a large number of parallel connections is prohibitively expensive.

The data bus is logically connected to one of its sources, which can be realized with a multiplexer. Another approach is to use a multi-input OR gate, merge the sources into one signal (the data bus), but enable only one source at a time. This is shown in Figure 10.4, where we have again used four 1-bit registers for demonstration. Of the four enable signals (EN_0–EN_3) only one is equal to Logic-1; all others must have 0 values. This arrangement of AND–OR circuits was demonstrated in Chapter 3 and Experiment 3. By changing the physical characteristics of the outputs of IC devices, the OR gates may be eliminated by connecting the outputs of the four data registers in a so-called wired-OR configuration. The outputs of the IC devices we have used until now could not be connected together, but there are classes of TTL devices

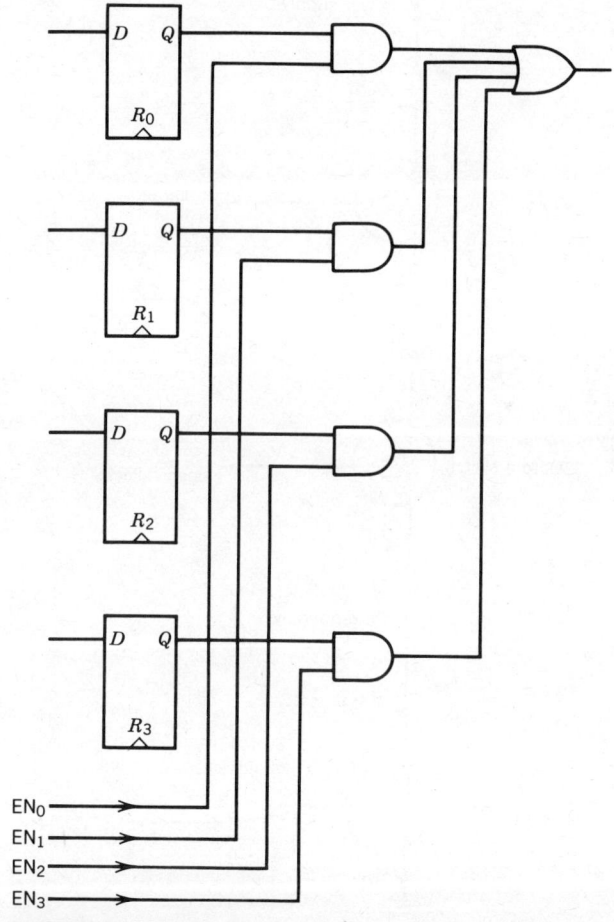

FIGURE 10.4 Using of AND–OR circuits to construct data bus from four 1-bit registers.

that are provided with special output circuits. The connection of device outputs is called either a wired-OR or a wired-AND connection depending on the logical function it performs and may be made by using IC devices that have either **open collector** or **tristate** outputs.

The differences between the electrical characteristics of ordinary, open-collector, and tristate outputs are demonstrated for the TTL class of IC devices in Figure 10.5. The output of an ordinary IC device is effectively a switch that has very low impedance when its output is equal to 0 and is around a hundred ohms when its output is in the Logic-1 state. If the outputs of two such ordinary IC devices were connected together and the two switches were in different positions, one of the 100-Ω resistors would be shorted to ground, and a large output current would flow in it. This could easily damage the IC device; therefore, outputs of ordinary TTL devices may not be connected together.

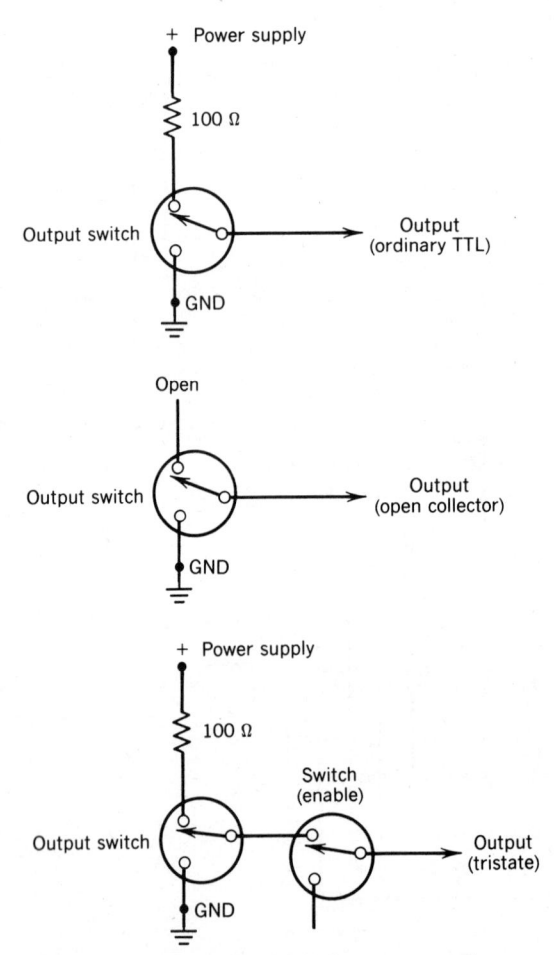

FIGURE 10.5 Output models for TTL normal, open-collector, and tristate IC devices.

10.2 REGISTER TRANSFERS, DATA BUS STRUCTURES, MEMORY SYSTEMS

As shown in Figure 10.5, the open-collector output contains only a switch, and no connection is provided to the power supply. For this reason, open-collector IC devices will not operate on their own, and for proper operation they require external resistors. In Figure 10.6 the operation of an open-collector, type 74170 storage device is demonstrated. In this diagram only those input and output connections are shown that are relevant to our current discussion of this device. The type 74170 IC device contains four 4-bit registers. It can store 16 bits, but only 4 of these are available at any one time. The outputs Q_0–Q_3 belong to the stored bits of one 4-bit register, and the binary values of the selection inputs S_0 and S_1 determine which register outputs are selected. When the values of S_1 and S_0 are 00, the selected register is R_0; when the values are 01, it is register R_1; and so on. In order for this device to operate, four "pull-up" resistors must be connected to its outputs. The other end of the resistors must be connected to the V_{cc} power supply, which, for TTL-type IC devices, is equal to 5 V. When an output is equal to Logic-1, the respective output switch of the open-collector device is opened, and the output connection voltage is raised to a positive value (it will be nearly 3.5 V if the output of the device is not loaded down by the connections to the inputs of other devices).

In addition to the selection inputs, the type 74170 storage device contains a READ ENABLE or OUTPUT ENABLE input as well. This "enable" input plays an essential role in the interconnection of open-collector devices. When the OUTPUT ENABLE signal is inactive (has value Logic-1 in this case), all open-collector output switches are opened (logically they are at the Logic-1 level when the positive-logic convention is used). This allows the connection of outputs of many open-collector devices as long as only one such device is enabled.

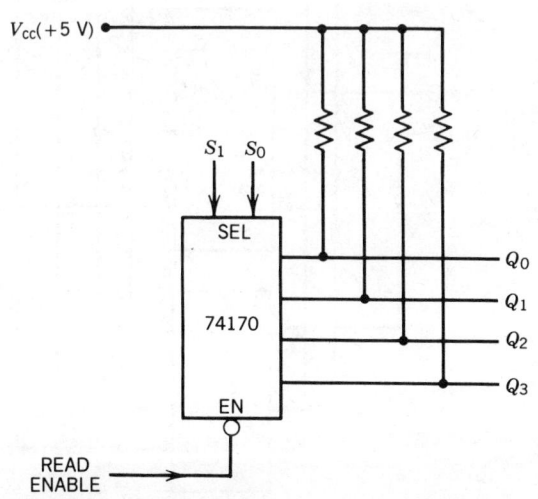

FIGURE 10.6 Use of resistors for open-collector outputs of the type 74170 memory device.

The interconnection of four type 74170 devices is shown in Figure 10.7, which provides storage for **sixteen** 4-bit registers. There are four selection inputs (E_0–E_3), two of which (E_0 and E_1) are connected to the selection inputs of all four storage devices. The other two selection inputs are connected to a type 74155 2-to-4 decoder/demultiplexer. With the two inputs of the demultiplexer enabled (at Logic-0 level), the 74155 behaves as a simple decoder with negative-logic outputs; that is, exactly one of four of its outputs is equal to 0 (active), whereas the other three outputs are equal to 1 (inactive). These

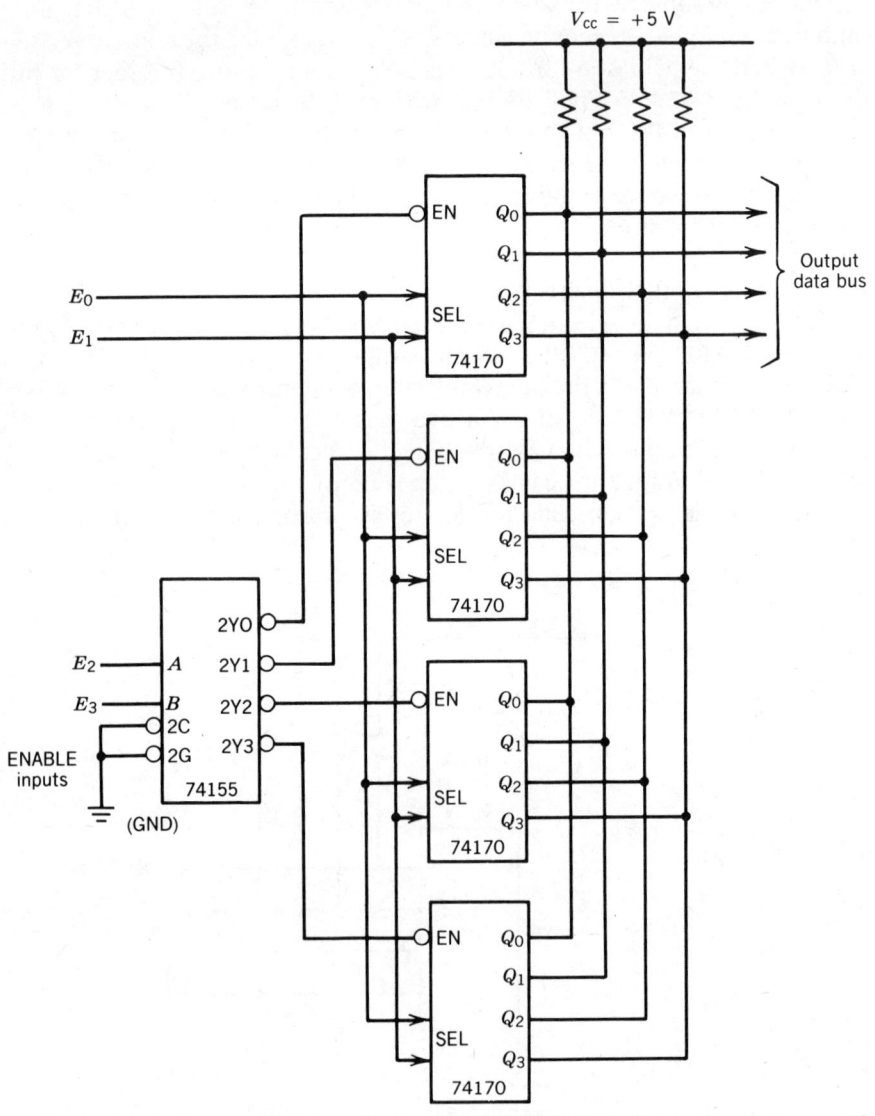

FIGURE 10.7 Hardware connections for four type 74170 devices that produce memory with sixteen 4-bit registers.

10.2 REGISTER TRANSFERS, DATA BUS STRUCTURES, MEMORY SYSTEMS

decoder outputs are connected to the enable inputs of the four type 74170 storage devices and ensure that only one open-collector device is enabled at any one time. The circuit in Figure 10.7 demonstrates how the outputs of open-collector devices may be connected together and their functions merged without the use of a multiplexer. The outputs of this circuit (containing 4 bits) is indicated as the output data bus.

In Figure 10.8 we show a considerably simpler and very useful schematic representation of this circuit, where the merging of the 16 register outputs is shown schematically and should be interpreted as a functional, rather than a wiring, feature of the circuit. Once the operation of the data bus is understood, many of the wiring details can be dropped from a functional or conceptual data flow diagram such as the one shown in Figure 10.8. In this diagram only the available data paths are shown. It is unnecessary to show the enable and control lines or the associated multiplexers and demultiplexers. As a further simplification, the data bus is shown as a single line with a small number indicating the number of bits it contains (4 bits in this case). The diagram in Figure 10.8 indicates that at a specified time the 4 bits of the output data bus may be made equal to the outputs of any one of the 16 registers. Similarly, the diagram shows that during a specified significant clock transition the 4 stored bits of any 1 of the 16 registers may be updated (written) according to the values carried by the input data bus.

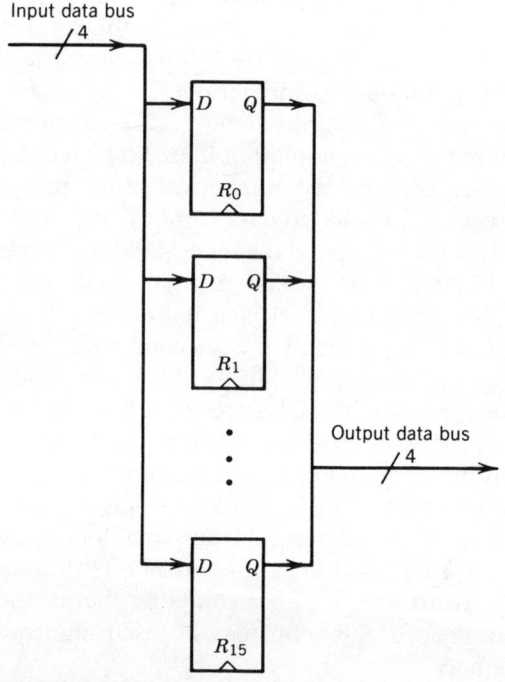

FIGURE 10.8 Schematic representation of data paths for input and output data bus.

Integrated-circuit devices provided with tristate outputs have the equivalent electrical characteristics of an additional switch in their output lines, as shown in Figure 10.5. Tristate devices always include enable inputs, and when such a device is disabled, its output impedance becomes very high (the serially connected switch is opened). When the device is enabled, it behaves as an ordinary IC device. Consequently, IC devices with tristate outputs can be connected as long as it is ensured that only one of them will be enabled at any one time. The advantage of tristate devices is that no external resistors are required for their operation. Their only disadvantage is that they may be damaged if more than one device is enabled at the same time due to a wiring error or faulty control circuits. Wired-OR or wired-AND circuits are not damaged when more than one of them are enabled by mistake. Apart from this difference, the operation of connected tristate and open-collector devices is the same.

So far we have been discussing the storage of information in computers in terms of registers. Other fundamental computer storage systems are often referred to as **memory systems, memory units,** or simply the **memory** of the computer. The internal structure of memory systems differ very little from the structure of the storage element constructed from sixteen 4-bit registers, shown in Figure 10.7. A hardware distinction may be made between registers and memories of a computer by the number of parallel output lines they provide. Registers are isolated storage devices, and their outputs are directly available. Memories contain a number of registers, but the output of only one register is available at any one time. In this sense, the four registers of the type 74170 IC device (Figure 10.6) form a small read–write memory unit called RAM (random-access memory).

Since memories contain a very large number of storage registers, they require a large number of selection input lines. The selection input values are collected into an unsigned binary number called the **memory address,** or simply the **address,** of the selected data word. In fact, we have already used this concept for the four registers of the type 74170 IC device shown in Figure 10.6, where the values of the selection inputs indicated the index of the selected register (e.g., values 10 selected register R_2). We can consider the unsigned binary number formed by the selection inputs as the address of the register and change our notation slightly by indicating the address as an integer enclosed in square brackets. Thus, register R_0 is indicated as $R[0]$, register R_1 as $R[1]$, and so on. (Pascal programmers will discover the obvious connection between this notation and arrays in Pascal.) The same notation may be used for memories when a single memory storage register is identified as MEMORY[address], where *address* is an unsigned binary value in the range of 0 to $2^k - 1$ for k selection lines. Designers often use the term *memories* (distinct from the word *memory*) for a computer that includes all stored bits of a computer; therefore, this term may be used interchangeably for both registers and memory.

A general memory unit for TTL devices is shown in Figure 10.9. As described in the preceding, the identification of one particular storage register

10.2 REGISTER TRANSFERS, DATA BUS STRUCTURES, MEMORY SYSTEMS

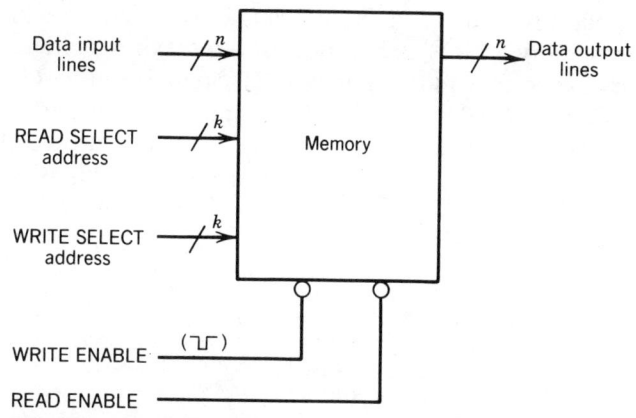

FIGURE 10.9 Block diagram of general memory unit containing 2^k n-bit storage registers.

among the large number of stored registers in the memory is made by a binary unsigned number, called an address. The number of bits contained in the address of a memory limits the number of available storage registers. The memory unit shown in Figure 10.9 has k address bits and thus 2^k storage registers. It has separate READ and WRITE SELECT address lines, two control lines called READ ENABLE and WRITE ENABLE, and both input and output data lines. As shown, every storage register contains n data bits.

Enable control lines are shown with negative-logic convention because this use is standard for TTL devices. The role of the READ ENABLE line has been demonstrated. It allows the connection of the outputs of multiple memory units as it enables only one of the connected units at any one time. The WRITE ENABLE input is equivalent to an inverted CLOCK signal of a D-type flip-flop. With the separate READ and WRITE SELECT lines, this general memory unit is capable of updating the contents of one register and at the same time enable its outputs from another register; in other words, it is capable of executing the general "memory-to-memory" register transfer operation:

$$\text{Memory[Destination]} \leftarrow \text{Memory[Source]}$$

In this case the n data output bits of the memory are connected to its DATA INPUT lines. In order to execute this register transfer operation, the bit values indicating the destination address are applied to the WRITE SELECT input lines, the values of the source address are applied to the READ SELECT lines, the READ ENABLE signal is set to 0 (active), and a negative clock pulse is applied at the WRITE ENABLE input line. The type 74170 storage device is equivalent to this general memory model with $n = 4$ and $k = 2$.

Such memory units that can be both written and read are called random-access memories, or RAMs. RAMs can be constructed from various types of memory units, some of which cannot be read and written simultaneously. Another type of RAM, called dynamic RAM, requires the writing of all storage registers periodically. Otherwise, the stored information does not remain constant in the memory. This is called **memory-refresh** operation.

For many types of memory units external circuits must be added for them to operate correctly (see Figure 10.10). Two additional registers are used. One of these, the memory address register (MAR), holds the address of the memory word, and another, the memory buffer register (MBR), holds the data. The following "memory cycles," or register transfer operations, are available, each one activated by pulsing the CLOCK input:

Memory cycle	Register transfer
READ	MBR ← MEMORY[MAR]
WRITE	MEMORY[MAR] ← MBR
LOAD MBR	MBR ← Input Data
LOAD MAR	MAR ← Input Address

Here MEMORY[MAR] indicates the storage register whose address is equal to the contents of the MAR register. In order to write a word into memory, the LOAD MAR cycle or instruction must be followed by the LOAD MBR and the WRITE cycles. In order to read the contents of the memory, the LOAD MAR cycle must be followed by a READ cycle. This memory unit is capable of "refreshing" its contents by the execution of a LOAD MAR, READ, and WRITE cycles executed in this order.

There are also memory units that contain a fixed set of stored bits and can only be read. These memory units are called read-only memories, or ROMs. Many ROM types do not contain a set of stored bits when purchased. The required contents of these ROMs are written once by the designer using a specially designed ROM programmer. This ROM type is called programmable read-only memory, or PROM. A popular type of PROM is called EPROM, that is, erasable PROM. The stored information in the EPROM may be erased by ultraviolet radiation and the PROM reprogrammed. A similarly programmable and erasable ROM device called EEPROM (electrically erasable and programmable read-only memory) is gaining rapid acceptance in modern digital systems. The EEPROM may be reprogrammed electrically without removing the device from the system. Electrically the EEPROM is similar to a RAM, but the information is written into the EEPROM only once, and then the WRITE ENABLE signal is disabled. When power is turned off, information stored in ROMs remains unchanged. Most RAM devices lose their stored information when power is turned off.

Digital computers use both constant and variable stored information. Therefore, both ROMs and RAMs play important roles in their design and construction. We have shown that the structures of memory and register storage elements are very similar. However, their use and implementation are very different. Most of the registers are used frequently; they are always RAMs and are usually constructed with the fastest hardware technology the manufacturer can afford. They are the main "workhorses" of computer data

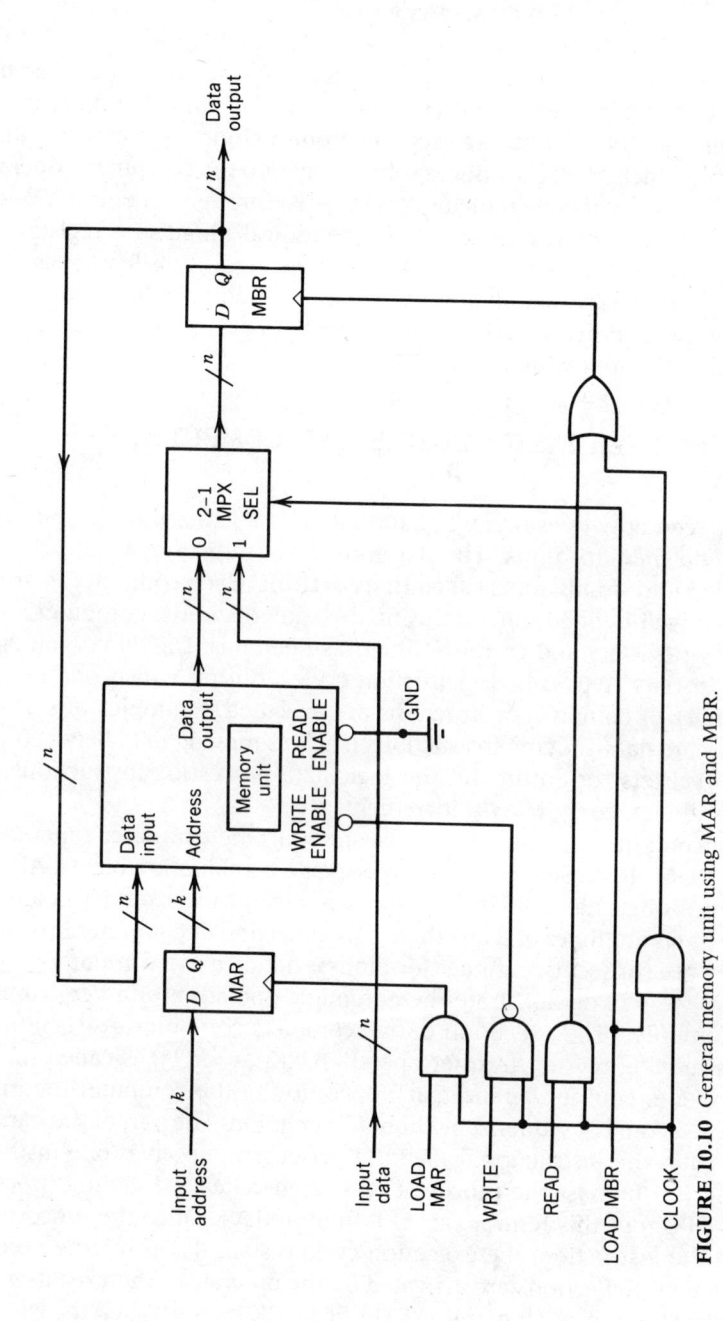

FIGURE 10.10 General memory unit using MAR and MBR.

processing. Their contents are available "instantly," without any time delay, in most working phases of the computer. On the other hand, in order to obtain the contents of a storage register from memory, a memory "READ" is required, which often introduces some delay into the computer's operation. A similar delay occurs when memory is updated through a memory "WRITE" operation. This difference in speed of operation distinguishes registers from memory in most data processors. Another distinguishing feature is that in most cases registers contain "data," whereas the memory of a computer stores computer instructions (i.e., programs) as well. Computer instructions will be discussed in the following section.

10.3 COMPUTER INSTRUCTIONS AND COMPUTER CYCLES

We have seen how the storage of information is organized in computers into register and memory units. The data-processing functions of the computer are divided into simple units called **instructions.** Each computer instruction executes a well-defined function using the contents of the computer's stored memories (registers and memory units) as operands. Digital computers operate in distinct steps called instruction cycles, during which one or a very few number of computer instructions are executed. Examples of simple instructions are data-moving instructions (such as moving data between memory and registers) or arithmetic and logic data-processing instructions (such as add, subtract, complement, increment, etc).

Any computer instruction may be stored in the computer's memory (in RAM or ROM) because it can be expressed as a combination of bits. Although instructions often have different number of bits, in most modern computers they are often multiples of 8 bits (bytes). A collection of computer instructions that often use consecutive storage locations is handled as a computer program.

First, we will demonstrate the automatic operation of a very simplified digital computer. The execution of one complete computer instruction (e.g., the data-moving register transfer operation $R[2] \leftarrow R[1]$) is called the **execution cycle.** During the automatic operation of the computer the instructions are executed in sequence without interruption. The part of the hardware that executes the instruction, called the **processor,** receives one instruction at a time (in the past the processor was often called the central processing unit, or CPU, but this term is out of fashion today). Once the processor has received the instruction, the execution cycle is started. During the execution cycle the required actions are activated by the processor. The execution cycle itself may extend through many SYSTEM CLOCK cycles. When the execution cycle is completed, the processor is ready to receive its next instruction. Since the instructions are stored in a memory unit, the execution cycle must be preceded by a memory READ function, which is called the **fetch cycle.** During the fetch cycle the "next" instruction is "fetched" from memory (this is the instruction that will be executed during the next execution cycle). During the execution cycle the instruction is executed, and the memory address of the "next" instruction is determined. The two cycles (fetch and execute) make up a complete computer cycle, and they alternate during the automatic operation period of the computer.

A very simple model of a working automatic computer is shown in Figure 10.11. Two special registers are added to the memories of the computer. The **program counter (PC)** contains the address of the next instruction. The bits contained in the PC select a particular memory storage register that contains the bits of a computer instruction. During the fetch cycle a MEMORY READ function is executed, and the instruction bits are transferred to the **instruction register (IR).** We have to store the instruction in the IR register during the entire execution cycle because during the execution of the instruction, other data values may be required from the memory, and thus, the outputs of the memory unit (normally held in the MBR register, as was shown in Figure 10.10) may change. The contents of the PC must be updated before the execution of an instruction is completed. In its simplest form the PC is incremented, and successive instructions will be fetched from consecutive storage locations. It is simple to incorporate the increment function into the fetch cycle (this occurs often in practice), in which case the fetch cycle may be described by two register transfer operations as follows:

$$IR \leftarrow M[PC] \qquad (FETCH)$$

$$PC \leftarrow PC \text{ plus } 1$$

During the execution cycle the bits stored in the IR control the operation of the computer. In this way different combinations of bits in the IR will initiate different actions during the execution cycle. This is how the execution of a variety of computer instruction types is handled by the processor.

Assuming that during the execution of the instruction the PC remains unchanged, this simple model of a digital computer will execute a computer

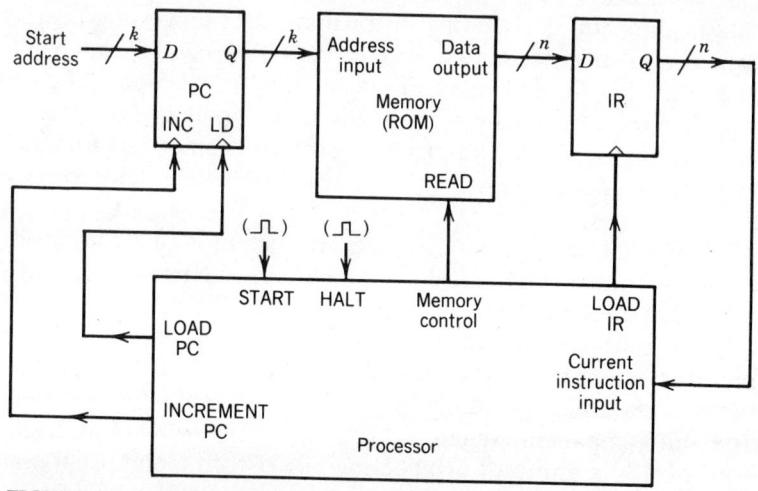

FIGURE 10.11 Model of simple automatic digital computer.

program whose instructions are placed in consecutive memory locations (since the PC is incremented during the fetch cycle). At the start of the execution of a program the PC must be loaded with the memory address where the first instruction of the program is stored (the start address). Similarly, some means must be provided by which the operation of the computer is stopped once the program is finished. These two facilities are indicated schematically in Figure 10.11 as START and HALT. We will postpone a detailed discussion on how the execution of a program is controlled until Chapter 11.

This computer model has unrealistically limited capabilities because the PC can only be incremented by the processor (in more general computer models it can be loaded with an arbitrary address as well), and the program memory can be read only (it is a ROM). However, this restricted model demonstrates the basic two cycles of automatic computer operation. The operation of the computer is summarized as follows: During the fetch cycle the next instruction is read from memory and loaded into the IR. After this loading operation the PC is incremented. The computer is now ready for the execution cycle. During the execution cycle the processor's control circuitry receives the bits of the instruction from the IR and executes the instruction according to the values of these bits. After completion of the execution of the instruction, the computer returns to its fetch cycle. Since the PC was incremented during the fetch cycle, the address for the next instruction is one larger than that of the previous instruction, and the computer automatically executes a program of consecutively stored instructions. Let us examine now how the simplest type of instructions, the data-handling instructions, are executed by the processor.

10.4 DATA-HANDLING INSTRUCTIONS

Computer instructions are separated into a number of instruction types. There are data-loading and data-storing instructions, data-processing instructions, program control instructions (similar to FORTRAN GOTO statements and Pascal WHILE loops), and input or output (I/O) instructions. The simplest types of computer instructions are of the data-handling (i.e., data-loading and data-storing) types. A data-loading instruction copies data from one register/memory to another register/memory without making changes to the data itself.

The collection of bits of a computer instruction is called an instruction **code**. The instruction code for data-handling instructions can be separated into three components. The first component specifies **what** the instruction does. This is called the **operation code,** or **opcode** for short. The second component specifies where the data originates from and is called the **source** component. The third specifies where the data is sent to and is called the **destination** component. When the separate components of an instruction are expressed by a group of neighboring bits (which is the usual practice), the groups are called instruction **fields.**

10.4 DATA-HANDLING INSTRUCTIONS

The bits within the opcode field determine what functions the instruction performs. These bits may distinguish different types of instructions or different instructions within one type. The bits within the source field specify a storage register. If the source field refers to computer registers, it identifies a specific register. Usually, the bits in this field can be interpreted as a register address. If the source field specifies a memory storage location, it contains the memory address of the data used as the "source" operand for the instruction. Just like the source field, the destination field may specify a register address or an address for a memory storage register.

The division of the instruction code into fields of bits can be best demonstrated with an example. The instruction examples will be described for a simple computer that has six 4-bit registers. Two registers are called register A and register B, and the other four registers are implemented by a type 74170 4×4-bit RAM or register file. The data paths of this simple 4-bit computer are shown in Figure 10.12. There are two data buses named R bus and A/B bus, and there are two types of data-moving instructions. The first type is called a LOAD instruction, and it writes the contents of one of the R registers into either the A or the B register. The R bus is used for this instruction. The other type is called a STORE instruction, which does the opposite; that is, it copies the contents of the A or B register into one of the R registers. The STORE instruction uses the A/B bus.

The number of bits belonging to each field is determined by the number of different combinations the field must express. There are two types of instructions (LOAD or STORE). Therefore, only one bit is needed for the operation code field. Two bits are needed to select one of the four R registers. Finally, 1 bit is sufficient for the field that controls the choice between the A and B registers. Since each instruction involves both the A and B registers (1 bit), the R registers (2 bits), and the specification of the instruction type (1 bit), a total of 4 bits are required.

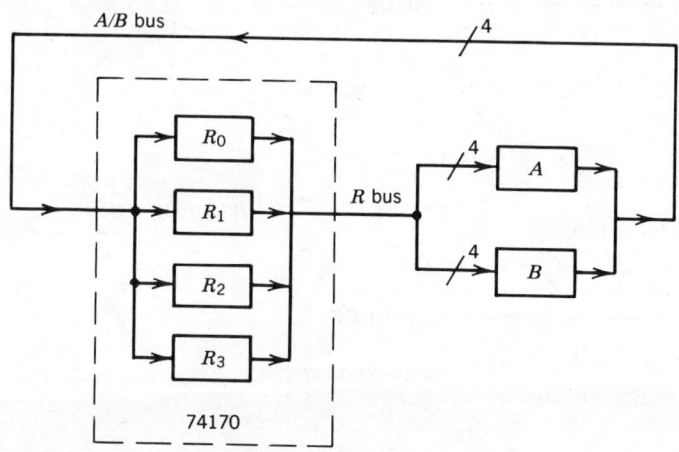

FIGURE 10.12 Data paths for simple 4-bit processor.

The number of bits in each field of the instruction code has been determined, but how these fields of bits are assigned within the instruction is yet unspecified. The assignment of 1's and 0's for each field is not specified either. The complexity of the circuits required for the execution of the instruction depends on these field and bit assignments, similar to the state assignments of sequential circuits. Therefore, the justification for a particular assignment can be made only when the computer circuits have been designed.

Arbitrary field and bit assignments for both the LOAD and the STORE instructions are shown in Figure 10.13. The 4 bits of the IR are designated from IR_0 (LSB) to IR_3 (MSB). The diagrams in Figure 10.13 clearly show the selections made by particular bit values. Figure 10.13 demonstrates the binary assignments for the instructions, which is called the **machine language** of the computer. A shorter and more easily interpretable description of the same bit assignments may be made with the help of the so-called **assembly language** notation. Assembly language expresses the encoding of instruction bits with the help of mneumonics, English words that express the action of the instruction. The operand fields are also indicated by more easily readable symbols. For example, the selection between the A and B registers is shown as A/B (either A or B, with the first choice for bit 0 and the second for bit 1). The selection for a particular R register is indicated by the word *address,* and the bits within this field form a binary address for the R registers. The instruction codes for the same two instructions are described in terms of assembly language mneumonics in Figure 10.14. In addition to the general form (e.g., LOAD address, A/B), which indicates values that may be substituted into the instructions, some examples of actual instructions (such as LOAD 2,A) and

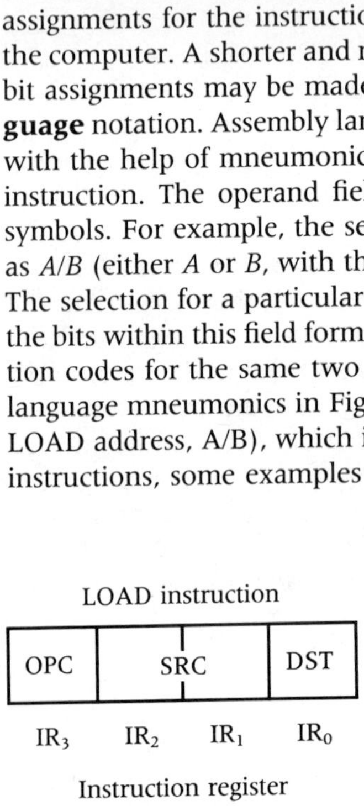

LOAD instruction

OPC	SRC	DST
IR_3	IR_2	IR_1 IR_0

Instruction register

	IR_3	IR_2	IR_1	IR_0	Selection
(OP. CODE)	0				LOAD
(SOURCE)		0	0		RAM[0]
		0	1		RAM[1]
		1	0		RAM[2]
		1	1		RAM[3]
(DESTINATION)				0	A
				1	B

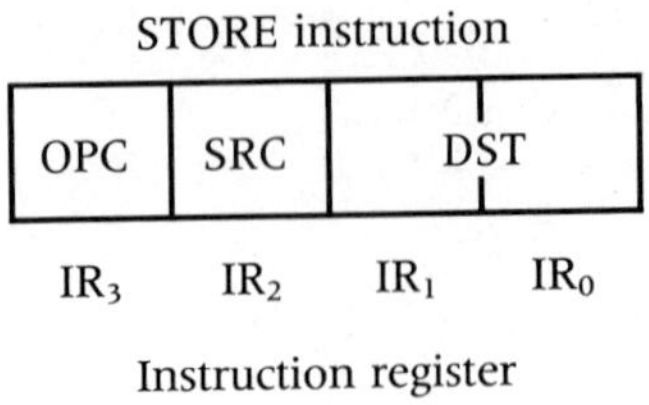

STORE instruction

OPC	SRC	DST
IR_3	IR_2	IR_1 IR_0

Instruction register

	IR_3	IR_2	IR_1	IR_0	Selection
(OP. CODE)	1				STORE
(SOURCE)		0			A
		1			B
(DESTINATION)			0	0	RAM[0]
			0	1	RAM[1]
			1	0	RAM[2]
			1	1	RAM[3]

FIGURE 10.13 Hardware instructions for 4-bit processor.

10.4 DATA-HANDLING INSTRUCTIONS

Instruction code	Assembly language form	Instruction examples	Register transfer
0 | address | A/B IR₃ | IR₂ IR₁ | IR₀	LOAD address,A/B	LOAD 2,A LOAD 3,B LOAD 1,A	A ← RAM[2] B ← RAM[3] A ← RAM[1]

Instruction code	Assembly language form	Instruction examples	Register transfer
1 | A/B | address IR₃ | IR₂ | IR₁ IR₀	STORE A/B,address	STORE A,0 STORE B,3 STORE B,1	RAM[0] ← A RAM[3] ← B RAM[1] ← B

FIGURE 10.14 Assembly language mneumonics for hardware instructions shown in Figure 10.13.

their equivalent register transfer functions are also shown. The general form (LOAD address, A/B) expresses the fact that this instruction "loads" either the A or B register with the contents of one of the R registers. The instruction has a variable address with the range 0–3. The actual instruction (LOAD 2,A) specifies the address for R as 2 (selects $R[2]$ or R_2) and selects the A register as the destination. The bit assignments for this instruction are 0100.

We are now ready to construct the actual circuitry that executes the LOAD and STORE instructions. One possible way of executing these two instructions is shown in Figure 10.15. The source of data for the STORE instruction is either from the A or the B register. As shown in Figure 10.15, this selection requires four 2-to-1 multiplexers (each line represents four data lines since all the registers contain 4 bits). The selection inputs of these multiplexers are directly connected to bit IR_2 of the IR, which indicates the source

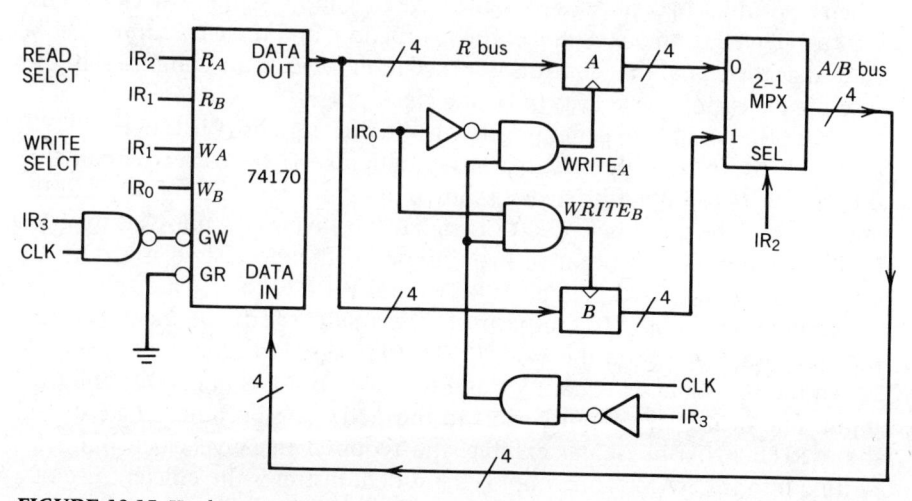

FIGURE 10.15 Hardware realization for simple 4-bit processor.

field for the STORE instruction. We can see that if IR_2 is equal to 0, then the A/B bus bits are equal to the contents of register A; otherwise, they are equal to the contents of register B. In this sense the IR_2 bit included in the STORE instruction selects the source operand of the instruction.

The source of data for the LOAD instruction is one of the R registers. The selection of a particular register from the type 74170 RAM is done by selection lines (READ SELECT). Therefore, a multiplexer for the RAM is not required. The selection of the RAM addresses for both the LOAD and STORE operations is applied directly to the RAM inputs and requires no additional circuitry. Bits IR_2 and IR_1 are connected to the READ SELECT inputs (the source operands for the LOAD instruction), whereas bits IR_1 and IR_0 are connected to the WRITE SELECT inputs (the destination operands for the STORE instruction) of the RAM.

The rest of the circuitry is needed for the generation of the correct clock signals that complete the register transfer operations. When the MSB of the instruction code (IR_3) is equal to Logic-0, either the A register or the B register has to be updated depending on the value of bit IR_0. The clock signals that update these registers are $WRITE_A$ and $WRITE_B$. Closer examination of the circuits used in Figure 10.15 reveals that the required circuitry is a 1-to-2 demultiplexer shown as an inverter and two AND gates. The input to the demultiplexer is a clock pulse generated by the output of the AND gate whose inputs are connected to bit IR_3 and the CLOCK signal. This clock pulse occurs only if $IR_3 = 0$ (LOAD instruction).

When the MSB of the instruction code is equal to Logic-1, the RAM is updated. The RAM WRITE ENABLE (GW) clock signal occurs only if IR_3 is equal to 1. A NAND gate is used since the RAM requires an enable signal with negative logic. The RAM READ ENABLE signal (GR) is always active, since only one RAM unit is used, and we do not require wired-OR connections. We can see that the write signals ($WRITE_A$, and $WRITE_B$) occur only when a positive clock pulse is present at the CLK input signal. The clock pulse may be viewed as the executor of the instruction, and the CLK input may be viewed as the signal that controls the execution cycle of the computer. In this case the execution cycle lasts only one clock pulse.

It is also possible to generate all three "write" signals ($WRITE_A$, $WRITE_B$, and GW) by using only one 2-to-4 demultiplexer. The correct circuit that uses a type 74155 demultiplexer is shown in Figure 10.16. Here, IR_3 and IR_0 are used as selection lines. The CLK signal (positive clock pulse) is inverted by the demultiplexer (negative-logic outputs); therefore, two inverters are needed to generate write signals $WRITE_A$ and $WRITE_B$. These are selected by selection inputs 00 and 01, respectively. Normally, the 1Q10 and 1Q11 outputs are equal to 1, and the WRITE ENABLE signal is equal to 1 (passive). When the IR_3 signal is equal to 1 and the clock pulse is applied to the CLK input, one of the two input signals to the AND gate becomes Logic-0, and the WRITE ENABLE signal exhibits the required negative clock pulse for writing into the RAM. This simple circuit demonstrates the efficient use of a demultiplexer for generating WRITE ENABLE signals.

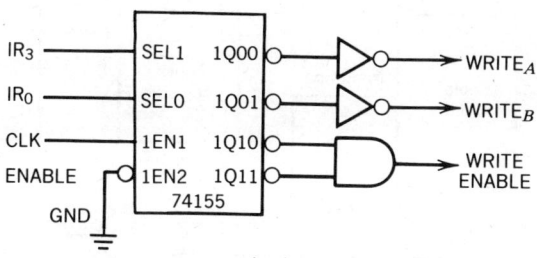

FIGURE 10.16 Write signals generated by type 74155 decoder.

10.5 EXTERNALLY PROGRAMMED COMPUTERS

The design of a simplified digital computer will be completed in two experiments (Experiments 10 and 11). Experiment 10 involves the design and construction of a simple externally programmed digital computer. The term *externally programmed* means that the computer does not have a program memory unit and has no need for an IR. The instruction bits (normally stored in the IR register) are supplied to the computer by external inputs, as shown in Figure 10.17.

The instruction bits are input to the computer, and the clock input is pulsed. The computer executes one instruction. A new instruction code is then applied to the computer, and the clock input is pulsed again. Data may be sent to this computer by external data input lines, and data output is received from it through data output lines. Many computer instruction types may be included in such an externally programmed computer. In addition to the data-handling and data-processing instructions we have discussed before, input and output instructions (I/O instructions) and conditional instructions are also possible. The only function that an externally programmed computer is unable to do is to change the sequence of instructions. This type of computer has no fetch cycles; therefore, it cannot execute program control (GOTO-type) instructions.

The usefulness of such a limited computer is enhanced by the so-called conditional instructions. We have seen that every computer instruction can be expressed as a register transfer operation (since each instruction operates on the stored bits of the computer). A general register transfer operation may

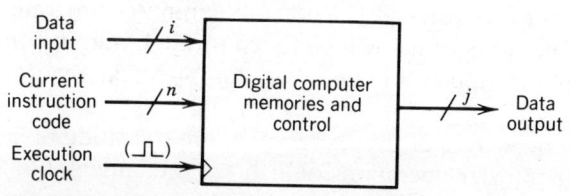

FIGURE 10.17 Block diagram for externally programmed digital computer.

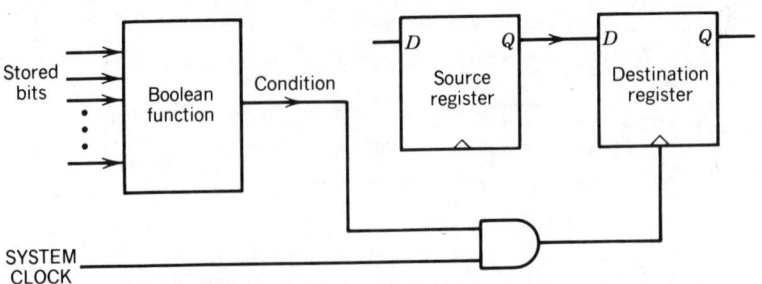

FIGURE 10.18 Hardware realization of conditional register transfer operation.

be executed conditionally and is indicated as

$$\text{DESTINATION} \leftarrow \text{SOURCE} : \text{CONDITION}$$

where the condition part is a Boolean function of the stored bits of the computer and can be true (Logic-1) or false (Logic-0). In the first case the register transfer is executed; in the second it is skipped, and the instruction behaves as a NOP (no operation). The hardware realization of such a conditional instruction is schematically shown in Figure 10.18. Such circuits are at the heart of "intelligent" data processors that can select their actions according to the data they possess.

Externally programmed computers are often used for numerically controlled milling machines and other tool-making heavy machinery. The machine is instructed to go through a constant set of steps while it makes a required machine part. In this case the sequence of instructions is never changed, and in fact, it is safer if it is not allowed to be changed by the computer program. The design of a simplified externally programmed computer of this type is the topic of Experiment 10.

EXPERIMENT 10
Externally Programmed Digital Computer

E10.1 Summary and Objectives

For this experiment a simple externally programmed digital computer will be designed and constructed. First, input instructions will be used to load the computer's memory with external data. After loading this initial data into the memory, a set of computer instructions (a computer program) will be executed, after which the results will be tested through output connections. The main task of the computer will be to sort four 4-bit data words in ascending order.

The main objectives of this experiment are the study of general (conditional) register transfer operations, which are fundamental to the control of digital computers; the use of memories (RAM and registers) within the processing elements of a digital computer; and the construction of machine language programs that execute a given processing function.

E10.2 Simple Data-Sorting Computer

The random-access memory (RAM) for this digital computer is a 4-by-4-bit register file (type 74170 IC). The four registers are named R_0, R_1, R_2, and R_3 but can be indicated as R[ADDRESS], where ADDRESS is a 2-bit unsigned binary number in the numerical range 0–3. Four 4-bit unsigned binary data words are loaded into the four registers, and after a number of computer instructions are executed, the four data words should be in ascending order; that is, the contents of $R[0]$ should be smaller (or equal) to the contents of $R[1]$, whose contents should be smaller or equal to those of $R[2]$, and so on.

In addition to the RAM, two 4-bit registers (called registers A and B) are also included in the computer's memories. These two registers help in the implementation of conditional transfer instructions necessary for the execution of a sorting program. The block diagram of the computer showing its memories and its external input and output connections are shown in Figure E10.1.

Four input lines are used as external data for the computer. A yet-unspecified number of input lines supply the bits for the "current computer instruction." When a clock pulse is applied, the current computer instruction is executed. Between clock pulses the contents of any one of the four registers may be tested by the examination of four output signals. The logic levels of two test input signals select one of the four registers and connect its four output lines to the output signals. Since the computer program is executed one step at a time, and between these time steps the contents of any one of the four registers may be examined, the correct operation of the computer can be verified after the execution of each instruction.

The computer is used for sorting only. Before the sorting program starts, four computer instructions load the RAM registers of the computer with external data. After the four 4-bit numbers are stored in registers $R[0]$–$R[3]$,

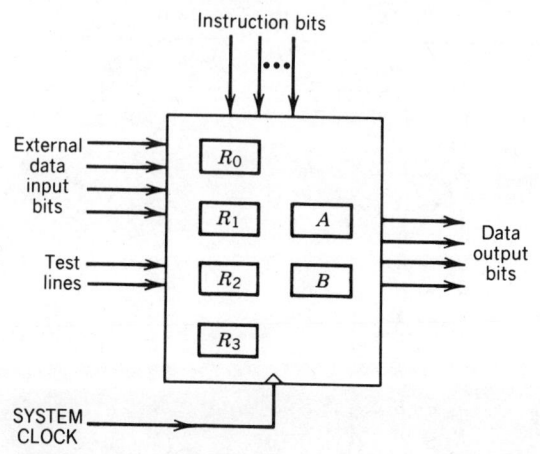

FIGURE E10.1 Block diagram of externally programmed 4-bit computer.

330 INTRODUCTION TO DIGITAL COMPUTERS

the complete sorting program is executed. At the conclusion of the execution of the sorting program the four data values are stored in an ascending order of their unsigned binary values. Register $R[0]$ contains the number with the smallest unsigned binary value, $R[1]$ the second smallest, $R[2]$ the larger, and $R[3]$ the largest number.

E10.3 Sorting Program

The operation of the computer is expressed in terms of general register transfer operations. Both unconditional and conditional register transfer operations are required. A relatively simple sorting program that uses two additional 4-bit registers are shown in Figure E10.2. Twenty-four instructions are required, out of which 12 are conditional instructions. The first 4 instructions exchange the contents of $R[0]$ and $R[1]$ if the magnitude of the number stored in $R[0]$ is larger than that of $R[1]$. These four program steps are equivalent to the following program statement:

Steps

    1–4   If $(R[0] > R[1])$ then exchange the contents of $R[0]$ and $R[1]$

If $R[0] > R[1]$ (the contents of these two registers are compared), then the following four register transfer operations are executed:

$$A \leftarrow R[0]$$
$$B \leftarrow R[1]$$
$$R[0] \leftarrow B$$
$$R[1] \leftarrow A$$

whereas if the contents of $R[0]$ is not larger than that of $R[1]$, only the first two transfers are executed, in which case the contents of both $R[0]$ and $R[1]$ remain unchanged. Thus, after the first 4 steps the smaller of the two data

Sorting Program

Program step	Operation	Program step	Operation
1	$A \leftarrow R_0$	13	$A \leftarrow R_1$
2	$B \leftarrow R_1$	14	$B \leftarrow R_2$
3	$R_0 \leftarrow B \; : A > B$	15	$R_1 \leftarrow B \; : A > B$
4	$R_1 \leftarrow A \; : A > B$	16	$R_2 \leftarrow A \; : A > B$
5	$A \leftarrow R_0$	17	$A \leftarrow R_1$
6	$B \leftarrow R_2$	18	$B \leftarrow R_3$
7	$R_0 \leftarrow B \; : A > B$	19	$R_1 \leftarrow B \; : A > B$
8	$R_2 \leftarrow A \; : A > B$	20	$R_3 \leftarrow A \; : A > B$
9	$A \leftarrow R_0$	21	$A \leftarrow R_2$
10	$B \leftarrow R_3$	22	$B \leftarrow R_3$
11	$R_0 \leftarrow B \; : A > B$	23	$R_2 \leftarrow B \; : A > B$
12	$R_3 \leftarrow A \; : A > B$	24	$R_3 \leftarrow A \; : A > B$

FIGURE E10.2 Sorting program in terms of register transfer operations.

words will be stored in R[0]. The next 20 steps are similar; only the roles of the registers change. They are as follows:

Steps

5–8	If $(R[0] > R[2])$ then exchange the contents of $R[0]$ and $R[2]$
9–12	If $(R[0] > R[3])$ then exchange the contents of $R[0]$ and $R[3]$
13–16	If $(R[1] > R[2])$ then exchange the contents of $R[1]$ and $R[2]$
17–20	If $(R[1] > R[3])$ then exchange the contents of $R[1]$ and $R[3]$
21–24	If $(R[2] > R[3])$ then exchange the contents of $R[2]$ and $R[3]$

Before the sorting program is executed, data are loaded into the four registers. These data-loading instructions need four initial clock periods:

$$R[0] \leftarrow \text{Input Data}$$
$$R[1] \leftarrow \text{Input Data}$$
$$R[2] \leftarrow \text{Input Data}$$
$$R[3] \leftarrow \text{Input Data}$$

After the loading operations 24 instructions are executed, and the four data words will be stored in registers R_0–R_3 in ascending order.

E10.4 Hardware Requirements

In order to execute both the sorting program and the initial loading of the RAM with external data, suitable hardware must be chosen for the computer. The first step is to construct the data path diagram of the computer that can be derived from examining the register transfer operations included in the required instructions. In addition to the type 74170 RAM, two 4-bit registers (A and B) are used. When the destination register is either the A or B register, the source register is any one of the four RAM registers. When the destination register is one of the RAM registers, the source may be the A or B register or external data. The data path diagram for this computer is shown in Figure E10.3.

In addition to the register transfer operations, a signal must be generated that indicates the relative magnitudes of the contents of the A and B registers. For this signal the commercially available TTL type 7485 magnitude comparator IC may be used. In Figure E10.4 the possible internal structure of this computer is shown. The only difference between this design and the indicated data paths in Figure E10.3 is the use of two 2-to-1 multiplexers in place of the one 3-to-1 multiplexer indicated in Figure E10.3. The required data select and "write" signals are also shown in Figure E10.4. The data select signals determine the data placed on the data buses (there are four such signals: SELECT$_{AB}$, SELECT$_{EX}$, and two READ SELECT lines). The write signals, when pulsed, execute register transfer operations (there are five write signals; three of these, WRITE$_A$, WRITE$_B$, and WRITE$_R$, require clock pulses to execute data transfers, whereas the two WRITE SELECT signals determine the RAM register that will be updated). With the proper selection of these nine control signals,

332 INTRODUCTION TO DIGITAL COMPUTERS

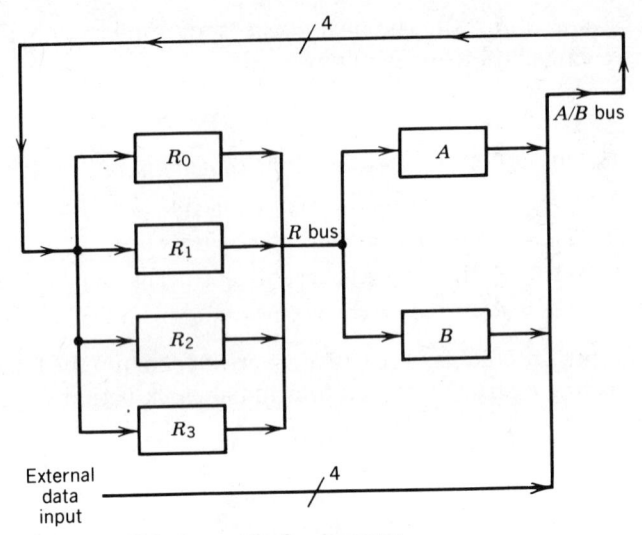

FIGURE E10.3 Data paths for computer.

any one of the register transfer operations in the loading or sorting program can be executed.

E10.5 Instruction Encoding

In the preceding section we have reduced the task of designing this sorting computer to the generation of nine control signals. The control circuitry required for these signals is shown schematically in Figure E10.5. The values of the control signals depend on the current instruction being executed and

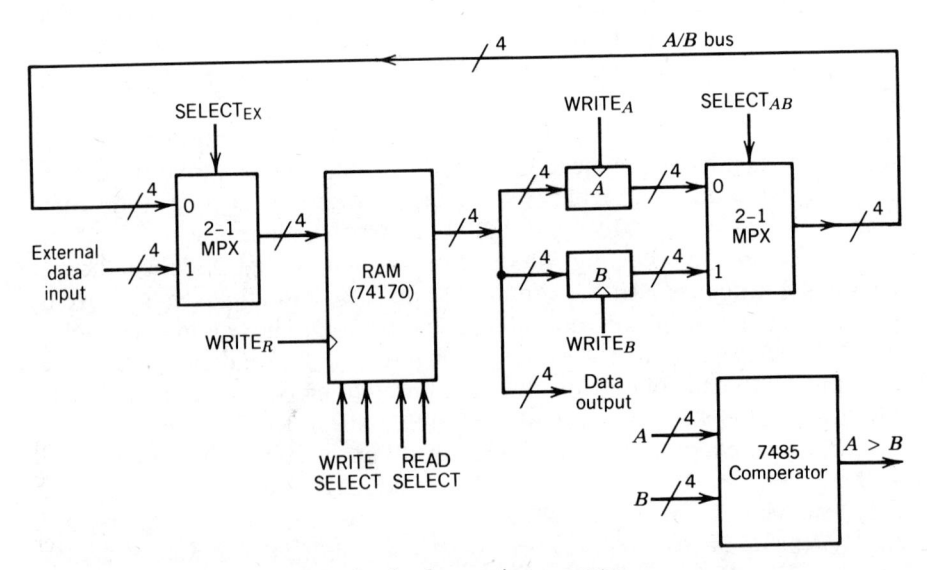

FIGURE E10.4 Internal organization for data-sorting computer.

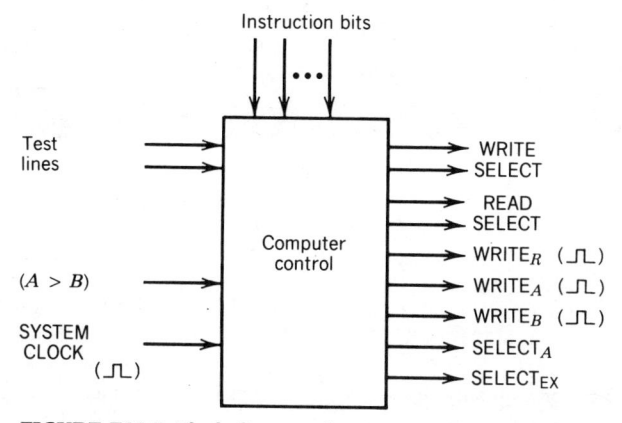

FIGURE E10.5 Block diagram showing generation of nine required control signals.

the output of the comparator. The SYSTEM CLOCK controls the pulsing of the write signals. Between write pulses the test lines control the READ SELECT signals, which determine the data on the external output lines. Now we are left with the task of determining the required number of instruction types and the encoding of bits into computer instructions.

Examining the proposed hardware structure of this computer in Figure E10.4, we find that there are three types of "address fields" derived from the three data buses in Figure E10.4. One bit is required to select either register A or B, and this address selection will be indicated as [A/B]. One bit is required to select between the external data and the A/B bus, for which we will use the symbol [EXT], and 2 bits are required to select the RAM register, which we will show as address [n], where n has the range 0–3. The loading of the registers is executed by the instruction type

$R[n] \leftarrow$ External Data ; where [EXT = 1] and [A/B = don't care]

The loading of the RAM needs a conditional instruction type:

$R[n] \leftarrow [A/B] : (A > B)$; where [EXT = 0]

And the loading of the A or B register is executed by the unconditional instruction type:

$[A/B] \leftarrow R[n]$; where [EXT = 0]

The minimum length of each instruction type may be deduced from the address and opcode fields. One bit must be assigned for the [EXT] field that distinguishes between the first and the other two instruction types. The first instruction type requires only 3 bits (one, the [EXT] bit, is equal to 1; the other 2 bits select the RAM register). The other two instruction types require 5 bits: the [EXT] bit (which is equal to 0), the [A/B] bit, two [n] bits, and one opcode bit that distinguishes between the load (A or B is updated) and the store (RAM is updated) instruction types.

Instruction type	Variable fields	Number of bits	Encoded bits
R[n] ← DATA	[n]	2	1 | [n] | X | X
R[n] ← A/B : (A > B)	[n] A/B	2 1	0 | [n] | 0 | A/B
A/B ← R[n]	[n] A/B	2 1	0 | [n] | 1 | A/B

FIGURE E10.6 Assembly language forms and encoded instruction bits for required computer instructions.

Assuming that all instructions have the same number of bits, the minimum number of bits is equal to 5, and a possible encoding of the bits for the three instruction types are shown in Figure E10.6. Note that once an instruction field is assigned for the instruction (e.g., the [n] field), its position is not changed when successive instruction types are designed. Since the bit assignments will determine where these inputs are used in the hardware, it is good practice to keep the address field assignments as constant as possible. The cross indicates a don't care input.

E10.6 Hardware Design

The minimum requirement for this experiment is that you design and construct your computer on the basis of the suggested hardware structure and the given sorting algorithm. If you select this minimum design route, then you must write a program on your own that sorts the numbers in descending order. You must demonstrate in the laboratory that you can re-sort the stored numbers in both ascending and descending order.

The hardware can be designed on the basis of a signal excitation table, shown in Table E10.1. In this table the nine control signals are set according to the bits of the current instruction. In this case specific instructions (not only instruction types) must be shown, but since the READ SELECT and the

TABLE E10.1 Generation of Control Signals

Current instruction	SEL_A	SEL_X	$\begin{pmatrix} WRITE \\ SEL \end{pmatrix}$	$\begin{pmatrix} READ \\ SEL \end{pmatrix}$	$WRITE_R$	$WRITE_A$	$WRITE_B$
R_n ← EXT	X	1	[n]	X	CLK	0	0
R_n ← A : (A > B)	0	1	[n]	X	CLK · (A > B)	0	0
R_n ← B : (A > B)	1	1	[n]	X	CLK · (A > B)	0	0
A ← R[n]	X	X	X	[n]	0	CLK	0
B ← R[n]	X	X	X	[n]	0	0	CLK

WRITE SELECT bits are always used as single units, only five different instructions are listed. Each entry in Table E10.1 represents four different hardware instructions. Consequently, this simple sorting computer has an instruction repertoire of 20 possible instructions.

The conditional nature of two instructions is indicated by the WRITE_R signal, which is pulsed only if the output of the comparator (signal $A > B$) is equal to Logic-1. The letter X is used when the value of the signal is immaterial for the correct operation of the computer (don't care cases). The incorporation of the test lines is not indicated in the excitation table because it is valid only for the time period when the SYSTEM CLOCK input is zero. First, design the computer by ignoring the test lines. Once the design is completed, incorporate the test lines into your design if you have time and enough hardware resources for this additional feature.

E10.7 Alternate Experiments

Our computer was defined with 11 input bits: 4 bits for EXTERNAL INPUT, 5 bits for the instruction, and 2 bits for the test lines. It is possible (with some additional hardware) to share different functions of some of these input bits. Design the computer with the minimum number of input bits you can manage and that still satisfies the specified sorting problem. (The requirement is that after loading four numbers into the computer it executes the given sorting algorithms, one for descending, the other for ascending order.)

The sorting algorithm given was very simple-minded and rather long. Design a similar sorting computer with the design criterium that the number of instructions for the sorting program should be kept to a minimum. For example, allow multiple register transfers to take place in one clock cycle. Show the algorithms for sorting the numbers in both descending and ascending order.

Design a sorting computer with the design criterium that the hardware should be kept to a minimum. This may lengthen your sorting algorithm. Try to eliminate multiplexers, one of the two registers, and try to use a different number of bits used for the instructions. Show that both sorting algorithms are possible with your new design.

The proposed computer uses 5 bits for its instructions and has 20 useful instructions. With 5 bits it is possible to encode 32 different instructions. Find useful additions to your hardware so that some of the unused bit combinations may be used for additional instruction types and new processing algorithms. Could you add data-processing (add, subtract, increment, decrement, logical functions, etc.) functions in this way? You may also consider using a shift register or a counter (with parallel load functions) for register A, register B, or both. Do not use more than 5 bits for the computer instructions.

Drill Problems

10.1. Using a 74170 type RAM and a crossbar switch as shown in Figure 10.1, design a data bus system that, in addition to the four general register transfer operations

336 INTRODUCTION TO DIGITAL COMPUTERS

$R_i \leftarrow R_j$ can clear or set any one of the 4-bit registers. (It can also perform the operations $R_i \leftarrow 0$ and $R_i \leftarrow -1$.)

10.2. What additional hardware is required for the switch of Problem 10.1 if four 16-bit registers are required?

10.3. Do Problems 10.1 and 10.2 using a time-multiplexed switch. How much hardware is saved?

10.4. What logical function of two input signals is generated when the outputs of two open-collector inverters are connected together?

10.5. Draw the data path diagram of the crossbar switch shown in Figure 10.1 and compare it to Figure 10.8.

10.6. Design the fetch–execute cycles of a digital computer for which two memory words are used for each computer instruction. The first word contains the bits of the instruction, which is transferred into the IR. The second word contains the memory address of the next instruction to be executed. What hardware additions are required to the circuits in Figure 10.11 for this computer?

10.7. Design the circuits of a general register transfer system for the type 74170 RAM and the two registers shown in Figure 10.15 that allow the clearing (set to Logic-0) of any one of the six registers in addition to the LOAD and STORE instructions shown in Figure 10.13. Design and encode the hardware instructions for this computer and suggest assembly language mneumonics for it.

10.8. Three additional register transfer operations are added to the circuits shown in Figure 10.5:
  (a) $A \leftarrow B$.
  (b) $B \leftarrow A$.
  (c) Exchange contents of A and B registers.
Show the data path diagram for the new, expanded circuits. Determine the additional hardware needed, encode the instruction bits, and suggest the assembly language mneumonics for this expanded system.

CHAPTER 11

Introduction to Programmable Digital Computers

11.1 INTRODUCTION

In Chapter 10 we described an externally programmed 4-bit digital computer that we used as a model of a very simple digital processor. We discussed the fetch–execute cycle of the computer, hardware instructions, assembly language, and encoding of instruction bits. In this chapter we will extend our model to programmable computers, that is, digital processors whose programs are stored within the computer's own memory. There are two fundamental changes: First, a memory with a large number of storage registers (memory that has a large address space) is added to the hardware, which stores both data and instructions. The large address space allows various forms of data and instruction address calculations. Second, the changing of the sequence in which the instructions are executed becomes a variable. When one can change the order in which the instructions are executed, programming blocks, such as IF, WHILE, REPEAT, or FOR loops may be constructed. Program loops allow the automatic operation of the computer.

In this chapter we will discuss new instruction types, such as data-processing instructions (arithmetic add, subtract, logical shift and complement, etc.), unconditional and conditional control instructions (SKIP, JUMP, and relative BRANCH), and I/O (input–output) instructions. We shall find that all instruction types and, indeed, the entire operation of the computer may be described by conditional register transfer operations. Our examples will be limited to the simplest forms of instructions and addressing types used today, but they will present the framework and methods by which more complex and practical microprocessors may be described, analyzed, and even designed.

A short discussion of simple machine language programs such as hardware diagnostics will also be given. Finally, Experiment 11 will involve the detailed design of a basic but usable 4-bit digital computer. This exercise should serve as an introduction to modern microprocessor architectures and should help in the understanding of how digital computers and their instruction sets have evolved in the past 30 years.

In the experimental model both instructions and data will be stored in the main memory of the computer, which contains two hundred fifty-six 4-bit storage locations. Thus, while data words contain only 4 bits, memory addresses will need 8 bits. Some of the computer instructions will also require

8 bits. This model will demonstrate how a computer using variable word lengths is organized. All modern microprocessors and larger processors are of this type, and the concepts of variable word length and addressing types are essential to the understanding of their operation.

11.2 MEMORY

The most central element of a modern digital computer is its memory. Basically, it is a simple collection of a large number of storage registers, but its organization (how many storage registers of how many bits), speed, structure (how many read–write operations can be done simultaneously), and type (RAM or ROM) are all very important factors in the efficiency and performance of the computer. By the 1970s a standardization had taken place, and most computers were built around bytes, or 8-bit words. Thus, memory addresses were expressed in terms of bytes.

At first, rather small address spaces were allowed, and 16-bit addresses became common. This allowed the addressing of 64K (approximately 64,000) bytes. Soon, programmers required more memory and with improved technology (VLSI devices) the hardware manufacturers could provide more memory at reasonable prices. Many of today's microprocessors can address up to 4M bytes (approximately four million bytes, which require 22-bit addresses), and even 32-bit addresses have become feasible, which provide approximately 4×10^9 (practically unlimited) addressable storage locations.

With the addition of memory to the computer system, the storage of both instructions and data are solved. Since, as we shall see shortly, a data-moving instruction must either contain the data itself or a memory address that "points" to the data (selects the storage register where the data is stored), logically, we have two types of addresses: an instruction address and a data address. Similarly, we have two types of memory contents: instruction bits and data bits. We have already introduced the concepts of the program counter (PC), instruction register (IR), memory address register (MAR), and memory buffer register (MBR), with which the separation of these two types of addresses and contents may be handled. Figure 11.1 shows the data paths of a common hardware structure for a simple memory element. The READ and WRITE clock pulses execute the so-called memory read and memory write cycles, which may be expressed by the following register transfer operations:

$$\text{Memory read cycle:} \quad \text{MBR} \leftarrow \text{M[Address]}$$

$$\text{Memory write cycle:} \quad \text{M[Address]} \leftarrow \text{MBR}$$

where Address is contained either in the PC (a program address) or in the MAR (a data address). In this model the "fetch" cycle contains a memory read cycle with the IR as the destination register, and the PC contains the address of the "next" instruction:

$$\text{Fetch cycle:} \quad \text{MEMORY READ (MBR} \leftarrow \text{M[PC])}$$

$$\text{IR} \leftarrow \text{MBR}$$

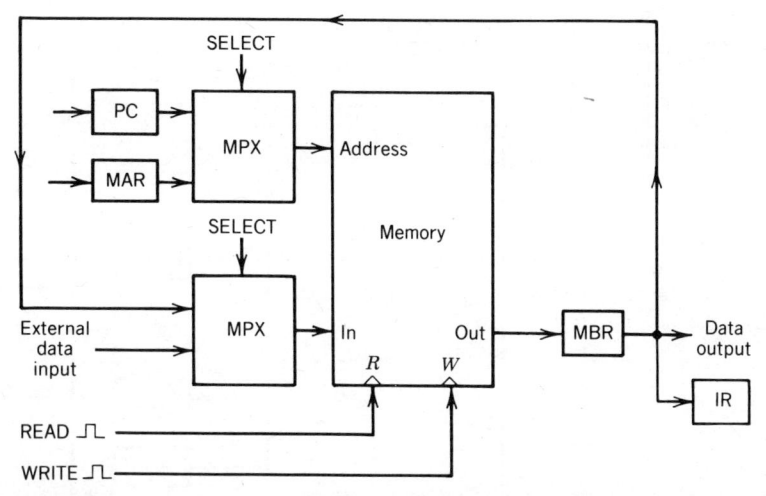

FIGURE 11.1 Hardware structure for simple memory element.

In simpler computers the MBR may be eliminated, in which case data may be directly written into and read from memory. For many practical microprocessors the number of bits for the data and address registers is different. Usually, the number of bits used is a multiple of some basic number (4, 8, or 16). In order to keep the number of instructions manageable, we will use 4 bits for instructions/data and 8 bits for addresses. The principles are similar regardless of the number of bits used. In our case the MBR and the IR are 4-bit registers, whereas the PC and the MAR contain 8 bits. As we shall see shortly, it is an advantage to be able to handle registers uniformly because this allows the transferring of data from any one of the registers to any other. Since the length of the PC and the MAR is twice that of the MBR and the IR, we define two registers in place of the PC, which will be called PCH and PCL with 4 bits each. Similarly, two registers are defined for the MAR, called MARH and MARL, where the letter H refers to "high," or the upper 4 bits, and L to "low," or the lower 4 bits of the address. In order to be able to store an address in the MBR, we extend it to contain 8 bits and similarly split it into two 4-bit registers: MBRH and MBRL.

Let us first consider a general model built around data bus structures. Since we defined all our registers as having the same number of bits (4), the buses, as shown in Figure 11.2, can be quite general. For example, the address bus is constructed from two 4-bit buses, with PCL, MARL, and MBRL contributing to the lower 4 bits and PCH, MARH, and MBRH to the upper 4 bits of the address. From this general data path diagram we can construct a number of different register transfer operations, which are discussed below.

The register–register transfer operations allow the transfer of data from any one of the 4-bit data or address registers, or external data to any data/address registers. The IR is handled differently because it is modern practice not to

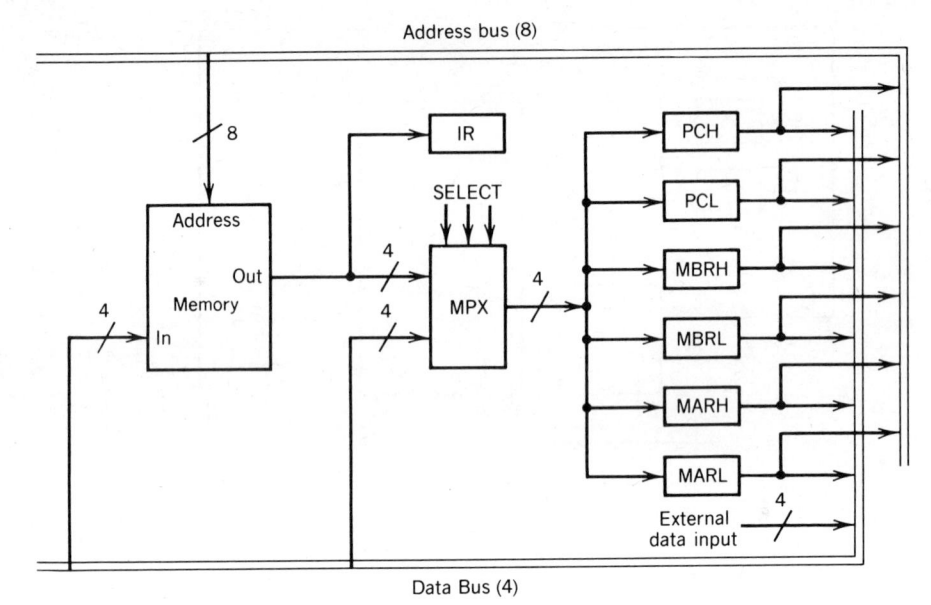

FIGURE 11.2 General model of 4-bit computer with memory.

consider the instruction bits as data, that is, not to allow the instruction bits to be changed by the processor. (In a nonconventional, nonstandard processor, there are no physical limitations, and the IR could be handled just like any other register in the system.)

Data transfer between external data and the PC register specifies a memory address of a program. This requires two register transfer operations. First the PCH register must be selected and updated followed by the PCL register. The transfer between the MBR and the MAR signifies something entirely different. If the contents of the MBR originated from memory, the transfer of the contents of the MBR to the MAR specifies the address of some data in memory with the address itself having been stored in a storage location.

The memory–register transfer operations allow the updating of any one of the seven registers from memory, with the address supplied by the PC or the MAR or MBR registers. Some care must be taken if the destination register is one of the active address registers since, in this case, the register that supplies the address for the read memory cycle changes when the register transfer is executed. Edge-triggered registers and careful timing of the transfer operation must be used to ensure the proper operation of this type of register transfer.

The register–memory transfer operations allow the storing of data from any one of the data/address registers or from an external data source. The data address is supplied by the PC or the MAR or MBR registers. In the next section we will examine how these general register transfer operations may be used for data transfer operations.

11.3 DATA TRANSFER OPERATIONS

We have already mentioned that the memory is used to store data as well as instructions. In order to transfer the data, a memory address must be specified that, in our case, contains 8 bits and must be stored in one of the three register pairs (PC, MAR, or MBR). Thus, an instruction that specifies data in memory must contain 8 bits in addition to the instruction bits (which specify the operation). If the number of bits for the instructions are 4, then the minimum length of the instruction is 12 bits (three 4-bit words). The most convenient way of storing these 12 bits is to store the instruction bits at a given address (the address contained in the PC) and store the 4 upper bits of the address in the next memory location and the lower 4 bits in the following memory location (Figure 11.3).

Assuming that the instruction and address bits are stored in memory, as shown in Figure 11.3, and that the 8-bit address stored in the PC "points" to this instruction, the following seven clock cycles execute a data transfer operation with a specified 8-bit memory address:

Clock cycle			
FETCH-1	IR	← M[PC]	(Fetch the instruction bits)
FETCH-2	PC	← PC plus 1	(Increment the PC)
EX-1	MARH	← M[PC]	(Get address/upper bits)
EX-2	PC	← PC plus 1	(Increment the PC)
EX-3	MARL	← M[PC]	(Get address/lower bits)
EX-4	PC	← PC plus 1	(Increment the PC)
MEMORY/READ	MBRL	← M[MAR]	(Get data)

where the data is transferred to the MBRL. The first two cycles belong to the normal FETCH operation. After the PC is incremented, it "points" to the next 4-bit word that contains the four upper bits of the memory address. These bits are transferred to the MARH (high) register. The PC is incremented again, and the lower order 4 bits are transferred to the MARL. Now the correct

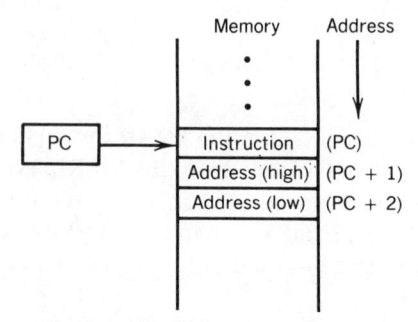

FIGURE 11.3 Storage for direct LOAD instruction in memory.

address is contained in the 8 bits of the MAR, and the data may be transferred from memory to the memory buffer register MBRL. The PC is incremented again in order to set its contents to the address of the next instruction.

This sequence of register transfer operations is a clever example of utilizing existing hardware. The incrementing of the program counter must be provided for the FETCH operation. The same hardware facility is used for setting up the memory address. Notice also that after the above seven clock cycles are executed, the PC contains the address of the next instruction, just as if the hardware instruction contained only 4 bits and the PC were incremented once.

For a register-to-memory operation the first six clock cycles are the same, and the last clock cycle is:

      MEMORY/WRITE   M[MAR] ← MBRL   (Store data)

These memory–register and register–memory instructions are called **direct** LOAD and STORE instructions because they contain the memory address of the data.

Let us consider now the problem of loading one of the data registers with a constant (say, 5). Let us assume that the execution of this instruction is required several times in a computer program. If we want to use the above direct LOAD instruction, then first we must store the constant (5) at a known memory location. We can do this simultaneously while loading the program into memory. Each time the same constant is required, a direct LOAD instruction has to be used (seven cycles). Fortunately, a much simpler way of executing the same function is possible by the use of the so-called **immediate** data. We have seen that an address may be attached to the instruction. Why should we not attach a constant? In this case, as shown in Figure 11.4, after the 4 bits of the instruction we require only 4 bits for the data. An immediate LOAD instruction requires only four cycles:

Clock cycle			
FETCH-1	IR	← M[PC]	(Fetch the instruction bits)
FETCH-2	PC	← PC plus 1	(Increment the PC)
EX-1	MBRL	← M[PC]	(Get immediate data)
EX-2	PC	← PC plus 1	(Increment the PC)

In this case the 4-bit word after the LOAD instruction contains the number 5, or binary 0101, which is loaded into the MBRL. We have saved memory locations and increased the speed of processing. We used only 8 bits for the immediate instruction and did not require an extra location to store the constant. At the same time we used only four clock cycles for the execution of the instruction.

These two types of LOAD instructions demonstrate that the instructions of a computer may contain one, two, or more basic data words (expressed in bytes for practical computers). The **type** of instruction is encoded into the

11.3 DATA TRANSFER OPERATIONS 343

```
              Memory    Address
                          |
                          v
   ┌────┐    ┌──────────┬──────┐
   │ PC │───▶│Instruction│ (PC) │
   └────┘    ├──────────┼──────┤
             │   Data    │(PC+1)│
             └──────────┴──────┘
```

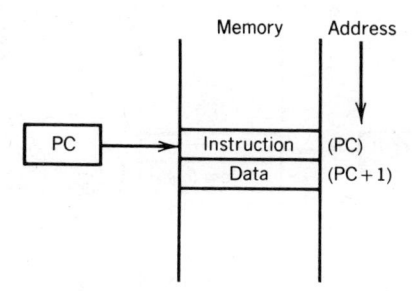

FIGURE 11.4 Storage for LOAD immediate data instruction.

bits of the instruction stored in the IR. The control circuitry of the computer must examine these bits and determine how many and which type of register transfer operations it must execute. Variable instruction lengths may simplify the program, but they introduce complexities in the control circuitry of the computer.

Logically, the immediate STORE instruction does not make sense and is rarely ever supplied. Since in many cases, computer programs are stored in ROM (read-only memory), it would be physically impossible to execute such an instruction.

Before concluding our discussion on data transfer operations, let us examine a different type of LOAD sequence when the contents of the MBR is transferred to the MAR:

Clock cycle			
FETCH-1	IR	← M[PC]	(Fetch the instruction bits)
FETCH-2	PC	← PC plus 1	(Increment the PC)
EX-1	MARH	← M[PC]	(Get address/upper bits)
EX-2	PC	← PC plus 1	(Increment the PC)
EX-3	MARL	← M[PC]	(Get address/lower bits)
EX-4	PC	← PC plus 1	(Increment the PC)
EX-5	MBRH	← M[MAR]	(Get address/upper bits)
EX-6	MAR	← MAR plus 1	(Increment MAR)
EX-7	MBRL	← M[MAR]	(Get address/lower bits)
MEMORY/READ	MBRL	← M[MBR]	(Get data into MBRL)

This sequence executes a so-called **indirect** LOAD operation. The memory address supplied with the 12-bit instruction points to a memory location that contains the upper 4 bits of an address. Thus, the instruction contains the address of a memory location that itself contains an address. This is why this addressing mode is called *indirect*. The 8 bits of the address are stored in two consecutive memory locations. The added register transfer operations load the MBR with the address (all 8 bits) and then use this address to find the data that is finally loaded into the MBRL. The addressing "chain" is shown symbolically in Figure 11.5.

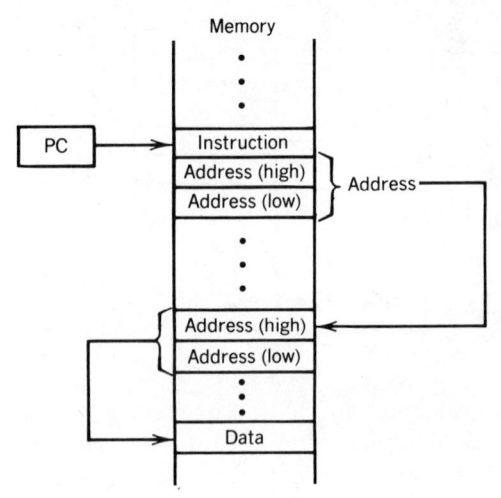

FIGURE 11.5 Storage for indirect LOAD instruction.

The preceding register transfer sequence is an example of one of those situations when the contents of the MBRL must be changed at the end of the clock pulse, whereas the contents of the same MBRL is used for the lower 4 bits of the address of the memory unit during the clock pulse. One possible hardware solution for solving this problem is schematically shown in Figure 11.6. The READ ENABLE signal uses negative logic so that the data output lines are enabled throughout the entire inverted clock pulse. The MBR must be constructed from edge-triggered or master–slave flip-flops. For edge-triggered flip-flops the transfer must take place during the positive (0-to-1) transition of the inverted clock pulse that is its trailing edge. For master–slave flip-flops the slave (output) flip-flop must be disabled during the clock pulse.

Usually, memory units have considerable time delays (hundreds of nanoseconds), and the length of the clock pulse in Figure 11.6 must be at least as long as this time delay. This ensures that at the time the MBR contents are updated (trailing edge), the memory data output lines have become stable.

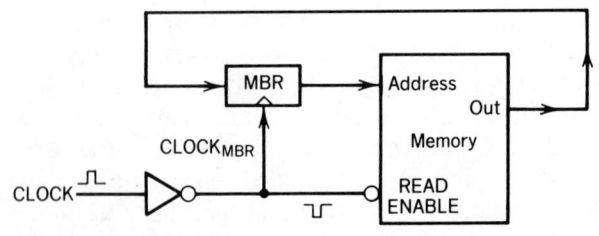

FIGURE 11.6 Simultaneous updating of MBR register and data in memory.

Delay should not be introduced into the $CLOCK_{MBR}$ line because the memory data output lines become invalid soon after the trailing edge of the clock pulse.

The indirect LOAD sequence of operations may not be used for a general indirect STORE operation because the MBRL contains the lower 4 bits of the address and cannot also contain the data. In this case we do not have a sufficient number of registers to implement an indirect STORE operation. This can be remedied by adding more registers to the computer.

The use of a large number of so-called **general** registers is a feature of most modern digital computers. A general register may be used to contain data or a component of a memory address. For example, in addition to the PC, the MAR, and the MBR, there could be a number of registers, say R_1, R_2, R_3, and so on, with data transfer instructions provided between all of them. The following register transfer operations execute general LOAD or STORE operations with the memory address contained in a general register.

Clock cycle

EX-1	MAR	$\leftarrow R_i$	(Transfer register contents)
EX-2	MBR	$\leftarrow$ M[MAR]	(Memory read)
or			
EX-2	M[MAR]	$\leftarrow$ MBR	(Memory write)

This is called indirect addressing through a register and is indicated by the register transfer operation Mem $\leftarrow (R_i)$, where the parentheses are used to indicate that R_i contains an address rather than data.

11.4 CONTROL INSTRUCTIONS

Control instructions change the sequence in which the instructions are executed. Adding two register–register transfer operations to the direct LOAD/STORE sequence, we can execute a **direct** JUMP (similar to the FORTRAN and BASIC GOTO) instruction:

Clock cycle

FETCH-1	IR	$\leftarrow$ M[PC]	(Fetch the instruction bits)
FETCH-2	PC	$\leftarrow$ PC plus 1	(Increment the PC)
EX-1	MARH	$\leftarrow$ M[PC]	(Get address/upper bits)
EX-2	PC	$\leftarrow$ PC plus 1	(Increment the PC)
EX-3	PCL	$\leftarrow$ M[PC]	(Load the PC with address/lower bits)
EX-4	PCH	$\leftarrow$ MARH	(Load address/high bits)

During the four "execute" cycles the PC is loaded with the "direct" address contained in the memory locations immediately following the four instruction bits. The instruction has the same format as the one shown in Figure 11.3. Since the PC is changed during the execution phase of the instruction, when the next instruction is fetched, the contents of the PC will be equal to the address specified by the previous instruction. Because all 8 bits are specified, the address of the next instruction may be at any location in memory. For the clock pulse EX-3 the PCL is changed while it holds 4 bits of the address of the memory unit; this is already a familiar problem, and we assume that it has been solved correctly in hardware.

Usually, a direct JUMP instruction is implemented by most processors. The large number of additional control instructions, which include conditional instructions as well, are more restricted. The problem is that computers with a large address space require a large number of bits for a direct JUMP instruction, whereas many control instructions are used for small program loops that do not require such large address spaces. Thus, many control instructions have only a limited range. For example, in our 4-bit computer an instruction may be supplied that loads only the PCL with a 4-bit address. The register transfer sequence for this type of instruction, which we will call a local BRANCH instruction, is given by:

Clock cycle		
FETCH-1	IR ← M[PC]	(Fetch the instruction bits)
FETCH-2	PC ← PC plus 1	(Increment the PC)
EX-1	PCL ← M[PC]	(Set PC/low)

The format of the instruction is shown in Figure 11.7. The PCL is changed by this instruction, but the upper 4 bits of the program counter remain the same. This has the effect of restricting the range of the BRANCH instruction to $2^4 = 16$ locations, which all have the same upper 4 bits in their memory addresses. For example, if the instruction is located at address 0–14, then the BRANCH instruction can select the next instruction from addresses 0–15. However, if the original instruction is located at address 15 (00001111), then after the FETCH-2 clock cycle the PC will be incremented to 16 (00010000), and the BRANCH instruction can select an address between 16 and 31 (00011111). In fact, the memory is divided into **segments.** Each segment contains 16 locations. The PCH register can be considered as a "segment

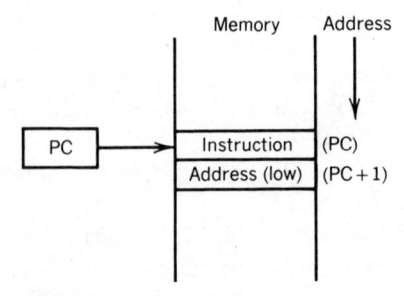

FIGURE 11.7 Storage for local BRANCH instruction.

register," that is, it contains the address of a segment (0–15), whereas the PCL register specifies a particular location within the selected memory segment. Computers with large memory address spaces (e.g., 32-bit addresses) usually use segmented addressing.

It is relatively simple to implement segmented addressing in hardware. As shown in the foregoing, loading either the PCL or PCH register with "immediate" data changes the individual location or the segment address of the next instruction. It is also useful to provide instructions that use so-called **relative** addressing. In this case the address is specified relative to a given location, which can point to anywhere in memory and does not depend on fixed memory segments. Relative BRANCH instructions require arithmetic processing and have the following sequence of register transfer operations:

Clock cycle		
FETCH-1	IR ← M[PC]	(Fetch the instruction bits)
FETCH-2	PC ← PC plus 1	(Increment the PC)
EX-1	PC ← PC plus M[PC]	(Set the PC)

The address of the next instruction is calculated by the arithmetic sum of the current contents of the PC and the address component stored in the memory location immediately after the instruction bits. For example, if the minimum word length of the processor is a byte, then the value added to the PC has the range of 0–255. It is usual practice to interpret this value as an 8-bit signed binary number. Thus, the number added to the contents of the PC has the range of -128 (10000000) to 127 (01111111). Notice that the addition is performed on the entire address, not only on the lower bits of the address; therefore, the BRANCH address is specified relative to the location of the instruction. If the address of the current instruction is equal to A, then the above register transfer operations implement a BRANCH instruction with the address range of $A - 127$ to $A + 128$.

We have already mentioned that control instructions become really useful when their execution depends on some condition. For example, a Pascal FOR or a FORTRAN DO loop is controlled by the comparison of the values of two variables. This may be implemented by the subtraction of two values and a conditional JUMP (or BRANCH) instruction executed only if the result of an arithmetic operation (subtraction in this case) is negative. This would be called a JUMP ON NEGATIVE or BRANCH ON NEGATIVE instruction. The consideration of all possible conditional control instructions is beyond the scope of this introductory text, but the understanding of the concept of the conditional instruction is fundamental to the understanding of the automatic operation of a digital computer. Therefore, we will discuss a simple SKIP instruction, which allows conditional program control by very simple means.

An unconditional SKIP instruction increments the PC (in addition to the incrementing required by the fetch cycle), which causes the next instruction in the sequence to be skipped. A conditional SKIP instruction executes the incrementing operation only if a given condition is true. This instruction is especially useful when it is used with an unconditional BRANCH instruction. Consider the end of a FOR (DO) loop, which may be coded in "pseudo"–assembly language as:

SUBTRACT two values (in order to compare)

SKIP ON NEGATIVE (if result is negative then skip the BRANCH)

BRANCH to start of loop

... continue ...

The advantage of the SKIP instruction is that it contains no address information; therefore, it has the shortest possible instruction length, and it is very easy to implement by register transfer operations. Notice that the number of words skipped must be the same as the length of the BRANCH or JUMP instruction. In our 4-bit case, if the BRANCH instruction is used, then the SKIP instruction must skip two 4-bit words. The register transfer operations become:

Clock cycle		
FETCH-1	IR ← M[PC]	(Fetch the instruction bits)
FETCH-2	PC ← PC plus 1	(Increment the PC)
EX-1	PC ← PC plus 1	: If condition is true
EX-2	PC ← PC plus 1	: If condition is true

11.5 DATA-PROCESSING INSTRUCTIONS

We discussed computer arithmetic and logic operations in Chapter 8. We studied an arithmetic-logic unit (ALU) that could produce a large number of arithmetic-logic functions such as add, subtract, increment, complement, negate (2's complement), AND, OR, XOR, and so on. In Chapter 9 we discussed shift operations and multiple precision arithmetic processing. Usually, these hardware elements are incorporated into the structure of the digital computer, and they provide a rich selection of data-processing instructions. Depending on the sources of data, we may find register–register instructions when both components of the source operands are in registers; memory–register operations when one of the source operands is fetched from memory; or even memory–memory instructions when both source operands are fetched from memory locations. The results of the operation may also be stored in registers or in memory.

In early digital computers, when hardware was very bulky, cumbersome, and expensive, only very few registers were used. In fact, in the earliest models there was only one data-processing register, called the **accumulator.** The data path diagram of a very simple one-register computer is shown in Figure 11.8. The accumulator (here, called the *A* register) can be loaded from the outputs of the ALU, or its contents can be stored in memory. All arithmetic operations that require only one source use the *A* and/or the *C* (carry bit)

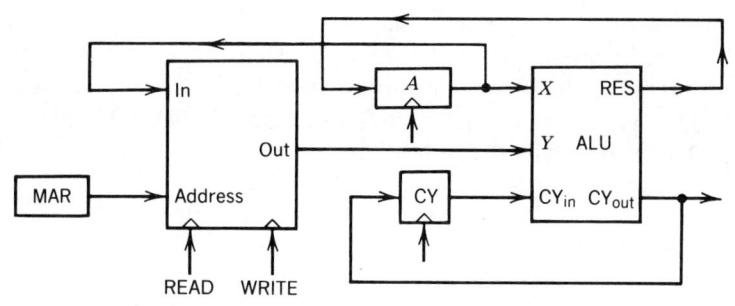

FIGURE 11.8 Data paths for computer with one register.

registers (such as clear, increment, decrement, shift, etc.). Operations with two source operands are of the memory/register type. The results of all data-processing operations are stored in the A and CY registers. Notice that there is no direct data path from memory to the A register because the A register can be loaded from memory by executing two instructions in sequence:

1. Clear the A register.
2. Add the contents of the A register and memory data and store the results in the A register.

The A register is called the accumulator because the following program loop "accumulates" (i.e., adds up) numbers stored in memory:

1. Clear the A register.
2. Add the contents of the A register and memory data using an address and store the results in the A register.
3. Change the memory address of the data.
4. JUMP to 2.

The first successful minicomputer, the 12-bit PDP-8 of Digital Equipment Corporation, had a similar structure.

The logical way of expanding the hardware capabilities of the circuits in Figure 11.8 is to add a second data-processing register, called the B register. At the early stages of computer hardware development the registers that contained data (A and B registers in this case) and those that contained a memory address or one of its components (called index registers) were separated. As hardware became less and less expensive, more and more registers were added to the processor. Also, registers became "general" registers. This meant that registers could be used for both data-processing operations and the calculation of memory addresses.

The advantage of having a large number of data-processing registers is that much of the temporary data may be kept in registers. By keeping a large amount of information in registers, the speed of processing increases as the required number of memory locations decreases. The speed of processing is increased because "reading" or "writing" memory takes much longer than updating a register. Normally, the contents of a register is always available to the processor; therefore, it is unnecessary to wait for a read or write operation. Thus, operations that involve registers are executed much faster than those that involve data in memory. The required storage for instructions that select the contents of registers and do not need data from memory is also reduced. This is normally the case because the selection of a register requires fewer bits than the specification of a memory address. For example, if we had two registers in our 4-bit example, then the selection of one of the registers requires 1 bit, whereas the specification of a full address (say, 16 bits) requires several bytes. Many small computers use eight registers and 3 bits to specify a particular register.

We have shown in Chapter 9 that multiple-precision arithmetic requires the storage of the carry bit, shift operations that involve the carry bit, and conditional instructions that test the carry bit. In most practical computers conditional instructions are of the control instruction type; that is, they are conditional "jump" or "branch" instructions. In addition to the testing of the carry bit, other conditions may be tested. For example, a conditional jump instruction is executed only if a register contains all zero bits. Other instructions may test a positive or negative number (testing the MSB), an even or odd number (testing the LSB), or parity (whether the number of 1's is even or odd). Another important test condition is whether an arithmetic operation produced an error (overflow or underflow). The variety of tests that can be applied helps the programmer's task but, of course, increases the complexity of the hardware. The number of possible computer instructions mainly affects the complexity of the computer's control circuitry.

11.6 CONTROL CIRCUITRY

We have seen several types of computer instructions needed for the construction of a useful computer program. Let us examine now how the execution of these instructions is controlled. There are three clearly identifiable groups of control signals:

1. signals that control the execution of different computer cycles,
2. signals that generate selection lines, and
3. signals that execute data transfers (clock signals).

The three types of signals are examined in what follows.

Signals that Control Computer Cycles

We have discussed the execution of computer instructions in terms of register transfer operations. The different clock periods, called clock cycles, were divided into two groups: fetch and execute. These groups may be viewed as the "states" of the computer, and the circuitry that controls these states is similar to the general state controller circuits discussed in Chapter 7. We have seen that different types of instructions may require different numbers of clock periods to execute, although they may all have the same number of fetch clock cycles. There are several methods that provide a variable number of execute clock periods for instructions. The STD for one of these methods is shown in Figure 11.9. The controller generates as many execute clock cycles as required by the instruction. Since the bits of the instruction are stored in the IR, the state sequence controller uses the bits of the IR as inputs. We can see that for a complex processor with several types of instructions, the controller circuitry may become very complex.

Another, much simpler solution is always to cycle through the same number of clock cycles (equal to the maximum number of cycles needed by any one of the computer instructions) and arrange the control signals in such a way that during the clock cycles not needed by the instruction, no register transfer operations are executed. This results in a trivial state controller (a counter) but wastes computer time. This method sacrifices computer efficiency for lower hardware costs.

It is also possible to identify unique computer states as collections of computer clock periods that execute a set of register transfer instructions. A specific computer instruction is executed by a collection of these states. In Figure 11.10 the execution of four different instructions is traced through the states of a computer. Most likely, this method will result in the most complex state controller circuitry but could produce the shortest execution times for the instructions.

Regardless of which state controller is used, signals are needed that indicate the "current" state of the computer. The register transfer operations are described in terms of instruction types and the state of the computer. The instruction type is determined from the stored bits in the IR, and we need a specific signal that is active when the computer is in a particular state, say, it is executing clock cycle EX-1. In case we need only two execution cycles, we may use the simple state controller circuit shown in Figure 11.11, where a decoder is connected to a 2-bit counter. The four output signals of the decoder are labeled according to the state of the computer, and their timing

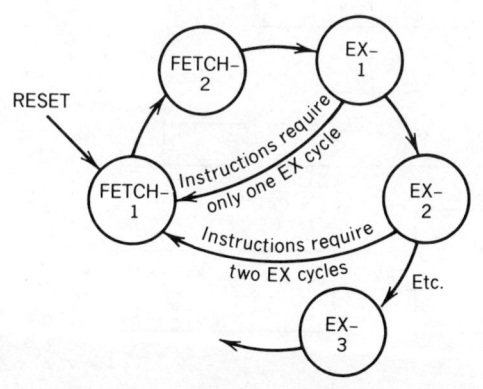

FIGURE 11.9 The STD for state control of automatic digital computer.

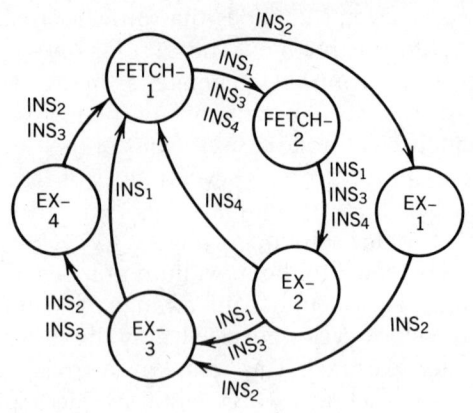

FIGURE 11.10 The STD using unique computing states.

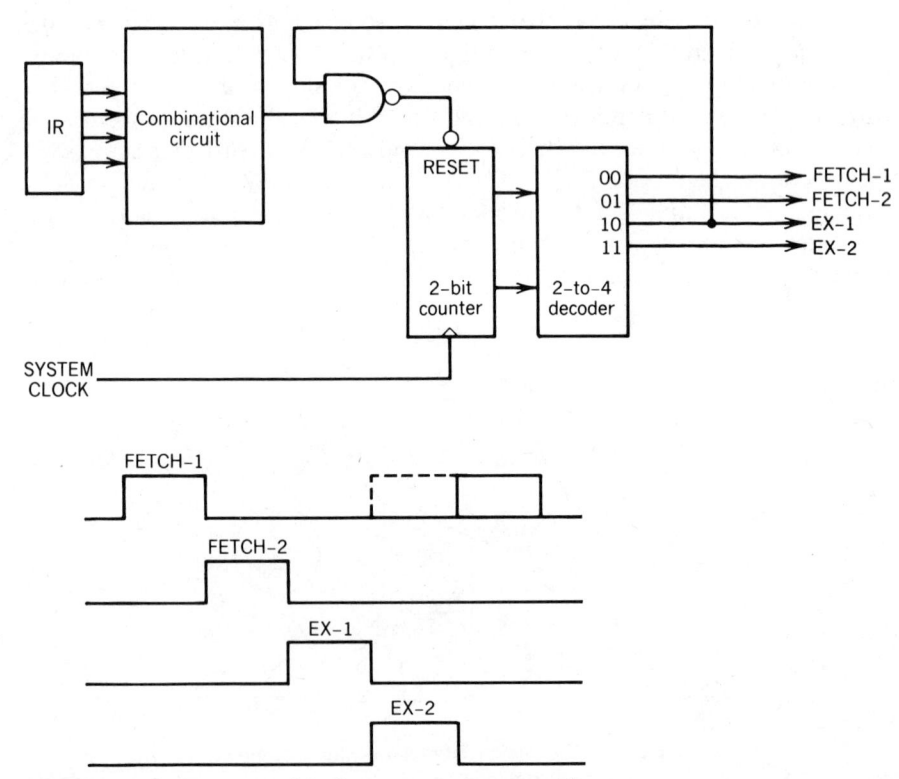

FIGURE 11.11 State controller of computer using resettable counter.

352

diagrams are also shown. The counter contains a RESET input that is activated by instructions that do not require state EX-2. The RESET input becomes active when the computer is in its EXEC-1 state and the instruction bits indicate that the EX-2 cycle is not needed. The state controller may be even simpler if the one-flip-flop-per-state method is used, in which case a decoder is not required.

In Figure 11.12 we show the control circuitry of a general multicycle computer that uses the parallel LOAD function of a counter to control its state transitions. The IR is used both to control the counter and to generate the control signals that execute the required register transfer instructions. The counter is controlled through its LOAD input, which becomes active if the computer must alter its normal state sequence. The combinational block that generates the control signals requires the system clock as input because, as will be shown later, the signals that update the registers must be derived from the system clock.

Signals for Selection Lines

When we examine the register transfer operations of several computer instructions, we find that, depending on the state of the computer and the type of instruction being executed, a register may receive its input from one of several possible sources. The hardware implementation for this requirement is a multiplexer, and the selection lines determine the source. Other selection lines determine the function of the ALU, the type of shift or LOAD operations executed by a general shift register, the address used for the memory, and so on. The binary values of the selection lines are derived from the bits of the instruction (stored in the IR) and the state of the computer.

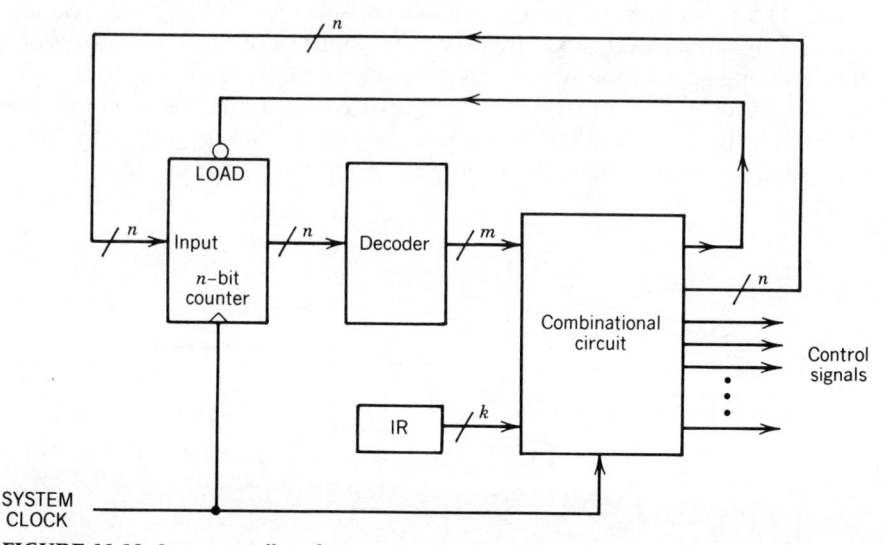

FIGURE 11.12 State controller of computer using loadable counter.

The generation of a typical selection line is shown in Figure 11.13. This line determines the operation of the PC. Assuming that the PC can be either incremented or loaded, this signal selects one or the other. It is active either if the computer is executing the FETCH-2 cycle or it is in the EXECUTE-2 cycle with either instruction bit IR_3 or IR_1 equal to 1. The instruction bits select all those instructions for which incrementing the PC is required. Note that the selection of this line does not necessarily mean the incrementing of the PC. The actual incrementing is controlled by a clock pulse (discussed in the next section). A select line may have a particularly simple logic function. For example, a select line may be connected to an instruction bit without any logic gates. In this case the computer instructions would be divided into two types, one for which the bit is equal to 1, the other for which it is equal to 0. It is no surprise then that computer hardware instructions naturally fall into instruction types selected by their bit values.

Signals for Updating Registers

Control signals belonging to this third group actually execute the register transfer operations. They should not be considered as signals with values of 1's or 0's but as having significant clock transitions. The generation of a typical UPDATE ($CLOCK_{PC}$) signal is shown in Figure 11.14. The logical selection of the signal depends on the instruction bits and the state of the computer, but the actual significant clock transition is controlled by the system clock.

The simplest form of computer control uses only one system clock. For a reliable system all selection lines should hold their correct values constant throughout the active period of the system clock, and all significant clock transitions must occur at the trailing edge of the system clock. A general solution that generates all the required positive clock signals is shown in Figure 11.15, where the unspecified block consists of a combinational circuit. Inverted signals (NAND gates) are used to provide positive (0-to-1) transitions at the trailing edge of the system clock pulse. If a negative (1-to-0) transition is required for the UPDATE signal, then an AND gate must be used in place of the NAND gate. Another method of providing clock signals is to use a demultiplexer with negative logic outputs (e.g., type 74156 IC device).

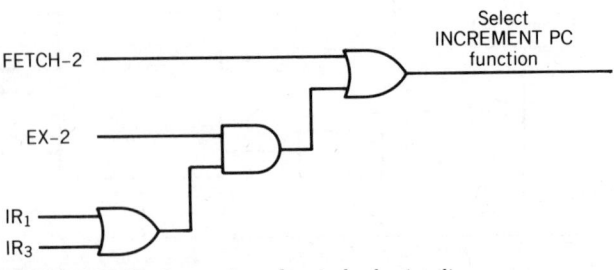

FIGURE 11.13 Generation of typical selection line.

11.6 CONTROL CIRCUITRY   355

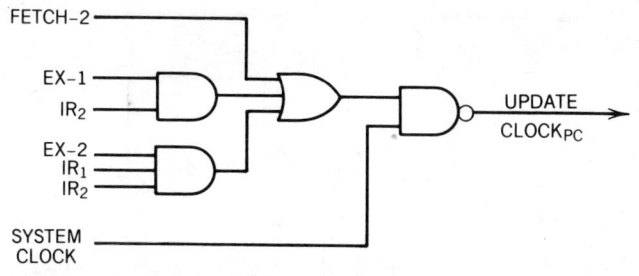

FIGURE 11.14 Generation of typical CLOCK line.

As long as only edge-triggered and/or master–slave flip-flops are used, the circuit shown in Figure 11.15 provides reliable operation. The clock period must be selected long enough to account for the longest delay in the system. This type of system should always work correctly in the single-clock-pulse mode. In this mode the system clock is operated manually by a properly debounced manual switch, and this implies infinitely long clock cycles.

If dc-latch-type flip-flops or memory elements are used, then extra care is required. For example, in the circuit shown in Figure 11.16, both registers A and B are updated by the same clock pulse. The selection lines can be set such that the contents of the A and B registers should be exchanged. The exchange operation works only if the flip-flops are of the edge-triggered or master–slave type. If dc latches are used, then the contents of the two registers become the same during the time period when the clock pulse is active. In practice, it is possible to use dc-latch-type flip-flops, but if exchanging the contents of the two registers is required, then two additional flip-flops and an additional clock signal are required. This is shown in Figure 11.17. The operation of this circuit is similar to that of a master–slave flip-flop. During the first clock pulse the contents of flip-flops A and B are transferred to the two additional flip-flops. During the second clock signal the contents of flip-flops A and B are updated to their correct values. Since there are two separate clock signals during one clock period, this technique uses the so-called **two-phase** clock. Many microprocessors provide two-phase clocks for reliable data transfers.

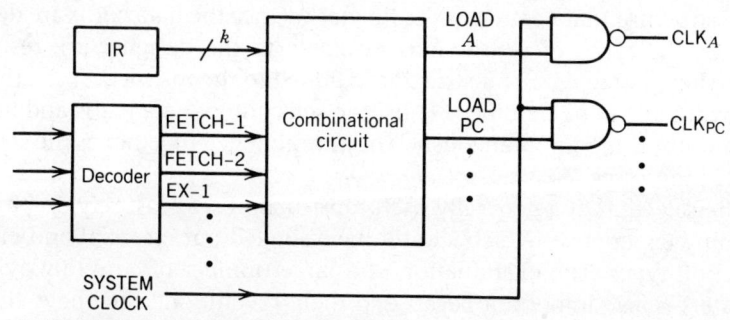

FIGURE 11.15 General solution to generation of all clock signals for computer.

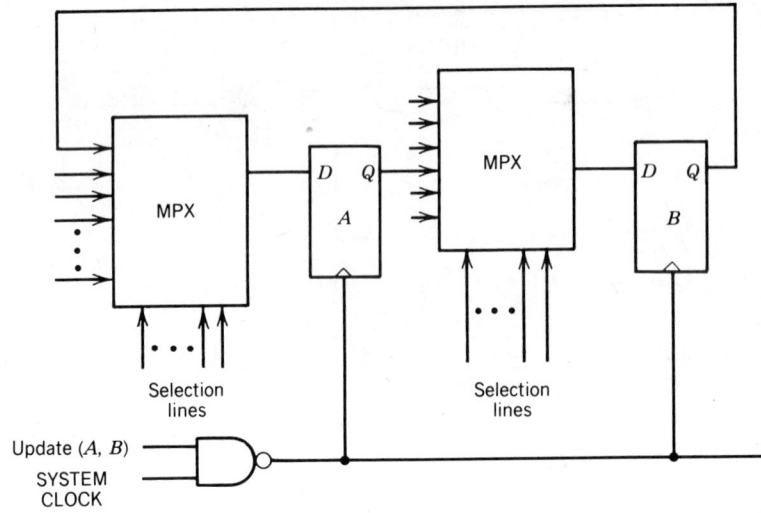

FIGURE 11.16 Data paths of computer that may lead to data exchange operation.

If only a single-phase clock is available, then the same exchange operation can be executed by using the two additional flip-flops and two computer cycles. If only one additional flip-flop is used, then three cycles are needed. This is the familiar exchange operation in three steps: during the first cycle TEMP ← A, during the second A ← B, and during the third B ← TEMP.

By generating the three types of control signals (state control, selection/function lines, and clock signals), all the required circuits of a computer may be designed and constructed. The selection of a particular set of computer instructions and the assignment of bit values to these instructions still have to be solved. The selection of an instruction set for a simple 4-bit computer will be discussed in the next section.

11.7 SELECTION OF COMPUTER'S INSTRUCTION SET

The selection of an instruction set for a yet undefined computer is not a straightforward design problem. One may consider the selection of the computer's structure (data paths) as the first task. Once the data paths are defined, however, the types of instructions available to the designer are restricted. Later, when a new type of instruction is added to the instruction set, the data paths will have to be modified. Thus, the selection of data paths and instructions must be done simultaneously. They are changed together as the selection process progresses.

The selection of a generally useful instruction set must rely on past programming experience, which should have singled out essential and efficient instruction types. The examination of a large number of currently available computers is also important because in their "evolution" they have retained

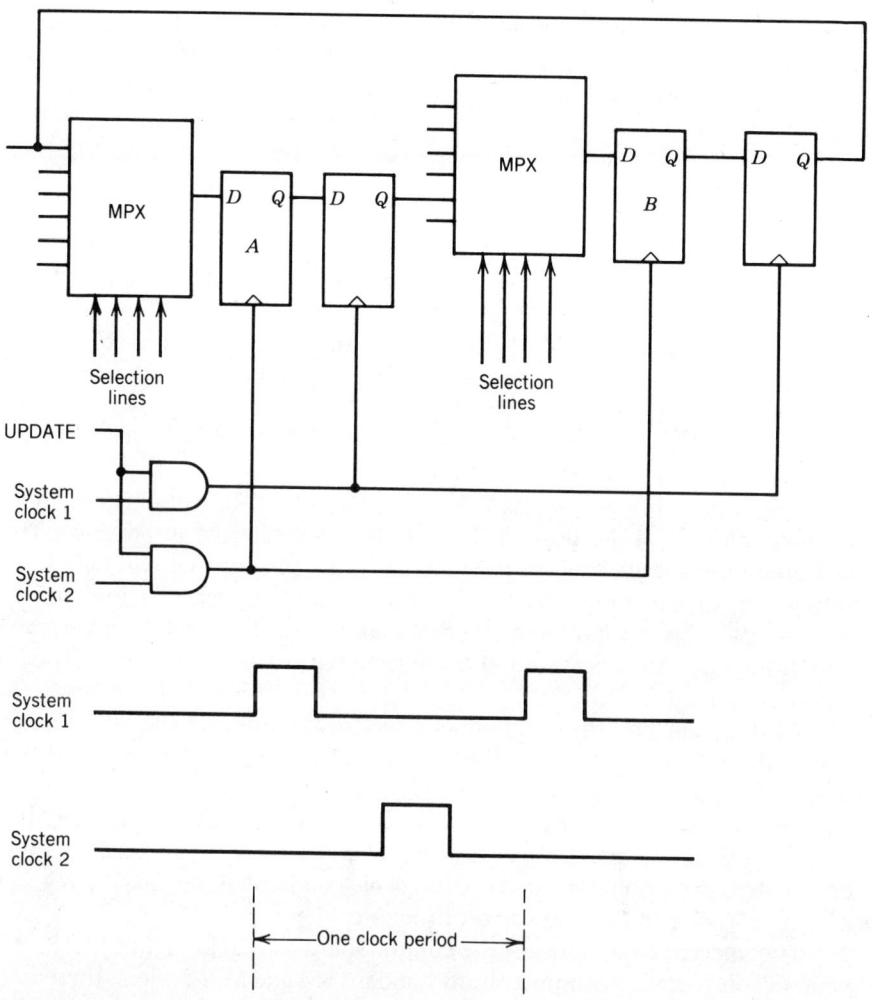

FIGURE 11.17 Solution of data exchange problem for dc latches using two additional flip-flops and two-phase clock.

features that have worked well in the past. It is difficult to combine all these factors into a coherent, "scientific" design procedure.

There are two major considerations in selecting the types of instructions supplied with a computer: hardware and software. Considering the hardware, one would like to use instructions that are easily implemented. We have seen in Chapter 10 how a number of bits in an instruction can select one particular source or destination register. In case these bits can perform the same selection for all instructions, the hardware becomes greatly simplified.

The selection of instructions on the basis of software, or programming efficiency, is a more difficult problem. There are two classes of instructions.

First, there are a number of instructions that are essential for constructing meaningful computer programs. Basic arithmetic functions (ADD, NEGATE, and SHIFT), basic logic functions (AND and COMPLEMENT), transfer instructions between registers and memory, input–output instructions, and finally, control instructions (with which control structures, such as REPEAT, WHILE/UNTIL, and FOR loops can be constructed) must be supplied at least as a minimum. Multiple-precision arithmetic and single-bit manipulations are also basic necessities.

Another class of instructions help the programmer to reduce the size of the program significantly. Examples of these are multiple-precision arithmetic instructions, MULTIPLY and DIVIDE instructions, a rich variety of conditional instructions, procedure or subroutine-handling instructions, and so on. When hardware was expensive, many of these additional instructions were treated as luxury. Today, programmers expect many of these instructions to be included even in the smallest of computers.

Our approach is to present a "case study" of a 4-bit computer with 14 instructions (out of the possible 16) having been selected in advance. This will allow the specification of the data paths at the start that will help us in subsequent explanations. We will comment on why particular instructions were selected, but we have to admit that many of the decisions were arbitrary, and in no way can we claim that the selected instruction set is optimal with respect to any given criteria.

At first, consider the computer's structure. Because of the very limited size of the instruction word (4 bits), a single data register is selected. When only one data register is used, no instruction bits are required to specify a particular data register for an instruction. An obvious way of extending this design for larger instruction words is to add register-addressing bits. This was demonstrated in Chapter 10. For this 4-bit computer we expect a register structure similar to the one shown in Figure 11.8.

The selected data paths for the computer are shown in Figure 11.18. We want to construct this computer from standard SSI and MSI devices; therefore, only the most essential hardware components are used. For example, the MBR is eliminated. The diagram does not show a direct data path from memory to the *A* register, and one may wonder how the loading of data from memory to the *A* register is implemented. The solution can be through the CLEAR A and ADD operations, as we have shown before, but can also be realized through one of the functions of the ALU. If one function of the ALU is to select its *Y* input for its output, then a direct data path from memory output to the *A* register input is established. We have selected this later solution.

Most circuit elements shown in Figure 11.18 have been used before and require no further explanations. The multiplexer connected to the carry flip-flop is a new element. Its role will become clear as we will discuss the selection of the essential instructions for this computer. For the arithmetic processing of 2's complement signed binary numbers, two arithmetic instructions (ADD

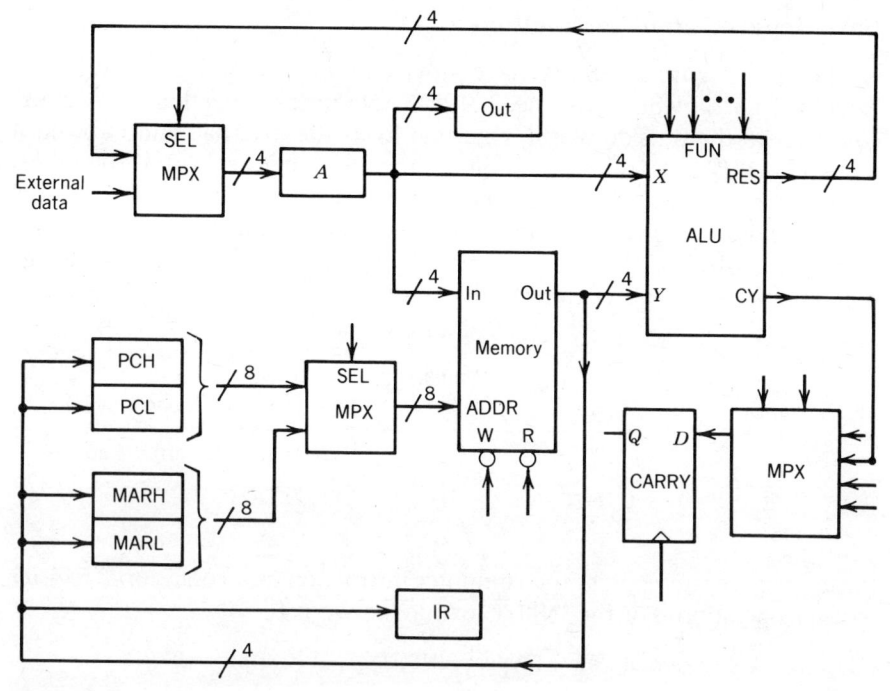

11.7 SELECTION OF COMPUTER'S INSTRUCTION SET   359

FIGURE 11.18 Selected data paths for 4-bit computer.

and INCREMENT) and the logical instruction COMPLEMENT are sufficient. Negation, that is, 2's complement of a binary number, is executed by the COMPLEMENT and INCREMENT operations. The advantage of selecting these three operations is that by adding one logical operation (e.g., logical AND), any arbitrarily complex logical expression may be calculated. These four data-processing instructions are sufficient for evaluating any arithmetic-logic expressions that involve single-word data.

Since the data word is very small, it is essential to provide multiword-processing capabilities. The carry bit is the key to multiword processing, as shown in Chapter 9. Programming experience has also shown that the carry bit can be very effectively used as a 1-bit flag, in which case it is advantageous to be able to set it, clear it, or complement it by hardware instructions. Another required function is the rotation of the A register with the carry bit (when the A register and the carry bit are considered to be a single 5-bit register), and it is also useful to be able to rotate the A register without the carry bit. All these "programming" considerations played an important role in the instruction selection process that yielded the instruction set described in the next section.

There are four types of instructions in a basic instruction set: input–output, data transfer, data processing, and program control. These four types are discussed next under separate headings.

Basic Input–Output Instructions

As shown in Figure 11.18, we have provided the simplest data paths for the external data. Obviously, we need at least two instructions that connect the computer to the outside world. One that loads the external data into the *A* register (INPUT), and another that loads the "out" register with the contents of the *A* register (OUTPUT). Since these instructions require no additional data or address information, one instruction word is all that is required for them. The instruction may be described by the following assembly language mneumonics and register transfer operations:

Number	Mneumonics	Number of words	Register transfer operations
1	INPUT	1	EX-1: A ← External Data
2	OUTPUT	1	EX-1: OUT ← A

In general, we can express the computer instruction as a conditional register transfer operation. For the INPUT instruction we get

$$A \leftarrow \text{External Data} : (\text{Type} = \text{INPUT}) \text{ AND } (\text{CYCLE} = \text{EX-1})$$

and

$$\text{OUT} \leftarrow A : (\text{Type} = \text{OUTPUT}) \text{ AND } (\text{CYCLE} = \text{EX-1})$$

We get similar conditional register transfer operations for all other instructions. The collection of all these register transfer operations completely describe the operation of the computer.

Basic Data Transfer

The choice for the basic data transfer instructions is not as simple as it was for the input–output instructions. For example, as we have seen before, the loading of the *A* register from memory may be executed by CLEAR A and ADD instructions or by an instruction that executes only the LOAD function. We will choose the latter method.

The possible choices for data transfer instructions are also influenced by the memory addressing methods used. The minimum requirement is that the contents of any memory location may be loaded into the *A* register (LOAD) or the contents of the *A* register may be loaded into any memory location (STORE). Since the memory address is stored in the MARH and the MARL, the contents of these two registers must be set up before the data transfer can take place. The LOAD/STORE instructions could contain the entire address, in which case these instructions would require three data words. We will limit the number of data words per instruction to 2 and supply an instruction that loads the higher order 4 bits of the MAR. Then, the LOAD and STORE instructions supply the four lower address bits in addition to the data transfer operations. The description of these three instructions follows:

Number	Mneumonics	Number of words	Register transfer operations
3	LOAD MARH (address)	2	EX-1: MARH ← M[PC] PC ← PC plus 1
4	LOAD A (address)	2	EX-1: MARL ← M[PC] EX-2: A ← M[MAR] PC ← PC plus 1
5	STORE A (address)	2	EX-1: MARL ← M[PC] EX-2: M[MAR] ← A PC ← PC plus 1

The word *address* (within parentheses) indicates that the instruction requires an additional word (4 bits) which is loaded into the MARH or MARL registers. Note that for all these instructions two register transfer operations are indicated for some clock cycles. The execution of more than one register transfer operation per computer cycle was selected in order to keep the number of computer cycles as small as possible. Another way of choosing clock cycles (which could produce a simpler overall hardware) is to use one cycle for each register transfer operation. If we used this method, then the number of execution cycles would have to be increased to 3.

Basic Data-Processing Instructions

In considering data-processing functions, we select a small number of instructions that will provide the basic arithmetic operations ADD and INCREMENT and the two logical operations COMPLEMENT and AND. With these basic operations arithmetic subtraction and all logical operations can be executed.

When data-processing instructions are selected, the role of the carry flip-flop becomes important. In addition to specifying the effect of the data-processing instructions on the contents of the *A* register (where the results of the operations are stored), their effect on the content of the carry flip-flop must be specified as well. The carry output of the ALU must be stored in the carry flip-flop for the arithmetic operations ADD and INCREMENT (this is essential for multiword data processing). The contents of the carry flip-flop may be set to any value for the logical operations AND and COMPLEMENT, since logical operations do not involve carries. We have selected to clear the carry flip-flop for the AND and complement its content for the COMPLEMENT operations. As shown in Figure 11.18, the content of the carry flip-flop can be controlled by a multiplexer. Four sources are used. As shown in Figure 11.19, the carry flip-flop may be updated by the carry output of the ALU

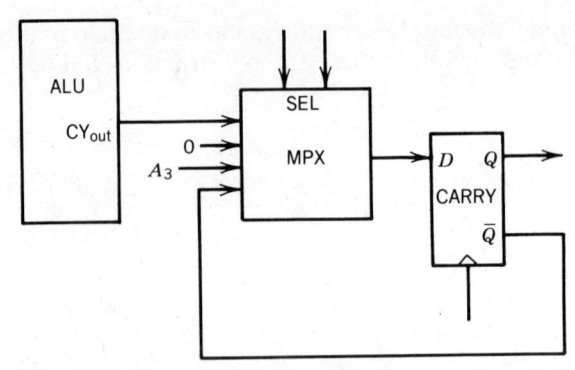

FIGURE 11.19 Selection of stored carry bit as implemented with multiplexer.

(ADD and INCREMENT operations), the inverted carry bit (COMPLEMENT), Logic-0 (AND), and the MSB of the A register, A_3 (ROTATE LEFT).

Figure 11.18 shows that all operations that require two source operands use the A register and memory. Keeping our instruction selection consistent, all operations that require data from memory will supply the four lower order address bits similarly to the LOAD and STORE operations. The following four instructions are selected:

Number	Mneumonics	Number of words	Register transfer operations
6	ADD (address)	2	EX-1: MARL ← M[PC] EX-2: A ← M[MAR] plus A CARRY ← CY_{out} PC ← PC plus 1
7	AND (address)	2	EX-1: MARL ← M[PC] EX-2: A ← M[MAR] AND A CARRY ← 0 PC ← PC plus 1
8	COMPL A	1	EX-1: A ← A' CARRY ← CARRY'
9	INC A	1	EX-1: A ← A plus 1 CARRY ← CY_{out}

The 2's complement operation is executed by the COMPLEMENT and the INCREMENT operations. In order to be able to write programs with multiple-precision arithmetic, shift operations that include the carry bit (and conditional control instructions that test the carry bit) are also needed. For general bit manipulations, a ROTATE type of instruction that does not include the carry bit is also very useful. Two such instructions are:

11.7 SELECTION OF COMPUTER'S INSTRUCTION SET

Number	Mneumonics	Number of words	Register transfer operations
10	ROT LEFT	1	EX-1: $A_i \leftarrow A_{i-1}$ $\quad i = 3, 2, 1$ $A_0 \leftarrow$ CARRY CARRY $\leftarrow A_3$
11	ROT RIGHT	1	EX-1: $A_i \leftarrow A_{i+1}$ $\quad i = 2, 1, 0$ $A_3 \leftarrow A_0$

Obviously, the selection of these two instructions is somewhat arbitrary. Experience with machine language programming tells us that with these two instructions most bit manipulations can be executed using a reasonably few number of instructions. The usefulness of these instructions also relies on a conditional control instruction that tests the content of the carry bit.

Basic Control Instructions

A minimum of two types of program control instructions are required. We need a "global" JUMP instruction with which the program can execute a GOTO instruction and the program can "jump" to any memory location. We also need a conditional control instruction. A third type of control instruction that "branches" within a limited range is also useful. We have discussed these instructions before and will demonstrate how they may be implemented using this example of a 4-bit computer.

For the global JUMP instruction both the PCH and the PCL registers must be updated. Keeping within the two-word limit per instruction, the PCH may be loaded with data, whereas the PCL register is cleared. This instruction would execute a JUMP to the starting address of one of 16 segments (each segment contains 16 instructions in memory). The BRANCH instruction with a limited range can be implemented by updating the PCL register only, whereas the conditional instruction will be of the SKIP type. These three control instructions are:

Number	Mneumonics	Number of words	Register transfer operations
12	JUMP (address)	2	EX-1: PCH $\leftarrow$ M[PC] PCL $\leftarrow$ 0000
13	BRANCH (address)	2	EX-1: PCL $\leftarrow$ M[PC]
14	SKIP ON C	1	EX-1: PC $\leftarrow$ PC plus 1 if $C = 1$ EX-2: PC $\leftarrow$ PC plus 1 if $C = 1$

With these 14 instructions a very small but practical and usable instruction set has been designed. There are two more possible but so-far unassigned instructions. What type of instructions should one choose? If a computer is designed for special calculations (like the sorting computer of Experiment 10), then the selection of additional instructions could be governed by the primary use of the computer. For a general computer either the minimization of hardware costs or the advice of experienced assembly language programmers could be the deciding factor.

Let us consider only the minimization of hardware here. Table 11.1, which shows all the selected instructions, is very useful for this discussion.

TABLE 11.1 Register Transfer Instructions for All Selected Computer Instructions

		Fetch cycle: IR ← M[PC], PC ← PC plus 1	
	Instruction	Execution cycle 1	Execution cycle 2
1	INPUT	A ← External Data	
2	OUTPUT	OUT ← A	
3	LOAD MARH	MARH ← M[PC] PC ← PC plus 1	
4	LOAD A	MARL ← M[PC] PC ← PC plus 1	A ← M[MAR]
5	STORE A	MARL ← M[PC] PC ← PC plus 1	M[MAR] ← A
6	ADD	MARL ← M[PC] PC ← PC plus 1	A ← M[MAR] plus A CARRY ← CY[ALU]
7	AND	MARL ← M[PC] PC ← PC plus 1	A ← M[MAR] AND A CARRY ← 0
8	COMPL A	A ← A' CARRY ← CARRY[1]	
9	INC A	A ← A plus 1 CARRY ← CY[ALU]	
10	ROT LEFT	A[i] ← A[i − 1] i = 3, 2, 1 A[0] ← CARRY CARRY ← A[3]	
11	ROT RIGHT	A[i] ← A[i + 1] i = 2, 1, 0 A[3] ← A[0]	
12	JUMP	PCH ← M[PC] PCL ← 0000	
13	BRANCH	PCL ← M[PC]	
14	SKIP ON C	PC ← PC plus 1 IF (C = 1)	PC ← PC plus 1 IF (C = 1)

For example, we find that we have selected five instructions that require two execution cycles and nine instructions that need only one. If we had divided the instructions into two equal groups (eight one-cycle and eight two-cycle instructions), then one instruction bit could be used to indicate the number of execution cycles required by the instruction (i.e., if this instruction bit is equal to 0, then the instruction requires one cycle; otherwise it requires two). This simple encoding of a bit translates into simple hardware since, in this case, the signal that controls the counter can be derived directly from the instruction bit (e.g., in Figure 11.11 the input to the NAND gate would be connected to 1 bit of the IR). It may be worth considering the rearrangement of the register transfer operations in order to simplify the hardware.

The collection of similar instructions (i.e., instructions with similar register transfer sequences) into groups of two, four, or eight instructions will also provide simplified hardware. For example, there are four instructions for which the MARL ← M[PC] register transfer instructions are executed. If the same two instruction bits are assigned for these four instructions, say, $IR_3 = 1$ and $IR_2 = 1$, then the Boolean equation for the clock signal that updates the MARL becomes very simple:

$$CLOCK_{MARL} = (IR_3) \text{ AND } (IR_2) \text{ AND } (EX-1)$$

These examples demonstrate important relationships between selected instructions, the encoding of instruction bits, and the complexity of the hardware.

The selection of a complete instruction set, the assignment of bit values to instructions, and the detailed design of this or a similar 4-bit computer are the subjects of Experiment 11. Further discussion on these topics may be found within the description of this experiment. Before we proceed to Experiment 11, however, a short discussion on program writing in assembly language is presented in the next section.

11.8 PROGRAMMING IN ASSEMBLY LANGUAGE

The subject of this course is hardware design. But the design of computer hardware cannot (and should not) be separated from the programming of the designed computer. Major faults in hardware design are found only after programs have been written for the computer. Often, major advancements in hardware design are suggested by programmers when they realize that some types of programs are very difficult to write or impossible to execute. We will introduce software by discussing two examples of the diagnostic program type.

Diagnostic programs are used to test the correct operation of the computer's hardware. Sometimes, hardware tests are made in single-step mode (the system clock signal is operated by a properly debounced switch). At other times, the program runs in a loop and the tests are repeated. Lights are connected to the flip-flop outputs of the output register and to the PC. The lights of the PC show that the program is running, and the output lights change according to the values of the external input data lines. Both following programs are designed to run in program loops.

The instruction codes of the indicated program are stored in the memory of the computer beginning at address 0000 0000 (memory addresses are shown as two groups of 4 bits with the left 4 bits stored in the PCH or MARH register and the right group of 4 bits stored in the PCL or MARL registers, respectively). The test lights are observed while the computer cycles are executed. It is assumed that there is a facility to "reset" the computer by setting the contents of both PC registers to binary zeros. Thus, all program examples start at memory location 0000 0000.

Diagnostic Program 1: Input–Output Test

The first program reads the external data input lines and updates the output register (OUT) as a function of the input bit values. This program runs in a loop. The hardware can be tested by connecting lights to the four outputs of the output register. When the logic levels of the four external data lines are changed, the lights will change according to the executed instructions in the program.

When the MSB of the input data is Logic-0, the output register is loaded with the input value. To make the program a little more interesting, the 2's complement of the input data is sent to the output register if the MSB of the input data is equal to 1. The program uses (therefore, it tests) the LOAD MARH, INPUT, OUTPUT, LOAD A, STORE A, BRANCH, ROTATE LEFT, SKIP (conditional), COMPLEMENT, and INCREMENT instructions.

The program listing shown in Figure 11.20 lists mneumonics and uses brief comments for explanation. The program requires fifteen 4-bit words plus one memory location for the storage of the input data.

Diagnostic Program 2: Test all Computer Instructions

The first program did not test four of the basic computer instructions: ROTATE RIGHT, ADD, AND, and JUMP. Instead of writing another program for these instructions, we will expand Program 1 to include the missing four instructions. The additional tests use the data values 0101 and 1010, which are stored in memory. Note that when these two values are added, the result is equal to 1111. The value of 1111 may be tested by incrementing it when the carry bit should become 1. (This is the only value that, when incremented, produces a carry value of 1.)

The idea is to produce the value 1111, test it, and if the test fails, use a BRANCH instruction that "jumps to itself." Thus, if any of the tests fail, the program will be "stuck" at a given location, will not run in a loop, and when the input data values are changed, the output lights will not follow.

Notice also that Program 2 (shown in Figure 11.21) continues beyond location 0000 1111; therefore, the storage locations used for data had to be moved beyond the end of the program.

Memory address	Mneumonics instruction/data	Comments
0000 0000	LOAD MARH	Set up MAR HIGH register so that it contains
0000 0001	0000	four higher order address bits: 0000.
0000 0010	INPUT	Load external data into A register.
0000 0011	STORE A	Store its contents into memory location:
0000 0100	1111	0000 1111.
0000 0101	COMPL A	Complement most significant bit
0000 0110	ROT LEFT	Shift complement of MSB into carry bit.
0000 0111	LOAD A	Load back contents of memory location
0000 1000	1111	0000 1111 into A register.
0000 1001	SKIP ON C	Skip two instructions if carry = 1 (MSB=0).
0000 1010	COMPL A	Complement contents of A register.
0000 1011	INC A	Increment contents of A register.
0000 1100	OUTPUT	Copy contents of A into output register.
0000 1101	BRANCH	Branch back to memory location:
0000 1110	0010	0000 0010 (INPUT instruction).
0000 1111	XXXX	Input data bits stored in this location.

FIGURE 11.20 Program listing for Diagnostic Program 1.

EXPERIMENT 11
Design of Four-Bit Computer

E11.1 Summary and Objectives

The main objective of this experiment is the detailed design of a small digital computer. The construction of a diagnostic assembly language program with which the instruction set of the computer may be tested is also required. If time and hardware resources permit, the computer should be constructed and the hardware tested. It may be more practical to construct this computer as a group effort, with several students working on one computer. In this case each subsystem should be the responsibility of one or two students.

Another objective of this experiment is to demonstrate how all the different hardware units of a computer work together to form a programmable digital computer. The concepts of instruction types, different forms of addressing, computer control instructions, and the problems of hardware testing of a relatively complex digital circuit will all be important elements of Experiment 11.

In Chapter 11 the operation and instruction set of a 4-bit computer have been discussed. When these ideas are used for a practical design, new problems arise. The selection of hardware devices and the minimization of the required hardware circuits will be two of these problems. The student is encouraged to change the hardware structure of the computer, its basic or extended instruction set, the recommended hardware components, or any other aspects

Memory address	Mneumonics instruction/data	Comments
0000 0000	LOAD MARH	Set up MAR HIGH register so that it contains
0000 0001	0010	four higher order address bits: 0010.
0000 0010	INPUT	Load external data into A register.
0000 0011	STORE A	Store its contents into memory location:
0000 0100	0100	0010 0100.
0000 0101	COMPL A	Complement A (including MSB).
0000 0110	ROT LEFT	Shift MSB into carry bit.
0000 0111	LOAD A	Load back contents of memory location
0000 1000	0100	0010 0100 into A register.
0000 1001	SKIP ON C	Skip two instructions if carry = 1 (MSB=0).
0000 1010	COMPL A	Complement contents of A register
0000 1011	INC A	Increment contents of A register
0000 1100	OUTPUT	Copy contents of A into output register
0000 1101	LOAD A	Load data 0101 from memory location:
0000 1110	0011	0010 0011 into A register.
0000 1111	ADD A	Add same value to it ("A" should contain:
0001 0000	0011	1010 with carry bit equal to 0).
0001 0001	ADD A	Add same value again ("A" should contain:
0001 0010	0011	1111 with carry bit equal to 0).
0001 0011	INC A	Test value by incrementing it and
0001 0100	SKIP ON C	then testing carry bit.
0001 0101	BRANCH	If test is O.K. then this BRANCH is skipped.
0001 0110	0101	Otherwise, program is stuck here.
0001 0111	LOAD A	Now test AND instruction with data 1010
0001 1000	0010	stored in location 0010 0010
0001 1001	AND A	and data 0101 stored in location:
0001 1010	0011	0010 0011.
0001 1011	COMPL A	Result should be 0000 which is complemented
0001 1100	INC A	incremented
0001 1101	SKIP ON C	and tested.
0001 1110	BRANCH	This is skipped if data is O.K.
0001 1111	1110	or program is stuck here on error.
0010 0000	JUMP	Jump back to beginning of program
0010 0001	0000	at location 0000 0000.
0010 0010	1010	Data value 1010 = -6
0010 0011	0101	Data value 0101 = $+5$
0010 0100	XXXX	Temporary location to hold value of A.

FIGURE 11.21 Program listing for Diagnostic Program 2.

of the described 4-bit computer either to minimize hardware or to provide a more convenient instruction set. Both the discussions in Chapter 11 and the recommendations made in the following sections should be considered only as a guide and, if possible, the examples used should be improved and not copied.

E11.2 Some Practical Aspects

We discussed the model of a 4-bit computer in Chapter 11. When such a computer is designed, constructed, and tested, new problems arise. The first problem that has to be solved is how to test the computer's operation. We showed hardware diagnostic programs but did not discuss how the instruction bits are first placed into the computer's memory. This may be done by a separate circuit connected to the memory before the testing starts, the required instructions loaded into the correct memory locations, and then the computer connected to the memory, which contains the correct program. Another approach is to build this initial memory-loading function into the computer's hardware. This can be implemented by a special switch that could control the mode of operation of the computer. When this special switch is set, the computer is in the "load memory" mode and executes the following two register transfer operations:

$$M[PC] \leftarrow \text{External Data}$$

$$PC \leftarrow PC \text{ plus } 1$$

We indicate this solution in the hardware diagram of our proposed 4-bit computer (Figure E11.1). When the test switch is set to Logic-1, the external data lines are applied to the memory data input lines. Two separate test clock

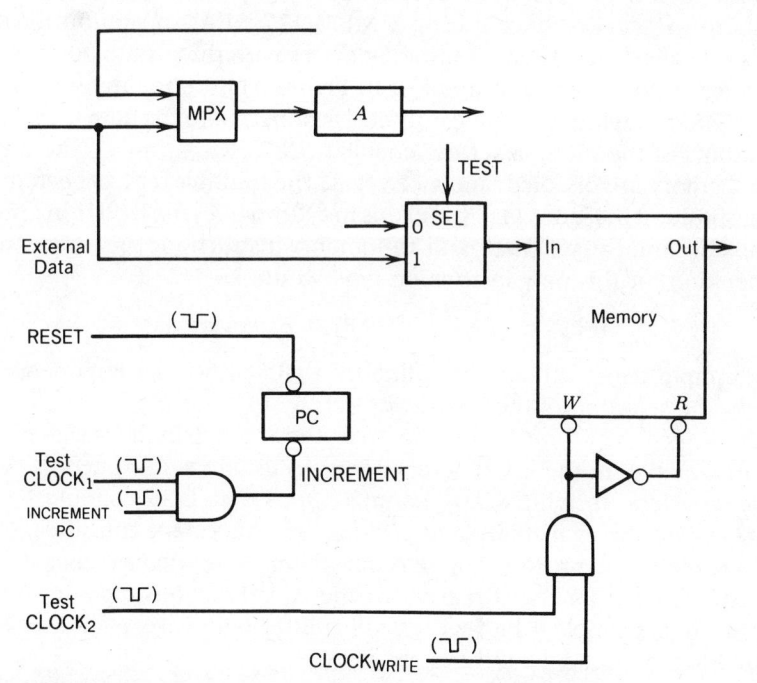

FIGURE E11.1 Addition of test signal and two test clocks to computer's hardware.

signals control the data transfer and the incrementing operations. It is assumed that the system clock signal is inactive, and updating occurs during the negative transition of the clock. Depending on the required clock signals for the memory and the PC, other types of gates may be required for a given hardware design.

We have mentioned in Chapter 11 that a "reset" function is required in order to be able to start the operation of the computer at address 0000 0000. This reset function is also required by the loading operation so that the first data stored in memory will be at PC address 0000 0000.

Other practical problems occur when the hardware components are selected. The selection of hardware components may influence the hardware design and even the preferred instruction set. The selection of the memory is very important. One may select the most convenient memory device, for example, SIGNETICS type-2101 RAM, which is very simple to operate. The assigned pins of this 22-pin device are shown in Figure E11.2. As indicated on this diagram, the outputs of the memory are active unless a write memory cycle is executed. All enabling signals may be kept active ($\overline{CE_1}$ = 0, CE_2 = 1, OD = 0) and the R/W signal can control the writing operation.

One may want to use a smaller IC device and choose the 18-pin type-2111 memory device. The difference is that in this case the data for the input and output are shared. The control of this memory (three-state outputs) is the same as for the type-2101 RAM, but the inputs cannot be connected directly to the device, since during a MEMORY READ operation the same signals act as outputs. Therefore, some devices with three-state outputs must be connected to these four data pins. In Figure E11.3 a multiplexer (similar to type 74S257) with three-state outputs is connected to the data lines. When the outputs of the multiplexer are enabled (CLOCK$_{WRITE}$ = 1), the outputs of the memory are disabled, and vice versa. The multiplexer has been moved from its position in Figure 11.18, and this move modifies the INPUT instruction so that it becomes an external-data-to-memory instruction. The register transfer operation for this new instruction type would be

INPUT   EX-1: M[MAR] ← External Data

This example demonstrates the influence that hardware components may have on the selection of the instruction set.

The selection of components should be the first step in the design process. For this experiment the list of component specifications in Appendix A may not be sufficient, in which case manufacturer's data books should be consulted. Obviously, availability and cost are two important considerations in the device selection process. The selection of the ALU will influence the type of register needed for the *A* register. If the ALU functions include shift operations, then a simple 4-bit register will suffice; otherwise, a shift register is required to store the bits of the *A* register.

The PC has to be built from two 4-bit binary counters, and the type of counter used will affect the complexity of the controller. It is worth considering the use of a counter for the MAR register as well, in which case, an additional

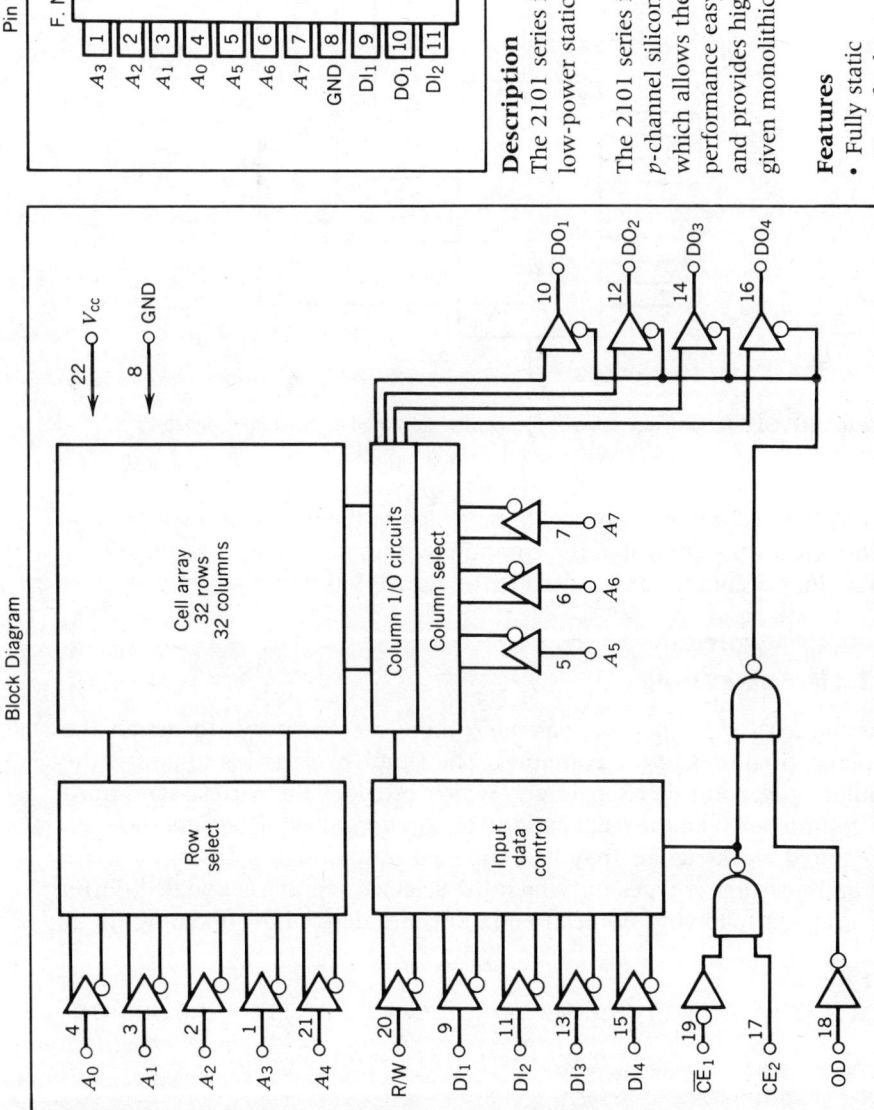

Pin configuration

F. N. PACKAGE

Description

The 2101 series is high-performance, low-power static read write RAMs.

The 2101 series is fabricated with p-channel silicon case technology, which allows the design of high-performance easy-to-use MOS circuits and provides high functional density on given monolithic chip.

Features

- Fully static
- No refresh operations, sense amps or clocks required
- All inputs and outputs are TTL compatible
- One 5-V power supply required

Block Diagram

FIGURE E11.2 A 256×4 static RAM by Signetics Corp.

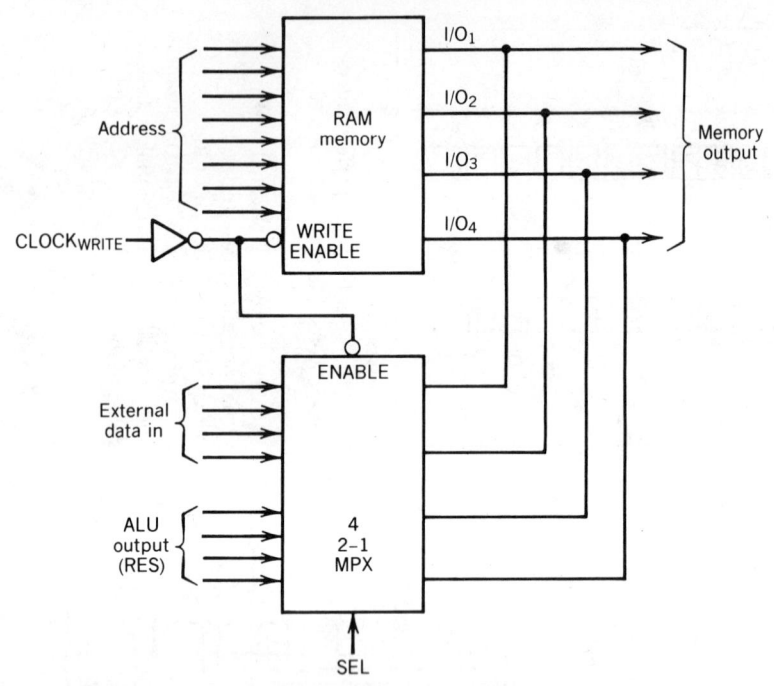

FIGURE E11.3 Hardware solution for RAM with shared input–output signals.

INCREMENT MAR instruction could be added to the instruction set. Finally, components for the control of the computer cycles have to be selected. These could include a counter and a decoder or simple flip-flops.

E11.3 Hardware Design

As mentioned in Chapter 11, basically, there are two ways to design the controlling circuitries of the computer. The simpler method is to design the computer cycle controller separately, which provides the correct state transition control between the fetch and the required number of execution cycles. The required clock signals may be expressed by the logical ORing of those states and instruction types for which the selected register is updated. From Table 11.1 the following Boolean equations are derived for updating the A register:

$$CLOCK_A = (EX\text{-}1) \cdot ((INPUT) + (COMPLA) + (INCA) + (ROTLEFT) + (ROTRIGHT))$$
$$+ (EX\text{-}2) \cdot ((LOADA) + (ADD) + (AND))$$

where the terms in parentheses refer to signals that indicate the computer cycle and the instruction type (from bits stored in the IR). The plus sign and centerdot signify logical OR and AND operations. Similarly, the updating of the PCL register is expressed by the following Boolean expression:

CLOCK$_{PCL}$ = (FETCH)

$\quad$ + (EX-1) · ((LOAD MARH) + (JUMP) + (BRANCH) + (SKIP · (C = 1)))

$\quad$ + (EX-2) · (SKIP' + (C = 1))

It is assumed that the selection lines control the function of the PC whether it is loaded, cleared, or incremented. The other clock signals can be similarly derived. The brute-force method of providing these clock signals is to use a 4-to-16 decoder for the 4 bits of the IR, which provides 16 individual signals with only one active signal for each instruction type, and to use AND and OR gates to provide the required CLOCK signals. Great simplifications can be achieved by grouping the instruction types together, as indicated in Chapter 11.

The design of the selection lines can also be completed on the basis of a table similar to Table 11.1. The necessary selection values for the instruction

TABLE E11.1 Values for Selection Lines of 4-Bit Computer

	Source of A	Source of address	Source of CARRY	ALU function	PC function
FETCH CYCLE	X	PC	X	X	INC
EXEC-1 CYCLE					
1 INPUT	EXT	X	X	X	X
2 OUTPUT	X	X	X	X	X
3 LOAD MARH	X	PC	X	X	INC
4 LOAD A	X	PC	X	X	INC
5 STORE A	X	PC	X	X	INC
6 ADD	X	PC	X	X	INC
7 AND	X	PC	X	X	INC
8 COMPL A	ALU	X	CARRY'	COMP	X
9 INC A	ALU	X	CY[ALU]	INC	X
10 ROT LEFT	ALU	X	A[3]	ROT/L	X
11 ROT RIGHT	ALU	X	X	ROT/R	X
12 JUMP	X	PC	X	X	LOAD
13 BRANCH	X	PC	X	X	LOAD
14 SKIP ON C	X	X	X	X	INC
EXEC-2 CYCLE					
4 LOAD A	ALU	MAR	X	Y[ALU]	X
5 STORE A	X	MAR	X	X	X
6 ADD	ALU	MAR	CY[ALU]	ADD	X
7 AND	ALU	MAR	0	AND	X
14 SKIP ON C	X	X	X	X	INC

set in Table 11.1 are shown in Table E11.1. The table shows the required source registers for multiplexers and the required functions for function selectors. The symbol X indicates that the values for the selection lines are don't care values. The particular binary values of these selection lines are assigned after the hardware devices are chosen and the instruction bits are encoded. Hardware savings can be made by finding convenient groups of instructions, especially since a large number of don't care cases are involved.

Thus, the hardware design that uses a simple computer cycle control is executed by the construction of the logic tables, the assignments of the instruction bits, and the design of the combinational circuits that provide the correct Boolean functions for the required signals. The design for the state model is based on the selection of specific states shared by many different types of instructions. The fetch cycle is an obvious state used by all instructions. As a demonstration of this method, let us select three states for the register transfer operations:

State E-1: MARL ← M[PC] ; PC ← PC plus 1.

State E-2: Update the A register.

State E-3: Update the carry flip-flop.

It is assumed that the selection lines provide the correct source when the A register is updated. The STD for the instructions that involve these states are shown in Figure E11.4. The numbers in the circles refer to the instruction numbers in Table 11.1. It is probable (but not necessarily obvious) that the state controller that produces these state transitions yields complex circits. The advantage is that the clock signals become very simple Boolean functions of these states.

E11.4 Requirements

The minimum requirement for this experiment is the detailed design of a 4-bit computer along the lines suggested by its description in Chapter 11. The two unassigned instructions should be selected and their testing included into the diagnostic program that tests all instructions. In addition to the diagnostic program, one more small program should be written in assembly language that performs some meaningful function. A list of suggested programs is given in the next section. The selection of the additional two instructions may be made by trying to minimize the length and execution time of this example program.

If sufficient time and hardware resources are available, then the computer should be constructed and tested. Especially if many students participate in the building of one computer, the integration and testing of this digital processor will provide valuable experience and insight into the difficulties digital designers face in practice.

It is highly recommended that the students go beyond the example shown in Chapter 11 and alter and/or expand the structure and instruction set of

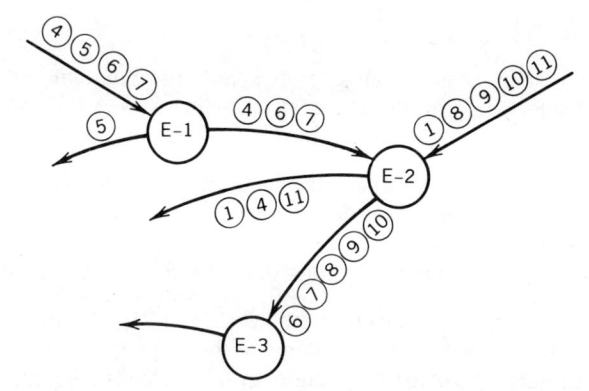

FIGURE E11.4 Control for execution of instructions for three selected unique system states.

the computer. Some suggestions are listed as drill problems after Experiment 11. In any case, a diagnostic program should be provided that tests all computer instructions. Facilities must be present for both loading computer programs into memory and for the starting (reset function) and step-by-step operation of the computer.

E11.5 Software Tasks

Write an assembly language program for your computer that executes one of the following data-processing functions:

1. Add (or subtract) two 8-bit unsigned binary numbers stored in memory and store the 8-bit result also in memory. The state of the carry bit should indicate the carry out of the MSB of the 8-bit addition process.
2. Multiply a 4-bit unsigned binary number stored in memory with external data supplied by the four input lines. (Assume that the input data is also given in unsigned binary form.) Store the 8 bits of the product back into memory.
3. Divide an unsigned 8-bit number stored in memory by an unsigned 4-bit binary number given by the external data input lines. Send the result to the output register. If overflow occurs, turn all output bits on (result equal to 1111).
4. Calculate the number of 1's appearing in the input data values and send this result to the output register. (Input data 1101, for example, should produce 0011 in the output register.)
5. Assume that the system clock operates at a rate of 1000 cycles per second. Write a program that "blinks" the bits of the output register (alternates between data values 0000 and 1111) at the rate of 10 cycles per second.
6. Compare two unsigned 8-bit numbers stored in memory and turn all the output bits on if the first one is larger than the second. Turn all output bits off if the reverse is true, and send the binary value 0101 to the output register if they are equal.

The preceding list of programs gives a good indication of the types of small programs that are instructive to code. Other programs that involve arrays of numbers could also be tried, but special instructions are needed to handle arrays with manageable-sized programs.

Drill Problems

The following problems suggest changes and expansions to the simple 4-bit computer described in Chapter 11. Consult Figure 11.18, which shows the data paths of the original design.

11.1. Add two more data-processing registers to the 4-bit computer. One is a second source register (*B*); the other is a "destination" register. Both are connected to the ALU (see Figure D11.1). Discuss the advantages of this computer over the original 4-bit computer as far as programming and instruction set are concerned.

11.2. Change the instruction set so that the LOAD and STORE instructions are one-word instructions and use the contents of the MAR (all 8 bits) as the address. Change the LOAD MARH instruction so that it loads all 8 bits of the MAR. Consider two additional instructions: one increments the MAR, and the other is a conditional skip instruction that tests the contents of the MARL and "skips" if the contents are equal to 0000. Discuss the advantages of this extended instruction set for array handling.

11.3. Implement the changes suggested in Drill Problem 11.2 and change all the data-processing instructions (including INPUT and OUTPUT) so that they all become register–memory instructions or just memory instructions (INC and COMPL) with the MAR holding the memory address. Examine how much additional hardware you need to implement these instructions. Discuss whether it is advantageous to store the results of the operations in the *A* register rather than in memory from both the hardware and software points of view. Design your instruction set with any mix of register–memory, register–register, or memory–memory instructions that would be most convenient for a programmer.

11.4. Extend the arithmetic-logic instructions (ADD, AND, INC, and COMPL) to contain two words. Use the additional 4 bits to specify the source and destination registers for the operations. Depending on the number of registers and the methods you use, there can be many different possibilities. For example, in the original model 2 bits may be used for the source and 2 for the destination registers and four addressable registers may be chosen from the six registers of the model (A, PCL, PCH, MARL, MARH, and IR). Or you may use 1 bit to select either the source or the destination register out of two registers (e.g., A

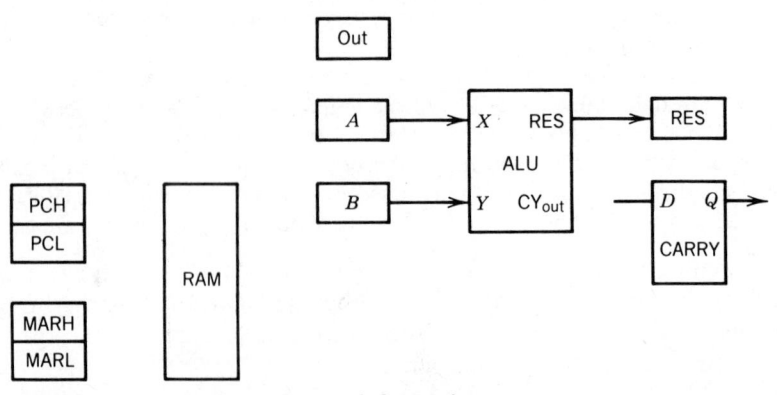

FIGURE D11.1 A 4-bit computer with four registers.

or *B*) and use the other 3 bits to select any one of eight registers for the other operand. Discuss the added programming conveniences introduced by this new class of instructions and the added hardware required to implement it.

11.5. Extend the testing capabilities of the computer by increasing the size of the SKIP instruction to two words. The additional 4 bits may be used to test four different conditions. Designate the extra 4 bits of this new SKIP instruction by C_0–C_3. Then, for example, the Boolean expression

$$\text{TEST} = C_3 \cdot (\text{CARRY} = 1) + C_2 \cdot (A = 0000) + C_1 \cdot (\text{OVERFLOW}) + C_0 \cdot (A < 0)$$

tests each condition (or a combination of conditions) when the condition bits C_i are specified by the instruction. Discuss the programming convenience supplied by this new instruction and the required hardware to implement it.

11.6. Discuss how "indirect" addressing could be introduced into such a small computer. One possible method is to decrease the addressing space to 128 words and use the MSB of the address as an indirect indication. The following sequence of register transfer instructions execute the indirect address calculation:

IND-1: MBR ← M[MAR] : IF ($MARH_3 = 1$)

MAR ← MAR plus 1 : IF ($MARH_3 = 1$)

IND-2: MARL ← M[MAR] : IF ($MARH_3 = 1$)

MARH ← MBR : IF ($MARH_3 = 1$)

IND-3: return to cycle IND-1: IF ($MARH_3 = 1$)

where one extra 4-bit register (MBR) was added to the original computer model. Discuss how these cycles could be implemented in hardware. Another possible way of introducing indirect addressing would be first to implement the changes suggested by Drill Problem 11.2 and then add an instruction that would unconditionally execute the first four register transfer operations of the preceding sequence.

11.7. Design a 4-bit computer with a different philosophy (and logic) than the one shown in Chapter 11. Add three more data-processing/address registers (registers *B*, *R*, and MBR) to the five contained in the original model (*A*, PCH, PCL, MARH, MARL) and design the computer around an ALU that includes the SHIFT, INCREMENT, DECREMENT, and ROTATE operations. Assume that the MEMORY READ and MEMORY WRITE operations involve the MBR register only and that the registers are constructed from simple flip-flops (not from shift registers or counters). The components of this computer are shown in Figure D11.2.

For simplicity's sake, assume that all computer instructions contain 8 bits (two words). The hardware includes a general switch so that it can perform the register transfer operations:

$$R_{\text{destination}} \leftarrow R_{\text{source}} \text{ FUNCTION } R_{\text{destination}}$$

or

$$R_{\text{destination}} \leftarrow \text{FUNCTION } R_{\text{destination}}$$

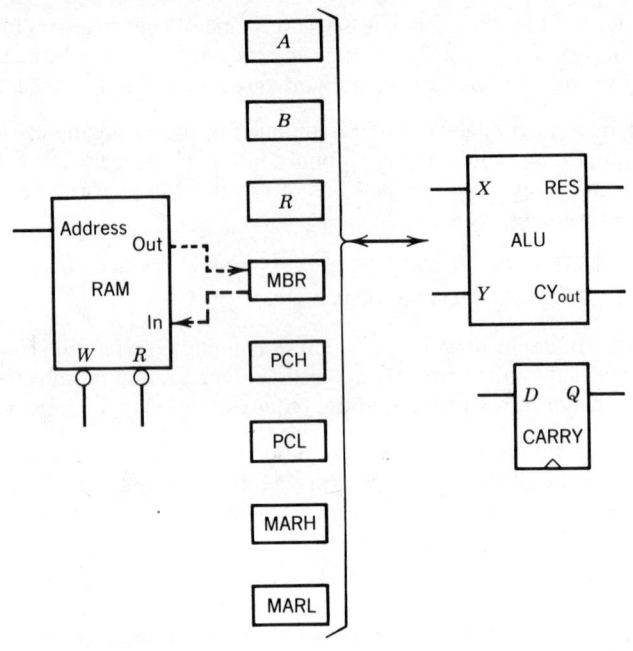

FIGURE D11.2 A 4-bit computer with eight registers and regular instruction set.

with 3 bits specifying both the source and destination registers. For example, in order to implement the fetch cycle, the following register transfer instructions are required:

> FETCH-1: IR ← M[PC]
>
> PCL ← PCL plus 1
>
> FETCH-2: PCH ← PCH plus 1 : IF (CARRY = 1)

Explore the required basic instruction set and consider extensions to it. Examine the types of instructions you may implement with such a "regular" hardware structure. Discuss the advantages and disadvantages of this computer from both the hardware and software points of view.

CHAPTER 12

Digital System Design

12.1 INTRODUCTION

In Chapter 1 we started with logic (binary) signals and gates, and in the following chapters we progressed through combinational and sequential circuit design until, in Chapter 11, we were ready to discuss the design problems associated with automatic digital computers. If we follow the history of the digital design discipline, we find the same natural progression. At the early stages the main preoccupation of digital designers was the minimization of circuits, that is, the search for a digital circuit that satisfied a set of requirements with the minimum number of gates. Today the majority of digital designers are concerned with the design of very large complex systems constructed from a relatively small number of well-defined (but still very complex) subsystems. Design at the gate level may be called "design in the small," whereas the integration of several subsystems into a well-organized whole may be called "design in the large." In the later chapters of this book we have been practicing "design in the middle," design with both gates and relatively simple digital units such as registers, multiplexers, demultiplexers, and so on.

It is our aim in this chapter to introduce some aspects of design in the large using practical digital building blocks such as ROMs, programmable logic arrays (PLAs), and universal logic modules as examples of the tools of this trade. At the same time the differences in the methods of design, whether they are used in the small or in the large, will also be emphasized. The two opposing methods, "design for the specific" and "design for the general," will also be discussed since they are very closely related to the problems of selecting the most appropriate design method for a given problem. We are going to base many of our discussions on the design examples covered in earlier chapters but will also use examples that will introduce a new concept, microprogramming.

Instead of constructing new experiments for this concluding chapter, we recommend that one or several of the earlier experiments be redesigned using the methods and philosophy of designing in the large. Suggestions as to how this may be done will be given under the heading of Experiment 12. It would be very instructional if the student could estimate and compare the amount of time spent in designing, constructing, and testing the circuits and the different costs associated with constructing either a small or a very large number of the same circuit using these two different design methods. Often

380 DIGITAL SYSTEM DESIGN

designing in the large is better suited for very complex systems or simple systems that are produced in very large quantities.

12.2 DESIGN IN THE LARGE AND IN THE SMALL

The development of digital design, like all other engineering disciplines, has been governed by changes in technology. In the early 1960s computers were still in operation that used vaccum tubes for their digital circuitry. One flip-flop could easily occupy a volume of 10 cm^3. By the early 1980s the gate density in VLSI devices reached a million flip-flops in 1 cm^3. This phenomenal progress of seven orders of magnitude in 20 years is unique to the area of digital electronics and had an overwhelming effect on its practicing designers. We assume here that 1 cm^3 is a convenient volume for a logically well-defined subsystem unit; therefore, the changes in technology have increased the number of digital gates per subunit by 10^7.

What are the major changes and problems that such an increase in the number of gates per module brings? A list of some of these follows:

1. Since a million gates may be packed into one module, the variety in the **types** of possible (even useful) modules is astronomical. An effective standardization takes place around a relatively few types of modules, but new modules will appear all the time (and old ones disappear). The designer has a very difficult time to get up-to-date information about all the latest modules since each type of module is an extremely complex device.
2. Since, ultimately, each module must be designed at the gate level, the design of VLSI modules becomes a highly complex and unique field. The role of the digital designer is to design systems with interconnecting modules; the role of the VLSI designer is to design modules. The separation of these two disciplines results in the development of two groups of professionals. The VLSI designers want to produce modules that are easy to to design, construct, and test, whereas the digital system designers want modules that are easy to connect together, work reliably in any system, and provide all the required system functions.
3. The system designer spends most of his or her time in studying the external specifications of complex modules. Testing and integration become major problems because any misunderstanding or erroneous information about a complex module will have to be corrected during the testing of the integrated system. This can be done successfully only if the designer understands the detailed operation of the module and not only its external specifications. This understanding can only come from knowledge and experience with designing in the small.
4. The digital system designer may not have the freedom to choose the best (most logical, most efficient, most reliable) design, since he or she has to rely on available system modules. This limitation will encourage designers to give up designing in the large selected parts of the system and to return to designing parts of the system in the small.
5. Because of the module size (1 cm^3), the number of external connections to a module is limited to around 100 (20–80 are typical). This introduces a new type of minimization problem: How can one minimize the number of external connections to

a module that performs very complex functions? Limits on the number of external connections forces the designer to "multiplex" signals, which slows down the operational speed of the module and complicates its interconnection to other modules. This also forces the system designer to understand the detailed, internal structure and operation of the module.

The preceding points reinforce the fact that although most of the time the modern system designer connects digital modules together, he or she must have a thorough understanding of the basic operation of digital gates and small digital modules as well as a thorough knowledge of the currently available digital subsystem elements.

We will demonstrate some of the techniques of designing in the large with the example of a small digital computer. The first important realization is that all good systematic design methods are **hierarchical.** At the lowest level there are the gates and flip-flops. Gates are organized into combinational building blocks: multiplexers, decoders, demultiplexers, comparators, and ALUs; flip-flops are organized into registers, counters, shift registers, and so on. At the next level there are memory subsystems, data-processing subsystems, input–output subsystems, controller subsystems, and so on. At the next level of hierarchy complete microprocessors, memory systems, other type of complex processors (e.g., graphics processors), and controllers may be connected together.

Designers have argued long about the correct way of designing systems. Should it be top down (from top to bottom) or bottom up? If we examine the example of the digital computer we designed in Chapter 11, what do we find? What type of design style did we use? The method we used, which is used most often in practice, is in fact neither one of the two methods mentioned. We used design by example. This involves the description of a representative model or prototype that demonstrates the fundamental features of the system. In addition to the description of the example system, an explanation of how various features of the system may be changed must also be given. In this way the designer learns about the principal features of the system as well as how to tailor the "general," or prototype, system to a specific application.

Designing in the large suggests a top-down approach. Starting at the very top is simple if one can specify what one wants. In the case of designing a small computer, we could state (the obvious):

1. I want to design the best small digital computer.

From the previous (obvious) statement one follows successive "refinement" steps that should lead to the correct system structure, the correct division into subsystems, after which each subsystem can be designed in a similar top-down manner.

Let us take a few steps of refinement. Obviously, in order to specify the product in more detail, we have to know how to specify it in general terms.

We know that two of the most important choices we have to make are the smallest word size of the computer and whether it is controlled by an external or a stored program. We choose a 4-bit automatic computer:

> 2. I want to design the best 4-bit automatic (stored-program) computer.

We may go one step further in general terms, since we know that the addressable memory space is another important feature of the computer. We choose two hundred fifty-six 4-bit words and state:

> 3. I want to design the best 4-bit automatic digital computer with a memory of two hundred fifty-six 4-bit words.

At this point it is very difficult to proceed further. The next step of refinement would be the specification of the types of computer instructions we would want that, in turn, could specify the number of data-processing, index, and/or general registers required and could functionally specify the ALU. At this point we would be ready to divide the computer into subsystem elements.

The specification of the instruction set is a very difficult problem. Experienced programmers will disagree on the type and number of instructions the computer should have. Historically, programmers wanted more and more, and the number of instructions increased rapidly with decreasing hardware costs. Today, there is a reverse trend with the introduction of RISC (reduced instruction set computers) type of processors. The main problem is that computers are designed for the "general" rather than a "specific" use. It would be relatively easy to specify the instruction set of a data processor that is used only for a specific task (such as the sorting computer in Chapter 10), but it requires a large amount of arbitrary choices (and guessing) to specify the instruction set of a general data processor.

We have to abandon now the strict top-down approach and continue the refinement of our design by choosing hardware (subsystem) components first. Since we are not certain about the instruction set, we would like to design the computer hardware independently from the instruction set. This is possible with the use of a ROM (read-only memory). We demonstrated in Chapter 11 that the signals within the processor fall into three groups: state control, selection, and clock signals; a design based on a general ROM can be constructed from the three subsystems shown in Figure 12.1.

The *System state* block consists of a collection of flip-flops that store the system state. The *Processor* block includes all the processing registers of the computer, its memory subsystem, the ALU, the input–output registers, and the data buses that establish the data paths and thus define the instruction set. The ROM receives its inputs (address) from the stored system state, the IR (the current instruction being executed), and special conditions generated by the processor (CARRY = 1, or A = 0000, etc.). The outputs of the ROM control all aspects of the processor. The next-state outputs select the "next" system state; the selection outputs control the multiplexers and function selectors of the processor; whereas signals ANDed with the system clock produce the clock signals that execute the required register transfer operations.

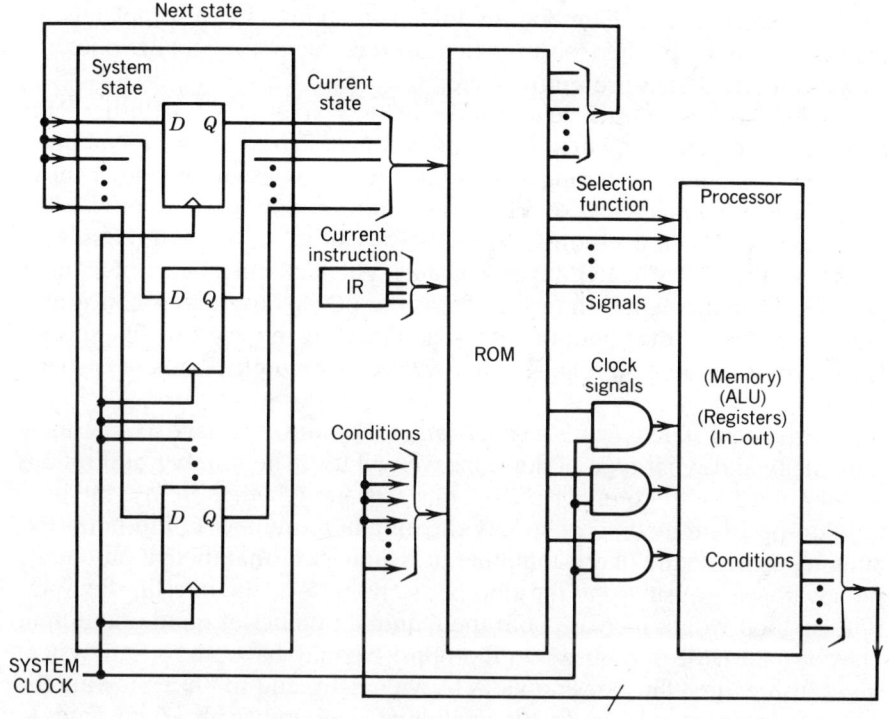

FIGURE 12.1 Block diagram of computer using ROM.

This model is a typical example of designing in the large since, by using a ROM to generate all the signals, there is no need to use individual logic gates at all. Since the cost of the ROM depends on its size (number of inputs and number of outputs), the factors that influence the overall cost of the hardware are

1. the number of system states that determine the number of flip-flops required, the number of ROM inputs used for the stored current state, and the number of ROM outputs needed to specify the next state of the computer;
2. the number of registers, type of ALU, number of data buses, and so on, that determine the cost of the processor subsystem and the number of ROM outputs used for both selection and clock signals; and
3. the variety of testable conditions that determines the number of ROM inputs assigned to "conditions."

Beyond these factors, the number and types of instructions provided (the details of the instruction set) does not influence the cost or the complexity of the hardware. Any instruction set may be specified, and the truth table for the ROM would follow. The price of the ROM depends only on the number of inputs, number of outputs, speed (maximum delay time), and type of construction, but it does not depend on its truth table. This example shows very clearly the advantage of designing in the large for the general case. The consideration of special cases for "useful" instructions, or easily implemented instructions (which often occur when we design in the small), does not complicate the design procedures here.

Before we discuss the details of designing with ROMs (and other large and general system elements), let us follow the design of our 4-bit computer example of Chapter 11 through its completion. As we shall see, other interesting facts will emerge as we follow our top-to-bottom design route.

The next design decisions involve the specifications of the three subsystems in Figure 12.1. The ROM is an available system module; hence, it requires no further "refinement." In fact, as far as the ROM subsystem is concerned, we have arrived at the "bottom" of our design. Only the selection of a specific hardware device will be required that will be made on the basis of cost and speed.

The system state block is a collection of flip-flops. We have to determine how many and what type of flip-flops we will use. The number of flip-flops is determined by the number of system states and whether we use the one-flip-flop-per-state method or specify the minimum number of flip-flops. Examining the structure of the computer, it is quite clear that there is absolutely no advantage in using one flip-flop per state; in fact, the one-flip-flop-per-state method would increase both the required number of input and output lines of the ROM. If we use two flip-flops, we may have up to four system states; using three flip-flops provides us with eight; and so on. Following the example in Chapter 11, at first we will limit the number of system states to 4 and specify two flip-flops for the system state subsystem. These states may be divided into two fetch and two execute or one fetch and three execute states. The choice of the number and types of system state becomes another variable that can be selected when a particular instruction set is designed without affecting the design of the hardware since the ROM contents will determine the specific state control functions.

Now we have to choose the type of flip-flops. It is interesting to note that when we designed circuits in the small, the selection of JK flip-flops often produced much simpler control circuits than those designed with D-type flip-flops. In this case it would be foolish to choose JK flip-flops since they would double the number of inputs to the system state block and increase the number of outputs of the ROM without any benefit to the hardware. Since only the number of inputs of the state system block is of concern, T-type (toggle) flip-flops could be used just as well as D-type flip-flops.

This concludes the "bottom" design of the state system block. The selection of the hardware components will be influenced by cost, availability, speed, and so on. Our next task is to specify the processor subsystem, at least as far as the number of inputs and outputs are concerned. We will use our experience in designing a 4-bit computer in Chapter 11 to arrive at a satisfactory model.

We use the 4-bit computer example in Chapter 11 as we select the components and the data paths of the processor. The main elements of the processor block are memory, the *A* register, the stored carry bit (CARRY), two address registers (PC and MAR), and the ALU. We have seen that we need three multiplexers: one to select between external and internal input data, one to select between the two addresses stored in the PC and MAR, and one that determines the input to the carry flip-flop. Function selection signals are needed for the ALU (we select three for eight possible functions) and for the INCREMENT/LOAD functions of the PC (two signals are sufficient).

With the components of the processor block specified, we may still find several ways to connect them; therefore, we have to rely on experience or else experiment with different data paths to arrive at an acceptable model. One such model was shown in Figure 11.18 which we adopt here. Our aim is to provide maximum flexibility (so that different ROM contents could provide a wide variety of different instruction sets) with a minimum number of required control signals (so that cost can be minimized by limiting the size of the ROM). Therefore, in Figure 12.2 we show only the components of the processor without their interconnections. The selection of specific interconnections can be left to the designer.

Requiring only one testable condition (CARRY = 1), the number of

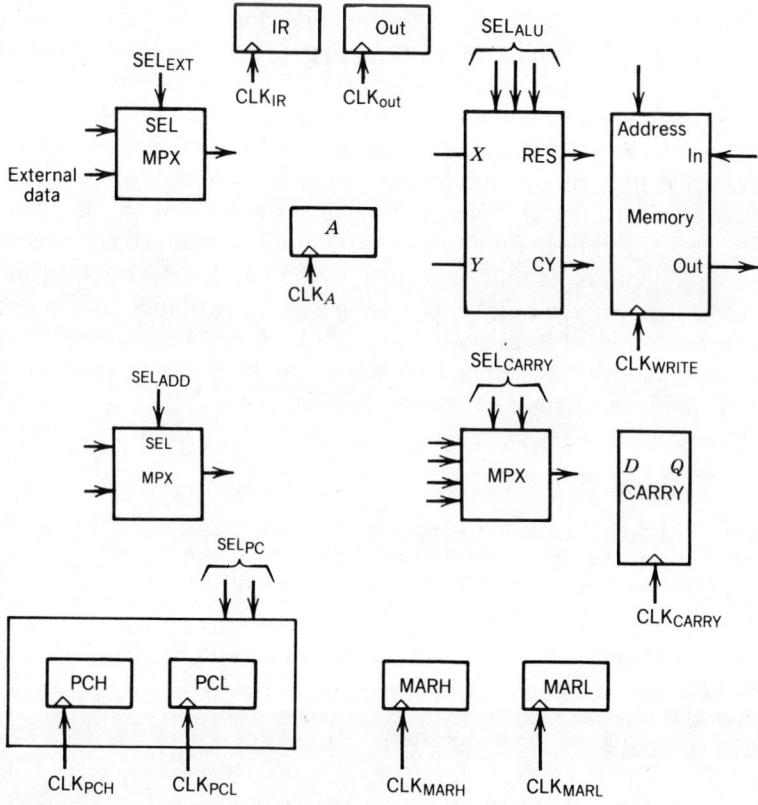

FIGURE 12.2 Components of 4-bit computer using ROM.

inputs for the ROM are specified. We have seven inputs (four from the IR, two from the state flip-flops, and one is the CARRY = 1 signal). Choosing a practical ROM, one realizes that most have been standardized to 6-, 8-, 9-, ... bit addresses (inputs) and 4 or 8 bits of output. Thus, we select 8 bits for input and note that the processor block has to be specified only so far as setting the required number of output bits for the ROM.

The required select and clock signals are summarized in Table 12.1. According to this list, we require two 256 × 8 and one 256 × 4 ROMs. These provide 20 output and 8 input signals. We have one spare input and two spare output signals that, obviously, could become very useful if we overlooked some required functions or would like to expand the capabilities of the computer for special applications.

Before we conclude this exercise in design in the large, let us examine our initial model shown in Figure 12.1 for possible reductions in the size of the ROM. We can see that this design uses AND gates for the generation of clock signals (a remnant of design in the small). In previous design examples we found that a demultiplexer can be used effectively for generating clock signals. Substituting a demultiplexer for the clock signals reduces the number of required ROM outputs to 4, with which 16 different clock signals may be generated. However, we have changed the fundamental operation of the processor block because, in this case, only one clock signal can be generated for one system clock period (only one of the 16 outputs of the demultiplexer may be selected). In our original design we tried to keep the number of system states as small as possible by executing multiple register transfer operations during one clock cycle. This resulted in an efficient (fast) processor. Here, we are interested in a general model that could provide a number of different processors, and the use of the demultiplexer provides us with a large number of clock signals with fewer number of ROM outputs (a total of 16 is required). Since the generation of only one clock signal per system clock period will necessarily increase the number of system states, the number of flip-flops for the system state block is increased to 3, and the number of possible system states is increased to 8. Figure 12.3 shows the final design with which the 4-bit computer of Chapter 11 may be constructed.

Because we have designed a general computer, the same hardware may be used for a variety of instruction sets, processor configurations, specialized

TABLE 12.1 Required Select Function and Clock Signals

Select signals	Number	Clock signals	Number
Select external data	1	Update A register	1
Select carry input	2	Update CARRY register	1
Select address	1	WRITE memory	1
Select ALU function	3	Update MAR	2
Select PC function	2	Update PC	2
		Update OUT register	1
		Update IR	1
Total signals	9		9

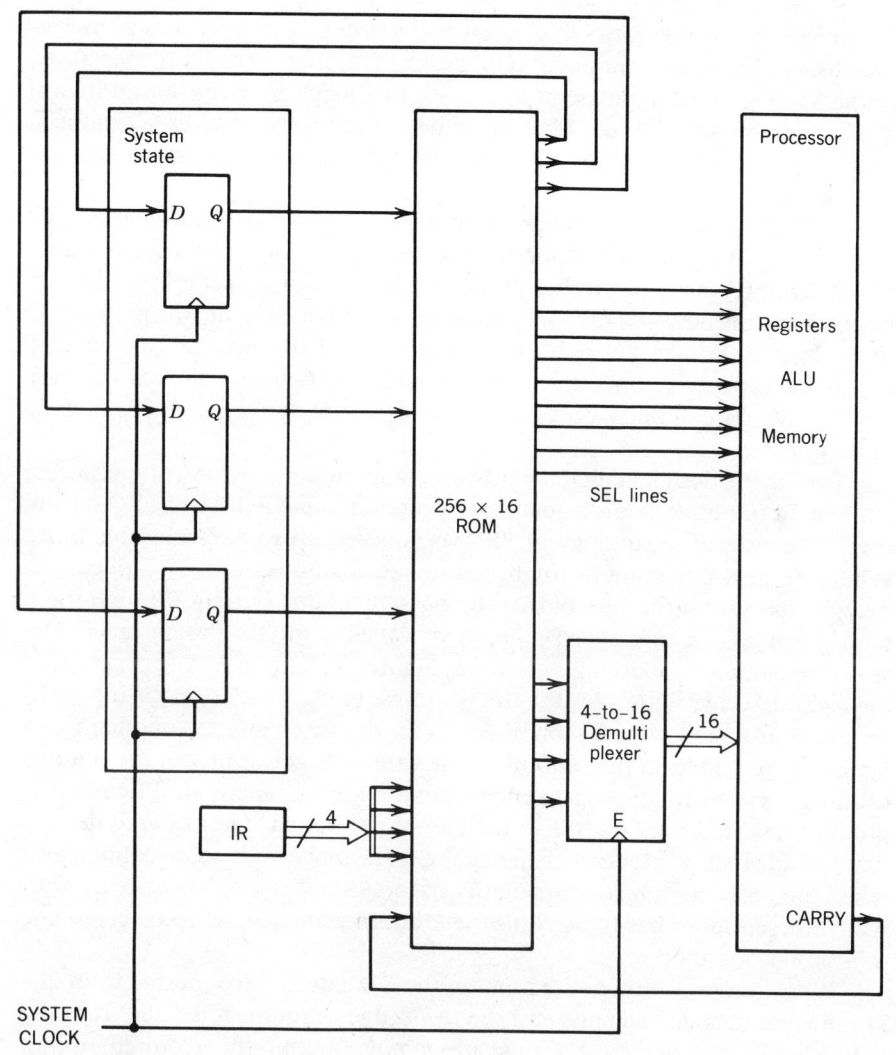

FIGURE 12.3 Final design of 4-bit computer with 256 × 16 ROM.

processors, and so on. It is also clear that the model can be easily expanded by providing more flip-flops for the system state block, more input and output (condition) signals for the processor block, and a larger demultiplexer (more clock signals), all of which would only increase the size of the ROM but would not alter the basic structure of the design. We have achieved a "generic" computer design with the approach of design in the large.

12.3 ROMS AND PROGRAMMABLE LOGIC MODULES

In the last section we showed how a ROM can be used to design a generic computer without specifying the details of its instruction set. Now, we will discuss how the ROM contents may be specified in practice. The technology of semiconductor memories in general and ROMs in particular has been well established. There are four basic ROM types. First, there are highly specialized ROMs such as character generators for video displays or drivers for numerical displays manufactured just like any other ordinary IC devices. Obviously, these ROMs can be used only for the particular applications for which they were designed.

Second, IC manufacturers will provide ROMs with customer-defined truth tables. The customer-defined ROMs are also manufactured like ordinary ICs, but the IC mask has to be specifically designed and made for each customer. The time between the specification of the truth table and the production of the first ROM device may be quite long, and the process can be very expensive. However, once the mask is made, subsequent devices are very inexpensive. The application of these ROMs are in systems that are already in production.

During the design of digital hardware, or in the construction of specialized equipment not built in large quantities, programmable ROMs, or PROMs, are used. The current technology of PROMs is based on so-called fusable links. When the device is manufactured, a number of connections is supplied that may be "broken" (the fuse blown) by passing a large current through them with the use of a specialized equipment called a PROM programmer. The structure of an 8×3 PROM is shown in Figure 12.4. Eight AND gates provide the eight decoded signals for the three address lines. All three eight-input OR gates are provided with fusable links. With the fuses intact (conducting), a Logic-1 is provided to the outputs of the three OR gates; hence, the outputs of this PROM in its unprogrammed state are all 1's. There are PROMs that produce inverted outputs (0's in their unprogrammed state) as well devices that provide logical connections when the fuses are blown and no connections when the fuses are intact. The polarity of the PROM is a practical detail with which the designer has to be concerned, but in principle, all PROMs operate in a similar manner.

When a fuse is blown, the input of the OR gate is disconnected from one of the eight decoded signals, and the respective minterm is deleted from the OR expression. Successive terms are removed until the required output expression is achieved. A standard shorthand notation (in terms of logic diagrams) has been developed for this type of AND–OR structure, and both the unprogrammed and programmed versions are shown in Figure 12.5. The unprogrammed device shows all connections (crosses where lines meet) intact; for the programmed PROM some of the connections have been removed.

A particular ROM output is selected by a number of intact fuses. Each single-input line of an OR gate represents the eight inputs shown in Figure 12.4. The single-input line shown for each AND gate represents three inputs in this case. A small cross indicates logical connection (a term in the Boolean expression). A logical product term, like $A \cdot B'$ requires two logical connections for the OR input, since both terms $A \cdot B' \cdot C$ and $A \cdot B' \cdot C'$ contribute to the output. The connections to the AND gates are fixed. Once the PROM is programmed, it behaves like any other ROM. The advantage of a PROM is

12.3 ROMS AND PROGRAMMABLE LOGIC MODULES

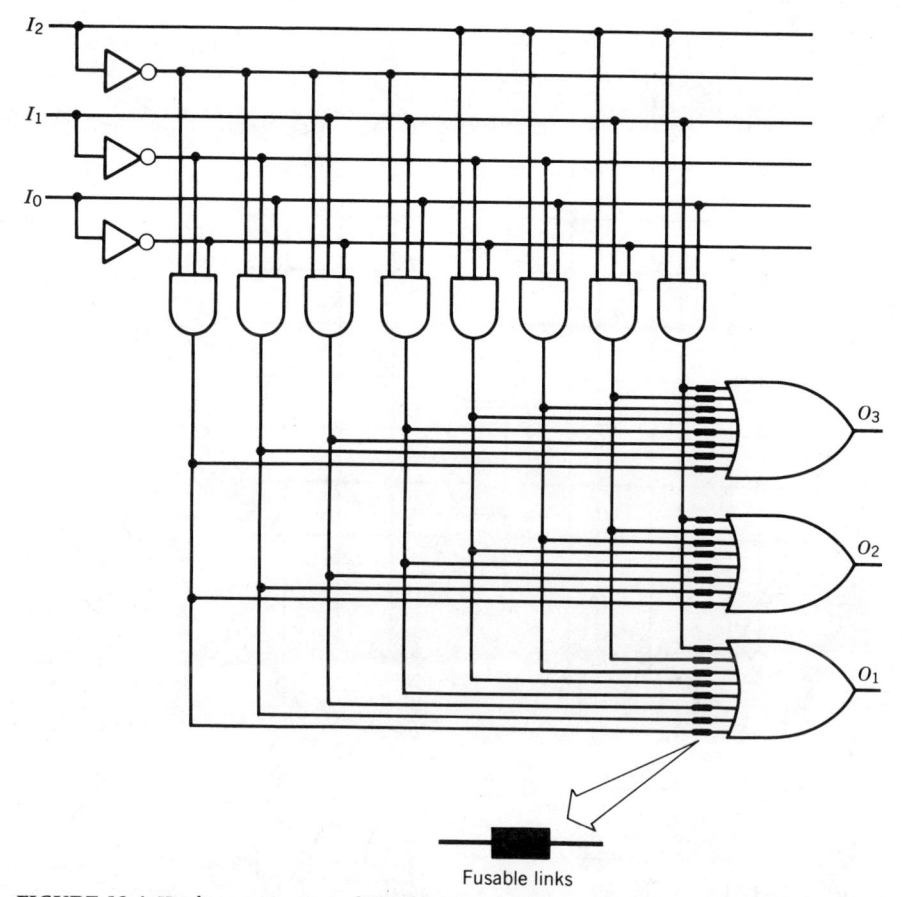

FIGURE 12.4 Hardware structure of PROM.

that its contents can be selected by the designer of the system and changed for different versions of the hardware. Its disadvantage is the higher cost when large quantities are used and the time the system builder must spend in programming (and testing) each PROM device.

The fourth type of ROM is the erasable PROM, or EPROM. EPROMs do not differ logically from PROMs. They are built with a different technology that allows the reconnecting of blown links either by radiation of ultraviolet light or by an electrical current (electrically erasable/programmable ROMs, or EEPROMs). They are very effective during the development of systems when the contents of the ROMs often change. They are more expensive than PROMs and less reliable for permanent use. In many systems EEPROMs are used for storing system variables that can be set by the computer but are set very rarely. Most EEPROMs have a limited lifetime; that is, only a number of memory updates may be made even though the number may be in the hundred thousands (nevertheless, they cannot be used as RAMs).

Let us return now to the design aspects of using PROMs in digital systems.

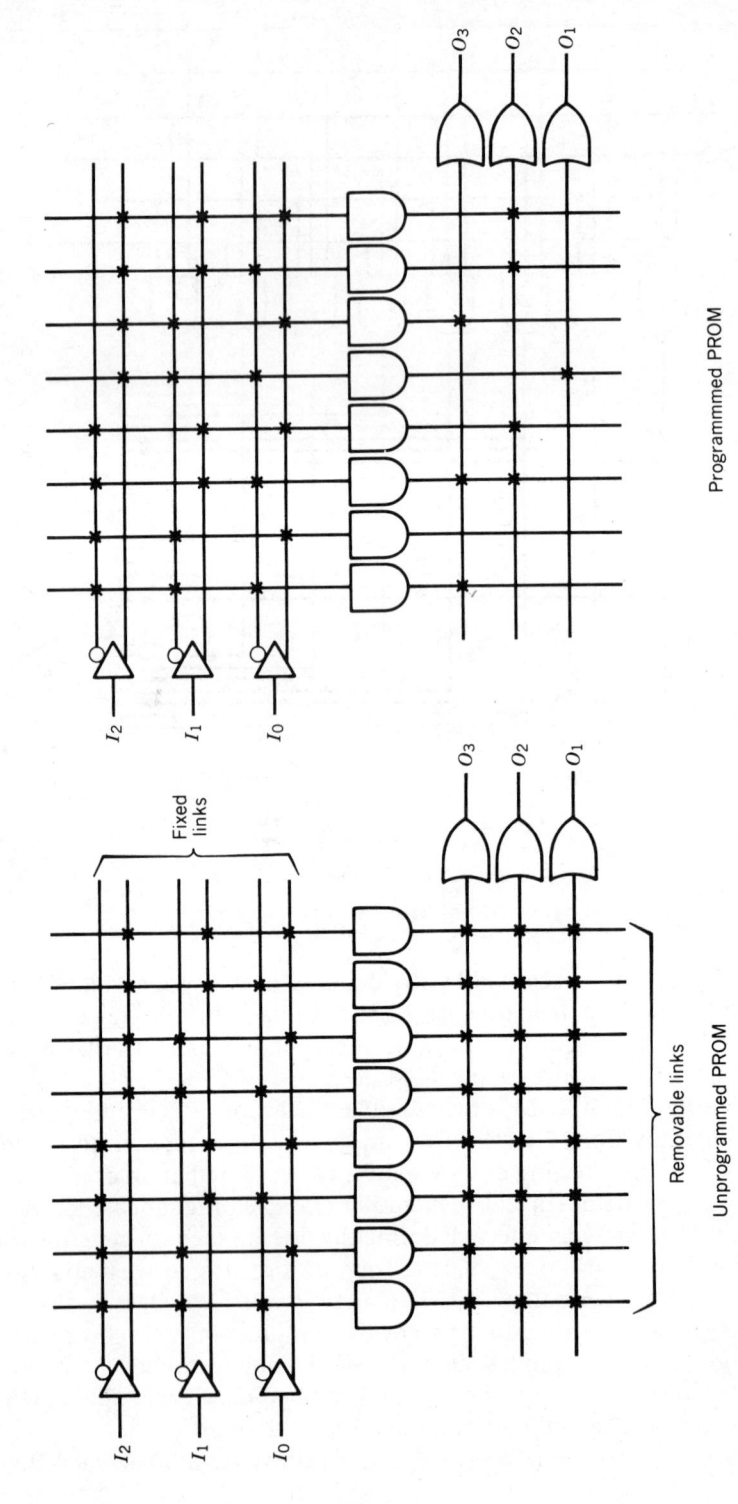

FIGURE 12.5 Simplified graphic diagram for PROM in its unprogrammed and programmed states.

12.3 ROMS AND PROGRAMMABLE LOGIC MODULES   391

As shown in Figure 12.5, an unprogrammed PROM is shown by having fixed connections for its AND gate and removable connections for its OR gate. Expanding a hardware design involves increasing the number of inputs and outputs of the ROM. We can see that the number of fusable links is equal to $M \times 2^N$ for N inputs and M outputs. Hence, the number of links of the PROM becomes very large when the number of inputs increases.

Expressing the required control signals of our 4-bit computer as Boolean expressions, we find that many signals require only a few terms, which very rarely include all inputs. For example, the O_2 output of the PROM in Figure 12.5 is equal to the logical ORing of four product terms ($I_2'I_1I_0' + I_2'I_1I_0' + I_2I_1I_0' + I_2I_1I_0$). The combined logic function is equal to I_2 which requires no logic gates at all. This indicates four three-input AND gates and one eight-input OR gate (the other four inputs are at Logic-0). This extreme example shows that the ROM, while being very general, is at the same time very wasteful, especially when the number of inputs becomes large and the output signals can be expressed as the logical sum of a few product terms.

The VLSI logic devices called PLAs (programmed logic arrays), PAL (gate arrays, or programmed array logic), and general logic modules were developed for logic functions that can be expressed with a limited number of logical sum expressions. Both mask-programmed (customer-specified) and field programmable logic arrays (sometimes called FPLAs) are available in practice. The shorthand graphical notation used in Figure 12.5 is convenient to demonstrate the various types of programmable logic devices. A very small version of the most general PLA is shown in Figure 12.6. The device has six inputs, three outputs, and a maximum of eight product terms. It is referred to as a $6 \times 8 \times 3$ programmable logic array. A maximum of eight different product terms may be generated, and each output can be expressed as a sum of any number of these product terms. There are $2 \times 6 \times 8 = 96$ fusable links to program the product terms and $3 \times 8 = 24$ fusable links for the logical sums (a total of 120 links).

A more typical PLA in practice would have 16 inputs, 48 product terms, and 8 outputs (size $16 \times 48 \times 8$) and would have $2 \times 16 \times 48 + 8 \times 48 = 1920$ fusable links. For the same logic function a PROM would have to have 16 inputs and 8 outputs, which would include 8×2^{16}, or approximately 500,000, fusable links (a factor of 250 to 1). Thus, if the number of product terms is limited, a considerable savings in size, cost, and complexity may be achieved by using PLAs.

It can be seen in Figure 12.6 that, even though any one of the product terms may be used by any one of the outputs, they are all shared by all the outputs. The structure of a simplified version of the PLA is shown in Figure 12.7. The fusable links for the AND–OR gates are removed, and a fixed number of inputs are provided for each output. This PLA has 6 inputs and $3 \times 4 = 12$ possible product terms, but each output can be expressed only as a logical sum of up to 4 product terms. The 4 product terms are completely arbitrary and are not shared between outputs. The number of fusable links are equal to $3 \times 2 \times 6 \times 4 = 144$ in this case. A practical 20-pin logic

392 DIGITAL SYSTEM DESIGN

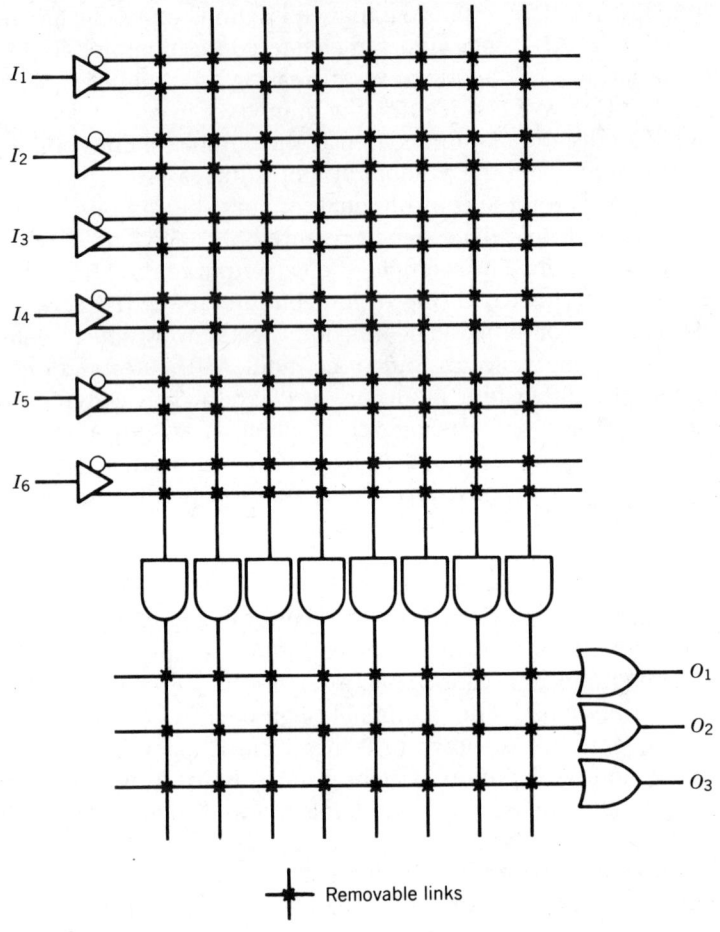

FIGURE 12.6 Simplified graphic diagram of general 6 × 8 × 3 PLA.

device (Monolithic Memories type 16L2) is shown in Figure 12.8 that contains 16 inputs and 2 outputs with 8 possible logical sum terms for each output. Similar 20-pin devices can be used to provide 4 outputs with 14 inputs or 6 outputs with 12 inputs.

Clearly, it is not our aim to give an exhaustive list of available PLA devices. We have chosen these examples to demonstrate how the tools of designing in the large are being developed at present. ROMs and PLAs are general Boolean function generators. Theoretically, only ROMs are needed, but when practical limitations arise, new devices are developed that are better suited for specialized applications. The number and types of programmable logic devices are growing, and soon the designer in the large will have the same problem of selecting the most suitable IC device for his or her application as the designer in the small had in choosing appropriate gates.

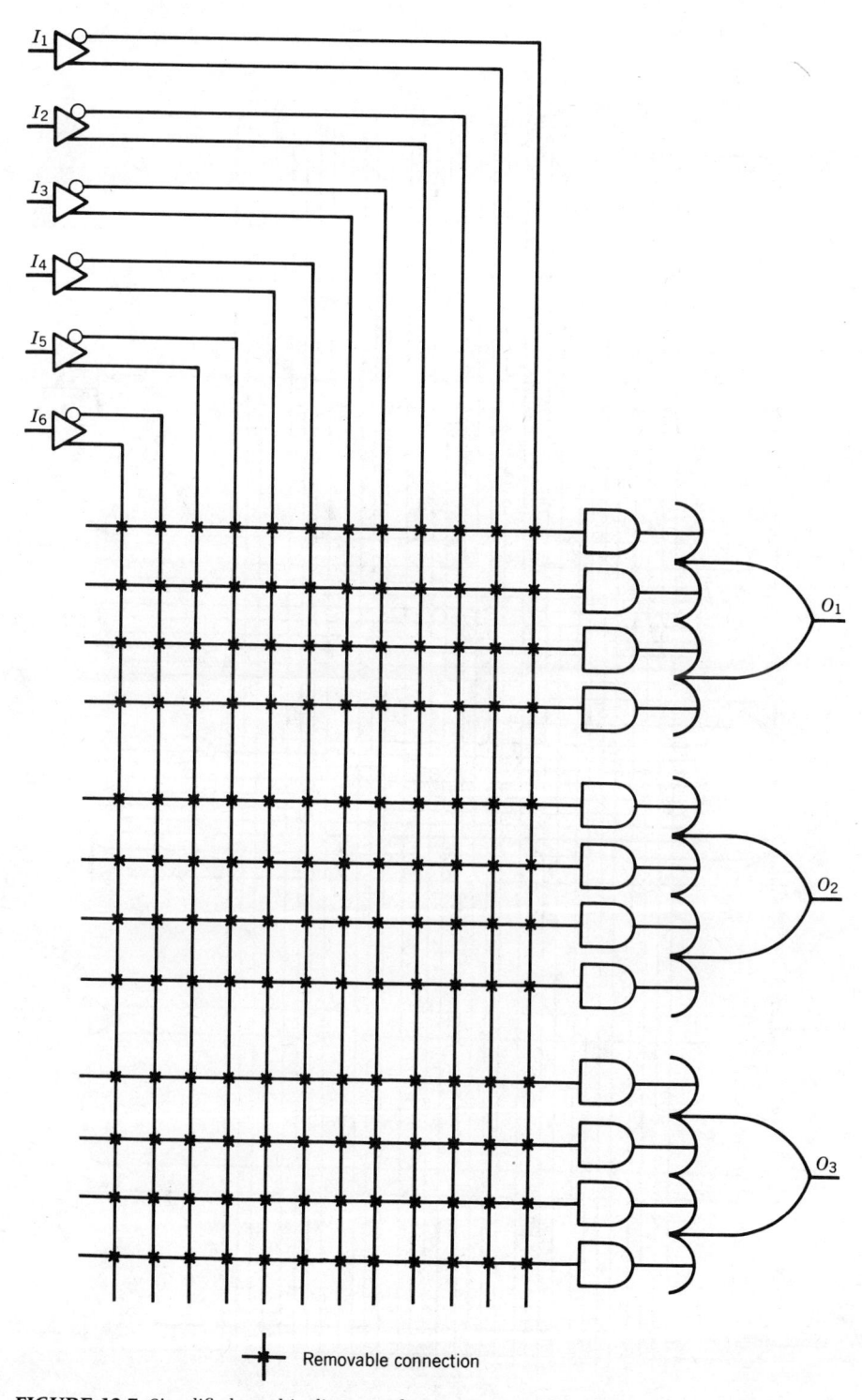

FIGURE 12.7 Simplified graphic diagram of unprogrammed fixed AND PLA.

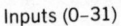

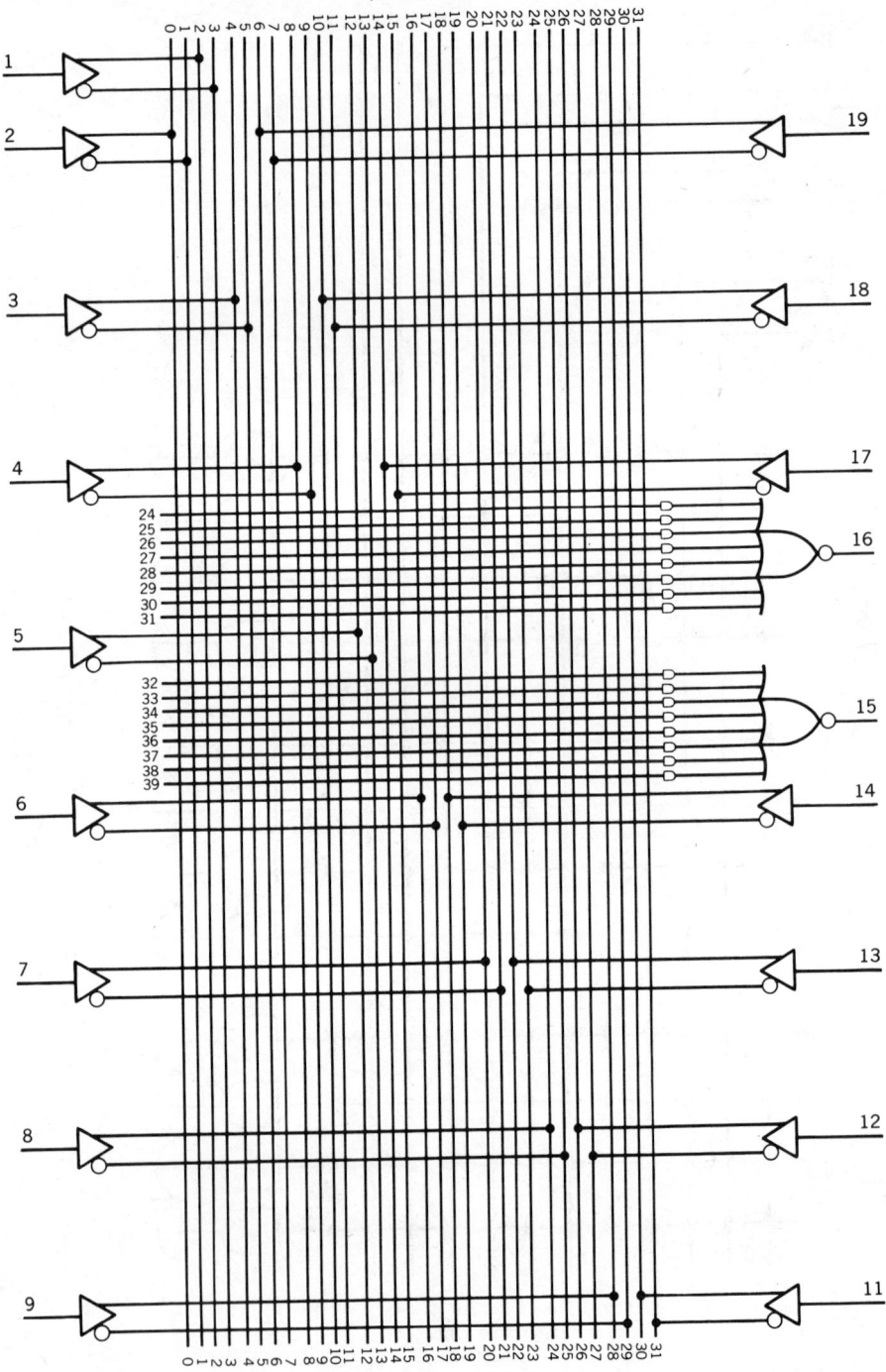

FIGURE 12.8 Monolithic Memories PLA type 16L2 device.

12.4 DESIGN EXAMPLES WITH ROMS AND PLAS

We have described how one may use ROMs and PLAs for the design of a computer without specifying the detailed instruction set of the computer. It is important to carry the design to completion in order to see the possible complications that may arise when we design digital systems in the large. We also want to demonstrate the practical problems one encounters in designing digital systems with ROMs and PLAs. For these design examples we will use the instruction set suggested in Chapter 11, shown in Table 11.1, and the data path diagram of the processor block shown in Figure 11.18. We will not show all the details but will present enough examples to indicate how the design may be carried out to completion.

First we will use a ROM. We have found that the smallest ROM we can use has 8 input and 16 output lines. Thus, we have to specify $2^8 = 256$ 16-bit binary numbers. This could be very tedious, but the organized structure of the processor allows the specification of a large number of outputs at the same time, and the large number of don't care outputs do not have to be specified at all.

In order to specify the ROM inputs as 1's and 0's, binary values must be assigned to the system states and to the instructions (the bits stored in the IR). These assignment are quite arbitrary, but if we list the instructions in order of their required number of states, the ROM assignments will become easier. The list of the 14 instructions along with the assignments of 1's and 0's are shown in Table 12.2. There are five states because only one register transfer operation may be executed during one clock period (we are using a demultiplexer for the clock signals). These are also shown in Table 12.2.

The ROM outputs must be described in terms of 1's and 0's; therefore, we have to encode the demultiplexer inputs (which select the clock signals), the function bits for the ALU and the PC, and the multiplexer select inputs. We will assume arbitrary choices for these binary values, and in most cases the actual bit values do not affect the circuits at all. In the case of the ALU or PC functions (LOAD, INCREMENT, etc.), the selected hardware may suggest the simplest bit combinations. In Table 12.3 we show these assignments.

Examining the register transfer operations in Table 12.2, we notice that for some instructions only the PCL or the PCH register is updated, whereas for the increment operation both registers may have to be updated. Since the number of available clock pulses (16) is larger than the required number of single clock pulses, we may assign an extra signal for updating both PC registers. The control of the two registers PCL and PCH are executed by three clock signals, and we have to translate these three signals into the two clock signals that operate the registers. The required circuit for this translation of three-to-two signals is shown in Figure 12.9. Only two additional OR gates are needed to provide either single- or double-register clock signals for the PC register. We used the Karnaugh map in Figure 12.9 as a demonstration that in this case we had to return to our design in the small to find a simple circuit. A similar clock signal is needed to update both the carry flip-flop and

TABLE 12.2 Register Transfer Operations for ROM Controlled 4-Bit Computer

		System states	
	Binary:	000	001
		FETCH-1	FETCH-2
(For all instructions)		IR ← M[PC]	PC ← PC plus 1

		System states		
		010	011	100
Code	Instruction	EXEC-1	EXEC-2	EXEC-3
0000	LDA	MARL ← M[PC]	PC ← PC plus 1	A ← M[MAR]
0001	STA	MARL ← M[PC]	PC ← PC plus 1	M[MAR] ← A
0010	ADD	MARL ← M[PC]	PC ← PC plus 1	A ← M[MAR] plus A
0011	AND	MARL ← M[PC]	PC ← PC plus 1	A ← M[MAR] AND A
0100	LOAD MARH	MARH ← M[PC]	PC ← PC plus 1	X
0101	JUMP	PCH ← M[PC]	PCL ← 0000	X
0110	SKIP	PC ← PC plus 1:CY	PC ← PC plus 1:CY	X
0111	BRANCH	PCL ← M[PC]	X	X
1000	INPUT	A ← EXT. INP.	X	X
1001	OUTPUT	OUT ← A	X	X
1010	INC A	A ← A plus 1, (CY)	X	X
1011	COMPL A	A ← A' (CY ← 1)	X	X
1100	ROT LEFT	[A,CY] ← ROTL[A,CY]	X	X
1101	ROT RIGHT	A ← ROTR[A]	X	X

TABLE 12.3 Encoding of Bits for Clock Signals and Select/Function Signals

Clock signals		ALU functions		PC functions	
Code	Clock signal	Code	Function	Code	Function
0000	IR	000	X plus Y	00	None
0001	OUT	001	X AND Y	01	INCREMENT
0010	PCL	010	X plus 1	10	LOAD
0011	PCH	011	X' (COMPLEMENT)	11	CLEAR
0100	MARL	100	ROT.LFT(X,CY)		
0101	MARH	101	ROT.RGHT(X)	Carry inputs	
0110	A	110	Y	Code	Input
0111	CARRY	111	Spare	00	CY$_{out}$
1000	MEM(WRITE)			01	A$_3$
1001	PCL and PCH	Address select		10	0
1010	A and CARRY	Code	Address	11	1
...				External select	
1111	NOP (none)	0	PC	Code	Input
		1	MAR	0	ALU output
				1	External inp

396

12.4 DESIGN EXAMPLES WITH ROMS AND PLAS

$CLOCK_1$	$CLOCK_2$	$CLOCK_3$	$CLOCK_{PCL}$	$CLOCK_{PCH}$
0	0	0	0	0
0	0	1	1	1
0	1	0	0	1
0	1	1	X	X
1	0	0	1	0
1	0	1	X	X
1	1	0	X	X
1	1	1	X	X

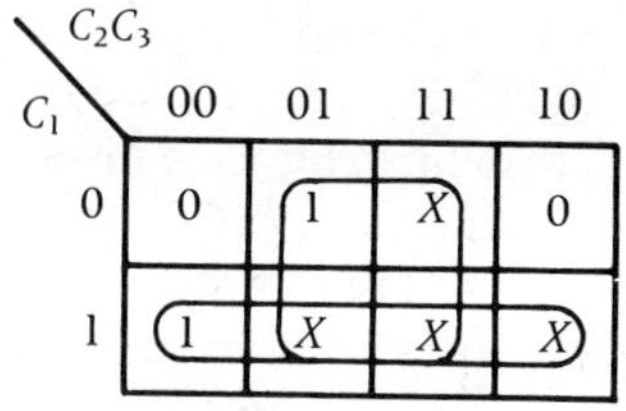

$$CLOCK_{PCL} = CLOCK_1 + CLOCK_3$$
$$CLOCK_{PCH} = CLOCK_2 + CLOCK_3$$

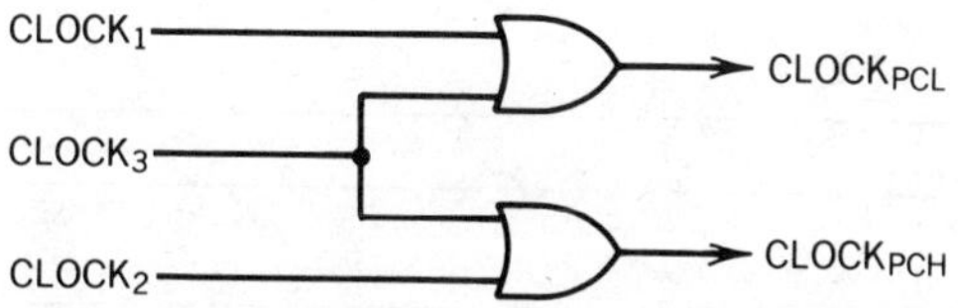

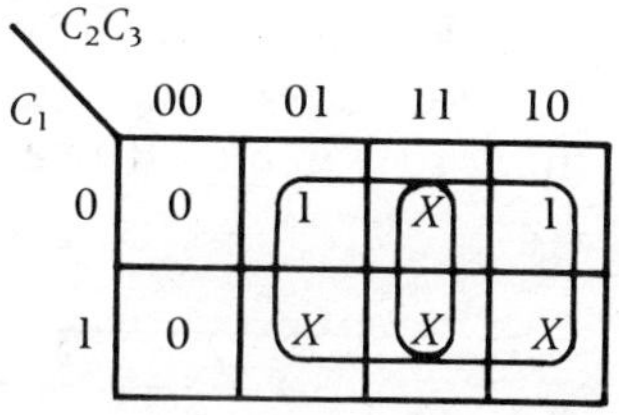

FIGURE 12.9 Generating both single and combined clock signals for PCL and PCH registers.

the A registers because some instructions (ADD, INCREMENT, etc.) affect both registers, whereas others (e.g., LOAD A) do not affect the carry register. Thus, the number of clock signals have to be increased from 9 to 11. Entries in Table 12.3 show the selected binary codes for all the required clock signals.

In addition to the extra two clock signals that operate two registers at the same time, we also have to provide for the case when no register transfer operation takes place at all. An extra clock signal is assigned for this NOP (no-operation) function; in other words, this clock signal is not connected to the circuit. This is required for the SKIP instruction in case the carry bit is equal to 0, when none of the registers is updated.

Now, we are ready to specify the ROM outputs. The inputs of the ROM are assigned (left to right) to the state flip-flop outputs (three signals), to the contents of the IR register (four signals), and finally, to the carry flip-flop output. Let us start with system state 000, which is used for the first fetch cycle. The assigned input and output bits are shown below:

Inputs			Outputs							
State	IR	Carry	New state	Clock	FUN ALU	SEL CY	FUN PC	SEL EXT	SEL ADDR
0 0 0	X X X X	X	0 0 1	0 0 0 0	X X X	X X	X X	X	0

A don't care value for the input means that the same output values are assigned for a number of ROM inputs. In this case all inputs between binary inputs 00000000 to 00011111 are assigned the same values. Thus, with one line, 32 ROM locations may be specified. In our case only 14 instructions are defined; therefore, the required number of entries is equal to 28. Two entries are defined for each instruction reflecting the two possible states of the CARRY signal input (which is ignored for all except the SKIP instruction).

A don't care output value means that the ROM does not have to be programmed for the indicated bit and can be left in its original (unprogrammed) state. For the two unassigned instructions all output bits may be left unprogrammed.

The second state is specified similarly. We have to assign bit values for the "increment PC" function which occurs during the second state of the FETCH cycle. As we mentioned before, the PC consists of two registers, the PCL and the PCH; therefore, the clock signal CLK_{1001} is selected. The bits for thirty-two (required twenty-eight) ROM locations 00100000 to 00111111 are:

	Inputs		Outputs						
State	IR	Carry	New state	Clock	FUN ALU	SEL CY	FUN PC	SEL EXT	SEL ADDR
0 0 1	X X X X	X	0 1 0	1 0 0 1	X X X	X X	0 1	X	X

We have assigned the first required 56 entries of the ROM. For the next, the first execute state, all 14 instructions must be separately encoded. For 13 instructions the carry bit is treated as a don't care input. For these instructions two entries use the same outputs. For the SKIP instruction both entries must be indicated, but if the carry bit is equal to 0, none of the registers should be updated (as mentioned before, this is called a NOP, or no-operation, instruction, and we call it CLK_{1111}). The encoding of a few instructions for this system state are:

	Inputs		Outputs							
State	IR	Carry	New state	Clock	FUN ALU	SEL CY	FUN PC	SEL EXT	SEL ADDR	
0 1 0	0 0 X X	X	0 1 1	0 1 0 0	X X X	X X	X X	X	0	(LDA-AND)
0 1 0	0 1 0 1	X	0 1 1	0 0 1 1	X X X	X X	X X	X	0	(JUMP)
0 1 0	0 1 1 0	0	0 0 0	1 1 1 1	X X X	X X	X X	X	X	(SKIP)
0 1 0	0 1 1 0	1	0 1 1	1 0 0 1	X X X	X X	0 1	X	X	(SKIP)
0 1 0	0 1 1 1	X	0 0 0	0 0 1 0	X X X	X X	X X	X	0	(BRANCH)
0 1 0	1 0 1 0	X	0 0 0	1 0 1 0	0 1 0	0 0	X X	0	X	(INC A)

The first line specifies eight entries because for the first four instructions the register transfer function is the same during state 010 and the carry input is ignored. For the INCREMENT A operation we have used clock signal CLK_{1010}, which updates both the A register and the carry flip-flop. This signal is treated

12.4 DESIGN EXAMPLES WITH ROMS AND PLAS 399

similarly to the signal used for both PCL and PCH registers. These few examples demonstrate how the entire truth table is completed for the third system state of the computer (28 entries for the 14 instructions).

A similar table must be constructed for the fourth or EXECUTE-2 state. In this case, the number of instructions which are effected is decreased to seven and the number of entries to be specified is equal to fourteen. We show the required outputs for all these instructions:

Inputs			Outputs							
State	IR	Carry	New state	Clock	FUN ALU	SEL CY	FUN PC	SEL EXT	SEL ADDR	
0 1 1	0 0 X X	X	1 0 0	1 0 0 1	X X X	X X	0 1	X	X	(LDA-AND)
0 1 1	0 1 0 0	X	0 0 0	1 0 0 1	X X X	X X	0 1	X	X	(LOAD MH)
0 1 1	0 1 0 1	X	0 0 0	0 0 1 0	X X X	X X	1 1	X	X	(JUMP)
0 1 1	0 1 1 0	1	0 0 0	1 0 0 1	X X X	X X	0 1	X	X	(SKIP)

Only four lines are required to specify 13 entries, since the outputs are the same for the first 8 entries, and the carry bit is ignored by most instructions. The SKIP instruction has to be specified only for the CARRY = 1 case because, when the carry bit is at Logic-0, this system state is skipped (see the outputs for the new state when the system is in state 010). Finally, the outputs for state 011 have to be specified, which requires only four lines:

Inputs			Outputs							
State	IR	Carry	New state	Clock	FUN ALU	SEL CY	FUN PC	SEL EXT	SEL ADDR	
1 0 0	0 0 0 0	X	0 0 0	0 1 1 0	1 1 0	X X	X X	0	1	(LDA)
1 0 0	0 0 0 1	X	0 0 0	1 0 0 0	X X X	X X	X X	X	1	(STA)
1 0 0	0 0 1 0	X	0 0 0	1 0 1 0	0 0 0	0 0	X X	0	X	(ADD)
1 0 0	0 0 1 1	X	0 0 0	1 0 1 0	0 0 1	1 0	X X	0	X	(AND)

This concludes the detailed specification of the ROM. Out of the possible 256 locations of the ROM, only 28 + 28 + 13 + 8 = 77 entries had to be specified, and there were only 1 + 1 + 12 + 4 + 4 = 22 distinct output bit combinations. This shows that the specification for the ROM outputs is greatly simplified. At the same time it becomes obvious that the ROM is not used very efficiently. The use of the ROM has the advantage that this technique is the most general design in the large; therefore, most likely, it can be executed automatically. It is possible to develop a computer-aided design (CAD) method that accepts the register transfer statements as input and generates the ROM contents automatically.

Let us see how designing the same computer with PLAs compares with using a ROM. PLAs are not as general as ROMs because they have a limited number of AND connections, and only a limited number of logical product terms can be used. Consequently, the design method is less general, and attention must be paid to the number of logical product terms needed for the outputs. This implies some form of minimization, possibly the use of Karnaugh maps. The requirements are not as severe as for designing with gates (where, as we know, the minimization and factoring of Boolean expressions are very important) because if the number of product terms are reduced to the number supplied by the PLA, there is no need for further reductions. Our ultimate goal is to use as few IC packages as possible and to utilize selected IC devices as much as possible.

The selection of the appropriate PLAs is made difficult by the growing number of available programmable logic modules. Since our computer requires 8 inputs and 16 outputs, we look for a device that matches these requirements. We find a PLA device with 10 input and 8 output connections, type HPL-16L8, whose logic diagram is shown in Figure 12.10. (The manufacturer refers to this device as PAL but we will use the generic term PLA for all programmable logic arrays regardless of their internal structure and commercially used name.) All 8 outputs are three-state outputs, and 7 output signals are fed back and are made available as inputs to the AND–OR circuits through fusable links. These signal connections allow the use of these seven pins either as input or output connections and make this programmable logic device configurable so that in addition to the dedicated output signal (pin 12) and 9 input signals, the other 7 signals may be used either for output or input (we may use up to 16 inputs, in which case only one output will be available). Since we require only 8 inputs, we need two of these devices, which provide 9 inputs and 16 outputs.

This PLA is of the fixed OR connection type. Each output may be generated from up to seven logical product terms, and each product term is given as an arbitrary combination of nine input signals or their inverse. Since all seven configurable signals are used as outputs, the feedback connections are not used at all. As long as we can limit our expressions to seven logical product terms, these two PLA devices will be sufficient for our computer.

From the three groups of output signals (state control, clock, and select) the state control signals will be designed first. As shown in Figure 12.3, there are three output signals that determine the next state of the computer. The next state of the computer depends on its current state, the current instruction stored in the IR, and possibly the carry bit. The nature of the PLA is such that Logical 1's are generated by the limited number of product terms, but all other don't care terms generate 000 outputs for the next state. Thus, the 000 output state should be chosen for the system state generated by the largest number of conditions. Examining the register transfer operations of the computer (Table 12.2), it is found that the FETCH-1 (or reset) state is the next state of all completed computer operations; therefore, it is assigned the binary value of 000. This assignment also solves the problem in case the computer finds itself in one of its "unused" states. (This could happen when PLAs are used because they contain a large number of don't care terms.) Only five of the eight possible states are used. Assigning outputs 000 to the FETCH-1 state

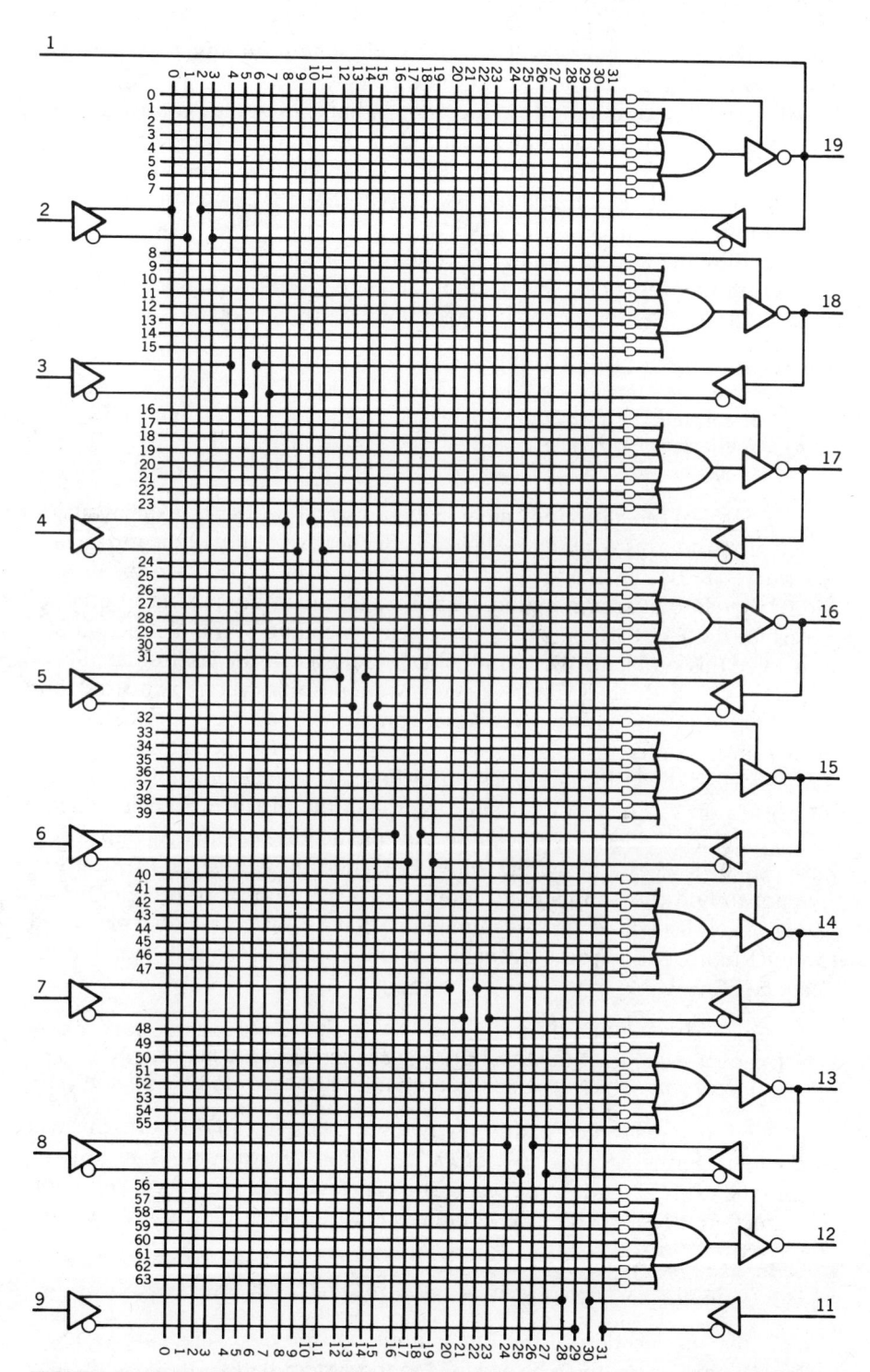

FIGURE 12.10 Monolithic Memories PLA type 16L8 device.

ensures that the computer will return to this state from any state that was not specifically programmed by the PLA.

The smallest number of product terms, that is, a single product term, is generated by those system states that follow the previous state without any conditions. For example, the FETCH-2 state always follows the FETCH-1 state, and the EXECUTE-1 state follows the FETCH-2 state. The outputs for these state transitions will be easily implemented with the PLA. The next state generated by the largest number of conditions is the EXECUTE-2 state. The register transfer instructions show that there are seven instructions that enter the EXECUTE-2 state. The logical conditions for this state are:

("next state" = EXECUTE-2 State) IF
    ("current state" = EXECUTE-1 State and Instruction = LOAD A)   OR
    ("current state" = EXECUTE-1 State and Instruction = STORE A)   OR
    ("current state" = EXECUTE-1 State and Instruction = ADD)   OR
    ... etc.

The logical functions indicate seven product terms. Using up all available seven product terms would not allow any further expansion of the instruction set (there are two unassigned instructions that may be used for this expansion); therefore, it seems a good idea to try to lower the number of product terms for the EXECUTE-2 state. The careful encoding of the instruction bits can solve this problem. We found when we designed the control circuits of the computer with gates that reductions were possible when the instructions were placed into groups or types that contained 2, 4, 8, ... individual but similar instructions. Instructions in one group were assigned consecutive binary codes. If the groups or types were chosen correctly, only a few number of instruction bits had to be used to represent the instructions belonging to one group. In Table 12.4 we show the instruction set divided into a few groups. Since we are designing with PLAs and not with gates, it is not important to find the groups that produce the least number of product terms. A reasonable list will reduce the numbers so that a sufficient number of unused product terms will become available. In this case the Boolean equation for state EXEC-2 is

("next state" = EXEC-2) IF
    ("current state" = EXEC-1 and $IR_3 = 0$ and Instruction is not SKIP) OR
    ("current state" = EXEC-1 and Instruction = SKIP and CARRY is not 0)

This Boolean equation contains only two product terms in this form but will yield more terms because the logical condition "Instruction is not SKIP" cannot be expressed with one term. The expression for the logical condition "Instruction is NOT SKIP" is given in terms of the other IR bits and contains three terms:

$$\text{Instruction is NOT SKIP} = IR_2' + IR_2 \cdot IR_1 + IR_2 \cdot IR_1' \cdot IR_0'$$

Before the specification of the PLA outputs for the next state can be determined, binary values must be assigned to the system states. These values may be arbitrarily chosen (except the 000 state, which already has been assigned), but these values will influence the required number of product terms. All states that share a binary 1 value with any one of the three outputs of the PLA will contribute logical product terms. For example, if three logical product terms are needed to generate the next state 100 and four product terms to generate the next state 101, then the output that represents the MSB of the next-state outputs will require seven product terms. Therefore, the state assignments should be chosen in such a way that next-state outputs with large numbers of product terms should contain disjoint 1's in their output values. For our very small and limited example the chosen PLA could be sufficient for an arbitrary state assignment; however, we present a general principle that can be frequently used with PLAs. In this case, system state EXEC-2 requires the largest number of product terms; therefore, it will be assigned state 100, and the MSB of the new-state outputs will be set to 0 for the other states. The state assignments for the system states are:

FETCH-1	FETCH-2	EXEC-1	EXEC-2	EXEC-3
000	001	010	100	011

With these binary assignments the Boolean expressions for the next-state outputs are as follows:

$$S_2 \text{ (next)} = S_2' \cdot S_1 \cdot S_0' \cdot IR_3' \cdot IR_2' + S_2' \cdot S_1 \cdot S_0' \cdot IR_3' \cdot IR_2 \cdot IR_1$$
$$+ S_2' \cdot S_1 \cdot S_0' \cdot IR_3' \cdot IR_2 \cdot IR_1' \cdot IR_0'$$
$$+ S_2' \cdot S_1 \cdot S_0' \cdot IR_3' \cdot IR_2 \cdot IR_1' \cdot IR_0 \cdot CARRY$$

$$S_1 \text{ (next)} = S_2' \cdot S_1' \cdot S_0 + S_2 \cdot S_1' \cdot S_0' \cdot IR_3' \cdot IR_2'$$

$$S_0 \text{ (next)} = S_2' \cdot S_1' \cdot S_0' + S_2 \cdot S_1' \cdot S_0' \cdot IR_3' \cdot IR_2'$$

where the conditions for the S_1 bit combine the conditions for the EXEC-1 and EXEC-3 states and for the S_0 bit the FETCH-2 and EXEC-3 states. So far, the largest number of product terms is 4, which is well within the capabilities of the PLA chosen.

The outputs for the selection/function lines and for the demultiplexer are similarly determined. We are not going to determine all the Boolean expressions but will examine whether the maximum of seven product terms is sufficient for this computer. The selection/function outputs consist of very few product terms since specific values for them are very infrequently required (there are many don't care terms). For example, the ADD function of the ALU is required only for the ADD operation occurring during system state EXEC-3. Similarly, selection line SEL_{EXT}, which controls the multiplexer with the external data inputs, is set to 1 (external data inputs) only when the INPUT instruction is executed during system state EXEC-1. At all other times the output of the ALU is selected. We expect that the function/selection output lines will not exceed the seven-product-term capacity of the PLA.

404 DIGITAL SYSTEM DESIGN

TABLE 12.4 Instructions Grouped into Similar Instruction Types

IR_3	IR_2	IR_1	IR_0	Instruction	Groups	
0	0	0	0	LOAD A	Instructions with three execution cycles	Instructions with more than one execution cycle
0	0	0	1	STORE A		
0	0	1	0	ADD		
0	0	1	1	AND		
0	1	0	0	LOAD MARH	Instructions with two execution cycles	
0	1	0	1	SKIP		
0	1	1	0	JUMP		
0	1	1	1	Unassigned		
1	0	0	0	INC A	Carry flip-flop is updated	
1	0	0	1	COMPL A		
1	0	1	0	ROTL A		
1	0	1	1	Unassigned		
1	1	0	0	ROTR A		
1	1	0	1	INPUT		
1	1	1	0	BRANCH		
1	1	1	1	OUTPUT		

Finally, we have to determine the Boolean equations for the four select output lines of the demultiplexer that generates the clock signals. Since a clock signal will be generated for each binary combination of the four signal values, we must assign the selection values 0000 (CLK_{0000}) to the NOP condition. This clock signal will be generated for all the don't care cases when no register transfer will take place. The binary assignments of the other clock signals are controlled by the number of conditions for each clock signal, which becomes the number of logical product terms for that clock signal. The number of conditions for a clock signal is determined from the register transfer operations (Table 12.2). For example, the PC ← PC plus 1 register transfer requires both the PCH and PCL clocks. In Table 12.2 there are eight instances of the INCREMENT PC operation; two of these are conditional depending on the output of the carry flip-flop. Fortunately, four instructions that require the PC ← PC plus 1 operation are in a group with consecutive binary codes and yield only one logical product term. The Boolean equation for this clock signal is

$$CLOCK_{PCL+PCH} = S_2' \cdot S_1' \cdot S_0 + S_2' \cdot S_1 \cdot S_0' \cdot IR_3' \cdot IR_2 \cdot IR_1 \cdot IR_0' \cdot CARRY$$
$$+ S_2 \cdot S_1' \cdot S_0' \cdot IR_1' \cdot IR_0' + S_2 \cdot S_1' \cdot S_0' \cdot IR_3' \cdot IR_2 \cdot IR_1' \cdot IR_0'$$
$$+ S_2 \cdot S_1' \cdot S_0' \cdot IR_3' \cdot IR_2 \cdot IR_1 \cdot IR_0' \cdot CARRY$$

The first term represents the FETCH-2 cycle, the second term the EXEC-1 cycle when only the SKIP instruction contributes to the incrementing of the PC. The last three terms represent the EXEC-2 cycle with the four three-cycle instructions (LOAD, STORE, ADD, and AND), the LOAD MARH, and the

12.4 DESIGN EXAMPLES WITH ROMS AND PLAS   405

SKIP instructions. Thus, the generation of the PCL + PCH clock pulse requires five product terms. The encoding of the demultiplexer's selection lines for the clock pulses can be done conveniently in a table where the number of required product terms is listed. The 1's and 0's of the selection lines are assigned so that the maximum number of product terms does not exceed 7. A possible set of assignments is shown in Table 12.5, which can be realized without exceeding the seven-terms limit. The binary assignments for the selection inputs followed techniques similar to those used for the assignments of the system state outputs. Clock signals with a large number of terms are assigned binary values with disjoint 1's. The PCH + PCL clock signal is assigned selection code 1000 (five terms), CLOCK$_A$ is assigned code 0100 (three terms), and so on.

With Table 12.5 we have completed the task of finding a feasible realization for our 4-bit computer using a PLA. The two PLA devices mentioned are sufficient for our example. The question remains, How can one design with PLAs when some of the hardware limits are exceeded? There may be too many inputs or too many logical product terms for a given PLA (any number of outputs may be realized by using a sufficient number of PLA devices). The physical limits of PLA devices can be extended either by cascading them or by ORing their outputs. Two four-input/four-output PLA devices are shown in Figure 12.11. The cascading of these two PLA devices provides Boolean expressions that may involve all the inputs (six in this case). The second PLA receives two of its inputs from the first PLA, which provides sums of logical product terms involving inputs I_1–I_4. The second PLA provides product terms that are combinations of these two inputs and inputs I_5 and I_6. As an example, let us assume that the two outputs B_3 and B_4 are equal to

$$B_3 = I_1 \cdot I_2' \quad \text{and} \quad B_4 = I_3 \cdot I_4'$$

TABLE 12.5 Encoding of CLOCK Signals for PLA Control

CLOCK	Number of product terms	S_3	S_2	S_1	S_0
IR	1	0	0	1	1
OUT	1	0	1	0	1
PCL	2	0	0	0	1
PCH	1	0	1	1	0
MARL	1	0	1	1	1
MARH	1	1	0	1	0
A	3	0	1	0	0
CARRY	0	—	—	—	—
WRITE (MEM)	1	1	0	0	1
PCH + PCL	5	1	0	0	0
A + CARRY	3	0	0	1	0
NOP		0	0	0	0
Total number of terms:		7	6	7	6

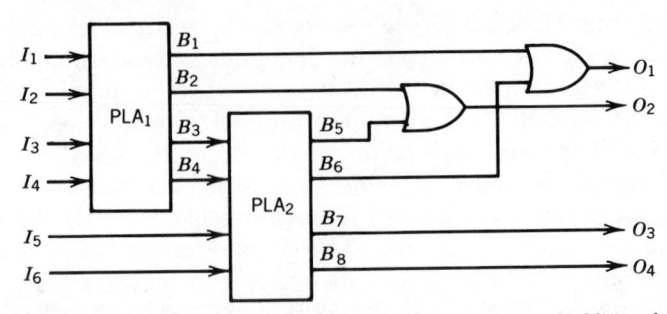

FIGURE 12.11 Expanding of PLA devices by cascading and additional OR gates.

Outputs B_5–B_8 can be expressed as the sum of product terms that include any combination of eight variables:

$$(I_5) \quad (I'_5) \quad (I_6) \quad (I'_6) \quad (I_1 \cdot I'_2)$$
$$(I_1 \cdot I'_2)' = (I'_1 + I_2) \quad (I_3 \cdot I'_4) \quad (I_3 \cdot I'_4)' = (I'_3 + I_4)$$

A Boolean expression implemented with the two four-input PLAs in Figure 12.11 must be expressed as a logical sum of the eight product terms shown. A general Boolean expression may have to be factored so that it is placed in the correct form and OR gates may be used to include more terms.

The connection of any number of PLA outputs by OR gates increases the number of product terms included in the output expression. In the example shown in Figure 12.11, outputs O_1 and O_2 include product terms generated by both PLAs. Of course, OR gates may be used for the outputs of one PLA. In Figure 12.12 we show an extreme case where the PLA device, with four outputs with n product term each, provides one output with $4n$ product terms. This arrangement is useful only for PLAs with fixed OR connections.

In conclusion, we have seen how computers may be designed in the large using ROMs or PLAs. The design methods are based on the understanding of Boolean relationships, binary gates, Boolean algebra, and minimization but often have to be adjusted to the specific structure of the programmable logic device. New, more powerful programmable logic devices are appearing continuously, which helps to decrease the number and the cost of required IC devices, but undoubtedly new design methods will be required for their efficient utilization. Before we conclude this chapter, we will examine one more general method of computer design that demonstrates the hierarchical nature of digital hardware.

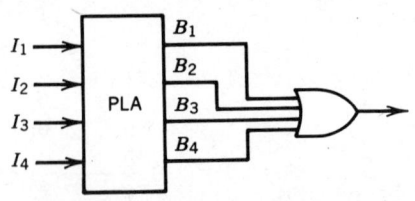

FIGURE 12.12 Increasing number of product terms for PLA device with fixed AND connections.

12.5 MICROPROGRAMMING

When we examine our 4-bit computer model, we find that there are two kinds of registers. Register A is used by a number of instructions and is considered a programmable register. This means that a computer program has total control over the contents of the A register (it can LOAD it with any data from memory). On the other hand, the IR cannot be accessed by a computer program. It is used by the system to hold the bits of the current instruction. The IR register is called a system register. The contents of other registers such as the PCH can be changed by the program but cannot be considered as a general programmable register. The distinction between *programmable* and *system* registers indicates two hierarchical levels. At the upper level computer instructions are expressed in terms of register transfer operations, which involve only programmable registers. At the lower, or system, level register transfer operations describe how the computer operates and how the register transfer operations of the instructions are executed by a number of system level operations.

The register transfer operations at the system level are often referred to as **microinstructions.** Each computer instruction may be expressed as a collection of microinstructions, some of which may be executed concurrently but will necessarily involve a number of sequential clock periods, which may be called **microsteps.** The collection of the microstep sequences for all available computer instructions may be thought of as a microprogram. In microprogrammed computers an automatic program executing system called a microsequencer is placed at the system level, and the execution of each computer instruction is specified by a sequence of microsteps stored in a ROM, which we will call the micro-ROM. The overwhelming advantage of the microprogrammed approach is that by reprogramming the micro-ROM the instruction set of the computer is redefined. Thus, one set of hardware may be made to behave as a number of different computer models or, depending on the application, the instruction set of the computer may be tailored to the task and its performance improved.

The difference between employing a ROM for the control element of the computer (demonstrated in Section 12.4) and using a micro-ROM for controlling a sequence of microinstructions is in the approach to the structure or data paths of the computer. The data paths for our 4-bit example computer was chosen rather arbitrarily, the selection being governed by the intended minimization of hardware components. The microprogrammed approach requires a much more logical and general structure where registers become general registers. This means that the registers may store instructions, data, addresses, or conditions. As mentioned before, some of these registers are available for computer instructions, whereas all registers can be used by microinstructions. One immediate effect of microprogramming on the instruction set of a computer is that the programmable registers become general registers and the instruction set becomes more logical. For example, in our example computer the ADD instruction was restricted to a register–memory instruction with the result of the ADD operation stored in the A register. In a microprogrammed computer we would expect the ADD instruction to be executed either between registers, register and memory, or even between memory and memory storage registers. This general and logical structure of the computer is demonstrated by the data path diagram shown in Figure 12.13.

Two bidirectional data buses (often called highways) are shown. One carries data and is connected to the data processor or ALU of the computer. The other carries address information. Both buses indicate their ability to accept information from or transmit information to external digital components. The general registers and the computer's main memory are connected to both buses.

For each microinstruction one source may be selected for each bus, and in the most general case, a number of destination registers may be updated if some specified conditions are met (normally, conditions are derived from the condition register, and if the conditions are not true, then no register transfers take place). In order to be able to implement such general microinstructions, the major subsystem elements, such as the ALU, memory, input, and output, must contain their own "private" registers. The data path diagram for these additional system registers is shown in Figure 12.14. We demonstrate the structure of a microprogrammed computer by choosing 16 registers (it is logical to select a number of registers equal to a power of 2, since the selection of a register can be decoded from a small number of bits). These include the 4 external registers (2 for data and 2 for addresses), 2 general programmable data registers (A and B), 2 general system registers (Q_1 and Q_2), the IR and

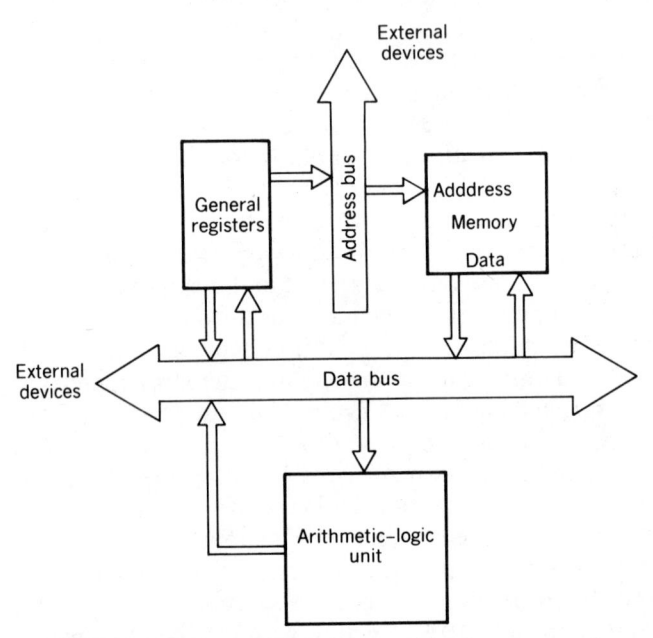

FIGURE 12.13 Data paths for computer with microprogramming.

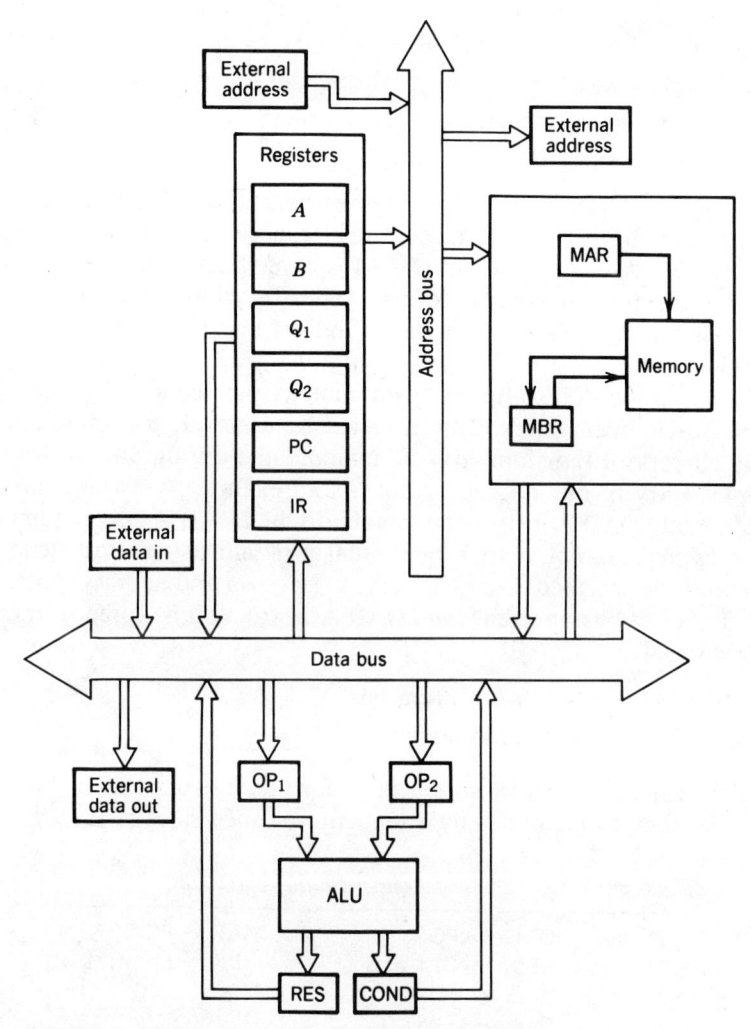

FIGURE 12.14 System registers for microprogramming.

PC registers, the MAR and MBR, which belong to the memory system, and the 4 registers of the ALU. We will also assume that the ALU has 16 functions, including ADD, SUBTRACT, INCREMENT, DECREMENT, COMPLEMENT, and logical and shift operations.

The next step in the design of the microinstructions is to break down the control of the possible register transfer operations into elementary select and control functions. In the model shown, there are 3 select functions,

1. select source for data bus,
2. select source for address bus, and
3. select ALU function,

and 18 control (clock) functions,

- 1–16   update a register,
- 17   memory write (M[MAR] ← MBR), and
- 18   memory read (MBR ← M[MAR]).

Since we have selected a model with 16 registers and 16 different ALU functions, the total number of bits that control all aspects of this general computer structure is 30. As shown in Figure 12.15, a microinstruction word, or microword, is divided into six fields of controlling functions. Since only 1 of the 16 registers may be selected as a source, only 4 bits are needed for the first two fields.

The execution of a computer instruction may be defined by a sequence of these 30-bit microwords. This is called **horizontal** microprogramming because all control functions may be manipulated during one microstep, a horizontal entry in our microprogram. As an example, we will express an unconditional LOAD A instruction (similar to the LOAD A instruction of our 4-bit computer example). In a horizontal program a microinstruction sequence must be specified. We assume here that data and address contain the same number of bits and that the LOAD A instruction is stored in memory in two words:

Memory location → LOAD A instruction

Next location   → Memory address

A possible sequence of microinstructions for this instruction (including the fetch cycle) may be expressed by a sequence of microsteps as follows:

Microstep	Register transfer operations	
1	OP1 ← PC	; MAR ← PC
2	RES ← OP1 plus 1	; MBR ← MEM[MAR]
3	IR ← MBR	
4	Q_1 ← RES	
5	MAR ← Q_1	; RES ← OP1 plus 1
6	PC ← RES	; MBR ← MEM[MAR]
7	A ← MBR	; MAR ← PC

The same sequence encoded into the various fields of the microword is shown in Figure 12.16. We can see that although computer instructions are made up of sequential operations, five out of the seven microsteps include multiple

Bit assignments

29 26	25 22	21 18	17 2	1	0
Data source	Address source	ALU function	Register clocks	Write memory	Read memory

FIGURE 12.15 The 30-bit microinstruction word.

12.5 MICROPROGRAMMING   411

Bit assignments

Micro step	29 26 Data source	25 22 Address source	21 18 ALU function	17 2 Register clocks	1 Write memory	0 Read memory	
1	PC	PC	X	MAR,OP1	0	0	
2	X	X	INC	RES	0	1	
3	MBR	X	X	IR	0	0	(fetch)
4	RES	X	X	Q1	0	0	(PC + 1)
5	X	Q1	INC	RES,MAR	0	0	(MAR←PC + 1)
6	RES	X	X	PC	0	1	(update PC)
7	MBR	PC	X	MAR,A	0	0	(load A and prepare MAR)

FIGURE 12.16 Encoding direct LOAD instruction into microinstruction word.

register transfer operations. Horizontal microprogramming is wasteful of hardware (large number of ROM outputs are required) but provides the most efficient (fastest) implementation of microsteps.

In order to reduce the hardware cost of the micro-ROM, the microprogram can be made more **vertical.** This means that fewer parallel operations are provided in one microstep. An obvious change would be to encode the 16 clock outputs into a field of 4 bits and decrease the required number of micro-ROM outputs by 12. This is similar to our design method in Section 12.3 when we used a demultiplexer for the clock signals. For a purely vertical program only one register transfer operation is allowed per microstep, in which case only one bus would be required. Even the bus could be eliminated by replacing it with a single register, in which case two separate microsteps would be required for a register transfer operation: one for transferring the contents of the source register to the bus register and the other to transfer the contents of the bus register to the destination register. Changing a horizontal microprogram into a vertical one is very similar at the system level to changing a bit-parallel process into a bit-serial one at the device level. We sacrifice speed to save hardware costs.

We have seen how one particular computer instruction is executed from a sequence of microinstructions. In order to implement a particular instruction set of a computer, a microprogram controller called a **microsequencer** is required, which selects the correct microprogram for each computer instruction and is capable of executing conditional instructions. The selection and control of the microprogram requires the introduction of a program counter for the microprogram (micro-PC) and the adding of several new microinstruction types. So far, we have shown only one type of microinstruction that controlled the computer through a 30-bit word. In order to control the microsequence, we require microinstructions that do the following:

1. unconditionally change the microsequence (JUMP),
2. change the microsequence according to the instruction bits stored in the IR (select microprogram),

3. execute register transfers with selection and/or clock lines specified by the instruction bits in the IR,
4. load a register with a constant (immediate data), and
5. execute microinstructions conditionally.

There are a number of commercially available microsequencers, but the structure of a microsequencer is simple enough so that it can be constructed from standard IC components. We will describe such a simple microsequencer circuit.

The microsequencer is connected to the computer as shown in Figure 12.17. The address of the current microinstruction is stored in the micro-PC. One bit of the microword determines whether the microinstruction operates the computer (through 30 bits of the microword) or changes the sequence of the microprogram. If the computer is operated, then the bits that determine the source and destination registers can be derived from the microword or from the IR. An additional bit in the microword selects either the IR or the microword. This is necessary because the instruction bits allow the selection of programmable registers for a computer instruction, whereas all registers can be manipulated by the microinstruction. The two additional bits in the microword are shown in Figure 12.18, where the bit value of 0 is selected for simple microsteps.

The structure of the microsequencer is shown in Figure 12.19, where a multiplexer determines whether the computer SEL/FUN lines are derived from

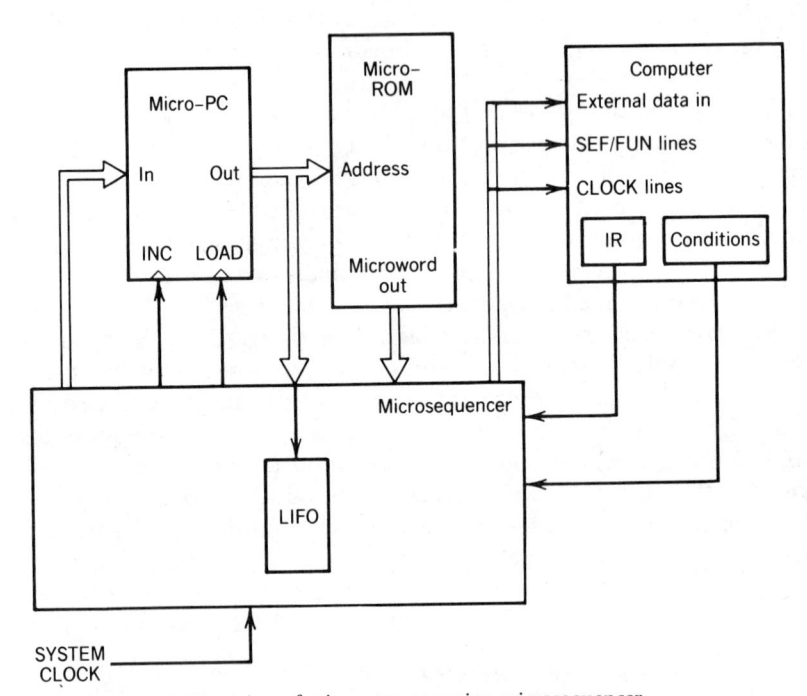

FIGURE 12.17 Execution of microprogram using microsequencer.

12.5 MICROPROGRAMMING 413

Bit assignments

31	30	29 26	25 22	21 18	17 2	1	0
0	SEL source	Data source	Address source	ALU function	Register clocks	Write memory	Read memory

FIGURE 12.18 Extened microinstruction word.

the microword or the contents of the IR. For each of these simple microsteps the contents of the PC are incremented, and a sequential microprogram is executed. When the MSB of the microword is equal to 1, a microprogram control instruction is executed. The contents of the micro-PC is changed with the parallel LOAD operation. There may be three sources for the address. If the address is derived from the bits of the microword, then a direct microJUMP (or procedure microCALL) instruction is executed. If the source is from the IR, then the address of the microJUMP (or microCALL) instruction is determined by the current instruction. This is how a microsequence is selected for a given computer instruction. The only difference between a microJUMP and

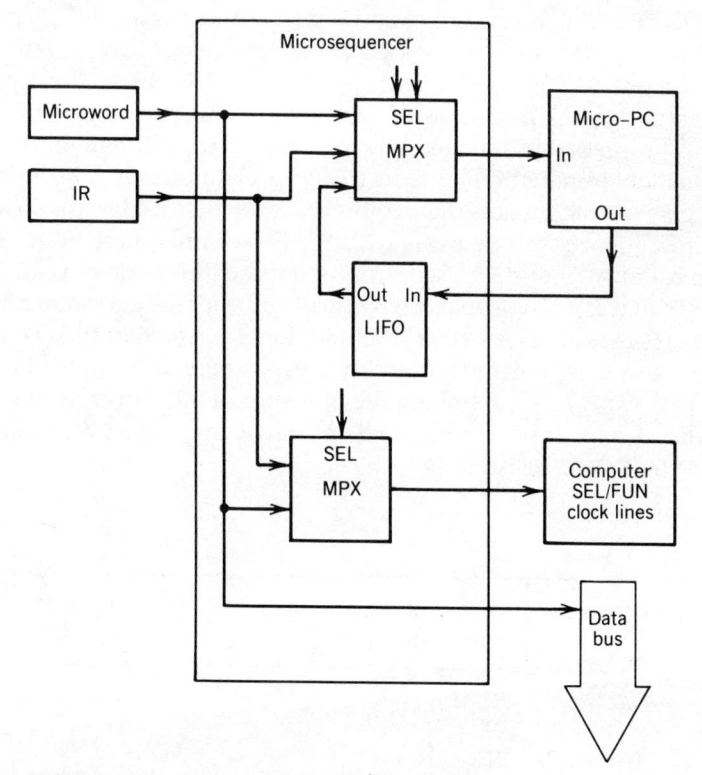

FIGURE 12.19 Structure of simple microsequencer.

a microCALL instruction is that for the microCALL the return address is pushed into the LIFO stack before the micro-PC is updated.

When the source for the micro-PC is from the LIFO stack, the microRETURN instruction (return from a procedure) is executed by the microsequencer. A stack POP operation must follow the updating of the micro-PC. So far, all instructions have been executed unconditionally. For the control instructions (microJUMP, microCALL, and microRETURN) conditional instructions must be provided. We assume that there are four condition signals from the computer. A convenient way of handling multiple conditions is to use so-called mask bits in the microword. Four mask bits are logically ANDed with the condition bits. The terms are ORed, and an additional mask bit controls the reversal of the condition. The following Boolean expression is evaluated:

$$\text{Condition} = M_5' \cdot (M_4 \cdot C_4 + M_3 \cdot C_3 + M_2 \cdot C_2 + M_1 \cdot C_1) \\ + M_5 \cdot (M_4 \cdot C_4 + M_3 \cdot C_3 + M_2 \cdot C_2 + M_1 \cdot C_1)'$$

where C_i are the condition signals from the computer and M_i are 5 bits in the microword. The microcontrol instruction is executed only if the value of the CONTROL expression is equal to Logic-1. The implementation of this equation is very simple. It requires four two-input AND gates, one four-input OR gate, and a two-input XOR gate.

In Figure 12.20 the encoding of the microword bits for the microcontrol instructions are shown. The address source bit chooses between the IR and the microword for the microJUMP instruction. The instruction type field chooses between the microJUMP, microCALL, and microRETURN instructions. Twenty-three bits remain, which is more than sufficient for the microaddress. A few hundred different instructions can be easily implemented by a few thousand microwords that require no more than 14 bits for an address. The microsequencer can be easily expanded since only three of the possible four control instructions are used. Notice that the direct microLOAD instruction is executed as a simple microstep instruction with the data source assigned to the EXTERNAL DATA INPUT. An additional instruction type is not necessary for this microinstruction. With these extended facilities the microsequencer is capable of controlling the sequence of microinstructions according to the current instruction stored in the IR and the system conditions stored in the COND register.

31	30	29 28	27 23	22 0
1	Address source	Instruction type	Condition mask	Direct address

Microword   JUMP
IR   CALL
                RETURN

FIGURE 12.20 Microcontrol instruction word.

In conclusion, we have seen that the design of a microprogrammed computer requires the specification of the micro-ROM. After the instructions have been defined in terms of microsteps, the microprogram is specified, which includes the microsteps of the computer instructions and special microinstructions that control the sequence of the microprogram. The complexity of this process is kept within reason by the logical and general specification of the computer's hardware and the very limited instruction set of the microsequencer.

EXPERIMENT 12
Experimental Designs with ROMs and PLAs

E12.1 Summary and Objectives

The main objective of this chapter was to introduce design in the large and apply it to the field of digital design. There is no specific experiment for this chapter because the main objective is to compare the two design methods (with ROMs and PLAs and with gates) and draw conclusions as to the differences between the two design philosophies. Comparisons between time spent on the design, the number of IC packages, costs, and testing methods are very important. We suggest that one (or a few) experiment is chosen from Experiment 6 to Experiment 11 and is redesigned using both ROMs and PLAs, after which comparisons can be made. When designing with ROMs and/or PLAs, try to structure the hardware so that it solves a general problem and can produce a class of hardware devices. Study the suggested alternative experiments and try to design the hardware so that it can be easily modified, expanded, and used for the suggested alternatives.

It is important to complete the design with the specification of the ROM and/or the graphical diagrams of the programmed PLAs. Comment on how much of the design had to be done in the small and how this may be avoided. Try to use realistic costing factors to arrive at the respective costs of the hardware. Especially for the earlier and simpler experiments, try to find modern programmable logic elements that include flip-flops so that no additional gates would be required for your design. After the design has been completed, evaluate how much of the total power of the logic devices are utilized in your design.

Drill Problems

12.1–12.7. For Drill Problems 11.1–11.7 determine and discuss the effects of suggested changes for a design if ROMs and/or PLAs were used. How different would the design be if microprogramming was used?

12.8. Complete the detailed design of the microsequencer shown in Figures 12.17 and 12.19. Could you (and should you) use a ROM or a PLA for the microsequencer controller?

APPENDIX 1

Integrated Circuit Devices

A1.1 INTRODUCTION

In the following pages a selected number of TTL IC devices are described. The information is derived from the *TTL Data Book for Design Engineers* by Texas Instruments Inc. The recommended number of selected devices have been used and found sufficient for the first 10 experiments in the text. Since it was suggested that Experiment 11 is constructed by a team of students, the number of devices available from three or four sets of ICs should be sufficient for Experiment 11. It is assumed that Experiment 12 is a paper exercise, and no practical devices are recommended for it.

With the detailed description of the following 18 IC device types, this book becomes self-sufficient. The alternative would have been to ask the student to obtain a copy of the Texas Instruments data book. Both approaches (using the exhaustive data book or a restricted list of devices) have their advantages and disadvantages. We have decided to build our experiments around a small set of devices.

The decision to use only a selected number of IC devices was made on both practical and educational grounds. The given number and types of devices are restrictive enough so that the student will be forced to face physical limitations. These limitations will have to be overcome by device type transformations, more careful minimization of the circuits, or the implementation of logic functions on a variety of logic devices. Thus, the restrictions force the student to practice the techniques discussed in the text. Since one of our aims is to teach systematic design principles in general, it is important to impose physical limitations on the student because limitations are very important factors in real design problems.

Obviously, the course can be taught equally well with an unrestricted list of IC devices. One disadvantage of this is that at the beginning, when the student should concentrate on digital design principles, he or she may have to spend a large amount of time to decipher the Texas Instruments data book. We attempted to help the student by using meaningful signal names and by supplying short explanations for the more complex devices.

A1.2 LIST OF AVAILABLE IC DEVICES

In Table A.1 the list of available IC devices are shown in order of IC numbers. In addition to their identification (type), their descriptive functions, number

418 INTEGRATED CIRCUIT DEVICES

TABLE A.1 List of Available IC Devices

IC type	Function	Number of pins	Price	Quantity
7400	Quad two-input NAND gates	14	0.20	4
7402	Quad two-input NOR gates	14	0.20	2
7404	Hex inverters	14	0.24	2
7408	Quad two-input AND gates	14	0.20	2
7420	Dual two-input NAND gates	14	0.20	3
7430	Single eight-input NAND gate	14	0.20	1
7474	Dual D-type edge-triggered flip-flops	14	0.26	2
7476	Dual JK master–slave flip-flops	16	0.24	2
7485	4-bit magnitude comperator	16	0.50	1
7486	Quad two-input XOR gates	14	0.24	2
74153	Dual 4-to-1 multiplexers	16	0.36	2
74155	Dual 2-to-4 decoders/demultiplexers	16	0.36	1
74157	Quad 2-to-1 multiplexers	16	0.44	2
74S163	4-bit binary counter with LOAD	16	0.40	1
74170	Open collector RAM, four 4-bit registers	16	0.90	1
74174	Hex edge-triggered D-type flip-flops	16	0.48	1
74181	4-bit arithmetic-logic unit (ALU)	24	1.50	1
74191	4-bit universal shift register	16	1.25	1
Total				31

of IC pins, and prices are also shown. The quantity entry is the suggested number of ICs found sufficient for the experiments in this book. The prices are shown in order that a total price for each completed circuit may be calculated. It is always useful to be able to calculate a measure of success for a design problem. One such measure could be the overall cost of the circuit. It is suggested that for each IC device used, additional "socket" and wiring costs are added to the total cost of the circuit. It would be difficult to evaluate these costs exactly, but the following costs are reasonable:

Number of pins	Socket plus wiring costs ($)
14	0.20
16	0.25
24	0.35

A healthy amount of competition may be introduced into a class when recognition is given to the student who could find the design solution with the lowest total cost.

A1.3 IC DEVICES

In the following pages the pin assignments for the available 18 IC device types are given. They are listed in order of their logic functions. No additional explanations are needed for the pin assignments of simple gates. For more complex devices signal names and short explanations are added. Signal names are chosen consistently. For example, inputs are named INP, IN, or I, outputs are OUT or Q. A number before the name of the signal indicates that the IC device contains a number of identical logic devices. For example, 1EN2 indicates the second enable input of the first device, whereas 2EN1 indicates the first enable input of the second device (see type 74155 demultiplexers). When no number is shown in front of the signal name (e.g., S1), the signal is common to all devices of the IC.

The circle convention is used to indicate signals with negative logic (Logic-0 for the active signal). Positive-logic clock signals operate during their 0-to-1, negative-logic clocks during their 1-to-0 transition.

A1.3.1 NAND Gates

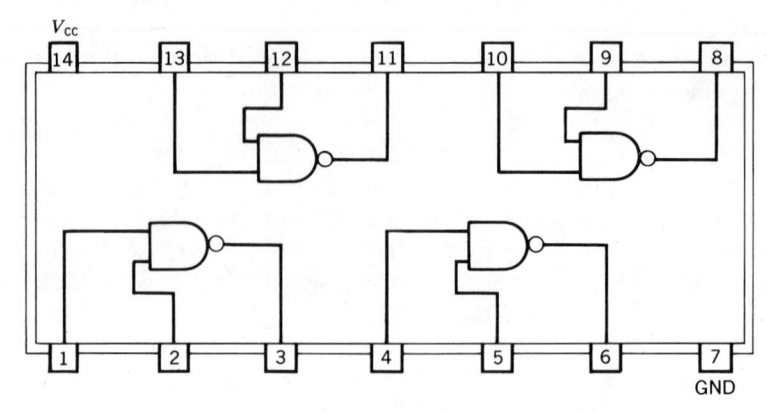

Type 7400

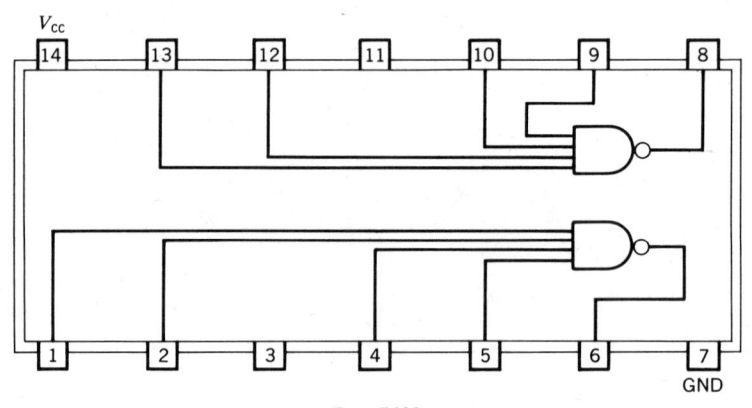

Type 7420

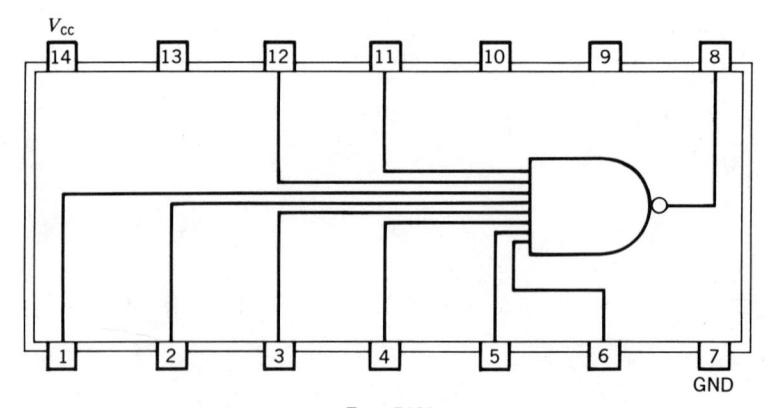

Type 7430

A1.3.2 AND Gates and Inverters

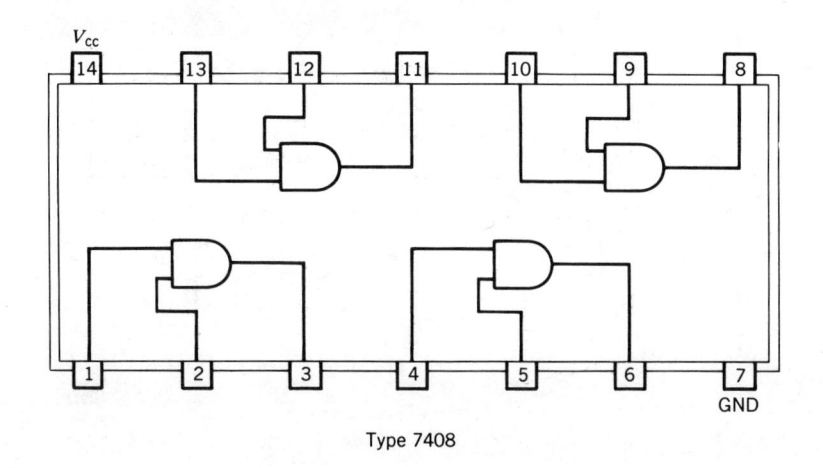

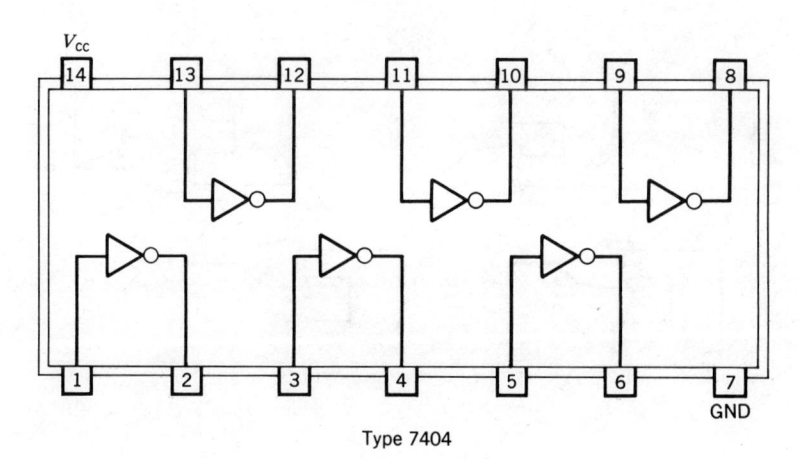

Type 7408

Type 7404

A1.3.3 NOR Gates and XOR Gates

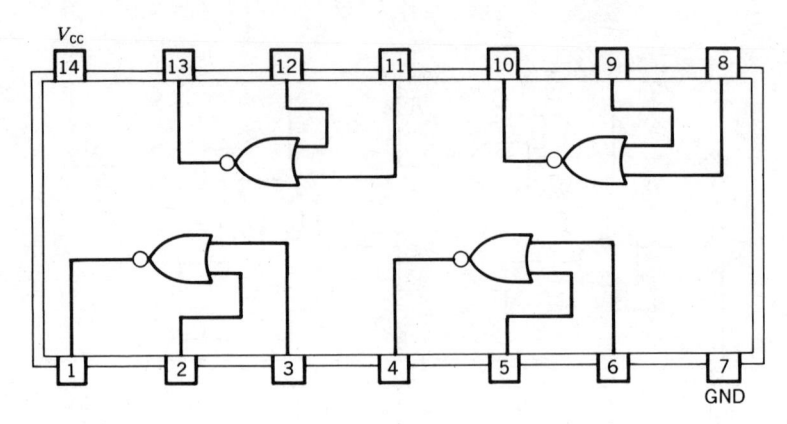

Type 7402

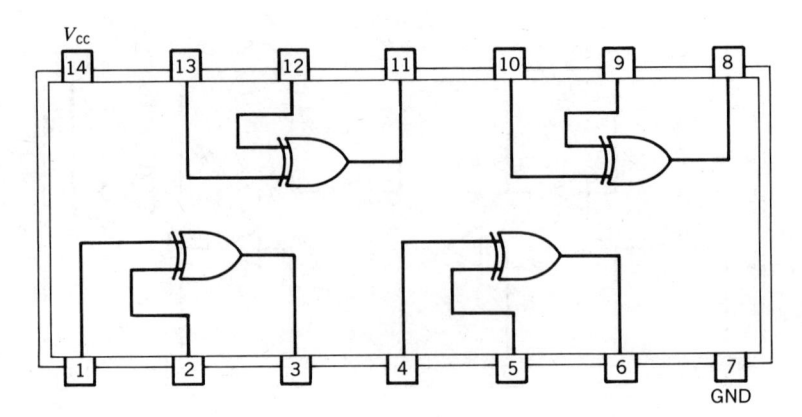

Type 7486

A1.3.4 Flip-Flops

The CLEAR and PRESET are asynchronous (dc) functions and use negative logic (active GND). Updating of the output occurs at the positive edge of the clock.

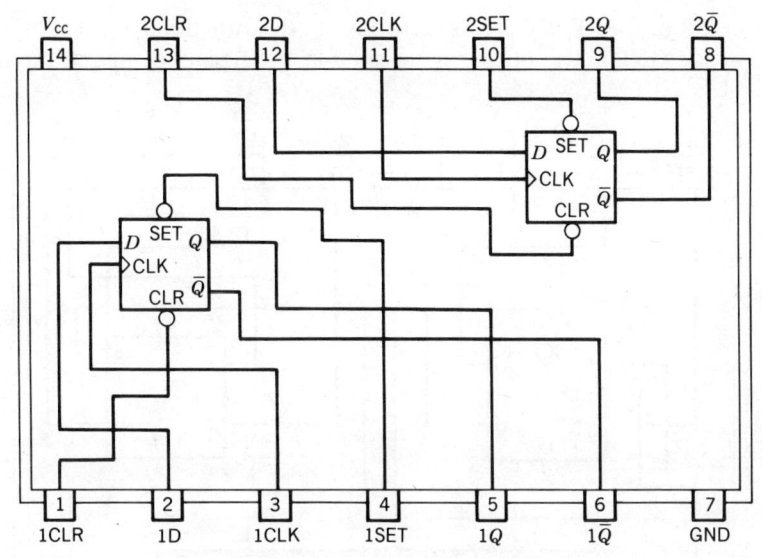

Type 7474, dual D-type edge-triggered flip-flops

The CLOCK and CLEAR inputs are common to all six flip-flops. The CLEAR function is asynchronous (dc) and uses negative logic (active GND). Updating of the outputs occur at the positive edge of the clock.

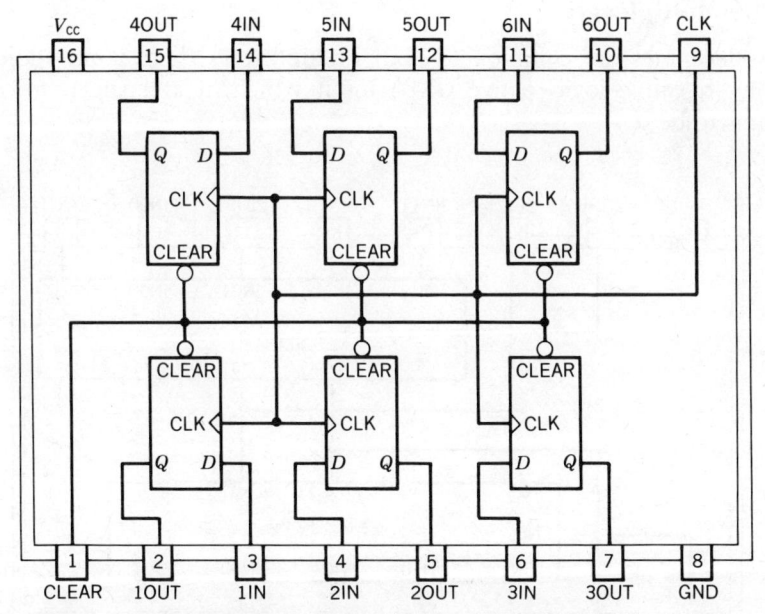

Type 74174, hex D-type edge-triggered flip-flops

The CLEAR and PRESET are asynchronous (dc) functions and use negative logic (active GND). The output is updated at the negative edge of the clock.

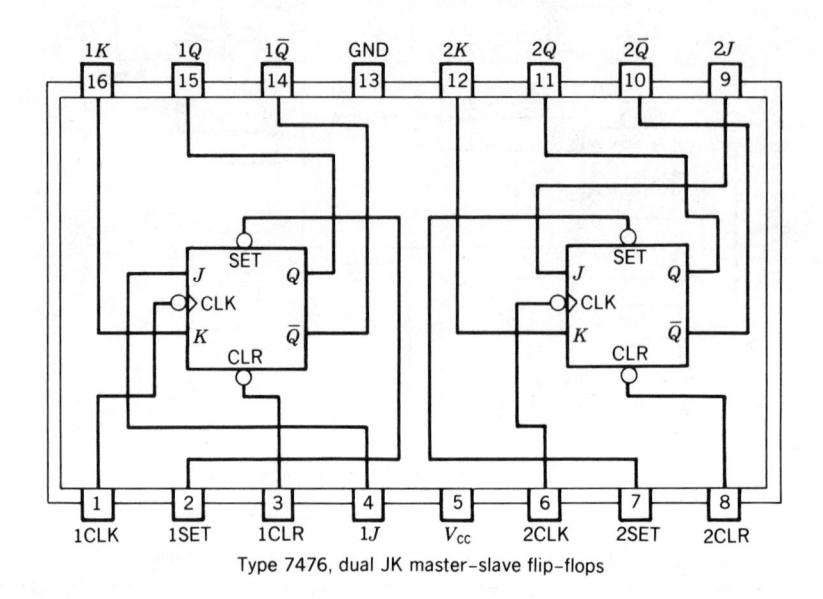

Type 7476, dual JK master–slave flip-flops

A1.3.5 Multiplexers

The select lines are common to both multiplexers. There is an additional enable (negative logic, active GND) input which, if at Logic-1, forces all outputs to be 0.

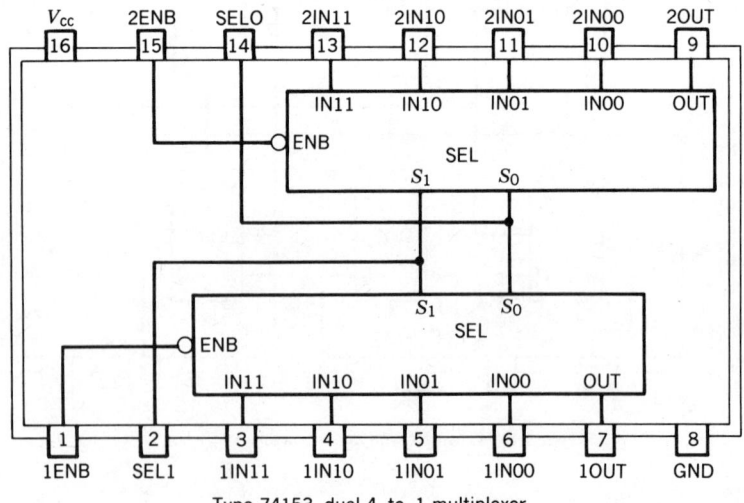

Type 74153, dual 4-to-1 multiplexer

The select line and the negative-logic enable line are common to all four multiplexers. When the enable line is at Logic-1 (passive), all outputs are at Logic-0.

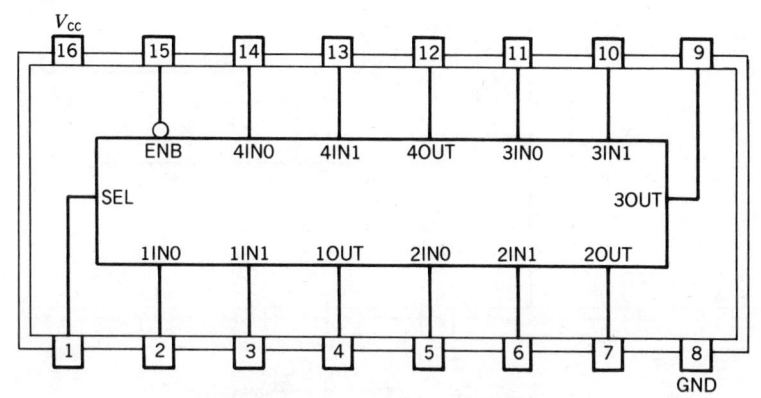

```
        V_CC
         16    15    14    13    12    11    10    9
         |     |     |     |     |     |     |     |
        ┌─────────────────────────────────────────────┐
        │     ○ENB  4IN0  4IN1  4OUT  3IN0  3IN1     │
        │ SEL                                    3OUT │
        │     1IN0  1IN1  1OUT  2IN0  2IN1  2OUT     │
        └─────────────────────────────────────────────┘
         |     |     |     |     |     |     |     |
         1     2     3     4     5     6     7     8
                                                    GND
```

Type 74157, quad 2–to–1 multiplexers

A1.3.6 Decoders/Demultiplexers

The selection lines are common to both demultiplexers. The outputs of the decoders are inverted and are at Logic-1 level when they are passive. There are two enable inputs for each multiplexer. The truth tables for this device follows:

Inputs						Outputs							
1ENB1	1ENB2	2ENB1	2ENB2	SEL1	SEL0	1Q00	1Q01	1Q10	1Q11	2Q00	2Q01	2Q10	2Q11
1	0	0	0	0	0	0	1	1	1	0	1	1	1
				0	1	1	0	1	1	1	0	1	1
				1	0	1	1	0	1	1	1	0	1
				1	1	1	1	1	0	1	1	1	0
All others						1	1	1	1	1	1	1	1

426 INTEGRATED CIRCUIT DEVICES

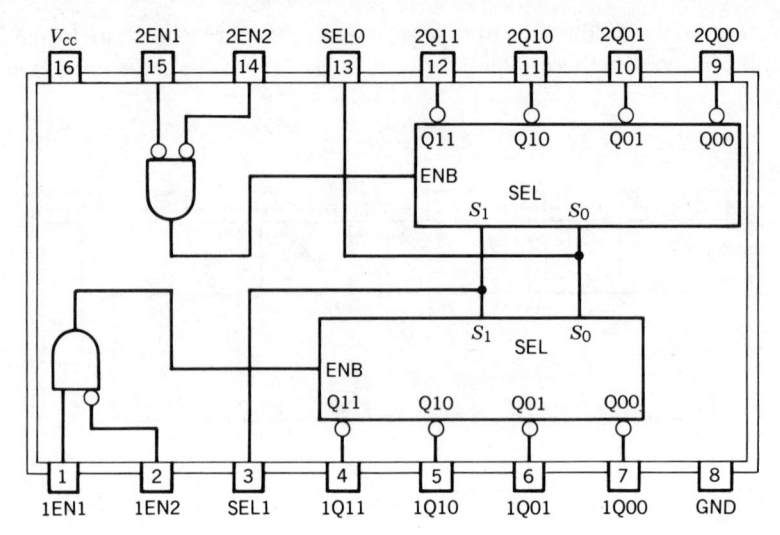

Type 74155, dual 2-to-4 decoder/demultiplexer

A1.3.7 Binary Counter

All functions of this 4-bit binary counter are executed synchronously (at the positive edge of the clock). The counter operates only when both enable inputs are active (at Logic-1). The function table for the counter follows:

Inputs				Function
ENB1	*ENB2*	*LOAD*	*CLEAR*	
1	1	1	1	INCREMENT
1	1	0	1	LOAD
1	1	1	0	CLEAR
1	1	0	0	Undefined

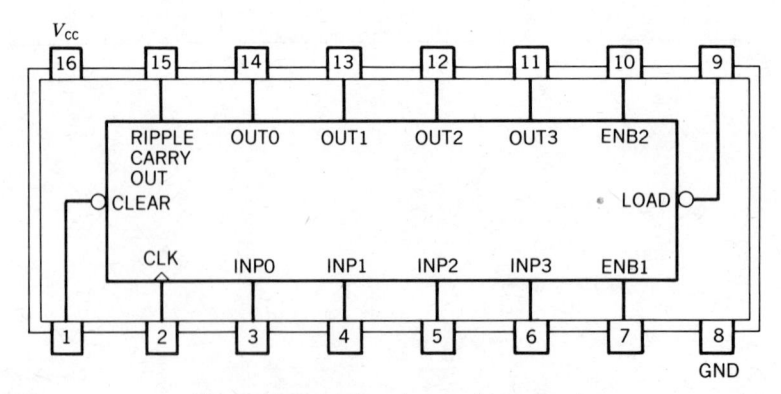

Type 74S163, 4-bit synchronous binary counter

A1.3 IC DEVICES

A1.3.8 Shift Register

The shift register operates during the positive edge of the CLOCK input according to the following function table:

SEL1	SEL0	Function
0	0	No operation
0	1	SHIFT RIGHT
1	0	SHIFT LEFT
1	1	LOAD

When the word is shifted right, OUT3 receives the value of the RIGHT SERIAL INPUT. For the left-shift function OUT0 receives the value of the LEFT SERIAL INPUT. The negative-logic CLEAR function sets all outputs to 0 asynchronously (independent of the CLOCK input).

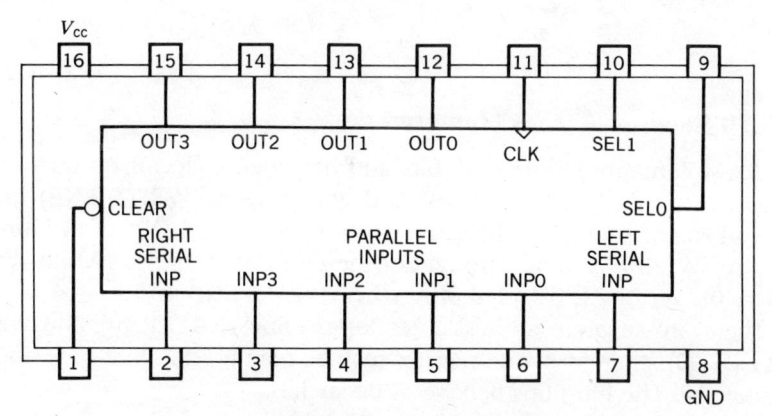

Type 74194, 4-bit universal shift register

A1.3.9 Comparator

This device compares two unsigned 4-bit binary numbers and generates a Logic-1 for one of its $A > B$, $A = B$, or $A < B$ outputs. The comparator has three similar inputs ($A > B$, $A = B$, and $A < B$), which it receives from higher order 4-bit comparators. To compare 4-bit numbers, set these inputs to:

$A < B$ INP	$A = B$ INP	$A > B$ INP
0	1	0

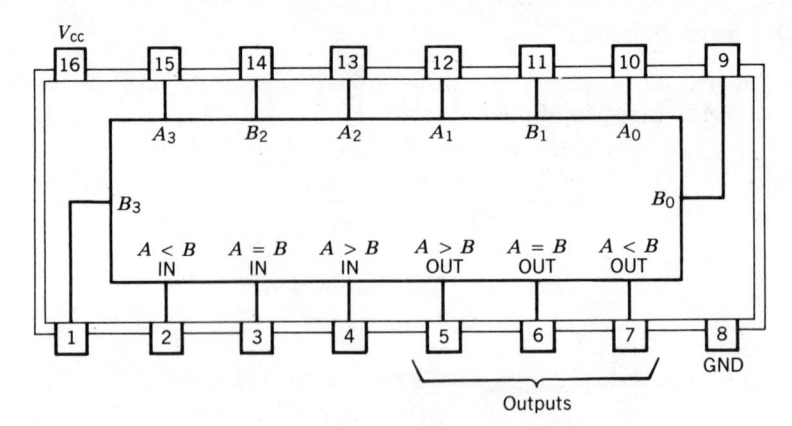

Type 7485, 4-bit magnitude comperator

A1.3.10 Random-Access Memory

This RAM contains 16 (4 × 4) bits and has open-collector outputs. It has separate read enable (READ ENB) and write enable (WRITE ENB) inputs. The read enable signal may be active at all times if wired-OR connections are not used. Writing is done throughout the entire write pulse (while WRITE ENB = 0). Writing is disabled only when WRITE ENB = 1.

There are separate address lines for reading (RA1, RA0) and writing (WA1, WA0); therefore, one register may be read at the same time another is written to. The flip-flops behave as dc latches.

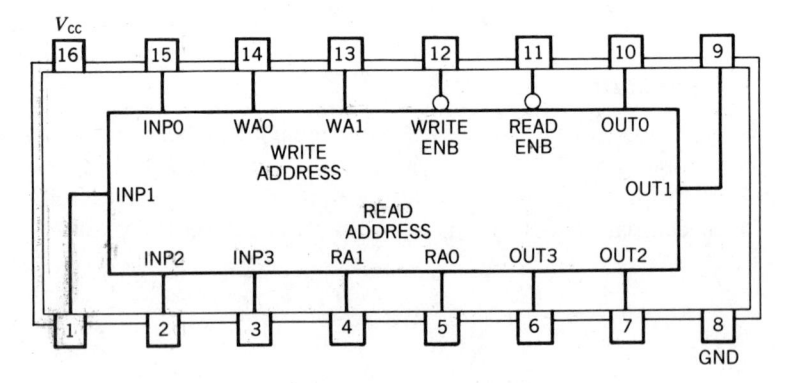

Type 74170, 4-by-4 bits register file

A1.3.11 Arithmetic-Logic Unit

This 4-bit ALU provides 16 logic and 32 arithmetic functions (some are repeated). There are six function selection lines (SEL0–SEL5). The CARRY OUT signal is inverted, and a comparator circuit generates the $A = B$ logic signal output. The functions are shown in Table A.2.

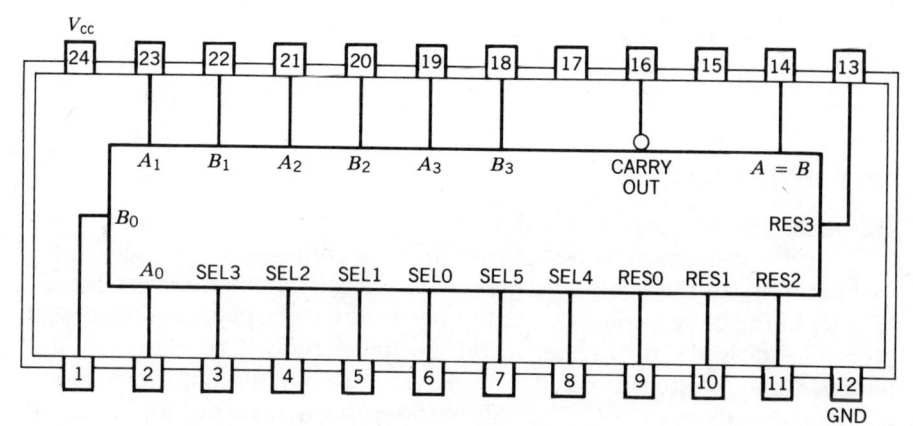

Type 74181, 4-bit arithmetic logic-unit

TABLE A.2 Function Table for Type 74181 Arithmetic-Logic Unit

$S_3S_2S_1S_0$	Logic $(S_4 = 0, S_5 = X)$	Arithmetic $(S_4 = 1)$ $S_5 = 0$	$S_5 = 1$
0000	A'	A	A plus 1
0001	$(A + B)'$	$A + B$	$(A + B)$ plus 1
0010	$A' \cdot B$	$A + B'$	$(A + B')$ plus 1
0011	0	-1	0
0100	$(A \cdot B)'$	A plus $A \cdot B'$	A plus $A \cdot B'$ plus 1
0101	B'	$(A + B)$ plus $A \cdot B'$	$(A + B)$ plus $A \cdot B'$ plus 1
0110	$A \oplus B$	$A - B - 1$	$A - B$
0111	$A \cdot B'$	$A \cdot B' - 1$	$A \cdot B'$
1000	$A' + B$	A plus $A \cdot B$	A plus $A \cdot B$ plus 1
1001	$A \odot B$	A plus B	A plus B plus 1
1010	B	$(A + B')$ plus $A \cdot B$	$(A + B')$ plus $A \cdot B$ plus 1
1011	$A \cdot B$	$A \cdot B - 1$	$A \cdot B$
1100	-1	A plus A	A plus A plus 1
1101	$A + B'$	$(A + B)$ plus A	$(A + B)$ plus A plus 1
1110	$A + B$	$(A + B')$ plus A	$(A + B')$ plus A plus 1
1111	A	$A - 1$	A

$+$ = OR, $\cdot$ = AND, $-1 = 1111$, X = don't care, $\oplus$ = XOR, $\odot$ = XNOR.

APPENDIX 2

Documentation

A2.1 INTRODUCTION

This section is directed toward those students and instructors who use this book in the same spirit as it was written. As mentioned in the preface, the main purpose of the book is to provide as realistic an environment for the practice of the design-construction-testing-maintenance phases of the design process as possible. In the real world, documentation is an extremely important (though often thoroughly neglected) responsibility of the designer. In this appendix we present our ideas about the written material required from our students while running this course.

Even in industry, it is not an easy task to force designers to produce good documentation unless there are strict standards and verification procedures that can determine that the standards have been adhered to. In an educational institution, especially among scientifically oriented students, many teachers dream of receiving good quality documentation; this dream, however, is not often realized. There are the conventional laboratory courses with severely limited supervised laboratory periods and required preparatory written work or neatly written up after-the-fact laboratory reports (or both). These reports often become a nightmare for both students and graders. I do not recall a single conventional laboratory course among those I have taken or have supervised for which the written material was anything but long, meaningless, deadly boring, and in most cases, copied from another source.

Since this course is an unconventional laboratory course, we tend to tailor the required written material to it so that it could help meet the course's objectives. As we have mentioned in the preface, the major requirement of this course is that the students themselves do the work, which includes design, construction, and testing. Since most of the students' work is completed outside of the laboratory, parts of the required documentation are used to check whether the students have fulfilled their obligations. Other parts are required that show the students' ability to communicate their results in writing in a concise and organized manner.

We have required two types of written work. The first type is called the **Engineering Notebook,** which the student keeps at all times. The second is called **Summary Report,** which is handed in and is graded.

A2.2 ENGINEERING NOTEBOOK

The Engineering Notebook should contain a written record of the design work as it progresses. It should be a notebook with numbered pages that are never removed (loose-leaf, ring-binder folders are not acceptable). The designer should always add the date and time to each entry. The purpose of the Notebook is twofold. First, it preserves the design work in a written form that should help in the construction and testing of the experiment and can later be looked up and reused when similar design problems are tackled. The page numbers help in referencing previous diagrams and circuits and eliminate the need to duplicate these diagrams in subsequent experiments.

The second purpose of the Notebook is to prove to the laboratory instructor that the work is original and was not simply copied from another student. When the wired circuit is checked by the instructor, the Notebook is also checked. A few probing questions should establish within a short time whether the student has produced or copied the work.

Beyond these two basic requirements the format and style of the Notebook will depend on individual students. The Notebook is never collected; it is treated as the property of the student, a tool for saving time. A well kept Notebook will be especially helpful to the student in performing the more complicated experiments.

A few general suggestions follow:

1. Use the first several pages of the Notebook as a table of contents. The description of an experiment should always start at the top of the page with its experiment number and title. After completing an experiment, extend the table of contents with significant entries, such as: The K-map, final circuit, gate transformation, checking, and so on.
2. Write a short description of the design problem as a reminder of what the experiment is about. Record the initial assumptions and indicate clearly the arbitrary choices you made throughout the design process.
3. Record all the simplifications and gate transformations you make. If your final circuit does not work according to specifications, these would be the first to check for errors.
4. It is a good practice to include a heading or a short explanation with every diagram, truth table, K-map, and so on. This makes the text much more readable and should help later if you need to locate some important information that is buried inside your Notebook.
5. The final wiring diagram should be included in a form that helps in its construction and testing. Make sure that the IC numbers, pin numbers, and connections correspond exactly to the wired circuit.
6. Include an independent checking procedure of your circuits and a table of expected results. Both of these should be extremely helpful in the debugging process if your circuit does not work.

Maintaining a well-organized, clearly written, and useful engineering notebook in this course will help you in your professional life since the keeping of such records is a very common requirement of engineering companies.

A2.3 SUMMARY REPORT

This report is treated as a formal description of the work done. It is collected and graded. It should be complete but very concise. It is intended for your manager whose valuable time should not be wasted with long-winded discussions of no consequence. The following sections of the report should be included in the same order as given here:

Title

This should include the number of the experiment and its title.

Design Specifications

This is often called the external specifications. A block diagram should show all required inputs and outputs that are labeled. The design specifications may be supplemented with a truth table or STD but should also include a written description of the specifications.

Design Choices

Many experiments include free choices for the designer. This section should list the choices made and also include the reasons the selected ones were preferred.

Expected Results

Normally, the expected results are given in a truth table. In addition to the required output signals, the results may be given for internal signals as well. These may help in the debugging process if the circuit does not work. For example, if J-K flip-flops are used, in addition to the outputs, the expected J and K input values should also be included in the table.

Even though there may be "don't care" values in the design specifications, there should not be any such values in the Expected Results tables. Since the Summary Report is a description of a designed circuit, the expected results are in terms of the expected signal values of the constructed circuit.

Circuit Diagram

This is the diagram used for construction, testing, and maintenance. It should correspond exactly to the actual circuit and contain IC numbers, pin numbers, list of IC types, and so on; the circuit diagram should be neatly drawn.

Circuit Cost and Time Spent

Overall circuit cost should be calculated and included in this section. Students should keep a record of (and list here) time spent for design, construction, and testing of the experiment.

Comments

This is an **optional** section and should be included only if there is something important to report. If the experiment cannot be completed because of reasons beyond the students' control, the circumstances should be reported here.

A2.4 EXAMPLE OF A SUMMARY REPORT

The following example of a Summary Report is presented for the design of a circuit that is similar (but not equivalent) to Experiment 7:

Experiment 7—Design of a Digital Lock

Description of the Design Problem

The object of this experiment is to design a digital lock that opens a safe door when the correct bit sequence (101) is entered at the input. If an incorrect bit sequence is used, the circuit turns on an alarm signal that may not be turned off by any digital input sequence. A special reset signal turns the alarm signal off and initializes the system.

The circuit has three input signals including the SYSTEM CLOCK and two output signals as shown in Figure A2.1. The LOCK SAFE signal remains **1** for all cases except when the safe is opened.

The detailed operation of the digital lock can be demonstrated by the STD shown in Figure A2.2. After the RESET signal is activated, the system is placed into its IDLE state and remains there as long as the input signal remains **0**. After the first **1** arriving at the input signal, the lock expects the

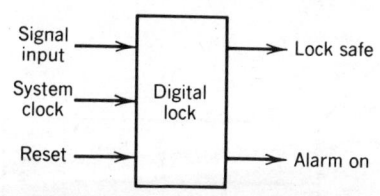

FIGURE A2.1 Block diagram of the digital lock.

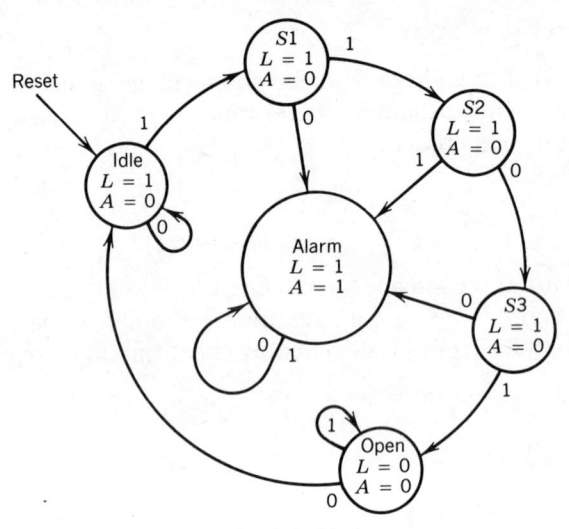

FIGURE A2.2 STD of the digital lock system.

bit sequence **101** after which the safe door opens (LOCK SAFE = 0). The door remains open as long as the input signal remains **1**. If the door is open, the first **0** input returns the system to its IDLE state (which locks the safe). If an incorrect bit sequence is input, the safe door remains locked and the alarm signal is turned on. This alarm signal remains in its ON state (ALARM ON = 1) until the RESET signal is reactivated.

Design Choices

A Moore-type bit sequence detector was chosen since reliability and fail-safe operations are both important in this case. Using the design rules for optimal state assignments, the assignments shown in Figure A2.3 satisfied most adjacency conditions for Rules 1 and 2.

Implementation was tried for both D-type and J-K type flip-flops and the number of ICs was found to be smaller for the J-K flip-flops; therefore, J-K flip-flops were chosen for construction.

Q_2 \ Q_1Q_0	00	01	11	10
0	Open	S_3	Alarm	S_2
1	Idle	S_1	X	X

FIGURE A2.3 State assignments.

A2.4 EXAMPLE OF A SUMMARY REPORT

Expected Results

The expected results are shown in the following truth table:

Old State	Q_2 Q_1 Q_0	Input	J_2 K_2	J_1 K_1	J_0 K_0	New State	Next Q_2 Q_1 Q_0
OPEN	0 0 0	0	1 0	0 1	0 0	IDLE	1 0 0
		1	0 0	0 0	0 1	OPEN	0 0 0
S3	0 0 1	0	0 1	1 0	0 0	ALRM	0 1 1
		1	0 1	0 0	0 1	OPEN	0 0 0
S2	0 1 0	0	0 0	0 1	1 0	S3	0 0 1
		1	0 0	0 0	1 0	ALRM	0 1 1
ALRM	0 1 1	0	0 1	1 0	1 0	ALRM	0 1 1
		1	0 1	0 0	1 0	ALRM	0 1 1
IDLE	1 0 0	0	1 0	0 1	0 0	IDLE	1 0 0
		1	0 0	0 0	1 1	S1	1 0 1
S1	1 0 1	0	0 1	1 0	0 0	ALRM	0 1 1
		1	0 1	1 0	1 1	S2	0 1 0
XXX	1 1 0	0	0 0	0 1	1 0	S1	1 0 1
		1	0 0	0 0	1 0	XXX	1 1 1
XXX	1 1 1	0	0 1	1 0	1 0	ALRM	0 1 1
		1	0 1	1 0	1 0	ALRM	0 1 1

Circuit Diagram

The circuit diagram is shown in Figure A2.4.

Circuit Cost and Time Spent

The following table summarizes the circuit costs:

Number of ICs	Type of IC	Cost/IC	Cost/Socket	Total
1	7400	.20	.20	0.40
1	7404	.24	.20	0.44
2	7408	.20	.20	0.80
2	7476	.24	.25	0.98
			Total circuit cost:	2.62

Total time spent on design: 3.4 hours

construction: 2.5 hours

testing: 0.5 hours (was lucky)

Comments

Two ICs (types 7404 and 7408) are under utilized. It seems possible that by gate transformations these two ICs may be reduced to one. Changing the

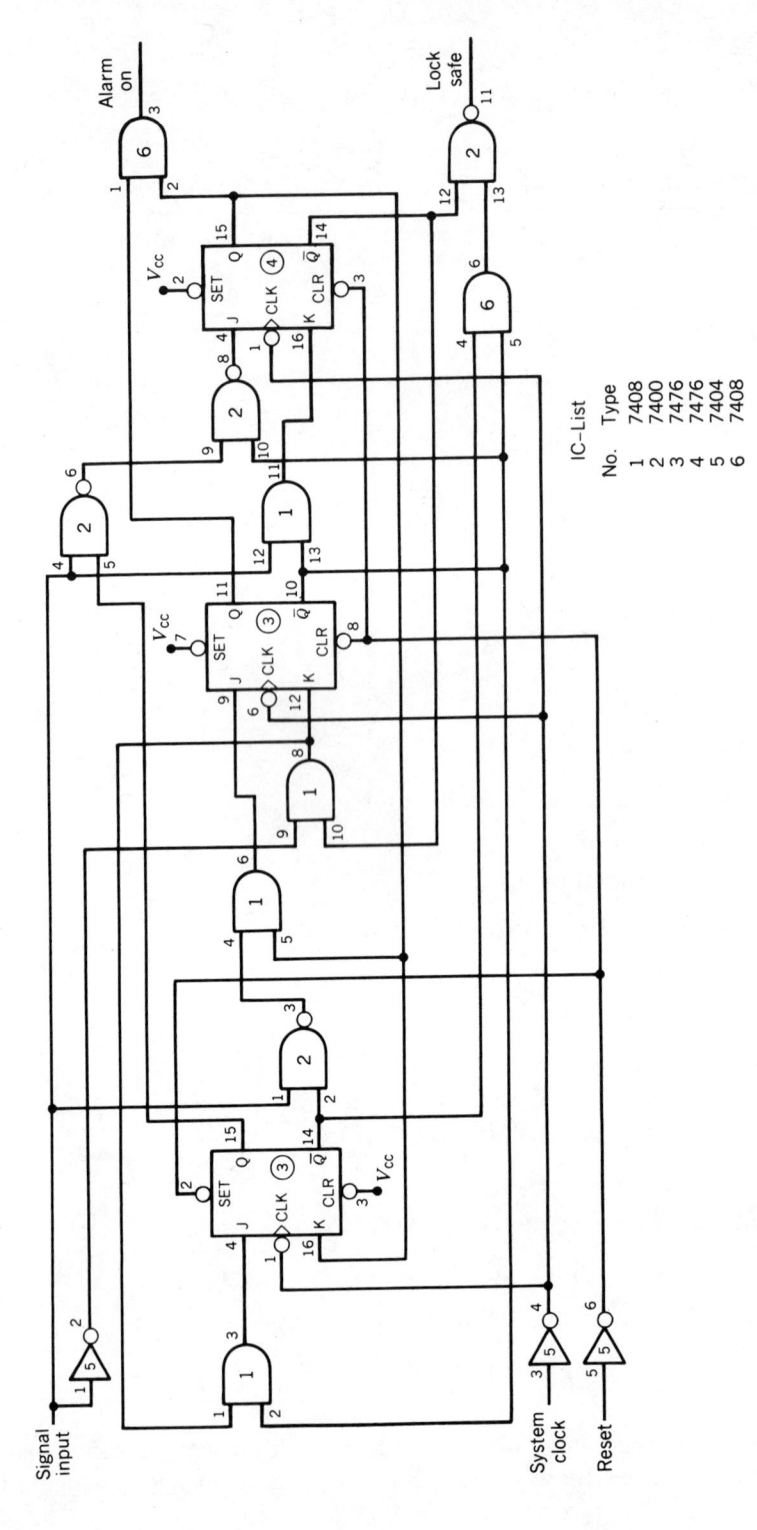

FIGURE A2.4 Circuit diagram.

AND gates to NOR gates seems worth a try. Unfortunately, time did not allow me to pursue this.

The unused states were checked. If at the time the power is turned on the system finds itself in one of its unused states, it will not open the safe. In fact, if the input signal is kept at Logic-0 then the system will proceed either to the IDLE or the ALARM states in a few clock periods. Thus, the system is reasonably fail-safe.

Index

Accumulator, 348
Addition
 parallel, 249–250
 serial, 250–252
 unsigned number representation and, 244–247
Address (memory address), 316, 347
AND gate, 11, 12, 421
 as variable Boolean function generator, 66
AND operator, 9–11
 Boolean algebra and, 17–22
 set theory and, 42
Arithmetic-logic unit (ALU), 429
Arithmetic operations, 239
 number representation and octal and hexadecimal representation of binary numbers, 260–263
  sign-plus-magnitude representation, 252–255
  two's complement representation, 254, 255–260
 parallel addition and subtraction, 249–250
 serial addition, 250–252
 unsigned number representation and, 242–249
Arithmetic shift operation, 288
Assembly language, 324–325
 programming in, 365–366, 367
Associative property, 21
Asynchronous control of practical counters, 194–197
Asynchronous (ripple) counters, 191–194
Asynchronous sequential circuits, 131

Binary (logic) circuits, 14–16
 analysis, construction, and testing of (experiment), 34–38
 design with minterms and maxterms, 83–88
 feedback connections in, 8, 122–132
 physical realization of, 23–28
Binary counter, 4-bit, 426
Binary (logic) gates, 11–14
 with one and two inputs, 50–53
 transformations, 53–58
Binary numbers, octal and hexadecimal representation of, 260–263

Bit-sequence detector (BSD) circuits, 217–219
Block diagrams, 2
Boolean algebra, 4, 16–23
Boolean function generators
 two-input, three-output (experiment), 58–63
 variable, 65–81
  design of (experiment), 77–80
  minimized, design of (experiment), 116–119
 multiplexers, 68–71
 multiplexers and decoders as, 73–77
 simple, 65–68
 using a decoder to build a multiplexer, 71–73
Boolean functions
 functions of one binary variable, 48
 functions of two binary variables, 49–50
 mapping, 93
Boolean simplification, 42–48
 using K-maps, 98–104
Borrow, unsigned number representation and, 247
BSD (bit-sequence detector) circuits, 217–219

Canonical forms for Boolean functions, 85
Carry flip-flop (carry register), 285
Carry generation, 291–294
 look-ahead, 292
Carry propagation, 290–294
Central process unit (CPU), *see* General data processor
CLEAR control, 196
Clock pulses, 133
Code (instruction code), 322
Combinational binary systems, 5–6, 82–88
Combination lock, serially controlled, design of (experiment), 233–237
Commutative operator, 20
Comparator, 427–428
Complement of a proposition, 9
Complement of a set, 42
Computer arithmetic, *see* Arithmetic operations
Contents of a register, 240
Control instructions, 345–348
Control signals (control circuitry), 350–356

INDEX **439**

440 INDEX

for computer cycles, 350–353
for selection lines, 353–354
for updating registers, 354–356
Control variables, 65
Counters, 164–200
   asynchronous (ripple) counters, 191–194
   down counter for excess-3 decimal digits using T flip-flops, 187–191
   five-state counter with JK flip-flops, 180–183
   five-state up-down counter, design of (experiment), 197–199
   JK flip-flop design using partitioned K-maps, 183–186
   synchronous and asynchronous control of practical counters, 194–197
   synchronous counters with D-type flip-flops, 165–176
   three-bit binary counter with SR flip-flops, 177–180
Cross-coupled NAND gates, 126–127, 128, 129–131
Custom-defined ROMs, 388
Cycles (computer cycles), 320–322
   control signals for, 350–353

Data bus structures, 306–316
Data-handling instructions, 322–326
DATA INPUTS, 196
Data-processing instructions, 348–350
   selection of, 361–363
Data rotate, 282
Data synchronization, 252
Data transfer instructions, 360–361
Data transfer operations, 341–345
Decoders, 65, 73–77
   for multiplexer construction, 71–73
Decoders/demultiplexers, 425–426
De Morgan's laws, 23
Denial of a proposition, 9
Design cycle, definition of, v
Design of digital computers, 379–415
   changes and problems due to increase in number of gates per module, 380–381
   designing in the large, 380–387
   microprogramming, 407–415
   with ROMs and PLAs, 395–406
      experiment, 415
   with ROMs and programmable logic modules, 387–394
Destination component of the instruction code, 322
Diagnostic programs, 365–366, 367
Digital circuits, 28–34
   logic gates with one and two inputs, 50–53
   practical gate transformations, 53–58
Digital computers, 306–336
   data bus structures, 306–316
   data-handling instructions, 322–326

externally programmed computers, 327–328
   design and construction of (experiment), 328–335
instructions and cycles, 320–322
memory systems, 316–320
programmable, 337–378
   control circuitry, 350–356
   control instructions, 345–348
   data-processing instructions, 348–350
   data transfer operations, 341–345
   four-bit computer, design of (experiment), 367–375
   memory and, 338–340
   programming in assembly language, 365–366, 367
   selection of basic instruction set, 356–365
register transfers, 306–316
Direct LOAD operation, 342
Distributive property, 20
Down counter for excess-3 decimal digits using T flip-flops, 187–191
D-type flip-flop circuits
   edge-triggered, 146–148
   master-slave, 145–146
   TTL 7474-type IC device, 150–152
D-type flip-flops, 136–138
   synchronous counters with, 165–176
Dual canonical forms for Boolean expressions, 85
Dual in-line IC device, 28, 29
Duality principle, 10, 21–22
Dual representation of basic logic operations, 12–13
Dynamic RAM, 317–318

Edge-triggered D-type flip-flop circuit, 146–148
Electrically erasable/programmable ROMs (EEPROMS), 318, 389
ENABLE input, 195
Engineering Notebook, 430, 431
Equivalence gate (EQU gate), 49
Erasable/programmable ROMs (EPROMs), 389
Excitation table for flip-flops, 164, 176–177
Exclusive NOR gate (XOR gate), 49, 422
Exclusive OR function (XOR function), 49
Execution cycle, 320–321
Experiments, 1
   binary circuits, analysis, construction, and testing of, 34–48
   externally programmed digital computer, design of, 328–335
   five-state up-down counter, design of, 197–199
   four-bit computer, design of, 367–375
   four-bit serial-parallel arithmetic processor, 294–304
   minimized variable Boolean function generator, design of, 116–119

INDEX 441

ROMs and PLAs, experimental designs with, 415
serial data processor, design of, 269–275
serially controlled combination lock, design of, 233–237
time-dependent digital signals, tracing and measuring, 153–160
two-input, three-output Boolean function generator, 58–63
variable Boolean function generator, design of, 77–80
Externally programmed digital computer, 327–328
   design of (experiment), 328–335

Fan-out, 34
Faulty master-slave flip-flop, 158–159
Feedback connections, 6, 7
   in binary circuits, 8, 122–132
Feed-forward connections, 6–8
Fetch cycle, 320
FETCH operation, 342
Fields of the instruction code, 322, 323–324
Finite-state machines, *see* Sequential systems
Five-state counter design with JK flip-flops, 180–183
Five-state up-down counter, design of (experiment), 197–199
Five-variable Karnaugh maps, 113–116
Flip-flop circuits, 140–152
   D-type
      edge-triggered, 146–148
      master-slave, 145–146
   elementary types, 148–150
   TTL 7474-type IC device, 150–152
Flip-flops, 140, 422–424
   D-type, 136–138
   synchronous counters with, 165–176
   excitation table for, 164, 176–177
   JK, 150, 162
      counter design using partitioned K-maps, 183–186
      five-state counter with, 180–183
      store state of, 201
   one-flip-flop-per-state method, 213–217
   SR, 149, 150
      three-bit binary counters with, 177–180
   state assignments for, 201–202
   T (toggle), 150, 151, 161
      down counter for excess-3 decimal digits with, 187–191
Flip-flop setup time, 219
Four-bit computer, design of (experiment), 367–375
Four-bit serial-parallel arithmetic processor (experiment), 294–304
Four-bit synchronous binary counter, 426
Four-variable Karnaugh maps, 89, 90, 91

Full adder, 246–247
Full subtractor, 249
Functional specification of the binary system, 8

General data processor, 263–269
   computer cycles and, 320
General synchronous one-bit memory, 135–136
Glitches (unwanted signal transitions), 154, 157

Half adder, 246
Half subtractor, 248–249
Hazards in synchronous sequential systems, 191–192
Hexadecimal representation of binary numbers, 260–263
Hierarchical design methods for digital computers, 381
High voltage state, 25, 26
Horizontal microprogramming, 410

Identity element, 18
Immediate data, 342
Implication functions, 50
Indirect LOAD operation, 343
Inhibition instructions, 50
INITIALIZE, 125
Input-output instructions, 360
Input-output signal synchronization in sequential systems, 219–222
Input variables, 2–3
Instruction (computer instruction), 320–322
   data-handling, 322–326
Instruction code, 322
Instruction fields, 322, 323–324
Instruction register (IR), 321–322, 338, 339, 340
Integer numbers, multiplication of, 289–290
Integrated circuit board and wires, 30
Integrated circuit (IC) devices, 417–429
   AND gates and inverters, 11, 12, 421
   arithmetic-logic unit (ALU), 429
   binary counters, 426
   comparators, 427–428
   decoders/demultiplexers, 65, 425–426
   flip-flops, 140, 422–424
   list of available devices, 417–419
   multiplexers, 65, 68–71, 424–425
   NAND gates, 14, 420
   NOR gates and XOR gates, 14, 49, 422
   random-access memory (RAM), 316–320, 428
   shift registers, 277–278, 427
   TTL family of, 26–28
Integrated circuit families, 25–26
Integrated circuits (ICs), 8
Intersection of two sets, 42

442　INDEX

INVERTER gates, 11, 12, 13, 14, 25–26, 421

JK flip-flops, 150, 162
  counter design using partitioned K-maps, 183–186
  five-state counter with, 180–183
  store state of, 201

Karnaugh maps (K-maps), 82–83
  for decimal decrementer using T flip-flops, 189, 190
  for five-state counter with JK flip-flops, 181–182
  for Mealy-type circuits, 227
  minimization with, 88–98
    with "don't care" terms, 104–106
    for five and six variables, 113–116
    with multiple outputs, 106–113
    of product-of-sum expressions, 96–98
    for simplification of Boolean functions, 98–104
    of sum-of-product expressions, 91–96
    for two, three, and four variables, 88–90, 91
  for one-flip-flop-per-state method, 216–217
  partitioned, JK flip-flop counter design using, 183–186
  state transition Karnaugh map (STKM), 136
  for synchronous counters with D-type flip-flops, 165–171, 173, 175
  for three-bit binary counters with SR flip-flops, 178, 179

Language
  assembly, 324–325
    programming in, 365–366
    machine, 324
Length of a register, 240
LOAD functions, 196
LOAD sequence of operations, 342–345
Logic-0, 5
Logic-1, 5
Logical shift operation, 287
Logical sum of logical products, 47
Logic gates, *see* Binary (logic) gates
Logic variables, 3
Look-ahead carry generation, 292
Low voltage state, 25, 26

Machine language, 324
Mapping Boolean functions, 93
MAR (memory address register), 318, 338, 339, 340, 341
Master-slave D-type flip-flop circuit, 145–146
  faulty, 158–159
Maxterms, binary circuit design with, 83–88
MBR (memory buffer register), 318, 338, 339, 340, 341

Mealy detector circuits, 220–221, 222
  for bit sequence 101, 225–228
  detecting nonoverlapped bit sequence 100–110, 230–233
Memory, 316–320, 338–340
  segments of, 346–347
  in sequential systems, 121–122, 124
Memory address (address), 316
Memory address register (MAR), 318, 338, 339, 340, 341
Memory buffer register (MBR), 318, 338, 339, 340, 341
Memory-refresh operation, 318
Microinstructions, 407–415
Microprogramming, 407–415
Micro-ROMS, 407, 411, 415
Microsequencer, 411–415
Microsteps, 407
Minimization, 82
  minimized variable Boolean function generation, design of (experiment), 116–119
  See also Karnaugh maps, minimization with
Minimized expression, 43
Minterms, binary circuit design with, 83–88
Moore detector circuits, 220, 221–222
  for bit sequence 101, 222–225
Multiplexers, 65, 68–71, 424–425
  for generating other Boolean functions, 73–77
  use of a decoder to build, 71–73
Multiplexing (time multiplexing), 310–312
Multiplication of integer numbers, 289–290

NAND gates, 14, 420
  cross-coupled, 126–127, 128, 129–131
  feedback connection and, 125–126, 128, 129–131, 132
  oscillations of, 126
N-bit general data-processor, 263–269
Negation, 253, 255
  for octal number representation, 262–263
Negative numbers, 252–253
Noise margin, 26
NOR gates, 14, 422
NOT operator, 9–11
  Boolean algebra and, 17–22
  set theory and, 42
Null element, 18
Null set, 42
Number representations, 239
  general data processor and, 263–269
  octal and hexadecimal representation of binary numbers, 260–263
  sign-plus-magnitude representation, 252–255
  two's complement representation, 255–260
  unsigned, arithmetic and, 242–249

INDEX 443

Octal representation of binary numbers, 260–263
One-bit memory, synchronous, 135–136
One-flip-flop-per-state design method, 213–217
One's complement number representation, 258
Operation code (opcode) of instructions, 322
OR gate, 11, 12, 13, 14
OR operator, 9–11
   Boolean algebra and, 17–22
   set theory and, 42
Oscillations of the NAND gate, 126
Output equation, 139
Output tables for sequential circuits, 130–131
Output transition tables, 176–177
Output variables, 2–3
Overflow errors, 244

PAL (programmed array logic), 391
Parallel addition, 249–250
Parallel data processor, thirty-two functions generated by, 266
Parallel data transmission, 277, 278
Parallel subtraction, 249–250
Parallel-to-serial converter, 277
   shift registers and, 279–285
Passive (quiescent) states, 141
PCBs (printed circuit boards), 32
Physical systems, 1–4
PLAs (programmed logic arrays), 391–395
   designing digital systems with ROMs and, 395–406, 415
Primitive elements of a system, 1
Printed circuit boards (PCBs), 32
Processor, 263–269
   computer cycles and, 320
Processor block using ROM, 382, 383, 385–386
Program control instructions, 363–365
Program counter (PC), 321–322, 338, 339, 340, 341
Programmable digital computers, 337–378
   control circuitry, 350–356
   control instructions, 345–348
   data-processing instructions, 348–350
   data transfer operations, 341–345
   four-bit computer, design of (experiment), 367–375
   memory and, 338–340
   programming in assembly language, 365–366, 367
   selection of basic instruction set, 356–365
      data processing, 361–363
      data transfer, 360–361
      input-output, 360
      program control, 363–365
Programmable read-only memory (PROM), 318, 388–391

Programmed array logic (PAL), 391
Programmed logic arrays (PLAs), 391–395
   designing digital systems with ROMs and, 395–406, 415
Propositions, 9

Quiescent (passive) states, 141

RAM (random-access memory), 316–320, 428
Read-only memories (ROMs), 318
   digital system design based on, 382–387
   micro-, 407, 411, 415
   PLAs and, designing digital systems with, 395–406, 415
   and programmable logic module, 387–394
Reduced instruction set computers (RISC), 382
Registers, 239, 240–242
Register transfer operations, 239, 240–242, 306–316
Relative addressing, 347
Representation of systems, 1–4
RESET function, 136
RIPPLE CARRY output, 195–196
Ripple counters, see Asynchronous (ripple) counters
ROMs, see Read-only memories (ROMs)

Segmented addressing, 347
Segments of memory, 346–347
Self-starting counters, 169–171
Semipermanent IC circuit, 30–31
Sequential systems, 8, 121–163
   asynchronous circuits, 131
   feedback connections in binary circuits, 8, 122–132
   input and output signal synchronization in, 219–222
   practical flip-flop circuits, 140–152
      edge-triggered D-type, 146–148
      elementary types, 148–150
      master-slave D-type, 145–146
      TTL 7474-type IC device, 150–152
   synchronous, 122, 131, 133–140
   time-dependent digital signals, tracing and measuring (experiment), 153–160
Serial addition, 250–252
Serial data processor, design of (experiment), 269–275
Serial data transmission, 277, 278, 279
Serially controlled combination lock, design of (experiment), 233–237
Serial-parallel arithmetic processor, four-bit (experiment), 294–304
Serial-to-parallel converter, 277
SET function, 136
Set-reset (SR) flip-flops, 149, 150
   three-bit binary counters with, 177–180
Set theory, 41–42

Shift registers, 277–278, 427
    parallel-serial data conversion and, 279–285
    shift operation of, 285–290
Signals (system variables), 5
Sign bit, 253
Sign-plus-magnitude number representation, 252–255
Simple variable Boolean function generator, 65–68
Six-variable Karnaugh maps, 113–116
Source component of the instruction code, 322
SR flip-flops, see Set-reset (SR) flip-flops
State assignments, 201–202
    selection of, 228–233
State controllers, 202–204
State transition diagrams, 127–129
State transition equation (STE), 136
State transition Karnaugh map (STKM), 136
State transition table (STT), 136
State variables, 2, 121, 127
STORE instruction, 342, 343
Subtraction
    parallel, 249–250
    unsigned number representation and, 247–249
Summary Report, 430, 432–437
Synchronous controllers, 201–228
    bit-sequence detection circuits, 217–219
    input and output signal synchronization in sequential systems, 219–222
    Mealy detector circuits for bit sequence 101, 225–228
    Moore detector circuits for bit sequence 101, 222–225
    one-flip-flop-per-state method, 213–217
    serially controlled combination lock, design of (experiment), 233–237
    state assignment selection, 228–233
    state controllers, 202–204
    traffic signal controller design, 204–213
Synchronous control of practical counters, 194–197
Synchronous counters with D-type flip-flops, 165–176
Synchronous one-bit memory, 135–136
Synchronous sequential systems, 122, 131, 133–140
    input and output signal, 219–222
SYSTEM CLOCK
    for asynchronous counters, 192–194
    counters and, 164
    for sequential system operation, 122, 133–135, 139, 219–220
Systems (physical systems), 1–4
System state block using ROM, 382, 383, 384–385

T flip-flops, see Toggle (T) flip-flops
Three-bit binary counter with SR flip-flops, 177–180
Three-variable Karnaugh maps, 89–90
Time (as variable in sequential systems), 121
Time-dependent digital signals, tracing and measuring (experiment), 153–160
Time multiplexing, 310–312
Toggle (T) flip-flops, 150, 151, 161
    down counter for excess-3 decimal digits using, 187–191
TOGGLE function, 136
Traffic signal controllers, design of, 204–213
Triggering, 159–160
Truth tables, 5
TTL (transistor-transistor logic) family of IC devices, 26–28
TTL 7474-type IC device, 150–152
TTL 7481-type IC device, 267–286
Two-input, three-output Boolean function generator (experiment), 58–63
Two-phase clocks for data transfer, 355–356
Two's complement number representation, 254, 255–260
Two-variable Karnaugh maps, 88

Underflow errors, 244
Union of two sets, 42
Universal set, 41–42
Universe (in Venn diagrams), 42
Unsigned number representation and arithmetic, 242–249
Unwanted signal transitions (glitches), 154, 157
Up-down counter, five-state, design of (experiment), 197–199

Variable Boolean function generators, 65–81
    design of (experiment), 77–80
    minimized, design of (experiment), 116–119
    multiplexers, 68–71
    multiplexers and decoders as, 73–77
    simple, 65–68
    using a decoder to build a multiplexer, 71–73
Veitch diagram, see Karnaugh maps
Venn diagrams, 41, 42
Vertical microprogramming, 411

Word size, 285

XOR function (exclusive OR function), 49
XOR gate (exclusive NOR gate), 49, 422
    as variable Boolean function generator, 67